AF094416

Mathematical Analysis and Analytic Number Theory 2019

Mathematical Analysis and Analytic Number Theory 2019

Editor

Rekha Srivastava

MDPI • Basel • Beijing • Wuhan • Barcelona • Belgrade • Manchester • Tokyo • Cluj • Tianjin

Editor
Rekha Srivastava
Department of Mathematics and Statistics,
University of Victoria,
Victoria, British Columbia V8W 3R4
Canada

Editorial Office
MDPI
St. Alban-Anlage 66
4052 Basel, Switzerland

This is a reprint of articles from the Special Issue published online in the open access journal *Mathematics* (ISSN 2227-7390) (available at: https://www.mdpi.com/journal/mathematics/special_issues/Mathematical_Analysis_Analytic_Number_Theory_2019).

For citation purposes, cite each article independently as indicated on the article page online and as indicated below:

LastName, A.A.; LastName, B.B.; LastName, C.C. Article Title. *Journal Name* **Year**, *Volume Number*, Page Range.

ISBN 978-3-0365-0032-4 (Hbk)
ISBN 978-3-0365-0033-1 (PDF)

© 2021 by the authors. Articles in this book are Open Access and distributed under the Creative Commons Attribution (CC BY) license, which allows users to download, copy and build upon published articles, as long as the author and publisher are properly credited, which ensures maximum dissemination and a wider impact of our publications.

The book as a whole is distributed by MDPI under the terms and conditions of the Creative Commons license CC BY-NC-ND.

Contents

About the Editor ix

Preface to "Mathematical Analysis and Analytic Number Theory 2019" xi

Hari Mohan Srivastava, Rekha Srivastava, Mahendra PalChaudhary and Salah Uddin
A Family of Theta-Function Identities Based upon Combinatorial Partition Identities Related to Jacobi's Triple-Product Identity
Reprinted from: *Mathematics* **2020**, *8*, 918, doi:10.3390/math8060918 1

Muhammad Naeem, Saqib Hussain, Shahid Khan, Tahir Mahmood, Maslina Darus and Zahid Shareef
Janowski Type q-Convex and q-Close-to-Convex Functions Associated with q-Conic Domain
Reprinted from: *Mathematics* **2020**, *8*, 440, doi:10.3390/math8030440 15

Davood Alimohammadi, Nak Eun Cho, Ebrahim Analouei Adegani and Ahmad Motamednezhad
Argument and Coefficient Estimates for Certain Analytic Functions
Reprinted from: *Mathematics* **2020**, *8*, 88, doi:10.3390/math8010088 29

Mohd Qasim, M. Mursaleen, Asif Khan and Zaheer Abbas
Approximation by Generalized Lupaş Operators Based on q-Integers
Reprinted from: *Mathematics* **2020**, *8*, 68, doi:10.3390/math8010068 43

Qaiser Khan, Muhammad Arif, Mohsan Raza, Gautam Srivastava, Huo Tang and Shafiq ur Rehman
Some Applications of a New Integral Operator in q-Analog for Multivalent Functions
Reprinted from: *Mathematics* **2019**, *7*, 1178, doi:10.3390/math7121178 59

Hari Mohan Srivastava, Asifa Tassaddiq, Gauhar Rahman, Kottakkaran Sooppy Nisar and Ilyas Khan
A New Extension of the τ-GaussHypergeometric Function and ItsAssociated Properties
Reprinted from: *Mathematics* **2019**, *7*, 996, doi:10.3390/math7100996 73

Nak Eun Cho, Ebrahim Analouei Adegani, Serap Bulut and Ahmad Motamednezhad
The Second Hankel Determinant Problem for a Class of Bi-Close-to-ConvexFunctions
Reprinted from: *Mathematics* **2019**, *7*, 986, doi:10.3390/math7100986 83

Zhaolin Jiang, Weiping Wang, Yanpeng Zheng, Baishuai Zuo and Bei Niu
Interesting Explicit Expressions of Determinants and Inverse Matrices for Foeplitz and Loeplitz Matrices
Reprinted from: *Mathematics* **2019**, *7*, 939, doi:10.3390/math7100939 93

Yunlan Wei, Yanpeng Zheng, Zhaolin Jiang and Sugoog Shon
A Study of Determinants and Inverses for Periodic Tridiagonal Toeplitz Matrices with Perturbed Corners Involving Mersenne Numbers
Reprinted from: *Mathematics* **2019**, *7*, 893, doi:10.3390/math7100893 111

Hari M. Srivastava, Qazi Zahoor Ahmad, Maslina Darus, Nazar Khan, Bilal Khan, Naveed Zaman and Hasrat Hussain Shah
Upper Bound of the Third Hankel Determinant for a Subclass of Close-to-Convex Functions Associated withthe Lemniscate of Bernoulli
Reprinted from: *Mathematics* **2019**, *7*, 848, doi:10.3390/math7090848 . . . . **123**

Namhoon Kim
Transformation of Some Lambert Series and Cotangent Sums
Reprinted from: *Mathematics* **2019**, *7*, 840, doi:10.3390/math7090840 . . . . **133**

Anthony Sofo and Amrik Singh Nimbran
Euler Sums and Integral Connections
Reprinted from: *Mathematics* **2019**, *7*, 833, doi:10.3390/math7090833 . . . . **143**

Pierpaolo Natalini and Paolo Emilio Ricci
Appell-Type Functions and Chebyshev Polynomials
Reprinted from: *Mathematics* **2019**, *7*, 679, doi:10.3390/math7080679 . . . . **167**

Lei Shi, Qaiser Khan, Gautam Srivastava, Jin-Lin Liu, and Muhammad Arif
A Study of Multivalent$q$-starlike Functions Connected with Circular Domain
Reprinted from: *Mathematics* **2019**, *7*, 670, doi:10.3390/math7080670 . . . . **175**

Konstantinos Kalimeris and Athanassios S. Fokas
A Novel Integral Equation for the Riemann Zeta Function and Large t-Asymptotics
Reprinted from: *Mathematics* **2019**, *7*, 650, doi:10.3390/math7070650 . . . . **187**

Aoen and Shuhai Li
The Application of Generalized Quasi-Hadamard Products of Certain Subclasses of Analytic Functions with Negative and Missing Coefficients
Reprinted from: *Mathematics* **2019**, *7*, 620, doi:10.3390/math7070620 . . . . **205**

Giuseppe Dattoli, Silvia Licciardi, Elio Sabia, Hari M. Srivastava
Some Properties and Generating Functions ofGeneralized Harmonic Numbers
Reprinted from: *Mathematics* **2019**, *7*, 577, doi:10.3390/math7070577 . . . . **215**

Lina Ma, Shuhai Li and Xiaomeng Niu
Some Classes of Harmonic Mapping with a Symmetric Conjecture Point Defined by Subordination
Reprinted from: *Mathematics* **2019**, *7*, 548, doi:10.3390/math7060548 . . . . **227**

Lei Shi, Izaz Ali,Muhammad Arif, Nak Eun Cho, Shehzad Hussain, and HassanKhan
A Study of Third Hankel Determinant Problem for Certain Subfamilies of Analytic Functions Involving Cardioid Domain
Reprinted from: *Mathematics* **2019**, *7*, 418, doi:10.3390/math7050418 . . . . **245**

Chenkuan Li, Changpin Li, Thomas Humphries and Hunter Plowman
Remarks on the Generalized Fractional Laplacian Operator
Reprinted from: *Mathematics* **2019**, *7*, 320, doi:10.3390/math7040320 . . . . **261**

Khurshid Ahmad, Saima Mustafa, Muhey U. Din, Shafiq Ur Rehman, Mohsan Raza and Muhammad Arif
On Geometric Properties of NormalizedHyper-Bessel Functions
Reprinted from: *Mathematics* **2019**, *7*, 316, doi:10.3390/math7040316 . . . . **279**

Shahid Mahmood, Janusz Sokół, Hari Mohan Srivastav and Sarfraz Nawaz Malik
Some Reciprocal Classes of Close-to-Convex and Quasi-Convex Analytic Functions
Reprinted from: *Mathematics* **2019**, 7, 309, doi:10.3390/math7040309 . . . . **291**

About the Editor

Rekha Srivastava was born in the small town of Chikati (District Ganjam) in the Province of Odisha in India. After completing her high school and intermediate college education, she received her B.Sc. degree in 1962 from Utkal University in the Province of Odisha. She then travelled to Banaras Hindu University in Varanasi in the Province of Uttar Pradesh in India, from where she first received her M.Sc. degree in 1965 and then her Ph.D. degree in 1967 at the age of 22. In fact, she happens to be the first woman in the entire Province of Odisha to have received a Ph.D. degree in Mathematics. She began her university-level teaching career in 1968 and, after having taught in India until 1972, she joined the University of Victoria in Canada, where she held the positions of Postdoctoral Research Fellow, Research Associate, and Visiting Scientist until the year 1977. Currently, she is a Retired Professor (Adjunct), having taught at the University of Victoria since 1977 until her retirement. In the year 1997, Professor Srivastava was nominated for the Best Teacher Award in the Faculty of Science at the University of Victoria. Professor Rekha Srivastava is currently associated with many scientific research journals as Editor or Guest Editor. She has published about 90 original papers in peer-reviewed international scientific research journals. Her current Google Scholar h-index is 24. She received the Distinguished Service Award, which was formally awarded to her on 14 December 2015 at the Eighteenth Annual Conference and the First International Conference of the Vijnana Parishad (Science Academy) of India at the Maulana Azad National Institute of Technology in Bhopal in the Province of Madhya Pradesh in India, which was held on 11–14 December 2015, for her outstanding contributions to Mathematics and for her distinguished services rendered to the Vijnana Parishad (Science Academy) of India. She also received the Fellowship Award (F.V.P.I.) on 26 November 2017 at the Twentieth Annual Conference of the Vijnana Parishad (Science Academy) of India at Manipal University in Jaipur in the Province of Rajasthan in India, which was held on 24–26 November 2017, for her outstanding professional contributions and scholarly achievements in Mathematics and Its Applications. Moreover, as a Member of the Organizing and Scientific Committees and also as an Invited Speaker, Professor Srivastava has participated in and delivered many lectures on her research work at a large number of international conferences held around the world. Professor Rekha Srivastava's research interests include several areas of pure and applied mathematical sciences, such as (for example) real and complex analysis, analytic number theory, fractional calculus and its applications, integral equations and integral transforms, and special functions and their applications. Further details about Professor Rekha Srivastava's professional achievements and scholarly accomplishments, as well as honors, awards, and distinctions, can be found at the following website: http://www.math.uvic.ca/~rekhas/.

Preface to "Mathematical Analysis and Analytic Number Theory 2019"

This volume contains a total of 22 peer-reviewed and accepted submissions (including several invited feature articles) from all over the world to the Special Issue of the MDPI journal, *Mathematics*, on the general subject area of "Mathematical Analysis and Analytic Number Theory".

This volume contains a total of 22 peer-reviewed and accepted submissions (including several invited feature articles) from all over the world to the Special Issue of the MDPI journal, Mathematics, on the general subject area of "Mathematical Analysis and Analytic Number Theory".

The suggested topics of interest for the call of papers for this Special Issue included, but were not limited to, the following themes:

- Theory and applications of the tools and techniques of mathematical analysis
- Theory and applications of the tools and techniques of analytic number theory
- Applications involving special (or higher transcendental) functions
- Applications involving fractional-order differential and differintegral equations
- Applications involving q-series and q-polynomials
- Applications involving special functions of mathematical physics and applied mathematics
- Applications involving geometric function theory of complex analysis
- Applications involving real analysis and operator theory

Finally, it gives me great pleasure to thank all of the participating authors in this Special Issue as well as the editorial personnel of the MDPI Editorial Office for *Mathematics* for their invaluable input and contributions toward the success of this Special Issue. The wholehearted support and dedication of one and all are indeed greatly appreciated.

Rekha Srivastava
Editor

Article

A Family of Theta-Function Identities Based upon Combinatorial Partition Identities Related to Jacobi's Triple-Product Identity

Hari Mohan Srivastava [1,2,3], Rekha Srivastava [1,*], Mahendra Pal Chaudhary [4] and Salah Uddin [5]

[1] Department of Mathematics and Statistics, University of Victoria, Victoria, BC V8W 3R4, Canada; harimsri@math.uvic.ca
[2] Department of Medical Research, China Medical University Hospital, China Medical University, Taichung 40402, Taiwan
[3] Department of Mathematics and Informatics, Azerbaijan University, 71 Jeyhun Hajibeyli Street, Baku AZ1007, Azerbaijan
[4] Department of Mathematics, Netaji Subhas University of Technology, Sector 3, Dwarka, New Delhi 110078, India; dr.m.p.chaudhary@gmail.com
[5] Department of Mathematics, PDM University, Bahadurgarh 124507, Haryana State, India; vsludn@gmail.com
* Correspondence: rekhas@math.uvic.ca

Received: 11 May 2020; Accepted: 3 June 2020; Published: 5 June 2020

Abstract: The authors establish a set of six new theta-function identities involving multivariable R-functions which are based upon a number of q-product identities and Jacobi's celebrated triple-product identity. These theta-function identities depict the inter-relationships that exist among theta-function identities and combinatorial partition-theoretic identities. Here, in this paper, we consider and relate the multivariable R-functions to several interesting q-identities such as (for example) a number of q-product identities and Jacobi's celebrated triple-product identity. Various recent developments on the subject-matter of this article as well as some of its potential application areas are also briefly indicated. Finally, we choose to further emphasize upon some close connections with combinatorial partition-theoretic identities and present a presumably open problem.

Keywords: theta-function identities; multivariable R-functions; Jacobi's triple-product identity; Ramanujan's theta functions; q-product identities; Euler's pentagonal number theorem; Rogers-Ramanujan continued fraction; Rogers-Ramanujan identities; combinatorial partition-theoretic identities; Schur's, the Göllnitz-Gordon's and the Göllnitz's partition identities; Schur's second partition theorem

1. Introduction and Definitions

Throughout this article, we denote by $\mathbb{N}$, $\mathbb{Z}$, and $\mathbb{C}$ the set of positive integers, the set of integers and the set of complex numbers, respectively. We also let

$$\mathbb{N}_0 := \mathbb{N} \cup \{0\} = \{0, 1, 2, \cdots\}.$$

In what follows, we shall make use of the following q-notations for the details of which we refer the reader to a recent monograph on q-calculus by Ernst [1] and also to the earlier works on the subject by Slater [2] (Chapter 3, Section 3.2.1), and by Srivastava et al. ([3] (pp. 346 et seq.) and [4] (Chapter 6)).

The q-shifted factorial $(a;q)_n$ is defined (for $|q|<1$) by

$$(a;q)_n := \begin{cases} 1 & (n=0) \\ \prod_{k=0}^{n-1}(1-aq^k) & (n \in \mathbb{N}), \end{cases} \quad (1)$$

where $a, q \in \mathbb{C}$, and it is assumed *tacitly* that $a \neq q^{-m}$ $(m \in \mathbb{N}_0)$. We also write

$$(a;q)_\infty := \prod_{k=0}^{\infty}(1-aq^k) = \prod_{k=1}^{\infty}(1-aq^{k-1}) \qquad (a, q \in \mathbb{C}; \ |q| < 1). \quad (2)$$

It should be noted that, when $a \neq 0$ and $|q| \geqq 1$, the infinite product in Equation (2) diverges. Thus, whenever $(a;q)_\infty$ is involved in a given formula, the constraint $|q| < 1$ will be *tacitly* assumed to be satisfied.

The following notations are also frequently used in our investigation:

$$(a_1, a_2, \cdots, a_m; q)_n := (a_1; q)_n (a_2; q)_n \cdots (a_m; q)_n \quad (3)$$

and

$$(a_1, a_2, \cdots, a_m; q)_\infty := (a_1; q)_\infty (a_2; q)_\infty \cdots (a_m; q)_\infty. \quad (4)$$

Ramanujan (see [5,6]) defined the general theta function $\mathfrak{f}(a,b)$ as follows (see, for details, in [7] (p. 31, Equation (18.1)) and [8,9]):

$$\begin{aligned} \mathfrak{f}(a,b) &= 1 + \sum_{n=1}^{\infty}(ab)^{\frac{n(n-1)}{2}}(a^n + b^n) \\ &= \sum_{n=-\infty}^{\infty} a^{\frac{n(n+1)}{2}} b^{\frac{n(n-1)}{2}} = \mathfrak{f}(b,a) \qquad (|ab|<1). \end{aligned} \quad (5)$$

We find from this last Equation (5) that

$$\mathfrak{f}(a,b) = a^{\frac{n(n+1)}{2}} b^{\frac{n(n-1)}{2}} \mathfrak{f}\big(a(ab)^n, b(ab)^{-n}\big) = \mathfrak{f}(b,a) \qquad (n \in \mathbb{Z}). \quad (6)$$

In fact, Ramanujan (see [5,6]) also rediscovered Jacobi's famous triple-product identity, which, in Ramanujan's notation, is given by (see [7] (p. 35, Entry 19)):

$$\mathfrak{f}(a,b) = (-a; ab)_\infty \, (-b; ab)_\infty \, (ab; ab)_\infty \quad (7)$$

or, equivalently, by (see [10])

$$\begin{aligned} \sum_{n=-\infty}^{\infty} q^{n^2} z^n &= \prod_{n=1}^{\infty}\left(1 - q^{2n}\right)\left(1 + zq^{2n-1}\right)\left(1 + \frac{1}{z}q^{2n-1}\right) \\ &= \left(q^2; q^2\right)_\infty \left(-zq; q^2\right)_\infty \left(-\frac{q}{z}; q^2\right)_\infty \qquad (|q|<1; \ z \neq 0). \end{aligned}$$

Remark 1. *Equation (6) holds true as stated only if n is any integer. In case n is not an integer, this result (6) is only approximately true (see, for details, [5] (Vol. 2, Chapter XVI, p. 193, Entry 18 (iv))). Moreover, historically speaking, the q-series identity (7) or its above-mentioned equivalent form was first proved by Carl Friedrich Gauss (1777–1855).*

Several q-series identities, which emerge naturally from Jacobi's triple-product identity (7), are worthy of note here (see, for details, (pp. 36–37, Entry 22) in [7]):

$$\varphi(q) := \sum_{n=-\infty}^{\infty} q^{n^2} = 1 + 2 \sum_{n=1}^{\infty} q^{n^2}$$
$$= \left\{(-q;q^2)_\infty\right\}^2 (q^2;q^2)_\infty = \frac{(-q;q^2)_\infty (q^2;q^2)_\infty}{(q;q^2)_\infty (-q^2;q^2)_\infty}; \tag{8}$$

$$\psi(q) := \mathfrak{f}(q,q^3) = \sum_{n=0}^{\infty} q^{\frac{n(n+1)}{2}} = \frac{(q^2;q^2)_\infty}{(q;q^2)_\infty}; \tag{9}$$

$$f(-q) := \mathfrak{f}(-q,-q^2) = \sum_{n=-\infty}^{\infty} (-1)^n q^{\frac{n(3n-1)}{2}}$$
$$= \sum_{n=0}^{\infty} (-1)^n q^{\frac{n(3n-1)}{2}} + \sum_{n=1}^{\infty} (-1)^n q^{\frac{n(3n+1)}{2}} = (q;q)_\infty. \tag{10}$$

Equation (10) is known as Euler's *Pentagonal Number Theorem*. Remarkably, the following q-series identity:

$$(-q;q)_\infty = \frac{1}{(q;q^2)_\infty} = \frac{1}{\chi(-q)} \tag{11}$$

provides the analytic equivalent form of Euler's famous theorem (see, for details, [11,12]).

Theorem 1. *(Euler's Pentagonal Number Theorem) The number of partitions of a given positive integer n into distinct parts is equal to the number of partitions of n into odd parts.*

We also recall the Rogers-Ramanujan continued fraction $R(q)$ given by

$$R(q) := q^{\frac{1}{5}} \frac{H(q)}{G(q)} = q^{\frac{1}{5}} \frac{\mathfrak{f}(-q,-q^4)}{\mathfrak{f}(-q^2,-q^3)} = q^{\frac{1}{5}} \frac{(q;q^5)_\infty (q^4;q^5)_\infty}{(q^2;q^5)_\infty (q^3;q^5)_\infty}$$
$$= \frac{q^{\frac{1}{5}}}{1+} \frac{q}{1+} \frac{q^2}{1+} \frac{q^3}{1+} \cdots \quad (|q|<1). \tag{12}$$

Here, $G(q)$ and $H(q)$, which are associated with the widely-investigated Roger-Ramanujan identities, are defined as follows:

$$G(q) := \sum_{n=0}^{\infty} \frac{q^{n^2}}{(q;q)_n} = \frac{f(-q^5)}{\mathfrak{f}(-q,-q^4)}$$
$$= \frac{1}{(q;q^5)_\infty (q^4;q^5)_\infty} = \frac{(q^2;q^5)_\infty (q^3;q^5)_\infty (q^5;q^5)_\infty}{(q;q)_\infty} \tag{13}$$

and

$$H(q) := \sum_{n=0}^{\infty} \frac{q^{n(n+1)}}{(q;q)_n} = \frac{f(-q^5)}{\mathfrak{f}(-q^2,-q^3)} = \frac{1}{(q^2;q^5)_\infty (q^3;q^5)_\infty}$$
$$= \frac{(q;q^5)_\infty (q^4;q^5)_\infty (q^5;q^5)_\infty}{(q;q)_\infty}, \tag{14}$$

and the functions $\mathfrak{f}(a,b)$ and $f(-q)$ are given by Equations (5) and (10), respectively.

For a detailed historical account of (and for various related developments stemming from) the Rogers-Ramanujan continued fraction (12) as well as the Rogers-Ramanujan identities (13) and (14), the interested reader may refer to the monumental work [7] (p. 77 et seq.) (see also [4,8]).

The following continued-fraction results may be recalled now (see, for example, (p. 5, Equation (2.8)) in [13]).

Theorem 2. *Suppose that $|q| < 1$. Then,*

$$A(q) := (q^2;q^2)_\infty (-q;q)_\infty$$
$$= \frac{(q^2;q^2)_\infty}{(q;q^2)_\infty} = \frac{1}{1-}\frac{q}{1+}\frac{q(1-q)}{1-}\frac{q^3}{1+}\frac{q^2(1-q^2)}{1-}\frac{q^5}{1+}\frac{q^3(1-q^3)}{1-}\cdots$$
$$= \cfrac{1}{1-\cfrac{q}{1+\cfrac{q(1-q)}{1-\cfrac{q^3}{1+\cfrac{q^2(1-q^2)}{1-\cfrac{q^5}{1+\cfrac{q^3(1-q^3)}{1-\cdots}}}}}}}, \qquad (15)$$

$$B(q) := \frac{(q;q^5)_\infty (q^4;q^5)_\infty}{(q^2;q^5)_\infty (q^3;q^5)_\infty} = \frac{1}{1+}\frac{q}{1+}\frac{q^2}{1+}\frac{q^3}{1+}\frac{q^4}{1+}\frac{q^5}{1+}\frac{q^6}{1+}\cdots$$
$$= \cfrac{1}{1+\cfrac{q}{1+\cfrac{q^2}{1+\cfrac{q^3}{1+\cfrac{q^4}{1+\cfrac{q^5}{1+\cfrac{q^6}{1+\cdots}}}}}}} \qquad (16)$$

and

$$C(q) := \frac{(q^2;q^5)_\infty (q^3;q^5)_\infty}{(q;q^5)_\infty (q^4;q^5)_\infty} = 1 + \frac{q}{1+}\frac{q^2}{1+}\frac{q^3}{1+}\frac{q^4}{1+}\frac{q^5}{1+}\frac{q^6}{1+}\cdots$$
$$= 1 + \cfrac{q}{1+\cfrac{q^2}{1+\cfrac{q^3}{1+\cfrac{q^4}{1+\cfrac{q^5}{1+\cfrac{q^6}{1+\cdots}}}}}}. \qquad (17)$$

By introducing the general family $R(s,t,l,u,v,w)$, Andrews et al. [14] investigated a number of interesting double-summation hypergeometric q-series representations for several families of partitions and further explored the rôle of double series in combinatorial-partition identities:

$$R(s,t,l,u,v,w) := \sum_{n=0}^\infty q^{s\binom{n}{2}+tn} \, r(l,u,v,w;n), \qquad (18)$$

where

$$r(l, u, v, w : n) := \sum_{j=0}^{[\frac{n}{u}]} (-1)^j \frac{q^{uv\binom{j}{2}+(w-ul)j}}{(q;q)_{n-uj}\,(q^{uv};q^{uv})_j}. \tag{19}$$

We also recall the following interesting special cases of (18) (see, for details, (p. 106, Theorem 3) in [14]; see also [8]):

$$R(2,1,1,1,2,2) = (-q;q^2)_\infty, \tag{20}$$

$$R(2,2,1,1,2,2) = (-q^2;q^2)_\infty \tag{21}$$

and

$$R(m,m,1,1,1,2) = \frac{(q^{2m};q^{2m})_\infty}{(q^m;q^{2m})_\infty}. \tag{22}$$

For the sake of brevity in our presentation of the main results, we now introduce the following notations:

$$R_\alpha = R(2,1,1,1,2,2),$$
$$R_\beta = R(2,2,1,1,2,2)$$

and

$$R_m = R(m,m,1,1,1,2) \quad (m \in \mathbb{N}).$$

Ever since the year 2015, several new advancements and generalizations of the existing results were made in regard to combinatorial partition-theoretic identities (see, for example, [8,15–24]). In particular, Chaudhary et al. generalized several known results on character formulas (see [22]), Roger-Ramanujan type identities (see [19]), Eisenstein series, the Ramanujan-Göllnitz-Gordon continued fraction (see [20]), the 3-dissection property (see [18]), Ramanujan's modular equations of degrees 3, 7, and 9 (see [16,17]), and so on, by using combinatorial partition-theoretic identities. An interesting recent investigation on the subject of combinatorial partition-theoretic identities by Hahn et al. [25] is also worth mentioning in this connection.

Here, in this paper, our main objective is to establish a set of six new theta-function identities which depict the inter-relationships that exist between the multivariable R-functions, q-product identities, and partition-theoretic identities.

Each of the following preliminary results will be needed for the demonstration of our main results in this paper (see [26] (pp. 1749–1750 and 1752–1754)):

I. If

$$P = \frac{\psi(q)}{q^{\frac{1}{2}}\psi(q^5)} \quad \text{and} \quad Q = \frac{\psi(q^3)}{q^{\frac{3}{2}}\psi(q^{15})},$$

then

$$PQ + \frac{5}{PQ} = \left(\frac{Q}{P}\right)^2 - \left(\frac{P}{Q}\right)^2 + 3\left(\frac{Q}{P} + \frac{P}{Q}\right). \tag{23}$$

II. If

$$P = \frac{\psi(q)}{q^{\frac{1}{4}}\psi(q^3)} \quad \text{and} \quad Q = \frac{\psi(q^5)}{q^{\frac{5}{4}}\psi(q^{15})},$$

then

$$(PQ)^2 + \left(\frac{3}{PQ}\right)^2 = \left(\frac{Q}{P}\right)^3 - \left(\frac{P}{Q}\right)^3 - 5\left(\frac{Q}{P} - \frac{P}{Q}\right) + 5\left(\frac{P}{Q}\right)^2 + 5\left(\frac{Q}{P}\right)^2. \tag{24}$$

III. If
$$P = \frac{\psi(q)}{q^{\frac{1}{4}}\psi(q^3)} \quad \text{and} \quad Q = \frac{\psi(q^7)}{q^{\frac{7}{4}}\psi(q^{21})},$$
then
$$(PQ)^3\left[\left(\frac{P}{Q}\right)^8 - 1\right] + 14P^5Q\left[\left(\frac{P}{Q}\right)^4 - 1\right] = P^6Q^2(7 - P^4) + \frac{7P^6}{Q^2}(P^4 - 3)$$
$$- \left\{27\left(\frac{P}{Q}\right)^4 - 7P^4\left[3 + 3\left(\frac{P}{Q}\right)^4 - P^4\right]\right\}. \tag{25}$$

IV. If
$$P = \frac{\psi(q)}{q^{\frac{1}{4}}\psi(q^3)} \quad \text{and} \quad Q = \frac{\psi(q^2)}{q^{\frac{1}{2}};\psi(q^6)},$$
then
$$\left(\frac{P}{Q}\right)^2 + \frac{3}{P^2} - P^2 + \left(\frac{Q}{P}\right)^2 = 0. \tag{26}$$

V. If
$$P = \frac{\psi(-q)}{q^{\frac{1}{4}}\psi(-q^3)} \quad \text{and} \quad Q = \frac{\psi(q^2)}{q^{\frac{1}{2}}\psi(q^6)},$$
then
$$\left(\frac{P}{Q}\right)^2 + \frac{3}{P^2} + P^2 - \left(\frac{Q}{P}\right)^2 = 0. \tag{27}$$

VI. If
$$P = \frac{\psi(-q)}{q^{\frac{1}{4}}\psi(-q^3)} \quad \text{and} \quad Q = \frac{\psi(q)}{q^{\frac{1}{4}}\psi(q^3)},$$
then
$$\left[\left(\frac{P}{Q}\right)^2 - \left(\frac{Q}{P}\right)^2\right] \cdot \left[\left(\frac{3}{PQ}\right)^2 - (PQ)^2\right] + \left(\frac{P}{Q}\right)^4 + \left(\frac{Q}{P}\right)^4 - 10 = 0. \tag{28}$$

2. A Set of Main Results

In this section, we state and prove a set of six new theta-function identities which depict inter-relationships among q-product identities and the multivariate R-functions.

Theorem 3. *Each of the following relationships holds true:*

$$\frac{R_1R_3}{R_5R_{15}} = \left(\frac{R_3R_5}{R_1R_{15}}\right)^2 - \left(\frac{q^2R_1R_{15}}{R_3R_5}\right)^2$$
$$+ \left(\frac{3qR_3R_5}{R_1R_{15}}\right) + \left(\frac{3q^3R_1R_{15}}{R_3R_5}\right) - \left(\frac{5q^4R_5R_{15}}{R_1R_3}\right) \tag{29}$$

and

$$\left(\frac{R_1R_5}{R_3R_{15}}\right)^2 = \left(\frac{R_3R_5}{R_1R_{15}}\right)^3 - \left(\frac{q^2R_1R_{15}}{R_3R_5}\right)^3 - \left(\frac{5q^2R_3R_5}{R_1R_{15}}\right) + \left(\frac{5q^4R_1R_{15}}{R_3R_5}\right)$$
$$+ 5q^5\left(\frac{R_1R_{15}}{R_3R_5}\right)^2 + 5q\left(\frac{R_3R_5}{R_1R_{15}}\right)^2 - \left(\frac{3q^3R_3R_{15}}{R_1R_5}\right)^2. \tag{30}$$

Equations (29) and (30) give inter-relationships between R_1, R_3, R_5 and R_{15}.

$$\left(\frac{R_1 R_7}{q^2 R_3 R_{21}}\right)^3 \cdot \left(\frac{q^{12}[R_1 R_{21}]^8}{[R_3 R_7]^8} - 1\right) = \frac{1}{q^5}\left(\frac{[R_1]^3 R_7}{[R_3]^3 R_{21}}\right)^2 \left(7 - \frac{[R_1]^4}{q[R_3]^4}\right)$$
$$+ \left(\frac{q[R_1]^3 R_{21}}{[R_3]^3 R_7}\right)^2 \left(\frac{[R_1]^4}{q[R_3]^4} - 3\right) - \frac{14}{q^3}\left(\frac{[R_1]^5 R_7}{[R_3]^5 R_{21}}\right) \cdot \left(\frac{q^6[R_1 R_{21}]^4}{[R_3 R_7]^4} - 1\right)$$
$$- 27 q^6 \left(\frac{R_1 R_{21}}{R_3 R_7}\right)^4 + \frac{21}{q}\left(\frac{R_1}{R_3}\right)^4 + 21 q^5 \left(\frac{[R_1]^2 R_{21}}{[R_3]^2 R_7}\right)^4 - \frac{7}{q^2}\left(\frac{R_1}{R_3}\right)^8. \tag{31}$$

Equation (31) gives inter-relationships between R_1, R_3, R_7, and R_{21}.

$$\left(\frac{R_1}{R_3}\right)^2 = \left(\frac{q^{\frac{1}{2}} R_1 R_6}{R_2 R_3}\right)^2 + \left(\frac{(3q)^{\frac{1}{2}} R_3}{R_1}\right)^2 + \left(\frac{R_2 R_3}{R_1 R_6}\right)^2. \tag{32}$$

Equation (32) gives inter-relationships between R_1, R_2, R_3, and R_6.

$$\left(\frac{R_\alpha R_2 (q^6; q^6)_\infty}{R_6 (q^2; q^2)_\infty (-q^3; q^6)_\infty}\right)^2 = \left(\frac{(3q)^{\frac{1}{2}} R_\alpha (q^6; q^6)_\infty}{(q^2; q^2)_\infty (-q^3; q^6)_\infty}\right)^2 + \left(\frac{(q^2; q^2)_\infty (-q^3; q^6)_\infty}{R_\alpha (q^6; q^6)_\infty}\right)^2$$
$$+ \left(\frac{q^{\frac{1}{2}} R_6 (q^2; q^2)_\infty (-q^3; q^6)_\infty}{R_\alpha R_2 (q^6; q^6)_\infty}\right)^2. \tag{33}$$

Equation (33) gives inter-relationships between R_2, R_6, and R_α. Furthermore, it is asserted that

$$\left(\frac{R_3 (q^2; q^2)_\infty (-q^3; q^6)_\infty}{R_1 R_\alpha (q^6; q^6)_\infty}\right)^4 + \left(\frac{R_1 R_\alpha (q^6; q^6)_\infty}{R_3 (q^2; q^2)_\infty (-q^3; q^6)_\infty}\right)^4$$
$$+ \left[\left(\frac{R_3 (q^2; q^2)_\infty (-q^3; q^6)_\infty}{R_1 R_\alpha (q^6; q^6)_\infty}\right)^2 - \left(\frac{R_1 R_\alpha (q^6; q^6)_\infty}{R_3 (q^2; q^2)_\infty (-q^3; q^6)_\infty}\right)^2\right]$$
$$\cdot \left[\left(\frac{3 q^{\frac{1}{2}} R_\alpha R_3 (q^6; q^6)_\infty}{R_1 (q^2; q^2)_\infty (-q^3; q^6)_\infty}\right)^2 - \left(\frac{R_1 (q^2; q^2)_\infty (-q^3; q^6)_\infty}{q^{\frac{1}{2}} R_\alpha R_3 (q^6; q^6)_\infty}\right)^2\right] - 10 = 0. \tag{34}$$

Equation (34) gives inter-relationships between R_1, R_3 and R_α.
It is assumed that each member of the assertions (29) to (34) exists.

Proof. First of all, in order to prove the assertion (29) of Theorem 3, we apply the identity (9) (with q replaced by q^3, q^5 q^{15}) under the given precondition of result (23). Thus, by using (20) and (21), and, after some simplifications, we get the values for P and Q as follows:

$$P = \frac{\psi(q)}{q^{\frac{1}{2}} \psi(q^5)} = \frac{R_1}{q^{\frac{1}{2}} R_5} \tag{35}$$

and

$$Q = \frac{\psi(q^3)}{q^{\frac{3}{2}} \psi(q^{15})} = \frac{R_3}{q^{\frac{3}{2}} R_{15}}. \tag{36}$$

Now, upon substituting from these last results (35) and (36) into (23), if we rearrange the terms and use some algebraic manipulations, we are led to the first assertion (29) of Theorem 3.

Secondly, we prove the second relationship (30) of Theorem 3. Indeed, if we first apply the identity (9) (with q replaced by q^3, q^5 and q^{15}) under the given precondition of the assertion (24), and then make use of (20) and (21), after some simplifications, the following values for P and Q would follow:

$$P = \frac{\psi(q)}{q^{\frac{1}{4}}\psi(q^3)} = \frac{R_1}{q^{\frac{1}{4}}R_3} \tag{37}$$

and

$$Q = \frac{\psi(q^5)}{q^{\frac{5}{4}}\psi(q^{15})} = \frac{R_5}{q^{\frac{5}{4}}R_{15}}. \tag{38}$$

Now, upon substituting from these last results (37) and (38) into (24), if we rearrange the terms and use some algebraic manipulations, we obtain the second assertion (30) of Theorem 3.

Thirdly, we prove the third relationship (31) of Theorem 3. For this purpose, we first apply the identity (9) (with q replaced by q^3, q^7 and q^{21}) under the given precondition of (25), and then use (20) and (21). We thus find for the values of P and Q that

$$P = \frac{\psi(q)}{q^{\frac{1}{4}}\psi(q^3)} = \frac{R_1}{q^{\frac{1}{4}}R_3} \tag{39}$$

and

$$Q = \frac{\psi(q^7)}{q^{\frac{7}{4}}\psi(q^{21})} = \frac{R_7}{q^{\frac{7}{4}}R_{21}}, \tag{40}$$

which, in view of (25) and after some rearrangements of the terms and the resulting algebraic manipulations, yields the third assertion (31) of Theorem 3.

Fourthly, we prove the identity (32) by applying the identity (9) (with the parameter q replaced by q^2, q^3 and q^6) under the given precondition of (26), we further use the assertions (20) and (21). Then, upon simplifications, we get the values for P and Q as follows:

$$P = \frac{\psi(q)}{q^{\frac{1}{4}}\psi(q^3)} = \frac{R_1}{q^{\frac{1}{2}}R_3} \tag{41}$$

and

$$Q = \frac{\psi(q^2)}{q^{\frac{1}{2}}\psi(q^6)} = \frac{R_2}{q^{\frac{1}{2}}R_6}. \tag{42}$$

Now, after using (41) and (42) in (26), if we rearrange the terms and and apply some algebraic manipulations, we get required result (32) asserted by Theorem 3.

We next prove the fifth identity (33). We apply the identity (9) (with the parameter q replaced by $-q$, $-q^3$, q^2 and q^6) under the given precondition of (27). We then further use the results (20) and (21). After simplification, we find the values for P and Q as follows:

$$P = \frac{\psi(-q)}{q^{\frac{1}{4}}\psi(-q^3)} = \frac{(q^2;q^2)_\infty (-q^3;q^6)_\infty}{q^{\frac{1}{4}} R_\alpha (q^6;q^6)_\infty} \tag{43}$$

and

$$Q = \frac{\psi(q^2)}{q^{\frac{1}{2}}\psi(q^6)} = \frac{R_2}{q^{\frac{1}{2}}R_6}. \tag{44}$$

Now, after using (43) and (44) in (27), we rearrange the terms and apply some algebraic manipulations. We are thus led to the required result (33).

Finally, we proceed to prove the last identity (34) asserted by Theorem 3. We make use of the identity (9) (with the parameter q replaced by $-q$, $-q^3$ and q^3) under the given precondition of (28). Then, by applying the identities (20) and (21), we obtain the values for P and Q as follows:

$$P = \frac{\psi(-q)}{q^{\frac{1}{4}}\psi(-q^3)} = \frac{(q^2;q^2)_\infty (-q^3;q^6)_\infty}{q^{\frac{1}{4}} R_\alpha(q^6;q^6)_\infty} \qquad (45)$$

and

$$Q = \frac{\psi(q)}{q^{\frac{1}{4}}\psi(q^3)} = \frac{R_1}{q^{\frac{1}{4}} R_3}. \qquad (46)$$

Thus, upon using (45) and (46) in (28), we rearrange the terms and apply some algebraic simplifications. This leads us to the required result (34), thereby completing the proof of Theorem 3. □

3. Applications Based upon Ramanujan's Continued-Fraction Identities

In this section, we first suggest some possible applications of our findings in Theorem 3 within the context of continued fraction identities. We begin by recalling that Naika et al. [27] studied the following continued fraction:

$$U(q) := \frac{q(1-q)}{(1-q^3)+} \frac{q^3(1-q^2)(1-q^4)}{(1-q^3)(1+q^6)+ \cdots +} \frac{q^3(1-q^{6n-4})(1-q^{6n-2})}{(1-q^3)(1+q^{6n})+ \cdots}, \qquad (47)$$

which is a special case of a fascinating continued fraction recorded by Ramanujan in his second notebook [5,28,29]. On the other hand, Chaudhary et al. (see p. 861, Equations (3.1) to (3.5)) developed the following identities for the continued fraction $U(q)$ in (47) by using such R-functions as (for example) $R(1,1,1,1,1,2)$, $R(2,2,1,1,2,2)$, $R(2,1,1,1,2,2)$, $R(3,3,1,1,1,2)$ and $R(6,6,1,1,1,2)$:

$$\frac{1}{U(q)} + U(q) = \frac{R(1,1,1,1,1,2)R(2,2,1,1,2,2)}{\{R(2,1,1,1,2,2)\}^2} \cdot \{(q^3;q^6)_\infty (q^6;q^{12})_\infty\}^3, \qquad (48)$$

$$\frac{1}{\sqrt{U(q)}} + \sqrt{U(q)} = \frac{R(2,1,1,1,2,2)}{R(2,2,1,1,2,2)} \left\{ \frac{R(1,1,1,1,1,2)R(2,2,1,1,2,2)}{q\, R(3,3,1,1,1,2)R(6,6,1,1,1,2)} \right\}^{\frac{1}{2}}, \qquad (49)$$

$$\frac{1}{\sqrt{U(q)}} - \sqrt{U(q)} = f(-q,q^3)$$

$$\cdot \left\{ \frac{R(1,1,1,1,1,2)\{R(2,2,1,1,2,2)\}^2}{q\, R(6,6,1,1,1,2)R(3,3,1,1,1,2)R(2,2,1,1,1,2)} \right\}^{\frac{1}{2}}, \qquad (50)$$

$$\frac{1}{\sqrt{U(q)}} + \sqrt{U(q)} + 2 = \frac{R(2,1,1,1,2,2)\{R(1,1,1,1,1,2)\}^2}{q\, R(6,6,1,1,1,2)R(3,3,1,1,1,2)R(2,2,1,1,2,2)} \qquad (51)$$

and

$$\frac{1}{\sqrt{U(q)}} + \sqrt{U(q)} - 2 = \frac{R(2,2,1,1,1,2)\{R(3,3,1,1,1,2)\}^3}{q\, R(1,1,1,1,1,2)\{R(6,6,1,1,1,2)\}^3}. \qquad (52)$$

By using the above formulas (48) to (52), we can express our results (29) to (34) in Theorem 3 in terms of Ramanujan's continued fraction $U(q)$ given here by (47).

Remark 2. *Even though the results of Theorem 3 are apparently considerably involved, each of the asserted theta-function identities does have the potential for other applications in analytic number theory and partition theory (see, for example, [30,31]) as well as in real and complex analysis, especially in connection with a significant number of wide-spread problems dealing with various basic (or q-) series and basic (or q-) operators (see, for example, [32,33]).*

Each of the theta-function identities (29) to (34), which are asserted by Theorem 3, obviously depict the inter-relationships that exist between q-product identities and the multivariate R-functions. Some corollaries and consequences of Theorem 3 may be worth pursuing for further research in the direction of the developments which we have presented in this article.

4. Connections with Combinatorial Partition-Theoretic Identities

Various extensions and generalizations of partition-theoretic identities and other q-identities, which we have investigated in this paper, as well as their connections with combinatorial partition-theoretic identities, can be found in several recent works (see, for example, [31,34,35]). The demonstrations in some of these recent developments are also based upon their combinatorial interpretations and generating functions (see also [25]).

As far as the connections with many different partition-theoretic identities are concerned, the existing literature is full of interesting findings and observations on the subject. In fact, in the year 2015, valuable progress in this direction was made by Andrews et al. [14], who established a number of interesting results including those for the q-series, q-products, and q-hypergeometric functions, which are associated closely with Schur's partitions, the Göllnitz-Gordon's partitions, and the Göllnitz's partitions in terms of multivariate R-functions. With a view to making our presentation to be self-sufficient, we choose to recall here some relevant parts of the developments in the remarkable investigation by Andrews et al. (see, for details, [14]).

We consider an integer partition of λ with parts $\lambda_1 \geqq \cdots \geqq \lambda_\ell$ and denote, as usual, its size by

$$|\lambda| := \lambda_1 + \cdots + \lambda_\ell$$

and its length (that is, the number of parts) by $\ell(\lambda)$ (see, for details, [36]).

Let us now assume that S denotes the set of Schur's partitions of λ such that

$$\lambda_j - \lambda_{j+1} > 3 \qquad (1 \leqq j \leqq \ell - 1),$$

with a strict inequality. We recall Schur's partitions as follows:

$$f_S(x;q) := \sum_{\lambda \in S} x^{\ell(\lambda)} q^{|\lambda|}, \qquad (53)$$

which is of special interest here due to the following strikingly important infinite-product identity known as *Schur's Second Partition Theorem* (see [37]):

$$f_S(1;q) = (-q;q^3)_\infty (-q^2;q^3)_\infty \qquad (54)$$

In fact, Equation (54) yields a double-series representation for the two-parameter generating function for Schur partitions, which is given below:

$$f_S(x;q) = \sum_{m,n \geqq 0} \frac{(-1)^n x^{m+2n} q^{(m+3n)^2 + \frac{m(m-1)}{2}}}{(q;q)_m (q^6;q^6)_n} \qquad (55)$$

or, alternatively, as follows (see [14] (p. 103)):

$$f_S(x;q) = (x;q^3)_\infty \sum_{n \geqq 0} \frac{x^n (-q, -q^2; q^3)_n}{(q^3; q^3)_n}.$$

We next suppose that GG denotes the set of the Göllnitz-Gordon partitions which satisfy the following inequality:

$$\lambda_j - \lambda_{j+1} \geqq 2 \qquad (1 \leqq j \leqq \ell - 1) \tag{56}$$

with strict inequality if either part is even. A direct combinatorial argument would now show that

$$f_{GG}(x;q) := \sum_{\lambda \in GG} x^{\ell(\lambda)} q^{|\lambda|} = \sum_{n \geq 0} \frac{x^n q^{n^2} (-q; q^2)_n}{(q^2; q^2)_n} \tag{57}$$

Hence, clearly, we have a new double-series representation of the generating function for the Göllnitz-Gordon partitions, which is given below:

$$f_{GG}(x;q) = \sum_{k,m \geq 0} \frac{(-1)^k x^{m++2k} q^{m^2+4mk+6k^2}}{(q;q)_m (q^4;q^4)_k}. \tag{58}$$

We also let G denote the set of the Göllnitz partitions which satisfy the following inequality:

$$\lambda_j - \lambda_{j+1} \geqq 2 \qquad (1 \leqq j \leqq \ell - 1)$$

with strict inequality if either part is odd. Then, the corresponding generating function for the Göllnitz partitions is given by

$$f_G(x;q) := \sum_{\lambda \in G} x^{\ell(\lambda)} q^{|\lambda|}. \tag{59}$$

We thus find the following double-series representation of the generating function for the Göllnitz partitions:

$$f_G(x;q) = \sum_{k,m \geq 0} \frac{(-1)^k x^{m++2k} q^{m^2+4mk+6k^2-2k}}{(q;q)_m (q^4;q^4)_k}. \tag{60}$$

Remark 3. *As pointed out by Andrews et al.* [14] *(p. 105, Equations (1.8) and (1.9)), alternative double-series representations for the double series in Equations (58) and (60) were given in an earlier publication by Alladi and Berkovich* [38].

In order to illustrate the connections of the above-mentioned partition-theoretic identities with the multivariable R-functions given by Equations (18) and (19), we note that the Schur's, the Göllnitz-Gordon and the Göllnitz partition identities can be expressed as follows:

$$R(3,t,0,2,3,4) = f_S(q^{t-1};q), \tag{61}$$

$$R(2,t,0,2,2,2) = f_{GG}(q^{t-1};q) \tag{62}$$

and

$$R(2,t,1,2,2,2) = f_G(q^{t-1};q). \tag{63}$$

5. An Open Problem

Based upon the work presented in this paper, we find it to be worthwhile to motivate the interested reader to consider the following related open problem.

Open Problem. Find inter-relationships between R_β and R_α, R_m ($m \in \mathbb{N}$), q-product identities and continued-fraction identities.

6. Concluding Remarks and Observations

The present investigation was motivated by several recent developments dealing essentially with theta-function identities and combinatorial partition-theoretic identities. Here, in this article, we have established a family of six presumably new theta-function identities which depict the inter-relationships that exist among q-product identities and combinatorial partition-theoretic identities. We have also considered several closely-related identities such as (for example) q-product identities and Jacobi's triple-product identities. In addition, with a view to further motivating research involving theta-function identities and combinatorial partition-theoretic identities, we have chosen to indicate rather briefly a number of recent developments on the subject-matter of this article.

The list of citations, which we have included in this article, is believed to be potentially useful for indicating some of the directions for further research and related developments on the subject-matter which we have dealt with here. In particular, the recent works by Adiga et al. (see [28,39]), Cao et al. [40], Chaudhary et al. (see [13,21,22]), Hahn et al. [25], and Srivastava et al. (see [23,29,33,41–45]), and by Yee [35] and Yi [31], are worth mentioning here.

Author Contributions: Conceptualization, M.P.C., H.M.S.; Formal analysis, H.M.S.; Funding acquisition, R.S.; Investigation, R.S., M.P.C. and S.U.; Methodology, H.M.S., M.P.C. and S.U.; Supervision, H.M.S. and R.S. All authors have read and agreed to the published version of the manuscript.

Funding: This research received no external funding.

Conflicts of Interest: All four authors declare that they have no conflict of interest.

References

1. Ernst, T. *A Comprehensive Treatment of q-Calculus*; Birkhäuser/Springer: Basel, Switzerland, 2012.
2. Slater, L.J. *Generalized Hypergeometric Functions*; Cambridge University Press: Cambridge, UK, 1966.
3. Srivastava, H.M.; Karlsson, P.W. *Multiple Gaussian Hypergeometric Series*; Halsted Press: Sydney, Australia; John Wiley and Sons: New York, NY, USA; Chichester, UK; Brisbane, Australia; Toronto, ON, Canada, 1985.
4. Srivastava, H.M.; Choi, J. *Zeta and q-Zeta Functions and Associated Series and Integrals*; Elsevier Science Publishers: Amsterdam, The Netherlands, 2012.
5. Ramanujan, S. *Notebooks*; Tata Institute of Fundamental Research: Bombay, India, 1957; Volumes 1 and 2.
6. Ramanujan, S. *The Lost Notebook and Other Unpublished Papers*; Narosa Publishing House: New Delhi, India, 1988.
7. Berndt, B.C. *Ramanujan's Notebooks*; Part III; Springer: Berlin/Heidelberg, Germany; New York, NY, USA, 1991.
8. Srivastava, H.M.; Chaudhary, M.P. Some relationships between q-product identities, combinatorial partition identities and continued-fraction identities. *Adv. Stud. Contemp. Math.* **2015**, *25*, 265–272.
9. Baruah, N.D.; Bora, J. Modular relations for the nonic analogues of the Rogers-Ramanujan functions with applications to partitions. *J. Number Theory* **2008**, *128*, 175–206. [CrossRef]
10. Jacobi, C.G.J. *Fundamenta Nova Theoriae Functionum Ellipticarum*; Regiomonti, Sumtibus Fratrum Bornträger: Königsberg, Germany, 1829; Reprinted in *Gesammelte Mathematische Werke* **1829**, *1*, 497–538; American Mathematical Society: Providence, RI, USA, 1969; pp. 97–239.
11. Hardy, G.H.; Wright, E.M. *An Introduction to the Theory of Numbers*, 6th ed.; Oxford University Press: London, UK; New York, NY, USA, 2008.
12. Apostol, T.M. *Introduction to Analytic Number Theory*; Undergraduate Texts in Mathematics; Springer: Berlin/Heidelberg, Germany, 1976.
13. Chaudhary, M.P. Generalization of Ramanujan's identities in terms of q-products and continued fractions. *Glob. J. Sci. Front. Res. Math. Decis. Sci.* **2012**, *12*, 53–60.
14. Andrews, G.E.; Bringman, K.; Mahlburg, K.E. Double series representations for Schur's partition function and related identities. *J. Combin. Theory Ser. A* **2015**, *132*, 102–119. [CrossRef]
15. Chaudhary, M.P. Some relationships between q-product identities, combinatorial partition identities and continued-fractions identities. III. *Pac. J. Appl. Math.* **2015**, *7*, 87–95.

16. Chaudhary, M.P.; Chaudhary, S. Note on Ramanujan's modular equations of degrees three and nine. *Pac. J. Appl. Math.* **2017**, *8*, 143–148.
17. Chaudhary, M.P.; Chaudhary, S.; Choi, J. Certain identities associated with 3-dissection property, continued fractions and combinatorial partition. *Appl. Math. Sci.* **2016**, *10*, 37–44. [CrossRef]
18. Chaudhary, M.P.; Chaudhary, S.; Choi, J. Note on Ramanujan's modular equation of degree seven. *Int. J. Math. Anal.* **2016**, *10*, 661–667. [CrossRef]
19. Chaudhary, M.P.; Choi, J. Note on modular relations for Roger-Ramanujan type identities and representations for Jacobi identities. *East Asian Math. J.* **2015**, *31*, 659–665. [CrossRef]
20. Chaudhary, M.P.; Choi, J. Certain identities associated with Eisenstein series, Ramanujan-Göllnitz-Gordon continued fraction and combinatorial partition identities. *Int. J. Math. Anal.* **2016**, *10*, 237–244. [CrossRef]
21. Chaudhary, M.P.; Choi, J. Certain identities associated with character formulas, continued fraction and combinatorial partition identities. *East Asian Math. J.* **2016**, *32*, 609–619. [CrossRef]
22. Chaudhary, M.P.; Uddin, S.; Choi, J. Certain relationships between q-product identities, combinatorial partition identities and continued-fraction identities. *Far East J. Math. Sci.* **2017**, *101*, 973–982. [CrossRef]
23. Srivastava, H.M.; Chaudhary, M.P.; Chaudhary, S. Some theta-function identities related to Jacobi's triple-product identity. *Eur. J. Pure Appl. Math.* **2018**, *11*, 1–9. [CrossRef]
24. Srivastava, H.M.; Chaudhary, M.P.; Chaudhary, S. A family of theta-function identities related to Jacobi's triple-product identity. *Russ. J. Math. Phys.* **2020**, *27*, 139–144. [CrossRef]
25. Hahn, H.-Y.; Huh, J.-S.; Lim, E.-S.; Sohn, J.-B. From partition identities to a combinatorial approach to explicit Satake inversion. *Ann. Combin.* **2018**, *22*, 543–562. [CrossRef]
26. Baruah, N.D.; Saikia, N. Two parameters for Ramanujan's theta-functions and their explicit values. *Rocky Mt. J. Math.* **2007**, *37*, 1747–1790. [CrossRef]
27. Naika, M.S.M.; Dharmendra, B.N.; Shivashankar, K. A continued fraction of order twelve. *Cent. Eur. J. Math.* **2008**, *6*, 393–404. [CrossRef]
28. Adiga, C.; Bulkhali, N.A.S.; Simsek, Y.; Srivastava, H.M. A continued fraction of Ramanujan and some Ramanujan-Weber class invariants. *Filomat* **2017**, *31*, 3975–3997. [CrossRef]
29. Srivastava, H.M.; Saikia, N. Some congruences for overpartitions with restriction. *Math. Notes* **2020**, *107*, 488–498. [CrossRef]
30. Liu, Z.-G. A three-term theta function identity and its applications. *Adv. Math.* **2005**, *195*, 1–23. [CrossRef]
31. Yi, J.-H. Theta-function identities and the explicit formulas for theta-function and their applications. *J. Math. Anal. Appl.* **2004**, *292*, 381–400. [CrossRef]
32. Srivastava, H.M. Operators of basic (or q-) calculus and fractional q-calculus and their applications in geometric function theory of complex analysis. *Iran. J. Sci. Technol. Trans. A Sci.* **2020**, *44*, 327–344. [CrossRef]
33. Srivastava, H.M.; Singh, S.N.; Singh, S.P. Some families of q-series identities and associated continued fractions. *Theory Appl. Math. Comput. Sci.* **2015**, *5*, 203–212.
34. Munagi, A.O. Combinatorial identities for restricted set partitions. *Discret. Math.* **2016**, *339*, 1306–1314. [CrossRef]
35. Yee, A.-J. Combinatorial proofs of generating function identities for F-partitions. *J. Combin. Theory Ser. A* **2003**, *102*, 217–228. [CrossRef]
36. Andrews, G.E. *The Theory of Partitions*; Cambridge University Press: Cambridge, UK; London, UK; New York, NY, USA, 1998.
37. Schur, I. *Zur Additiven Zahlentheorie*; Sitzungsber. Preuss. Akad. Wiss. Phys.-Math. Kl.: Berlin, Germany, 1926.
38. Alladi, K.; Berkovich, A. Göllnitz–Gordon partitions with weights and parity conditions. In *Zeta Functions, Topology and Quantum Physics*; Springer Series on Developments in Mathematics; Springer: New York, NY, USA, 2005; Volume 14, pp. 1–17.
39. Adiga, C.; Bulkhali, N.A.S.; Ranganatha, D.; Srivastava, H.M. Some new modular relations for the Rogers–Ramanujan type functions of order eleven with applications to partitions. *J. Number Theory* **2016**, *158*, 281–297. [CrossRef]
40. Cao, J.; Srivastava, H.M.; Luo, Z.-G. Some iterated fractional q-integrals and their applications. *Fract. Calc. Appl. Anal.* **2018**, *21*, 672–695. [CrossRef]
41. Srivastava, H.M. Some formulas of Srinivasa Ramanujan involving products of hypergeometric functions. *Indian J. Math. (Ramanujan Centen. Vol.)* **1987**, *29*, 91–100.

42. Srivastava, H.M. A note on a generalization of a q-series transformation of Ramanujan. *Proc. Jpn. Acad. Ser. A Math. Sci.* **1987**, *63*, 143–145. [CrossRef]
43. Srivastava, H.M. Srinivasa Ramanujan and generalized basic hypergeometric functions. *Serdica (Academician Ljubomir G. Iliev Dedication Vol.)* **1993**, *19*, 191–197.
44. Srivastava, H.M.; Zhang, C.-H. A certain class of identities of the Rogers–Ramanujan type. *Pan Am. Math. J.* **2009**, *19*, 89–102.
45. Srivastava, H.M.; Arjika, S.; Kelil, A.S. Some homogeneous q-difference operators and the associated generalized Hahn polynomials. *Appl. Set-Valued Anal. Optim.* **2019**, *1*, 187–201.

© 2020 by the authors. Licensee MDPI, Basel, Switzerland. This article is an open access article distributed under the terms and conditions of the Creative Commons Attribution (CC BY) license (http://creativecommons.org/licenses/by/4.0/).

Article

Janowski Type q-Convex and q-Close-to-Convex Functions Associated with q-Conic Domain

Muhammad Naeem [1,*], Saqib Hussain [2], Shahid Khan [3], Tahir Mahmood [1], Maslina Darus [4,*] and Zahid Shareef [5]

[1] Department of Mathematics and Statistics, International Islamic University Islamabad, Islamabad 44000, Pakistan; tahirbakhat@iiu.edu.pk
[2] Department of Mathematics, COMSATS University Islamabad, Abbottabad Campus 22060, Pakistan; saqib_math@yahoo.com
[3] Department of Mathematics, Riphah International University Islamabad, Islamabad 44000, Pakistan; shahidmath761@gmail.com
[4] Faculty of Science and Technology, University Kebangsaan Malaysia, Bangi 43600, Selangor, Malaysia
[5] Mathematics and Natural Science, Higher Colleges of Technology, Fujairah Men's, Fujairah 4114, UAE; zshareef@hct.ac.ae
* Correspondence: naeem.phdma75@iiu.edu.pk (M.N.); maslina@ukm.edu.my (M.D.)

Received: 19 January 2020; Accepted: 11 March 2020; Published: 18 March 2020

Abstract: Certain new classes of q-convex and q-close to convex functions that involve the q-Janowski type functions have been defined by using the concepts of quantum (or q-) calculus as well as q-conic domain $\left(\Omega_{k,q}[\lambda,\alpha]\right)$. This study explores some important geometric properties such as coefficient estimates, sufficiency criteria and convolution properties of these classes. A distinction of new findings with those obtained in earlier investigations is also provided, where appropriate.

Keywords: analytic functions; Janowski functions; conic domain; q-convex functions; q-close-to-convex functions

1. Introduction

The mathematical study of q-calculus, particularly q-fractional calculus and q-integral calculus, q-transform analysis has been a topic of great interest for researchers due to its wide applications in different fields (see [1,2]). Some of the earlier work on the applications of the q-calculus was introduced by Jackson [3,4]. Later, q-analysis with geometrical interpretation was turned into identified through quantum groups. Due to the applications of q-analysis in mathematics and other fields, numerous researchers [3,5–14] did some significant work on q-calculus and studied its several other applications. Recently, Srivastava [15] in his survey-cum-expository article, explored the mathematical application of q-calculus, fractional q-calculus and fractional q-differential operators in geometric function theory. Keeping in view the significance of q-operators instead of ordinary operators and due to the wide range of applications of q-calculus, many researchers comprehensively studied q-calculus such as Srivastava et al. [16], Muhammad and Darus [17], Kanas and Reducanu [18] and Muhammad and Sokol [19]. Motivated by [15–21], we consider subfamilies of q-convex functions and q-close to convex functions with respect to Janowski functions connected with q-conic domain.

Let $\mathcal{A}$ be the class of functions of the form

$$f(z) = z + \sum_{n=2}^{\infty} a_n z^n, \quad a_n \in \mathbb{C}, z \in U, \qquad (1)$$

which are analytic in the open unit disk $U = \{z : z \in \mathbb{C}, |z| < 1\}$. Let $\mathcal{A} \supseteq \mathcal{S}$, where $\mathcal{S}$ represents the set of all univalent functions in U. The classes of starlike (S^*) and convex (C) functions in U are the well known subclasses of S. Moreover, the class K of close to convex functions in U consists of normalized functions $f \in \mathcal{A}$ that satisfy the following conditions:

$$f \in \mathcal{A} \quad \text{and} \quad Re\left(\frac{zf'(z)}{g(z)}\right) > 0, \quad \text{where } g(z) \in S^*.$$

Now, for $\kappa \geq 0$, the classes κ-uniformly convex mappings ($\kappa - UCV$) and κ-starlike mappings ($\kappa - UST$), explored by Kanas and Wiśniowska, see [22–28]. Kanas and Wiśniowska [22,23] also initiated the study of analytic functions on conic domain Ω_κ, $\kappa \geq 0$ as:

$$\Omega_\kappa = \left\{u + iv : u > \kappa\sqrt{(u-1)^2 + v^2}\right\}.$$

See [22,23] for geometric interpretation of Ω_κ. These conic regions are images of the unit disk under the extremal functions $h_\kappa(z)$ given by:

$$h_\kappa(z) = \begin{cases} \frac{1+z}{1-z} & \kappa = 0, \\ 1 + \left(\log \frac{\sqrt{z}+1}{1-\sqrt{z}}\right)^2 \frac{2}{\pi^2} & \kappa = 1, \\ 1 + \sinh^2\left\{\arctan h\sqrt{z}\left(\frac{2}{\pi}\arccos \kappa\right)\right\} \frac{2}{1-\kappa^2} & 0 < \kappa < 1, \\ 1 + \frac{1}{\kappa^2-1}\sin\left(\frac{\pi}{2R(y)}\int_0^{\frac{u(z)}{\sqrt{y}}} \frac{dx}{\sqrt{1-x^2}\sqrt{1-y^2x^2}}\right) + \frac{1}{\kappa^2-1} & \kappa > 1, \end{cases} \quad (2)$$

where

$$u(z) = \frac{z - \sqrt{y}}{1 - \sqrt{y}z}, \quad z \in U.$$

Here, $\kappa = \cosh(\pi R'(y)/(4R(y))) \in (0,1)$, where $R(y)$ is Legendre's complete elliptic integral of first kind and $R'(y) = R(\sqrt{1-y^2})$ is its complementary integral, see [22,23,29–34]. If $h_\kappa(z) = 1 + \delta(\kappa)z + \delta_1(\kappa)z^2 + \cdots$ is taken from [23] for (2), then

$$\delta(\kappa) = \begin{cases} \frac{8(\arccos \kappa)^2}{\pi^2(1-\kappa^2)} & 0 \leq \kappa < 1, \\ \frac{8}{\pi^2} & \kappa = 1, \\ \frac{\pi^2}{4\sqrt{y}(\kappa^2-1)R^2(y)(1+y)} & \kappa > 1, \end{cases} \quad (3)$$

$$\delta_1(\kappa) = \delta_2(\kappa)\delta(\kappa),$$

where

$$\delta_2(\kappa) = \begin{cases} \frac{\mathcal{T}_1^2 + 2}{3} & 0 \leq \kappa < 1, \\ \frac{2}{3} & \kappa = 1, \\ \frac{4R^2(y)(y^2+6y+1)-\pi^2}{24R^2(y)(1+y)\sqrt{y}} & \kappa > 1, \end{cases} \quad (4)$$

where $\mathcal{T}_1 = \frac{2}{\pi}\arccos \kappa$, and $y \in (0,1)$.

Definition 1. ([35]) Let $p \in \mathcal{A}$ and $p(0) = 1$ be in the class $\mathcal{P}(\lambda, \alpha)$ if and only if

$$p(z) \prec \frac{1 + \lambda z}{1 + \alpha z}, \quad (-1 \leq \alpha < \lambda \leq 1),$$

where $\prec$ stands for subordination.

Janowski [35] initiated the class $\mathcal{P}(\lambda, \alpha)$ by showing that $p \in \mathcal{P}(\lambda, \alpha)$ if and only if there exists a mapping $p \in \mathcal{P}$ such that
$$\frac{p(z)(\lambda+1) - (\lambda-1)}{p(z)(\alpha+1) - (\alpha-1)} \prec \frac{1+\lambda z}{1+\alpha z},$$
where $\mathcal{P}$ is class the of mappings with non-negative real parts.

Definition 2. ([36]) *Let function $f \in \mathcal{A}$ be in the class $\mathcal{S}^*(\lambda, \alpha)$ if and only if*
$$\frac{zf'(z)}{f(z)} = \frac{p(z)(\lambda+1) - (\lambda-1)}{p(z)(\alpha+1) - (\alpha-1)}, \quad (-1 \leq \alpha < \lambda \leq 1).$$

Definition 3. ([36]) *Let function $f \in \mathcal{A}$ is in the class $\mathcal{C}(\lambda, \alpha)$ if and only if*
$$\frac{\left(zf'(z)\right)'}{(f(z))'} = \frac{p(z)(\lambda+1) - (\lambda-1)}{p(z)(\alpha+1) - (\alpha-1)}, \quad (-1 \leq \alpha < \lambda \leq 1).$$

Definition 4. ([7]) *Let function $f \in \mathcal{A}$, $n \in \mathbb{N}_0$ and $q \in (0,1)$, the q-difference (or $q-$derivative) operator D_q is defined as:*
$$D_q f(z) = -\frac{f(z) - f(qz)}{(q-1)z}.$$

Note that
$$D_q z^n = [n]_q z^{n-1}, \quad D_q \left\{ \sum_{n=1}^{\infty} a_n z^n \right\} = \sum_{n=1}^{\infty} [n]_q a_n z^{n-1},$$
where
$$[n]_q = \frac{1-q^n}{1-q}.$$

Definition 5. ([37]) *Let function $f \in \mathcal{A}$ is in the class $\mathcal{S}_q^*(\lambda, \alpha)$ if and only if*
$$\frac{zD_q f(z)}{f(z)} = \frac{(\lambda+1)\tilde{p}(z) - (\lambda-1)}{(\alpha+1)\tilde{p}(z) - (\alpha-1)}, \quad (-1 \leq \alpha < \lambda \leq 1), \quad q \in (0,1).$$

By principle of subordination we can be written as follows:
$$\frac{zD_q f(z)}{f(z)} \prec \frac{(\lambda+1)z + 2 + (\lambda-1)qz}{(\alpha+1)z + 2 + (\alpha-1)qz},$$
where
$$\tilde{p}(z) = \frac{1+z}{1-qz}.$$

Definition 6. ([37]) *Let function $f \in \mathcal{A}$ is in class $\mathcal{C}_q(\lambda, \alpha)$ if and only if*
$$\frac{D_q(zD_q f(z))}{D_q f(z)} = \frac{\tilde{p}(z)(\lambda+1) - (\lambda-1)}{\tilde{p}(z)(\alpha+1) - (\alpha-1)}, \quad (-1 \leq \alpha < \lambda \leq 1), \quad q \in (0,1).$$

Similarly, by principle of subordination, we can be written as follows:
$$\frac{D_q(zD_q f(z))}{D_q f(z)} \prec \frac{z(\lambda+1) + (\lambda-1)qz + 2}{z(\alpha+1) + (\alpha-1)qz + 2}.$$

Mahmood et al. [38] introduced the class $k - \mathcal{P}_q(\lambda, \alpha)$ as:

Definition 7. ([38]) A function $h \in k - \mathcal{P}_q(\lambda, \alpha)$, if and only if

$$h(z) \prec \frac{(\lambda O_1 + O_3) h_k(z) - (\lambda O_1 - O_3)}{(\alpha O_1 + O_3) h_k(z) - (\alpha O_1 - O_3)}, \quad k \geq 0, \, q \in (0,1),$$

where

$$O_1 = 1 + q \text{ and } O_3 = 3 - q.$$

In addition, $h_k(z)$ is defined in Label (2). Geometrically, the mapping $h \in k - \mathcal{P}_q(\lambda, \alpha)$ takes all domain values $\Omega_{k,q}(\lambda, \alpha)$, $1 \leq \alpha < \lambda \leq 1, k \geq 0$, which is definable as:

$$\Omega_{k,q}(\lambda, \alpha) = \{r = u + iv : \Re(\Psi) > k |\Psi - 1|\},$$

where

$$\Psi = \frac{(\alpha O_1 - O_3) r(z) - (\lambda O_1 - O_3)}{(\alpha O_1 + O_3) r(z) - (\lambda O_1 + O_3)}.$$

This domain describes the conic type domain; for details, see [38].
Note that

(i) When $q \to 1$, then domain $\Omega_{\kappa,q}(\lambda, \alpha)$ reduces to the domain $\Omega_\kappa(\lambda, \alpha)$ (see [39]).
(ii) When $q \to 1$, then the class $\kappa - \mathcal{P}_q(\lambda, \alpha)$ reduces to the class $\kappa - \mathcal{P}(\lambda, \alpha)$ (see [39]).
(iii) When $q \to 1$, and $\kappa = 0$, then $\kappa - \mathcal{P}_q(\lambda, \alpha) = \mathcal{P}(\lambda, \alpha)$ also $\kappa - \mathcal{P}(1, -1) = \mathcal{P}(h_\kappa)$ (see ([35]).

Definition 8. ([38]) Let $f \in \mathcal{A}$ be in the class $k - \mathcal{ST}_q(\beta, \gamma)$, if and only if

$$\Re \left(\frac{(\gamma O_1 - O_3) \frac{z D_q f(z)}{f(z)} - (\beta O_1 - O_3)}{(\gamma O_1 + O_3) \frac{z D_q f(z)}{f(z)} - (\beta O_1 + O_3)} \right)$$

$$> k \left| \frac{(\gamma O_1 - O_3) \frac{z D_q f(z)}{f(z)} - (\beta O_1 - O_3)}{(\gamma O_1 + O_3) \frac{z D_q f(z)}{f(z)} - (\beta O_1 + O_3)} - 1 \right|,$$

or, equivalently,

$$\frac{z D_q f(z)}{f(z)} \in k - \mathcal{P}_q(\beta, \gamma),$$

where $k \geq 0, -1 \leq \gamma < \beta \leq 1$.

We can see that, when $q \to 1$, then $\kappa - \mathcal{ST}_q(\beta, \gamma)$ diminishes to the renowned class which is stated in [39].

Motivated by the definition above, we introduced new classes $\kappa - \mathcal{UCV}_q(\beta, \gamma)$, $\kappa - \mathcal{UK}_q(\lambda, \alpha, \beta, \gamma)$ and $\kappa - \mathcal{UQ}_q(\lambda, \alpha, \beta, \gamma)$ of analytic functions.

Definition 9. Let $f \in \mathcal{A}$, be in the class $k - \mathcal{UCV}_q(\beta, \gamma)$ if and only if

$$\Re \left(\frac{(\gamma O_1 - O_3) \frac{D_q(z D_q f(z))}{D_q f(z)} - (\beta O_1 - (O_3))}{(\gamma O_1 + O_3) \frac{D_q(z D_q f(z))}{D_q f(z)} - (\beta O_1 + O_3)} \right)$$

$$> k \left| \frac{(\gamma O_1 - O_3) \frac{D_q(z D_q f(z))}{D_q f(z)} - (\beta O_1 - O_3)}{(\gamma O_1 + O_3) \frac{D_q(z D_q f(z))}{D_q f(z)} - (\beta O_1 + O_3)} - 1 \right|,$$

or, equivalently,
$$\frac{D_q\left(zD_qf(z)\right)}{D_qf(z)} \in k - \mathcal{P}_q(\beta,\gamma),$$

where $k \geq 0$, $-1 \leq \gamma < \beta \leq 1$.

One can clearly see that
$$f \in \kappa - \mathcal{UCV}_q(\beta,\gamma) \Leftrightarrow zD_q(z) \in \kappa - \mathcal{ST}_q(\beta,\gamma). \qquad (5)$$

Note that, when $q \to 1$, then the class $\kappa - \mathcal{UCV}_q(\beta,\gamma)$ reduces to a well-known class defined in [39].

Definition 10. *Let $f \in \mathcal{A}$, be in the class $k - \mathcal{UK}_q(\lambda,\alpha,\beta,\gamma)$ if and only if there exists $g \in k - \mathcal{ST}_q(\beta,\gamma)$, such that*

$$\Re\left(\frac{(\alpha O_1 - O_3)\frac{zD_qf(z)}{g(z)} - (\lambda O_1 - O_3)}{(\alpha O_1 + O_3)\frac{zD_qf(z)}{g(z)} - (\lambda O_1 + O_3)}\right)$$

$$> k\left|\frac{(\alpha O_1 - O_3)\frac{zD_qf(z)}{g(z)} - (\lambda O_1 - O_3)}{(\alpha O_1 + O_3)\frac{zD_qf(z)}{g(z)} - (\lambda O_1 + O_3)} - 1\right|.$$

We can write equivalently
$$\frac{zD_qf(z)}{g(z)} \in k - \mathcal{P}_q(\lambda,\alpha),$$

where $k \geq 0$, $-1 \leq \gamma < \beta \leq 1$, $-1 \leq \alpha < \lambda \leq 1$.

Note that, when $q \to 1$, then, the class $k - \mathcal{UK}_q(\lambda,\alpha,\beta,\gamma)$ reduces into the well-known class that is defined in (see [40]).

Definition 11. *Let $f \in \mathcal{A}$, belong to the class $k - \mathcal{UQ}_q(\lambda,\alpha,\beta,\gamma)$ if and only if there exist $g \in k - \mathcal{CV}_q(\beta,\gamma)$, such that*

$$\Re\left(\frac{(\alpha O_1 - O_3)\frac{D_q(zD_qf(z))}{D_qg(z)} - (\lambda O_1 - O_3)}{(\alpha O_1 + O_3)\frac{D_q(zD_qf(z))}{D_qg(z)} - (\lambda O_1 + O_3)}\right)$$

$$> k\left|\frac{(\alpha O_1 - O_3)\frac{D_q(zD_qf(z))}{D_qg(z)} - (\lambda O_1 - O_3)}{(\alpha O_1 + O_3)\frac{D_q(zD_qf(z))}{D_qg(z)} - (\lambda O_1 + O_3)} - 1\right|,$$

or, equivalently,
$$\frac{D_q\left(zD_qf(z)\right)}{D_qg(z)} \in k - \mathcal{P}_q(\lambda,\alpha),$$

where, for $k \geq 0$, $-1 \leq \gamma < \beta \leq 1$, $-1 \leq \alpha < \lambda \leq 1$.

It is simple to verify this
$$f \in \kappa - \mathcal{UQ}_q(\lambda,\alpha,\beta,\gamma) \Leftrightarrow zD_qf \in \kappa - \mathcal{UK}_q(\lambda,\alpha,\beta,\gamma). \qquad (6)$$

A special case arises when $q \to 1$, then the class $\kappa - \mathcal{UQ}_q(\lambda, \alpha, \beta, \gamma)$ reduces to a well known class defined in [40].

2. Set of Lemmas

Lemma 1. *([41]) Suppose $1 + \sum_{n=1}^{\infty} c_n z^n = d(z) \prec H(z) = 1 + \sum_{n=1}^{\infty} C_n z^n$. If $H(U)$ is convex and $H(z) \in \mathcal{A}$, then*

$$|c_n| \leq |C_1|, \quad n \geq 1.$$

Lemma 2. *([38]) Suppose $d(z) = 1 + \sum_{n=1}^{\infty} c_n z^n \in k - \mathcal{P}_q(\lambda, \alpha)$, then*

$$|c_n| \leq |\delta(k, \lambda, \alpha)| = \frac{O_1(\lambda - \alpha)}{4} \delta(k),$$

where $\delta(k)$ is given by (3).

Lemma 3. *([38]) Suppose $d \in k - \mathcal{ST}_q(\beta, \gamma)$, $k \geq 0$ is given by*

$$d(z) = z + \sum_{n=2}^{\infty} b_n z^n, \quad z \in U,$$

then

$$|b_n| \leq \prod_{m=0}^{n-2} \left(\frac{\left| \delta(k) O_1 (\beta - \gamma) - 4q [m]_q \gamma \right|}{4q [m+1]_q} \right),$$

where $\delta(k)$ is given by (3).

Lemma 4. *([42]) Suppose $d \in \mathcal{S}^*$, $f \in \mathcal{C}$ and $G \in \mathcal{S}$, then we have*

$$\frac{f(z) * d(z) G(z)}{f(z) * d(z)} \in \overline{co}(G(U)), \quad z \in U.$$

Here, "*" means convolution and $\overline{co}(G(U))$ means the closed convex hull $G(U)$.

Lemma 5. *([38]) The function $f \in \mathcal{A}$ will belong to the class $k - \mathcal{ST}_q(\beta, \gamma)$, if the following inequality holds:*

$$\sum_{n=2}^{\infty} \left\{ 2O_3(1+k)q[n-1]_q + \left| (\gamma O_1 + O_3)[n]_q - (\beta O_1 + (O_3)) \right| \right\} |a_n|$$
$$\leq O_1 |\gamma - \beta|.$$

Throughout this paper, we assume that $k \geq 0$, $-1 \leq \gamma < \beta \leq 1$, $-1 \leq \alpha < \lambda \leq 1$, and $q \in (0,1)$, unless otherwise specified.

3. Main Results

Theorem 1. *Let $f \in \mathcal{A}$; then, f is in the class $k - \mathcal{UCV}_q(\beta, \gamma)$, if the following inequality holds:*

$$\sum_{n=2}^{\infty} [n]_q \left\{ 2O_3(k+1)q[n-1]_q + \left| (\gamma O_1 + O_3)[n]_q - (\beta O_1 + O_3) \right| \right\} |a_n|$$
$$\leq O_1 |\gamma - \beta|.$$

Proof. By Lemma 5 and relation (5), the proof is straightforward. □

For $q \to 1^-$, in Theorem 1, then we obtained following corollary, proved by Malik and Noor [39].

Corollary 1. Let $f \in \mathcal{A}$; then, f belongs to $k - \mathcal{UCV}(\beta, \gamma)$, if the following inequality holds

$$\sum_{n=2}^{\infty} n \left\{ 2(k+1)(n-1) + |n(\gamma+1) - (\beta+1)| \right\} |a_n| \leq |\gamma - \beta|.$$

Theorem 2. Let $f \in \mathcal{A}$, then f is in the class $k - \mathcal{UK}_q(\lambda, \alpha, \beta, \gamma)$, if the condition (7) holds

$$\sum_{n=2}^{\infty} \left\{ 2O_3(k+1) \left| b_n - [n]_q a_n \right| + \left| (\alpha O_1 + O_3)[n]_q a_n - (\lambda O_1 + O_3) b_n \right| \right\}$$

$$\leq O_1 |\alpha - \lambda|. \tag{7}$$

Proof. Presuming that (7) holds, then it is enough to show that

$$k \left| \frac{(\alpha O_1 - O_3) \frac{z D_q f(z)}{g(z)} - (\lambda O_1 - O_3)}{(\alpha O_1 + O_3) \frac{z D_q f(z)}{g(z)} - (\lambda O_1 + O_3)} - 1 \right|$$

$$- \operatorname{Re} \left\{ \frac{(\alpha O_1 - O_3) \frac{z D_q f(z)}{g(z)} - (\lambda O_1 - O_3)}{(\alpha O_1 + O_3) \frac{z D_q f(z)}{g(z)} - (\lambda O_1 + O_3)} - 1 \right\}$$

$$< 1.$$

We have

$$k \left| \frac{(\alpha O_1 - O_3) \frac{z D_q f(z)}{g(z)} - (\lambda O_1 - O_3)}{(\alpha O_1 + O_3) \frac{z D_q f(z)}{g(z)} - (\lambda O_1 + O_3)} - 1 \right|$$

$$- \operatorname{Re} \left\{ \frac{(\alpha O_1 - O_3) \frac{z D_q f(z)}{g(z)} - (\lambda O_1 - O_3)}{(\alpha O_1 + O_3) \frac{z D_q f(z)}{g(z)} - (\lambda O_1 + O_3)} - 1 \right\},$$

$$\leq (k+1) \left| \frac{(\alpha O_1 - O_3) \frac{z D_q f(z)}{g(z)} - (\lambda O_1 - O_3)}{(\alpha O_1 + O_3) \frac{z D_q f(z)}{g(z)} - (\lambda O_1 + O_3)} - 1 \right|, \tag{8}$$

$$= 2O_3(k+1) \left| \frac{g(z) - z D_q f(z)}{(\alpha O_1 + O_3) z D_q f(z) - (\lambda O_1 + O_3) g(z)} \right|,$$

$$= 2O_3(k+1) \left| \frac{\sum_{n=2}^{\infty} \left\{ b_n - [n]_q a_n \right\} z^n}{O_1 (\alpha - \lambda) z + \sum_{n=2}^{\infty} \left\{ (\alpha O_1 + O_3)[n]_q a_n - (\lambda O_1 + O_3) b_n \right\} z^n} \right|,$$

$$\leq \frac{2O_3(k+1) \sum_{n=2}^{\infty} \left\{ \left| b_n - [n]_q a_n \right| \right\}}{O_1 |\alpha - \lambda| - \sum_{n=2}^{\infty} \left| (\alpha O_1 + O_3)[n]_q a_n - (\lambda O_1 + O_3) b_n \right|}.$$

The expression (8) is bounded above by 1 if

$$\sum_{n=2}^{\infty} \left[2O_3(k+1) \left| b_n - [n]_q a_n \right| + \left| (\alpha O_1 + O_3)[n]_q a_n - (\lambda O_1 + O_3) b_n \right| \right]$$
$$\leq (O_1) |\alpha - \lambda|.$$

□

Corollary 2. ([40]) Let $f \in \mathcal{A}$. Then, f is in the class $k - \mathcal{UK}_{q \to 1}(\lambda, \alpha, \beta, \gamma) = k - \mathcal{UK}(\lambda, \alpha, \beta, \gamma)$, if the following condition holds:

$$\sum_{n=2}^{\infty} \{ 2(k+1) |b_n - na_n| + |(\alpha+1)na_n - (\lambda+1)b_n| \} \leq |\alpha - \lambda|.$$

Here, $q \to 1$ represents the limiting value of q as it approaches 1.

Theorem 3. Let $f \in \mathcal{A}$. Then, f is in the class $k - \mathcal{UQ}_q(\lambda, \alpha, \beta, \gamma)$, if the following condition holds:

$$\sum_{n=2}^{\infty} [n]_q \left[2O_3(k+1) \left| b_n - [n]_q a_n \right| + \left| (\alpha O_1 + O_3)[n]_q a_n - (\lambda O_1 + O_3) b_n \right| \right]$$
$$\leq O_1 |\alpha - \lambda|.$$

Proof. By Theorem 2 and relation (6), the proof is straightforward. □

Corollary 3. ([40]) Let $f \in \mathcal{A}$. Then, f is in the class $k - \mathcal{UK}_{q \to 1}(\lambda, \alpha, \beta, \gamma) = k - \mathcal{UQ}(\lambda, \alpha, \beta, \gamma)$, if

$$\sum_{n=2}^{\infty} n \{ 2(k+1) |b_n - na_n| + |(\alpha+1)na_n - (\lambda+1)b_n| \} \leq |\alpha - \lambda|.$$

Corollary 4. ([43]) Let $f \in \mathcal{A}$. Then, f is in the class $1 - \mathcal{UK}_{q \to 1}(1 - 2\tau, -1, 1, -1) = \mathcal{UK}(\tau)$ if

$$\sum_{n=2}^{\infty} n^2 |a_n| \leq \frac{1-\tau}{2}.$$

Theorem 4. Let $f \in k - \mathcal{UCV}_q(\beta, \gamma)$, is of the form (1). Then,

$$|a_n| \leq \frac{1}{[n]_q} \prod_{m=0}^{n-2} \left(\frac{|\delta(k) O_1(\beta - \gamma) - 4q[m]_q \gamma|}{4q[m+1]_q} \right),$$

where $\delta(k)$ is given by (3).

Proof. By Lemma 3 and relation (5), the proof is straightforward. □

For $q \to 1^-$, Theorem 4 brings to the following corollary, proved by Noor [39].

Corollary 5. Let $f \in k - \mathcal{UCV}(\beta, \gamma)$. Then,

$$|a_n| \leq \frac{1}{n} \prod_{m=0}^{n-2} \left(\frac{|\delta(k)(\beta - \gamma) - 2m\gamma|}{2(m+1)} \right),$$

where $\delta(k)$ is given by (3).

Theorem 5. If $f \in k-\mathcal{UK}_q(\lambda,\alpha,\beta,\gamma)$ and $g \in k-\mathcal{ST}_q(\beta,\gamma)$, then,

$$|a_n| \leq \begin{cases} \frac{1}{[n]_q}\prod_{m=0}^{n-2}\frac{\left|\delta(k)O_1(\beta-\gamma)-4q[m]_q\gamma\right|}{4q[m+1]_q} \\ +\frac{\delta(k)O_1(\lambda-\alpha)}{4[n]_q}\sum_{j=1}^{n-1}\prod_{m=0}^{j-2}\frac{\left|\delta(k)O_1(\beta-\gamma)-4q[j]_q\gamma\right|}{4q[j+1]_q}, \quad n \geq 2, \end{cases}$$

where $\delta(k)$ is given in (3).

Proof. Let us take
$$\frac{zD_qf(z)}{g(z)} = h(z), \tag{9}$$

where
$$h \in k-\mathcal{P}_q(\lambda,\alpha) \text{ and } g \in k-\mathcal{ST}_q(\beta,\gamma).$$

Now, from (9), we have
$$zD_qf(z) = g(z)h(z),$$

which implies that

$$z + \sum_{n=2}^{\infty}[n]_q a_n z^n = \left(1+\sum_{n=1}^{\infty}c_n z^n\right)\left(z+\sum_{n=2}^{\infty}b_n z^n\right).$$

By equating z^n coefficients
$$[n]_q a_n = b_n + \sum_{j=1}^{n-1} b_j c_{n-j}, \quad a=1, \ b_1 = 1.$$

This implies that

$$[n]_q |a_n| \leq |b_n| + \sum_{j=1}^{n-1}|b_j||c_{n-j}|. \tag{10}$$

Since $h \in k-\mathcal{P}_q(\lambda,\alpha)$, therefore, by using Lemma 2 on (10), we have

$$[n]_q |a_n| \leq |b_n| + \frac{\delta(k)O_1(\lambda-\alpha)}{4}\sum_{j=1}^{n-1}|b_j|. \tag{11}$$

Again $g \in k-\mathcal{ST}_q(\beta,\gamma)$, therefore, by using Lemma 3 on (11), we have

$$|a_n| \leq \begin{cases} \frac{1}{[n]_q}\prod_{m=0}^{n-2}\left(\frac{\left|\delta(k)O_1(\beta-\gamma)-4q[m]_q\gamma\right|}{4q[m+1]_q}\right) \\ +\frac{\delta(k)O_1(\lambda-\alpha)}{4[n]_q}\sum_{j=1}^{n-1}\prod_{m=0}^{j-2}\left(\frac{\left|\delta(k)O_1(\beta-\gamma)-4q[m]_q\gamma\right|}{4q[m+1]_q}\right). \end{cases}$$

□

Corollary 6. ([40]) If $f \in k-\mathcal{UK}_{q\to 1}(\lambda,\alpha,\beta,\gamma) = k-\mathcal{UK}(\lambda,\alpha,\beta,\gamma)$, then

$$|a_n| \leq \begin{cases} \frac{1}{n}\prod_{m=0}^{n-2}\left(\frac{|\delta(k)(\beta-\gamma)-2m\gamma|}{2(m+1)}\right) \\ +\frac{\delta(k)(\lambda-\alpha)}{2n}\sum_{j=1}^{n-1}\prod_{m=0}^{j-2}\left(\frac{|\delta(k)(\beta-\gamma)-2m\gamma|}{2(m+1)}\right), \quad n \geq 2, \end{cases}$$

where $\delta(k)$ is defined by (3).

Corollary 7. ([26]) *If* $f \in k - \mathcal{UK}_{q \to 1}(1, -1, 1, -1) = k - \mathcal{UK}$, *then*

$$|a_n| \leq \frac{(\delta(k))_{n-1}}{n!} + \frac{\delta(k)}{n} \sum_{j=0}^{n-1} \frac{(\delta(k))_{j-1}}{(j-1)!}, \quad n \geq 2.$$

Corollary 8. ([44]) *If* $f \in 0 - \mathcal{UK}_{q \to 1}(1, -1, 1, -1) = \mathcal{K}$, *then*

$$|a_n| \leq n, \quad n \geq 2.$$

Theorem 6. *If* $f \in k - \mathcal{UQ}_q(\lambda, \alpha, \beta, \gamma)$, *then*

$$|a_n| \leq \begin{cases} \frac{1}{\left([n]_q\right)^2} \prod_{m=0}^{n-2} \left| \frac{\delta(k) O_1(\beta - \gamma) - 4q[m]_q \gamma}{4q[m+1]_q} \right| \\ + \frac{\delta(k) O_1(\lambda - \alpha)}{4\left([n]_q\right)^2} \sum_{j=1}^{n-1} \prod_{m=0}^{j-2} \left| \frac{\delta(k) O_1(\beta - \gamma) - 4q[j]_q \gamma}{4q[j+1]_q} \right|, \quad n \geq 2, \end{cases}$$

where $\delta(k)$ *is defined by* (3).

Proof. By Theorem 5 and relation (6), the proof is straightforward. □

Corollary 9. ([40]) *If* $f \in k - \mathcal{UQ}_{q \to 1}(\lambda, \alpha, \beta, \gamma) = \mathcal{UQ}(\lambda, \alpha, \beta, \gamma)$ *and is of the form* (1), *then*

$$|a_n| \leq \begin{cases} \frac{1}{n^2} \prod_{m=0}^{n-2} \left(\frac{|\delta(k)(\beta - \gamma) - 2m\gamma|}{2(m+1)} \right) \\ + \frac{\delta(k)(\lambda - \alpha)}{2n^2} \sum_{j=1}^{n-1} \prod_{m=0}^{j-2} \left(\frac{|\delta(k)(\beta - \gamma) - 2m\gamma|}{2(m+1)} \right), \quad n \geq 2. \end{cases}$$

Theorem 7. *If* $f \in k - \mathcal{P}_q(\beta, \gamma)$ *and* $\chi \in \mathcal{C}$, *then* $f * \chi \in k - \mathcal{P}_q(\beta, \gamma)$.

Proof. Here, we prove that

$$\frac{zD_q(\chi(z) * f(z))}{(\chi(z) * f(z))} \in k - \mathcal{P}_q(\beta, \gamma).$$

Consider

$$\frac{zD_q(\chi(z) * f(z))}{(\chi(z) * f(z))} = \frac{\chi(z) * f(z) \left(\frac{zD_q f(z)}{f(z)} \right)}{\chi(z) * f(z)},$$

$$= \frac{\chi(z) * f(z) \Psi(z)}{\chi(z) * f(z)},$$

where $\frac{zD_q f(z)}{f(z)} = \Psi(z) \in \mathcal{P}_q(\beta, \gamma)$. By using Lemma 4, we obtain the required result. □

Theorem 8. *If* $f \in k - \mathcal{UK}_q(\lambda, \alpha, \beta, \gamma)$ *and* $\chi \in \mathcal{C}$, *then* $f * \chi \in k - \mathcal{UK}_q(\lambda, \alpha, \beta, \gamma)$.

Proof. Since $f \in k - \mathcal{UK}_q(\lambda, \alpha, \beta, \gamma)$, there exist $g \in k - \mathcal{ST}_q(\beta, \gamma)$, such that $\frac{zD_q f(z)}{g(z)} \in k - \mathcal{P}_q(\lambda, \alpha)$.
It follows from Lemma 4 that $\chi * g \in k - \mathcal{ST}_q(\beta, \gamma)$.

Consider

$$\frac{zD_q\left(\chi(z)*f(z)\right)}{(\chi(z)*g(z))} = \frac{\chi(z)*(zD_qf(z))}{(\chi(z)*g(z))},$$

$$= \frac{\chi(z)*\left(\frac{zD_qf(z)}{g(z)}\right)g(z)}{\chi(z)*f(z)},$$

$$= \frac{\chi(z)*F(z)g(z)}{\chi(z)*g(z)},$$

where $F \in k-\mathcal{ST}_q(\lambda,\alpha)$. By using Lemma 4, we obtain the required result. □

4. Conclusions

In this paper, we use Quantum Calculus to define new subclasses $k-\mathcal{CV}_q(\beta,\gamma)$, $k-\mathcal{UK}_q(\lambda,\alpha,\beta,\gamma)$ and $k-\mathcal{UQ}_{q\to1}(\lambda,\alpha,\beta,\gamma)$ of analytic functions involving conic domain and associated with Janowski type function. We then investigate many geometric properties and characteristics of each of these families such as coefficient inequalities, sufficient condition, necessary condition, and convolution properties. For verification and validity of our main results, we have also pointed out relevant connections of our main results with those in several earlier related works on this subject.

For further investigation, we can make connections between the q-analysis and (p,q)-analysis, and the results for q-analogues which we have included in this article for $0 < q < 1$ can be possibly be translated into the relevant findings for the (p,q)-analogues with $(0 < q < p \le 1)$ by adding some parameter.

Author Contributions: Conceptualization, S.H.; Formal analysis, T.M. and M.D.; Funding acquisition, M.D.; Investigation, M.N., S.K. and Z.S.. All authors have read and agreed to the published version of the manuscript.

Funding: M.D. is thankful to MOHE grant: FRGS/1/2019/STG06/UKM/01/1.

Acknowledgments: The authors would like to thank the referees for their valuable comments and suggestions, which was essential to improve the quality of this paper.

Conflicts of Interest: The authors declare no conflict of interest.

References

1. Gauchman, H. Integral inequalities in q-calculus. *Comput. Math. Appl.* **2004**, *47*, 281–300.
2. Tang, Y.; Zhang, T. A remark on the q-fractional order differential equations. *Appl. Math. Comput.* **2019**, *350*, 198–208. [CrossRef]
3. Jackson, F.H. On q-definite integrals. *Quart. J. Pure Appl. Math.* **1910**, *41*, 193–203. [CrossRef]
4. Jackson, F.H. q-difference equations. *Amer. J. Math.* **1910**, *32*, 305–314.
5. Adams, C.R. On the linear partial q-difference equation of general type. *Trans. Am. Math. Soc.* **1929**, *31*, 360–371.
6. Carmichael, R.D. The general theory of linear q-difference equations. *Am. J. Math.* **1912**, *34*, 147–168. [CrossRef]
7. Jackson, F.H., XI. On q-functions and a certain difference operator. *Earth Environ. Sci. Trans. R. Soc. Edinb.* **1909**, *46*, 253–281.
8. Khan, Q.; Arif, M.; Raza, M.; Srivastava, G.; Tang, H. Some applications of a new integral operator in q-analog for multivalent functions. *Mathematics* **2019**, *7*, 1178. [CrossRef]
9. Mahmood, S.; Raza, N.; AbuJarad, E.S.; Srivastava, G.; Srivastava, H.M.; Malik, S.N. Geometric properties of certain classes of analytic functions associated with a q-integral operator. *Symmetry* **2019**, *11*, 719. [CrossRef]
10. Shi, L.; Khan, Q.; Srivastava, G.; Liu, J.L.; Arif, M. A study of multivalent q-starlike functions connected with circular domain. *Mathematics* **2019**, *7*, 670. [CrossRef]
11. Srivastava, H.M.; Tahir, M.; Khan, B.; Ahmad, Q.Z.; Khan, N. Some general families of q-starlike functions associated with the Janowski functions. *Filomat* **2019**, *33*, 2613–2626. [CrossRef]

12. Srivastava, H.M.; Raza, N.; AbuJarad, E.S.; Srivastava, G.; AbuJarad, M.H. Fekete-Szegö inequality for classes of (p,q)-Starlike and (p,q)-convex functions. *Revista de la Real Academia de Ciencias Exactas, Físicas y Naturales. Serie A Matemáticas* **2019**, *113*, 3563–3584. [CrossRef]
13. Yuan, Y.; Srivastava, R.; Liu, J.L. The order of strongly starlikeness of the generalized α-convex functions. *Symmetry* **2019**, *11*, 76. [CrossRef]
14. Zhang, H.Y.; Srivastava, R.; Tang, H. Third-order Hankel and Toeplitz determinants for starlike functions connected with the sine function. *Mathematics* **2019**, *7*, 404. [CrossRef]
15. Srivastava, H.M. Operators of basic (or q-) calculus and fractional q-calculus and their applications in geometric function theory of complex analysis. *Iran J. Sci. Technol. Trans. Sci.* **2020**, *44*, 327–344. [CrossRef]
16. Srivastava, R.; Zayed, H.M. Subclasses of analytic functions of complex order defined by g- derivative operator. Studia Universitatis Babes-Bolyai. *Mathematica* **2019**, *64*, 223–242. [CrossRef]
17. Aldweby, H.; Darus, M. Some subordination results on q-analogue of Ruscheweyh differential operator. *Abstr. Appl. Anal.* **2014**, *2014*. [CrossRef].
18. Kanas, S.; Răducanu, D. Some class of analytic functions related to conic domains. *Math. Slovaca* **2014**, *64*, 1183–1196.
19. Mahmood, S.; Sokół, J. New subclass of analytic functions in conical domain associated with Ruscheweyh q-differential operator. *Res. Math.* **2017**, *71*, 1345–1357. [CrossRef]
20. Goodman, A.W. *Univalent Functions*; Polygonal Publishing House: Washington, NJ, USA, 1983; Volume 1. [CrossRef]
21. Goodman, A.W. *Univalent Functions*; Polygonal Publishing House: Washington, NJ, USA, 1983; Volume 2. [CrossRef]
22. Kanas, S.; Wisniowska, A. Conic regions and k-uniform convexity. *J. Comput. Appl. Maths.* **1999**, *105*, 327–336.
23. Kanas, S.; Wisniowska, A. Conic domains and starlike functions. *Revue Roumaine de Mathématiques Pures et Appliquées* **2000**, *45*, 647–658.
24. Kanas, S.; Srivastava, H.M. Linear operators associated with k-uniformly convex functions. *Integr. Transforms Spec. Funct.* **2000**, *9*, 121–132. [CrossRef]
25. Kanas, S. Techniques of the differential subordination for domains bounded by conic sections. *Int. J. Math. Math. Sci.* **2003**, *2003*, 2389–2400.
26. Noor, K.I.; Arif, M.; Ul-Haq, W. On k-uniformly close-to-convex functions of complex order. *Appl. Math. Comput.* **2009**, *215*, 629–635. [CrossRef]
27. Noor, K.I. Applications of certain operators to the classes related with generalized Janowski functions. *Integr. Transforms Spec. Funct.* **2010**, *21*, 557–567. [CrossRef]
28. Thomas, D.K. Starlike and close-to-convex functions. *J. Lond. Math. Soc.* **1967**, *42*, 427–435. [CrossRef]
29. Akhiezer, N.I. *Elements of the Theory of Elliptic Functions*; American Mathematical Society: Providence, RI, USA, 1970; Volume 79. [CrossRef].
30. Fan, L.L.; Wang, Z.G.; Khan, S.; Hussain, S.; Naeem, M.; Mahmood, T. Coefficient bounds for certain subclasses of q-starlike functions. *Mathematics* **2019**, *7*, 969. [CrossRef]
31. Khan, S.; Hussain, S.; Zaighum, M.A.; Khan, M.M. New subclass of analytic functions in conical domain associated with Ruscheweyh q-differential operator. *Int. J. Anal. Appl.* **2018**, *16*, 239–253.
32. Naeem, M.; Hussain, S.; Mahmood, T.; Khan, S.; Darus, M. A new subclass of analytic functions defined by using Salagean q-differential operator. *Mathematics* **2019**, *7*, 458. [CrossRef]
33. Sakar, F.M.; Aydogan, S.M. Subclass of m-quasiconformal harmonic functions in association with Janowski starlike functions. *Appl. Math. Comput.* **2018**, *319*, 461–468.
34. Ul-Haq, W.; Mahmood, S. Certain properties of a class of close-to-convex functions related to conic domains. *Abstr. Appl. Anal.* **2013**, *2013*, 847287. [CrossRef]
35. Janowski, W. Some extremal problems for certain families of analytic function I. In *Annales Polonici Mathematici*; Institute of Mathematics Polish Academy of Sciences: Warsaw, Poland, 1973: Volume 28, pp. 297–326. [CrossRef]
36. Srivastava, H.M.; Tahir, M.; Khan, B.; Ahmad, Q.Z.; Khan, N. Some general classes of q-starlike functions associated with the Janowski functions. *Symmetry* **2019**, *11*, 292. [CrossRef]
37. Srivastava, H.M.; Bilal, K.; Nazar, K.; Ahmad, Q.Z. Coefficient inequalities for q-starlike functions associated with the Janowski functions. *Hokkaido Math. J.* **2019**, *48*, 407–425.

38. Mahmood, S.; Jabeen, M.; Malik, S.N.; Srivastava, H.M.; Manzoor, R.; Riaz, S.M. Some coefficient inequalities of q-starlike functions associated with conic domain defined by q-derivative. *J. Funct. Sp.* **2018**, *2018*, 8492072. [CrossRef]
39. Noor, K.I.; Malik, S.N. On coefficient inequalities of functions associated with conic domains. *Comput. Math. Appl.* **2011**, *62*, 2209–2217. [CrossRef]
40. Mahmood, S.; Arif, M.; Malik, S.N. Janowski type close-to-convex functions associated with conic regions. *J. Inequal. Appl.* **2017**, *2017*, 259. [CrossRef]
41. Rogosinski, W. On the coefficients of subordinate functions. *Proc. Lond. Math. Soc.* **1943**, *48*, 48–82.
42. Miller, S.S.; Mocanu, P.T. *Differential Subordinations: Theory and Applications*; CRC Press: Boca Raton, FL, USA, 2000. [CrossRef]
43. Subramanian, K.G.; Sudharsan, T.V.; Silverman, H. On uniformly close-to-convex functions and uniformly quasiconvex functions. *Int. J. Math. Math. Sci.* **2003**, *2003*, 3053–3058. [CrossRef]
44. Kaplan, W. Close-to-convex schlicht functions. *Mich. Math. J.* **1952**, *1*, 169–185. [CrossRef]

© 2020 by the authors. Licensee MDPI, Basel, Switzerland. This article is an open access article distributed under the terms and conditions of the Creative Commons Attribution (CC BY) license (http://creativecommons.org/licenses/by/4.0/).

Article

Argument and Coefficient Estimates for Certain Analytic Functions

Davood Alimohammadi [1], Nak Eun Cho [2,*], Ebrahim Analouei Adegani [3] and Ahmad Motamednezhad [3]

[1] Department of Mathematics, Faculty of Science, Arak University, Arak 38156-8-8349, Iran; d-alimohammadi@araku.ac.ir
[2] Department of Applied Mathematics, College of Natural Sciences, Pukyong National University, Busan 608-737, Korea
[3] Faculty of Mathematical Sciences, Shahrood University of Technology, P.O. Box 316-36155 Shahrood, Iran; analoey.ebrahim@gmail.com (E.A.A.); a.motamedne@gmail.com (A.M.)
* Correspondence: necho@pknu.ac.kr

Received: 3 December 2019; Accepted: 2 January 2020; Published: 5 January 2020

Abstract: The aim of the present paper is to introduce a new class $\mathcal{G}(\alpha, \delta)$ of analytic functions in the open unit disk and to study some properties associated with strong starlikeness and close-to-convexity for the class $\mathcal{G}(\alpha, \delta)$. We also consider sharp bounds of logarithmic coefficients and Fekete-Szegö functionals belonging to the class $\mathcal{G}(\alpha, \delta)$. Moreover, we provide some topics related to the results reported here that are relevant to outcomes presented in earlier research.

Keywords: starlike function; subordinate; univalent function

MSC: Primary 30C45; Secondary 30C80

1. Introduction and Preliminaries

Let $\mathbb{U}$ denote the open unit dick in the complex plane $\mathbb{C}$. A function $\omega : \mathbb{U} \to \mathbb{C}$ is called a *Schwarz function* if ω is a analytic function in $\mathbb{U}$ with $\omega(0) = 0$ and $|\omega(z)| < 1$ for all $z \in \mathbb{U}$. Clearly, a Schwarz function ω is the form

$$\omega(z) = w_1 z + w_2 z^2 + \cdots.$$

We denote by Ω the set of all Schwarz functions on $\mathbb{U}$.

Let $\mathcal{A}$ be consisting of all analytic functions of the following normalized form:

$$f(z) = z + \sum_{n=2}^{\infty} a_n z^n, \qquad (1)$$

in the open unit disk $\mathbb{U}$. An analytic function f is said to be *univalent* in a domain if it provides a one-to-one mapping onto its image: $f(z_1) = f(z_2) \Rightarrow z_1 = z_2$. Geometrically, this means that different points in the domain will be mapped into different points on the image domain. Also, let $\mathcal{S}$ be the class of functions $f \in \mathcal{A}$ which are univalent in $\mathbb{U}$. A domain D in the complex plane $\mathbb{C}$ is called *starlike* with respect to a point $w_0 \in D$, if the line segment joining w_0 to every other point $w \in D$ lies in the interior of D. In other words, for any $w \in D$ and $0 \le t \le 1$, $tw_0 + (1-t)w \in D$. A function $f \in \mathcal{A}$ is starlike if the image $f(D)$ is starlike with respect to the origin.

For two analytic functions f and F in $\mathbb{U}$, we say that the function f is *subordinate to* the function F in $\mathbb{U}$ and we write $f(z) \prec F(z)$, if there exists a Schwarz function ω such that $f(z) = F(\omega(z))$ for all $z \in \mathbb{U}$. Specifically, if the function F is univalent in $\mathbb{U}$, then we have the next equivalence:

$$f(z) \prec F(z) \iff f(0) = F(0) \quad \text{and} \quad f(\mathbb{U}) \subset F(\mathbb{U}).$$

The logarithmic coefficients γ_n of $f \in \mathcal{S}$ are defined with the following series expansion:

$$\log\left(\frac{f(z)}{z}\right) = 2\sum_{n=1}^{\infty} \gamma_n(f) z^n, \ z \in \mathbb{U}. \tag{2}$$

These coefficients are an important factor in studying diverse estimates in the theory of univalent functions. Note that we use γ_n instead of $\gamma_n(f)$. The concept of logarithmic coefficients inspired Kayumov [1] to solve Brennan's conjecture for conformal mappings. The importance of the logarithmic coefficients follows from Lebedev-Milin inequalities [2] (Chapter 2), see also [3,4], where estimates of the logarithmic coefficients were used to find bounds on the coefficients of f. Milin [2] conjectured the inequality

$$\sum_{m=1}^{n}\sum_{k=1}^{m}\left(k|\gamma_k|^2 - \frac{1}{k}\right) \leq 0 \quad (n = 1, 2, 3, \cdots),$$

which implies Robertson's conjecture [5], and hence, Bieberbach's conjecture [6]. This is the famous coefficient problem in univalent function theory. L. de Branges [7] established Bieberbach's conjecture by proving Milin's conjecture.

Definition 1. *Let $q, n \in \mathbb{N}$. The qth Hankel determinant is denote by $H_q(n)$ and defined by*

$$H_q(n) = \begin{vmatrix} a_n & a_{n+1} & \cdots & a_{n+q-1} \\ a_{n+1} & a_{n+2} & \cdots & a_{n+q} \\ \vdots & \vdots & \ddots & \vdots \\ a_{n+q-1} & a_{n+q} & \cdots & a_{n+2q-2} \end{vmatrix}, \tag{3}$$

where a_k $(k = 1, 2, \ldots)$ are the coefficients of the Taylor series expansion of a function f of the form (1). *Note that $a_1 = 1$.*

The Hankel determinant $H_q(n)$ was defined by Pommerenke [8,9] and for fixed q, n the bounds of $|H_q(n)|$ have been studied for several subfamilies of univalent functions. Different properties of these determinants can be observed in [10] (Chapter 4). The Hankel determinants $H_2(1) = a_3 - a_2^2$ and $H_2(2) = a_2 a_4 - a_3^2$, are well-known as *Fekete-Szegö* and *second Hankel determinant functionals*, respectively. In addition, Fekete and Szegö [11] introduced the generalized functional $a_3 - \lambda a_2^2$, where λ is a real number. Recently, Hankel determinants and other problems for various classes of bi-univalent functions have been studied, see [12–16].

For $\alpha \in [0, 1)$, we denote by $\mathcal{S}^*(\alpha)$ the subclass of $\mathcal{A}$ including of all $f \in \mathcal{A}$ for which f is a *starlike function of order α* in $\mathbb{U}$, with

$$\text{Re}\frac{zf'(z)}{f(z)} > \alpha \quad (z \in \mathbb{U}).$$

Also, for $\alpha \in (0, 1]$, we denote by $\widetilde{\mathcal{S}}^*(\alpha)$ the subclass of $\mathcal{A}$ consisting of all $f \in \mathcal{A}$ for which f is a *strongly starlike function of order α* in $\mathbb{U}$, with

$$\left|\text{Arg}\left(\frac{zf'(z)}{f(z)}\right)\right| < \frac{\alpha\pi}{2} \quad (z \in \mathbb{U}).$$

Note that $\widetilde{\mathcal{S}}^*(1) = \mathcal{S}^*(0) = \mathcal{S}^*$, the class of *starlike functions* in $\mathbb{U}$.

For $\alpha \in (0,1]$, we denote by $\widetilde{C}(\alpha)$ the subclass of $\mathcal{A}$ including all of $f \in \mathcal{A}$ for which

$$\left|\mathrm{Arg}\left(f'(z)\right)\right| < \frac{\alpha\pi}{2} \qquad (z \in \mathbb{U}).$$

Note that $\widetilde{C}(1) = C$, the subclass of *close-to-convex functions* in $\mathbb{U}$. Here we understand that $\mathrm{Arg}\, w$ is a number in $(-\pi, \pi]$.

For $\alpha \in (0,1]$, Nunokawa and Saitoh in [17] defined the more general class $\mathcal{G}(\alpha)$ consisting of all $f \in \mathcal{A}$ satisfying

$$\mathrm{Re}\left(1 + \frac{zf''(z)}{f'(z)}\right) < 1 + \frac{\alpha}{2} \qquad (z \in \mathbb{U}).$$

They proved that $\mathcal{G}(\alpha)$ is a subclass of $\mathcal{S}^*$. Ozaki in [18] showed that every function $\mathcal{G}(1)$ is univalent in the unit disk $\mathbb{U}$. In the following, Umezawa [19], Sakaguchi [20] and Singh and Singh [21] obtained some geometric properties of $\mathcal{G}(1)$ including, convex in one direction, close-to-convex and starlike, respectively. Obradović et al. in [22] proved the sharp coefficient bounds for the moduli of the Taylor coefficients a_n of $f \in \mathcal{G}(\alpha)$ and determined the sharp bound for the Fekete-Szegö functional for functions in $\mathcal{G}(\alpha)$ with complex parameter λ. Also, Ponnusamy et al. [22,23] studied bounds for the logarithmic coefficients for functions in $\mathcal{G}(\alpha)$.

Here, we introduce a class as follows:

Definition 2. *For $\alpha, \delta \in (0,1]$, we define the subclass $\mathcal{G}(\alpha,\delta)$ of $\mathcal{A}$ as the following:*

$$\mathcal{G}(\alpha,\delta) := \left\{ f \in \mathcal{A} : \left|\mathrm{Arg}\left(\frac{2+\alpha}{\alpha} - \frac{2}{\alpha}\left(1 + \frac{zf''(z)}{f'(z)}\right)\right)\right| < \frac{\delta\pi}{2} \quad (z \in \mathbb{U})\right\}.$$

It is clear that $\mathcal{G}(\alpha,1) = \mathcal{G}(\alpha)$ for $\alpha \in (0,1]$. Let $\alpha, \delta \in (0,1]$, identity function on $\mathbb{U}$ belongs to $\mathcal{G}(\alpha,\delta)$ which implies that $\mathcal{G}(\alpha,\delta) \neq \varnothing$. By means of the principle of subordination between analytic functions, we deduce

$$\mathcal{G}(\alpha,\delta) := \left\{ f \in \mathcal{A} : 1 + \frac{zf''(z)}{f'(z)} \prec -\frac{\alpha}{2}\left(\frac{1+z}{1-z}\right)^\delta + \frac{2+\alpha}{2} := \phi(z) \quad (z \in \mathbb{U})\right\}. \qquad (4)$$

Since the function f defined by

$$f(z) = \int_0^z \exp\left(\int_0^x \frac{-\frac{\alpha}{2}\left(\frac{1+t}{1-t}\right)^\delta + \frac{\alpha}{2}}{t}\,dt\right)dx \qquad (z \in \mathbb{U})$$

satisfies

$$1 + \frac{zf''(z)}{f'(z)} = \phi(z) \prec \phi(z),$$

we deduce $f \in \mathcal{G}(\alpha,\delta)$.

The aim of the present paper is to study some geometric properties for the class $\mathcal{G}(\alpha,\delta)$ such as strongly starlikeness and close-to-convexity. Also we investigate sharp bounds on logarithmic coefficients and Fekete-Szegö functionals for functions belonging to the class $\mathcal{G}(\alpha,\delta)$, which incorporate some known results as the special cases.

2. Some Properties of the Class $\mathcal{G}(\alpha, \delta)$

We denote by Q the class of all complex-valued functions q for which q is univalent at each $\overline{\mathbb{U}} \setminus E(q)$ and $q'(\zeta) \neq 0$ for all $\zeta \in \partial \mathbb{U} \setminus E(q)$ where

$$E(q) = \left\{ \zeta \in \partial \mathbb{U} : \lim_{z \to \zeta} q(z) = \infty \right\}.$$

The following lemmas will be required to establish our main results.

Lemma 1 ([24] (Lemma 2.2d (i))). *Let $q \in Q$ with $q(0) = a$ and let $p(z) = a + p_n z^n + \ldots$ be analytic in $\mathbb{U}$ with $p(z) \not\equiv 1$ and $n \geq 1$. If p is not subordinate to q in $\mathbb{U}$ then there exist $z_0 \in \mathbb{U}$ and $\zeta_0 \in \partial \mathbb{U} \setminus E(q)$ such that $\{p(z) : z \in \mathbb{U}, |z| < |z_0|\} \subset q(\mathbb{U})$,*

$$p(z_0) = q(\zeta_0).$$

Lemma 2. (see [25,26]) *Let the function p given by*

$$p(z) = 1 + \sum_{n=1}^{\infty} c_n z^n$$

be analytic in $\mathbb{U}$ with $p(0) = 1$ and $p(z) \neq 0$ for all $z \in \mathbb{U}$. If there exists a point $z_0 \in \mathbb{U}$ with

$$\left| \arg \left(p(z) \right) \right| < \frac{\beta \pi}{2} \qquad (|z| < |z_0|)$$

and

$$\left| \arg \left(p(z_0) \right) \right| = \frac{\beta \pi}{2},$$

for some $\beta > 0$, then

$$\frac{z_0 p'(z_0)}{p(z_0)} = ik\beta \qquad (i = \sqrt{-1}),$$

where

$$k \geq \frac{a + a^{-1}}{2} \geq 1 \quad \text{when} \quad \arg \left(p(z_0) \right) = \frac{\beta \pi}{2} \tag{5}$$

and

$$k \leq -\frac{a + a^{-1}}{2} \leq -1 \quad \text{when} \quad \arg \left(p(z_0) \right) = -\frac{\beta \pi}{2}, \tag{6}$$

where

$$[p(z_0)]^{1/\beta} = \pm ia \quad \text{and} \quad a > 0.$$

Theorem 1. *Let $\alpha, \beta \in (0, 1]$. If $f \in \mathcal{A}$ satisfies the condition*

$$\left| \text{Arg} \left(\frac{2 + \alpha}{\alpha} - \frac{2}{\alpha} \left(1 + \frac{z f''(z)}{f'(z)} \right) \right) \right| < \text{Arctan} \left(\frac{4\beta}{2 + \alpha} \right), \tag{7}$$

then

$$\left| \text{Arg} \left(\frac{z f'(z)}{f(z)} \right) \right| < \frac{\beta \pi}{2} \quad (z \in \mathbb{U}).$$

Proof. Let $f \in \mathcal{A}$ and define the function $p : \mathbb{U} \to \mathbb{C}$ by

$$p(z) = \frac{z f'(z)}{f(z)} = 1 + \sum_{n=1}^{\infty} c_n z^n \quad (z \in \mathbb{U}).$$

Then it follows that p is analytic in $\mathbb{U}$, $p(0) = 1$,

$$1 + \frac{zf''(z)}{f'(z)} = p(z) + \frac{zp'(z)}{p(z)} \quad (z \in \mathbb{U})$$

and $p(z) \neq 0$ for all $z \in \mathbb{U}$. In fact, if p has a zero $z_0 \in \mathbb{U}$ of order m, then we may write

$$p(z) = (z - z_0)^m p_1(z) \quad (m \in \mathbb{N} = 1, 2, 3, \cdots),$$

where p_1 is analytic in $\mathbb{U}$ with $p_1(z_0) \neq 0$. Then

$$\frac{2+\alpha}{\alpha} - \frac{2}{\alpha}\left(p(z) + \frac{zp'(z)}{p(z)}\right) = \frac{2+\alpha}{\alpha} - \frac{2}{\alpha}\left(p(z) + \frac{zp_1'(z)}{p_1(z)} + \frac{mz}{z-z_0}\right).$$

Thus, choosing $z \to z_0$, suitably the argument of the right-hand of the above equality can take any value between $-\pi$ and π, which contradicts (7).

Define the function $q : \overline{\mathbb{U}} \setminus \{1\} \to \mathbb{C}$ by

$$q(z) = \left(\frac{1+z}{1-z}\right)^\beta \quad (z \in \overline{\mathbb{U}} \setminus \{1\}).$$

Then $q \in Q$, $q(0) = 1$ and $E(q) = \{1\}$. It is clear that $|\operatorname{Arg}(p(z))| < \frac{\beta\pi}{2}$ for all $z \in \mathbb{U}$ if and only if $p \prec q$ on $\mathbb{U}$. Let $|\operatorname{Arg}(p(z_1))| \geq \frac{\beta\pi}{2}$ for some $z_1 \in \mathbb{U}$. Then p is not subordinate to q. By Lemma 1 there exists $z_0 \in \mathbb{U}$ and $\xi_0 \in \partial\mathbb{U} \setminus \{1\}$ such that $\{p(z) : z \in \mathbb{U}, |z| < |z_0|\} \subset q(\mathbb{U})$ and $p(z_0) = q(\xi_0)$. Therefore,

$$|\operatorname{Arg}(p(z))| < \frac{\beta\pi}{2},$$

for all $z \in \mathbb{U}$ with $|z| < |z_0|$ and

$$|\operatorname{Arg}(p(z_0))| = \frac{\beta\pi}{2}.$$

Then, Lemma 2, gives us that

$$\frac{z_0 p'(z_0)}{p(z_0)} = ik\beta,$$

where $[p(z_0)]^{\frac{1}{\beta}} = \pm ia$ ($a > 0$) and k is given by (5) or (6).

Define the function $g : (0, a) \to \mathbb{R}$ by

$$g(t) = \frac{\frac{2}{2+\alpha}\left(t^\beta \sin(\frac{\beta\pi}{2}) + \beta\right)}{1 - \frac{2}{2+\alpha}t^\beta \cos\frac{\beta\pi}{2}} \quad t \in (0, a).$$

Then g is a differentiable function on $(0, a)$ and $g'(t) > 0$ for all $t \in (0, a)$. This implies that the function $h : (0, a) \to \mathbb{R}$ defined by

$$h(t) = \operatorname{Arctan}(g(t)) \quad t \in (0, a),$$

is a non-decreasing function on $(0, a)$. Thus

$$h(a) \geq \lim_{t \to 0^+} h(t) = \operatorname{Arctan}\left(\frac{2\beta}{2+\alpha}\right).$$

Therefore, we have

$$\operatorname{Arctan}\left(\frac{\frac{2}{2+\alpha}\left(a^\beta \sin \frac{\beta\pi}{2}+\beta\right)}{1-\frac{2}{2+\alpha}a^\beta \cos \frac{\beta\pi}{2}}\right) \geq \operatorname{Arctan}\left(\frac{2\beta}{2+\alpha}\right). \tag{8}$$

Now we consider six cases for estimation of $\operatorname{Arg}(p(z_0))$ as follows:

Case 1. $\operatorname{Arg}(p(z_0)) = \frac{\beta\pi}{2}$ and $1 - \frac{2}{2+\alpha}a^\beta \cos \frac{\beta\pi}{2} > 0$. In this case we have $[p(z_0)]^{\frac{1}{\beta}} = ia\ (a > 0)$, and $k \geq 1$. Therefore,

$$\operatorname{Arg}\left(\frac{2+\alpha}{\alpha}\left(1 - \frac{2}{2+\alpha}\left(p(z_0) + \frac{z_0 p'(z_0)}{p(z_0)}\right)\right)\right) = \operatorname{Arg}\left(1 - \frac{2}{2+\alpha}a^\beta \cos \frac{\beta\pi}{2} - i\frac{2}{2+\alpha}\left(a^\beta \sin \frac{\beta\pi}{2} + k\beta\right)\right)$$

$$= \operatorname{Arctan}\left(\frac{-\frac{2}{2+\alpha}\left(a^\beta \sin \frac{\beta\pi}{2} + k\beta\right)}{1 - \frac{2}{2+\alpha}a^\beta \cos \frac{\beta\pi}{2}}\right)$$

$$\leq \operatorname{Arctan}\left(\frac{-\frac{2}{2+\alpha}\left(a^\beta \sin \frac{\beta\pi}{2} + \beta\right)}{1 - \frac{2}{2+\alpha}a^\beta \cos \frac{\beta\pi}{2}}\right)$$

$$= -\operatorname{Arctan}\left(\frac{\frac{2}{2+\alpha}\left(a^\beta \sin \frac{\beta\pi}{2} + \beta\right)}{1 - \frac{2}{2+\alpha}a^\beta \cos \frac{\beta\pi}{2}}\right)$$

$$= -h(a)$$

$$\leq -\operatorname{Arctan}\left(\frac{2\beta}{2+\alpha}\right). \tag{9}$$

Now applying (8) and (9) we get

$$\operatorname{Arg}\left(\frac{2+\alpha}{\alpha}\left(1 - \frac{2}{2+\alpha}\left(p(z_0) + \frac{z_0 p'(z_0)}{p(z_0)}\right)\right)\right) = \operatorname{Arg}\left(1 - \frac{2}{2+\alpha}\left(p(z_0) + \frac{z_0 p'(z_0)}{p(z_0)}\right)\right)$$

$$= \operatorname{Arg}\left(1 - \frac{2}{2+\alpha}\left(1 + \frac{z_0 f''(z_0)}{f'(z_0)}\right)\right)$$

$$\leq -\operatorname{Arctan}\left(\frac{\frac{2}{2+\alpha}\left(a^\beta \sin \frac{\beta\pi}{2} + \beta\right)}{1 - \frac{2}{2+\alpha}a^\beta \cos \frac{\beta\pi}{2}}\right)$$

$$\leq -\operatorname{Arctan}\left(\frac{2\beta}{2+\alpha}\right),$$

which contradicts (7).

Case 2. $\operatorname{Arg}(p(z_0)) = \frac{\beta\pi}{2}$ and $1 - \frac{2}{2+\alpha}a^\beta \cos \frac{\beta\pi}{2} = 0$. In this case, we have $p(z_0) = a^\beta(\cos \frac{\beta\pi}{2} + i \sin \frac{\beta\pi}{2})$ and $k \geq 1$. Thus $-\frac{2}{2+\alpha}\left(a^\beta \sin \frac{\beta\pi}{2} + k\beta\right) < 0$ and so

$$\operatorname{Arg}\left(\frac{2+\alpha}{\alpha}\left(1 - \frac{2}{2+\alpha}\left(p(z_0) + \frac{z_0 p'(z_0)}{p(z_0)}\right)\right)\right) = \operatorname{Arg}\left(-i\frac{2}{2+\alpha}\left(a^\beta \sin \frac{\beta\pi}{2} + k\beta\right)\right)$$

$$= -\frac{\pi}{2} < -\operatorname{Arctan}\left(\frac{2\beta}{2+\alpha}\right),$$

which contradicts (7).

Case 3. $\operatorname{Arg}(p(z_0)) = \frac{\beta\pi}{2}$ and $1 - \frac{2}{2+\alpha}a^\beta \cos \frac{\beta\pi}{2} < 0$. In this case, we have $p(z_0) = a^\beta(\cos \frac{\beta\pi}{2} + i \sin \frac{\beta\pi}{2})$ and $k \geq 1$. Thus

$$\frac{-\frac{2}{2+\alpha}\left(a^\beta \sin \frac{\beta\pi}{2} + k\beta\right)}{1 - \frac{2}{2+\alpha}a^\beta \cos \frac{\beta\pi}{2}} > 0.$$

Therefore,

$$\text{Arg}\left(\frac{2+\alpha}{\alpha}\left(1-\frac{2}{2+\alpha}\left(p(z_0)+\frac{z_0 p'(z_0)}{p(z_0)}\right)\right)\right) = \text{Arg}\left(1-\frac{2}{2+\alpha}a^\beta\cos\frac{\beta\pi}{2}-i\frac{2}{2+\alpha}\left(a^\beta\sin\frac{\beta\pi}{2}+k\beta\right)\right)$$

$$= -\pi + \text{Arctan}\left(\frac{-\frac{2}{2+\alpha}\left(a^\beta\sin\frac{\beta\pi}{2}+k\beta\right)}{1-\frac{2}{2+\alpha}a^\beta\cos\frac{\beta\pi}{2}}\right)$$

$$< -\pi + \frac{\pi}{2}$$

$$= -\frac{\pi}{2}$$

$$< -\text{Arctan}\left(\frac{2\beta}{2+\alpha}\right),$$

which contradicts (7).

Case 4. $\text{Arg}(p(z_0)) = -\frac{\beta\pi}{2}$ and $1-\frac{2}{2+\alpha}a^\beta\cos\frac{\beta\pi}{2} > 0$. In this case we have $p(z_0) = a^\beta(\cos\frac{\beta\pi}{2} - i\sin\frac{\beta\pi}{2})$ and $k \leq -1$. Thus $-\frac{2}{2+\alpha}\left(-a^\beta\sin\frac{\beta\pi}{2}+k\beta\right) < 0$. Now, applying (8) we get

$$\text{Arg}\left(\frac{2+\alpha}{\alpha}\left(1-\frac{\alpha}{2+\alpha}\left(p(z_0)+\frac{z_0 p'(z_0)}{p(z_0)}\right)\right)\right) = \text{Arg}\left(1-\frac{2}{2+\alpha}\left(a^\beta e^{\frac{-i\beta\pi}{2}}+ik\beta\right)\right)$$

$$= \text{Arctan}\left(\frac{-\frac{2}{2+\alpha}\left(-a^\beta\sin\frac{\beta\pi}{2}+k\beta\right)}{1-\frac{2}{2+\alpha}a^\beta\cos\frac{\beta\pi}{2}}\right)$$

$$\geq \text{Arctan}\left(\frac{-\frac{2}{2+\alpha}\left(-a^\beta\sin\frac{\beta\pi}{2}-\beta\right)}{1-\frac{2}{2+\alpha}a^\beta\cos\frac{\beta\pi}{2}}\right)$$

$$= \text{Arctan}\left(\frac{\frac{2}{2+\alpha}\left(a^\beta\sin\frac{\beta\pi}{2}+\beta\right)}{1-\frac{2}{2+\alpha}a^\beta\cos\frac{\beta\pi}{2}}\right)$$

$$\geq \text{Arctan}\left(\frac{2\beta}{2+\alpha}\right),$$

which contradicts (7).

For other cases applying the same method in Case 2. and Case 3. with $k \leq -1$ we obtain

$$\text{Arg}\left(\frac{2+\alpha}{\alpha}\left(1-\frac{2}{2+\alpha}\left(p(z_0)+\frac{z_0 p'(z_0)}{p(z_0)}\right)\right)\right) \geq \text{Arctan}\left(\frac{2\beta}{2+\alpha}\right),$$

which contradicts (7). Hence the proof is completed. □

Corollary 1. *Let $\alpha,\beta \in (0,1]$ and $\delta = \frac{2}{\pi}\text{Arctan}\left(\frac{2\beta}{2+\alpha}\right)$. If $f \in \mathcal{G}(\alpha,\delta)$, then $f \in \widetilde{\mathcal{S}}^*(\beta)$.*

Theorem 2. *Let $\alpha,\beta \in (0,1]$. If $f \in \mathcal{A}$ and*

$$\left|\text{Arg}\left(\frac{2+\alpha}{\alpha}-\frac{2}{\alpha}\left(1+\frac{zf''(z)}{f'(z)}\right)\right)\right| < \text{Arctan}\left(\frac{2\beta}{\alpha}\right), \qquad (10)$$

then

$$|\text{Arg}(f'(z))| < \frac{\beta\pi}{2} \quad (z \in \mathbb{U}).$$

Proof. Define the function $p : \mathbb{U} \to \mathbb{C}$ by

$$p(z) = f'(z) = 1 + \sum_{n=1}^{\infty} c_n z^n \quad (z \in \mathbb{U}).$$

Then p is analytic in $\mathbb{U}$, $p(0) = 1$,

$$1 + \frac{zf''(z)}{f'(z)} = 1 + \frac{zp'(z)}{p(z)}.$$

and $p(z) \neq 0$ for all $z \in \mathbb{U}$. If there exists a point $z_0 \in \mathbb{U}$ such that

$$|\operatorname{Arg}(p(z))| < \frac{\beta\pi}{2},$$

for all $z \in \mathbb{U}$ with $|z| < |z_0|$ and

$$|\operatorname{Arg}(p(z_0))| = \frac{\beta\pi}{2}.$$

Then, Lemma 2, gives us that

$$\frac{z_0 p'(z_0)}{p(z_0)} = ik\beta,$$

where $[p(z_0)]^{\frac{1}{\beta}} = \pm ia$ $(a > 0)$ and k is given by (5) or (6).

For the case $\operatorname{Arg}(p(z_0)) = \frac{\alpha\pi}{2}$ when

$$p(z_0)]^{\frac{1}{\beta}} = ia \quad (a > 0)$$

and $k \geq 1$, we have

$$\operatorname{Arg}\left(\frac{2+\alpha}{\alpha}\left(1 - \frac{2}{2+\alpha}\left(1 + \frac{z_0 p'(z_0)}{p(z_0)}\right)\right)\right) = \operatorname{Arg}\left(1 - \frac{2}{2+\alpha}\left(1 + \frac{z_0 p'(z_0)}{p(z_0)}\right)\right)$$

$$= \operatorname{Arg}\left(1 - \frac{2}{2+\alpha}(1 + ik\beta)\right)$$

$$= \operatorname{Arctan}\left(\frac{-2k\beta}{\alpha}\right)$$

$$\leq -\operatorname{Arctan}\left(\frac{2\beta}{\alpha}\right),$$

which contradicts (10).

Next, for the case $\operatorname{Arg}(p(z_0)) = -\frac{\alpha\pi}{2}$ when

$$p(z_0) = -ia \quad (a > 0)$$

and $k \leq -1$, using the same method as before, we can obtain

$$\operatorname{Arg}\left(\frac{2+\alpha}{\alpha}\left(1 - \frac{2}{2+\alpha}\left(1 + \frac{z_0 p'(z_0)}{p(z_0)}\right)\right)\right) = \operatorname{Arg}\left(1 - \frac{2}{2+\alpha}\left(1 + \frac{z_0 p'(z_0)}{p(z_0)}\right)\right)$$

$$= \operatorname{Arg}\left(1 - \frac{2}{2+\alpha}(1 + ik\beta)\right)$$

$$= \operatorname{Arctan}\left(\frac{-2k\beta}{\alpha}\right)$$

$$\geq \operatorname{Arctan}\left(\frac{2\beta}{\alpha}\right),$$

which is a contradicts (10).

Consequently, from the two above-discussed contradictions, it follows that

$$\left| \operatorname{Arg}\left(f'(z)\right) \right| < \frac{\beta \pi}{2} \quad (z \in \mathbb{U}).$$

and hence the proof is completed. □

Corollary 2. *Let $\alpha, \beta \in (0,1]$ and $\delta = \frac{2}{\pi} \operatorname{Arctan}\left(\frac{2\beta}{\alpha}\right)$. If $f \in \mathcal{G}(\alpha, \delta)$, then $f \in \widetilde{\mathcal{C}}(\beta)$. In other words, if $f \in \mathcal{G}(\alpha, \delta)$, then $f(z)$ is close-to-convex (univalent) in $\mathbb{U}$.*

3. Coefficient Bounds

In this section, we give a the general problem of coefficients in the class $\mathcal{G}(\alpha, \delta)$ like the estimates of coefficients for membership of this, bounds of logarithmic coefficients and the Fekete-Szegö problem with sharp inequalities. In order to achieve our aim we need to establish some knowledge.

Lemma 3 ([27] (p. 172)). *Let $\omega \in \Omega$ with $\omega(z) = \sum\limits_{n=1}^{\infty} w_n z^n$ for all $z \in \mathbb{U}$. Then $|w_1| \leq 1$ and*

$$|w_n| \leq 1 - |w_1|^2 \quad \text{for all } n \in \mathbb{N} \text{ with } n \geq 2.$$

Lemma 4 ([28] (Inequality 7, p. 10)). *Let $\omega \in \Omega$ with $\omega(z) = \sum\limits_{n=1}^{\infty} w_n z^n$ for all $z \in \mathbb{U}$. Then*

$$|w_2 - t w_1^2| \leq \max\{1, |t|\} \quad \text{for all } t \in \mathbb{C}.$$

The inequality is sharp for the functions $\omega(z) = z^2$ or $\omega(z) = z$.

Lemma 5 ([29]). *If $\omega \in \Omega$ with $\omega(z) = \sum\limits_{n=1}^{\infty} w_n z^n$ ($z \in \mathbb{U}$), then for any real numbers q_1 and q_2, we have the following sharp estimate:*

$$|p_3 + q_1 w_1 w_2 + q_2 w_1^3| \leq H(q_1; q_2),$$

where

$$H(q_1; q_2) = \begin{cases} 1 & \text{if } (q_1, q_2) \in D_1 \cup D_2 \cup \{(2,1)\}, \\ |q_2| & \text{if } (q_1, q_2) \in \cup_{k=3}^{7} D_k, \\ \frac{2}{3}(|q_1|+1)\left(\frac{|q_1|+1}{3(|q_1|+1+q_2)}\right)^{\frac{1}{2}} & \text{if } (q_1, q_2) \in D_8 \cup D_9, \\ \frac{q_2}{3}\left(\frac{q_1^2-4}{q_1^2-4q_2}\right)\left(\frac{q_1^2-4}{3(q_2-1)}\right)^{\frac{1}{2}} & \text{if } (q_1, q_2) \in D_{10} \cup D_{11} \setminus \{(2,1)\}, \\ \frac{2}{3}(|q_1|-1)\left(\frac{|q_1|-1}{3(|q_1|-1-q_2)}\right)^{\frac{1}{2}} & \text{if } (q_1, q_2) \in D_{12}, \end{cases}$$

and the sets D_k, $k = 1, 2, \ldots, 12$ are stated as given below:

$$D_1 = \left\{(q_1, q_2) : |q_1| \leq \frac{1}{2}, \ |q_2| \leq 1\right\},$$

$$D_2 = \left\{(q_1, q_2) : \frac{1}{2} \leq |q_1| \leq 2, \ \frac{4}{27}\left((|q_1|+1)^3\right) - (|q_1|+1) \leq q_2 \leq 1\right\},$$

$$D_3 = \left\{(q_1, q_2) : |q_1| \leq \frac{1}{2}, \ q_2 \leq -1\right\},$$

$$D_4 = \left\{(q_1, q_2) : |q_1| \geq \frac{1}{2}, |q_2| \leq -\frac{2}{3}(|q_1|+1)\right\},$$

$$D_5 = \{(q_1, q_2) : |q_1| \leq 2, q_2 \geq 1\},$$

$$D_6 = \left\{(q_1, q_2) : 2 \leq |q_1| \leq 4, q_2 \geq \frac{1}{12}(q_1^2 + 8)\right\},$$

$$D_7 = \left\{(q_1, q_2) : |q_1| \geq 4, q_2 \geq \frac{2}{3}(|q_1|-1)\right\},$$

$$D_8 = \left\{(q_1, q_2) : \frac{1}{2} \leq |q_1| \leq 2, -\frac{2}{3}(|q_1|+1) \leq q_2 \leq \frac{4}{27}\left((|q_1|+1)^3\right) - (|q_1|+1)\right\},$$

$$D_9 = \left\{(q_1, q_2) : |q_1| \geq 2, -\frac{2}{3}(|q_1|+1) \leq q_2 \leq \frac{2|q_1|(|q_1|+1)}{q_1^2 + 2|q_1| + 4}\right\},$$

$$D_{10} = \left\{(q_1, q_2) : 2 \leq |q_1| \leq 4, \frac{2|q_1|(|q_1|+1)}{q_1^2 + 2|q_1| + 4} \leq q_2 \leq \frac{1}{12}(q_1^2 + 8)\right\},$$

$$D_{11} = \left\{(q_1, q_2) : |q_1| \geq 4, \frac{2|q_1|(|q_1|+1)}{q_1^2 + 2|q_1| + 4} \leq q_2 \leq \frac{2|q_1|(|q_1|-1)}{q_1^2 - 2|q_1| + 4}\right\},$$

$$D_{12} = \left\{(q_1, q_2) : |q_1| \geq 4, \frac{2|q_1|(|q_1|-1)}{q_1^2 - 2|q_1| + 4} \leq q_2 \leq \frac{2}{3}(|q_1|-1)\right\}.$$

We assume that φ is a univalent function in the unit disk $\mathbb{U}$ satisfying $\varphi(0) = 1$ such that it has the power series expansion of the following form

$$\varphi(z) = 1 + B_1 z + B_2 z^2 + B_3 z^3 + \ldots, \quad z \in \mathbb{U}, \quad \text{with} \quad B_1 \neq 0. \tag{11}$$

Lemma 6 ([30] (Theorem 2)). *Let the function $f \in \mathcal{K}(\varphi)$. Then the logarithmic coefficients of f satisfy the inequalities*

$$|\gamma_1| \leq \frac{|B_1|}{4}, \tag{12}$$

$$|\gamma_2| \leq \begin{cases} \dfrac{|B_1|}{12} & \text{if} \quad |4B_2 + B_1^2| \leq 4|B_1|, \\ \dfrac{|4B_2 + B_1^2|}{48} & \text{if} \quad |4B_2 + B_1^2| > 4|B_1|, \end{cases} \tag{13}$$

and if B_1, B_2, and B_3 are real values,

$$|\gamma_3| \leq \frac{|B_1|}{24} H(q_1; q_2), \tag{14}$$

where $H(q_1; q_2)$ is given by Lemma 5, $q_1 = \dfrac{B_1 + \frac{4B_2}{B_1}}{2}$ and $q_2 = \dfrac{B_2 + \frac{2B_3}{B_1}}{2}$. The bounds (12) and (13) are sharp.

Theorem 3. *Let $f \in \mathcal{G}(\alpha, \delta)$. Then*

$$|a_2| \leq \frac{\alpha \delta}{2}, \quad |a_3| \leq \frac{\alpha \delta}{6}, \quad |a_4| \leq \frac{\alpha \delta}{12} H(q_1; q_2),$$

where $H(q_1; q_2)$ is given by Lemma 5,

$$q_1 = \frac{-3\alpha\delta}{2} + 2\delta \quad \text{and} \quad q_2 = \delta^2\left(\frac{-3\alpha}{2} + \frac{\alpha^2}{2} + \frac{2}{3}\right) + \frac{1}{3}.$$

The first two bounds are sharp.

Proof. Set $g(z) =: zf'(z)$, where $f \in \mathcal{G}(\alpha, \delta)$ and suppose that $g(z) = z + \sum_{n=2}^{\infty} b_n z^n$. Hence $b_n = n a_n$ for $n \geq 1$. Then from (4), it follows that

$$\frac{zg'(z)}{g(z)} \prec -\frac{\alpha}{2}\left(\frac{1+z}{1-z}\right)^\delta + \frac{2+\alpha}{2} =: \phi(z)$$

$$= 1 - \alpha\delta z - \alpha\delta^2 z^2 - \frac{1}{3}\alpha\delta(2\delta^2 + 1)z^3 + \cdots$$

$$:= 1 + B_1 z + B_2 z^2 + B_3 z^3 + \cdots.$$

Now, by the definition of the subordination, there is a $\omega \in \Omega$ with $\omega(z) = \sum_{n=1}^{\infty} w_n z^n$ so that

$$\frac{zg'(z)}{g(z)} = \phi(\omega(z))$$

$$= 1 + B_1 w_1 z + (B_1 w_2 + B_2 w_1^2)z^2 + (B_1 w_3 + 2w_1 w_2 B_2 + B_3 w_1^3)z^3 + \cdots.$$

From the above equality, it concludes that

$$\begin{cases} b_2 = B_1 w_1 \\ 2b_3 - b_2^2 = B_1 w_2 + B_2 w_1^2 \\ 3b_4 - 3b_2 b_3 + b_2^3 = B_1 w_3 + 2w_1 w_2 B_2 + B_3 w_1^3. \end{cases}$$

First, for b_2, from Lemma 3 we get $|b_2| \leq \alpha\delta$, and so $|a_2| \leq \frac{\alpha\delta}{2}$. Next, utilizing Lemma 3 for b_3 and using $|B_2 + B_1^2| \leq |B_1|$, we have

$$|b_3| \leq \frac{|B_1|(1 - |w_1|^2) + |B_2 + B_1^2||w_1|^2}{2}$$

$$= \frac{|B_1| + [|B_2 + B_1^2| - |B_1|]|w_1|^2}{2}$$

$$\leq \frac{|B_1|}{2} = \frac{\alpha\delta}{2}.$$

Ultimately, utilizing Lemma 5 for a_4, we have

$$|b_4| \leq \frac{B_1}{3}\left|c_3 + \left(\frac{3}{2}B_1 + \frac{2B_2}{B_1}\right)w_1 w_2 + \left(\frac{3}{2}B_2 + \frac{1}{2}B_1^2 + \frac{B_3}{B_1}\right)w_1^3\right|$$

$$\leq \frac{B_1}{3}H(q_1; q_2),$$

where

$$q_1 = \frac{3}{2}B_1 + \frac{2B_2}{B_1} = \frac{-3\alpha\delta}{2} + 2\delta \quad \text{and} \quad q_2 = \frac{3}{2}B_2 + \frac{1}{2}B_1^2 + \frac{B_3}{B_1} = \delta^2\left(\frac{-3\alpha}{2} + \frac{\alpha^2}{2} + \frac{2}{3}\right) + \frac{1}{3}.$$

The extremal functions for the initial coefficients a_n ($n = 2, 3$) are of the form:

$$f_n(z) = \int_0^z \exp\left(\int_0^x \frac{\phi(t^n) - 1}{t} dt\right) dx = z - \frac{\alpha\beta}{n(n+1)} z^{n+1} + \frac{\alpha\beta^2(\alpha/n - 1)}{2n(2n+1)} z^{2n+1} + \cdots,$$

obtained by taking $\omega(z) = z^n$ in (4). Therefore, this completes the proof. □

Theorem 4. *Let $f \in \mathcal{G}(\alpha, \delta)$. Then*

$$|\gamma_1| \leq \frac{\alpha\delta}{4}, \quad |\gamma_2| \leq \frac{\alpha\delta}{12}, \quad |\gamma_3| \leq \frac{\alpha\delta}{24} H(q_1; q_2),$$

where $H(q_1; q_2)$ is given by Lemma 5, $q_1 = \frac{-\alpha\delta+4\delta}{2}$, and $q_2 = \frac{-\alpha\delta^2 + \frac{2(2\delta^2+1)}{3}}{2}$. The first two bounds are sharp.

Proof. The results are concluded from Theorem 6 by setting $\varphi := \phi$. Also, two first bounds are sharp for $f_n(z)$ for $n = 1, 2$, respectively. Therefore, this completes the proof. □

Theorem 5. *Let $f \in \mathcal{G}(\alpha, \delta)$. Then we have sharp inequalities for complex parameter μ*

$$\left|a_3 - \mu a_2^2\right| \leq \begin{cases} \frac{\alpha\delta^2}{6}\left|1 - \alpha + \frac{3\mu}{2}\alpha\right| & \text{for } \left|\mu + \frac{2}{3\alpha}(1-\alpha)\right| \geq \frac{2}{3\alpha\delta}, \\ \frac{\alpha\delta}{6} & \text{for } \left|\mu + \frac{2}{3\alpha}(1-\alpha)\right| < \frac{2}{3\alpha\delta}. \end{cases}$$

Proof. Let $f \in \mathcal{G}(\alpha, \delta)$, then from (4), by the definition of the subordination, there is a $\omega \in \Omega$ with $\omega(z) = \sum_{n=1}^{\infty} w_n z^n$ so that

$$1 + \frac{zf''(z)}{f'(z)} = \phi(\omega(z)) = 1 + B_1 w_1 z + (B_1 w_2 + B_2 w_1^2)z^2 + \cdots.$$

Therefore, we get that

$$2a_2 = B_1 w_1 \quad \text{and} \quad 6a_3 - 4a_2^2 = B_1 w_2 + B_2 w_1^2.$$

Form the above equalities, we have

$$\left|a_3 - \mu a_2^2\right| = \frac{1}{6}|B_1|\left|w_2 + \nu w_1^2\right|.$$

The results are obtained by the application of Lemma 4 with $\nu = \left[\frac{B_2}{B_1} + B_1(1 - \frac{3\mu}{2})\right]$, where $B_1 = -\alpha\delta$ and $B_2 = -\alpha\delta^2$. Equality is attained in the first inequality by the function $f = f_1$ and in the second inequality for $f = f_2$. □

Remark 1.

(i) Taking into account $\delta = 1$ in Theorem 3, we get the result obtained in [31] (Theorem 1) for $n = 2, 3, 4$.
(ii) Setting $\delta = 1$ in Theorem 3, we have the result obtained in [23] (Theorem 2.10).
(iii) Letting $\delta = 1$ in Theorem 4, we obtain a correction of the result presented in [31] (Theorem 2).

Author Contributions: Investigation, D.A., N.E.C., E.A.A. and A.M.; Writing—original draft, E.A.A.; Writing—review and editing, N.E.C. The authors contributed equally to this work. All authors have read and agreed to the published version of the manuscript.

Funding: The second author was supported by the Basic Science Research Program through the National Research Foundation of Korea (NRF) funded by the Ministry of Education, Science and Technology (No. 2019R1I1A3A01050861).

Acknowledgments: The authors would like to express their gratitude to the referees for many valuable suggestions regarding a previous version of this paper.

Conflicts of Interest: The authors declare no conflict of interest.

References

1. Kayumov, I.R. On Brennan's conjecture for a special class of functions. *Math. Notes* **2005**, *78*, 498–502. [CrossRef]
2. Milin, I.M. *Univalent Functions and Orthonormal Systems*; Amer Mathematical Society, Translations of Mathematical Monographs: Proovidence, RI, USA, 1977; Volume 49.

3. Milin, I.M. On a property of the logarithmic coefficients of univalent functions. In *Metric Questions in the Theory of Functions*; Naukova Dumka: Kiev, Ukraine, 1980; pp. 86–90. (In Russian)
4. Milin, I.M. On a conjecture for the logarithmic coefficients of univalent functions. *Zap. Nauch. Semin. Leningr. Otd. Mat. Inst. Steklova* **1983**, *125*, 135–143. (In Russian) [CrossRef]
5. Robertson, M.S. A remark on the odd-schlicht functions. *Bull. Am. Math. Soc.* **1936**, *42*, 366–370. [CrossRef]
6. Bieberbach, L. Über die Koeffizienten derjenigen Potenzreihen, welche eine schlichte Abbildung des Einheitkreises vermitteln. *Sitz. Preuss. Akad. Wiss.* **1916**, *138*, 940–955.
7. De Branges, L. A proof of Bieberbach conjecture. *Acta Math.* **1985**, *154*, 137–152. [CrossRef]
8. Pommerenke, C. On the coefficients and Hankel determinant of univalent functions. *J. Lond. Math. Soc.* **1966**, *41*, 111–112. [CrossRef]
9. Pommerenke, C. On the Hankel determinants of univalent functions. *Mathematika* **1967**, *14*, 108–112. [CrossRef]
10. Vein, R.; Dale, P. Determinants and Their Applications in Mathematical Physics. In *Applied Mathematical Sciences*; Springer: New York, NY, USA, 1999; Volume 134.
11. Fekete, M.; Szegö, G. Eine Bemerkung Über Ungerade Schlichte Funktionen. *J. Lond. Math. Soc.* **1933**, *8*, 85–89. [CrossRef]
12. Analouei Adegani, E.; Cho, N.E.; Motamednezhad, A.; Jafari, M. Bi-univalent functions associated with Wright hypergeometric functions. *J. Comput. Anal. Appl.* **2020**, *28*, 261–271.
13. Cho, N.E.; Analouei Adegani, E.; Bulut, S.; Motamednezhad, A. The second Hankel determinant problem for a class of bi-close-to-convex functions. *Mathematics* **2019**, *7*, 986. [CrossRef]
14. Deniz, E.; Çağlar, M.; Orhan, H. Second Hankel determinant for bi-starlike and bi-convex functions of order β. *Appl. Math. Comput.* **2015**, *271*, 301–307. [CrossRef]
15. Kanas, S.; Analouei Adegani, E.; Zireh, A. An unified approach to second Hankel determinant of bi-subordinate functions. *Mediterr. J. Math.* **2017**, *14*, 233. [CrossRef]
16. Motamednezhad, A.; Bulboacă, T.; Adegani, E.A.; Dibagar, N. Second Hankel determinant for a subclass of analytic bi-univalent functions defined by subordination. *Turk. J. Math.* **2018**, *42*, 2798–2808. [CrossRef]
17. Nunokawa, M.; Saitoh, H. On certain starlike functions. *Srikaisekikenkysho Kkyroku* **1996**, *963*, 74–77.
18. Ozaki, S. On the theory of multivalent functions, II. *Sci. Rep. Tokyo Bunrika Daigaku. Sect. A* **1941**, *4*, 45–87.
19. Umezawa, T. Analytic functions convex in one direction. *J. Math. Soc. Jpn.* **1952**, *4*, 194–202. [CrossRef]
20. Sakaguchi, K. A property of convex functions and an application to criteria for univalence. *Bull. Nara Univ. Ed. Nat. Sci.* **1973**, *22*, 1–5.
21. Singh, R.; Singh, S. Some sufficient conditions for univalence and starlikeness. *Colloq. Math.* **1982**, *47*, 309–314. [CrossRef]
22. Obradović, M.; Ponnusamy, S.; Wirths, K.-J. Logarithmic coeffcients and a coefficient conjecture for univalent functions. *Monatshefte Math.* **2017**. [CrossRef]
23. Ponnusamy, S.; Sharma, N.L.; Wirths. K.-J. Logarithmic coefficients problems in families related to starlike and convex functions. *J. Aust. Math. Soc.* **2019**, in press. [CrossRef]
24. Miller, S.S.; Mocanu, P.T. *Differential Subordinations. Theory and Applications*; Marcel Dekker Inc.: New York, NY, USA, 2000; ISBN 0-8247-0029-5.
25. Nunokawa, M. On properties of non-Carathéodory functions. *Proc. Jpn. Acad. Ser. A Math. Sci.* **1992**, *68*, 152–153. [CrossRef]
26. Nunokawa, M. On the order of strongly starlikeness of strongly convex functions. *Proc. Jpn. Acad. Ser. A Math. Sci.* **1993** *69*, 234–237. [CrossRef]
27. Nehari, Z. *Conformal Mapping*; McGraw-Hill: New York, NY, USA, 1952.
28. Keogh, F.R.; Merkes, E.P. A coefficient inequality for certain classes of analytic functions. *Proc. Am. Math. Soc.* **1969**, *20*, 8–12. [CrossRef]
29. Prokhorov, D.V.; Szynal, J. Inverse coefficients for $(\alpha; \beta)$-convex functions. *Ann. Univ. Mariae Curie-Sklodowska Sect. A* **1984**, *35*, 125–143.

30. Analouei Adegani, E.; Cho, N.E.; Jafari M. Logarithmic coefficients for univalent functions defined by subordination. *Mathematics* **2019**, *7*, 408. [CrossRef]
31. Obradović, M.; Ponnusamy, S.; Wirths, K.-J. Coefficient characterizations and sections for some univalent functions. *Sib. Math. J.* **2013**, *54*, 679–696. [CrossRef]

© 2020 by the authors. Licensee MDPI, Basel, Switzerland. This article is an open access article distributed under the terms and conditions of the Creative Commons Attribution (CC BY) license (http://creativecommons.org/licenses/by/4.0/).

Article

Approximation by Generalized Lupaş Operators Based on q-Integers

Mohd Qasim [1], M. Mursaleen [2,3,4,*], Asif Khan [3] and Zaheer Abbas [1]

[1] Department of Mathematical Sciences, Baba Ghulam Shah Badshah University, Rajouri 185234, Jammu and Kashmir, India; bgsbuqasim@gmail.com (M.Q.); az1129200@yahoo.co.in (Z.A.)
[2] Department of Medical Research, China Medical University Hospital, China Medical University (Taiwan), Taichung 40402, Taiwan
[3] Department of Mathematics, Aligarh Muslim University, Aligarh 202002, India; asifjnu07@gmail.com
[4] Department of Computer Science and Information Engineering, Asia University, Taichung 41354, Taiwan
* Correspondence: mursaleenm@gmail.com; Tel.: +91-9411491600

Received: 6 December 2019; Accepted: 27 December 2019; Published: 2 January 2020

Abstract: The purpose of this paper is to introduce q-analogues of generalized Lupaş operators, whose construction depends on a continuously differentiable, increasing, and unbounded function ρ. Depending on the selection of q, these operators provide more flexibility in approximation and the convergence is at least as fast as the generalized Lupaş operators, while retaining their approximation properties. For these operators, we give weighted approximations, Voronovskaja-type theorems, and quantitative estimates for the local approximation.

Keywords: generalized Lupaş operators; q analogue; Korovkin's type theorem; convergence theorems; Voronovskaya type theorem

1. Introduction

Approximation theory rudimentarily deals with the approximation of functions by simpler functions or more easily calculated functions. Broadly, it is divided into theoretical and constructive approximation. In 1912, S.N. Bernstein [1] was the first to construct a sequence of positive linear operators to provide a constructive proof of the prominent Weierstrass approximation theorem [2] using a probabilistic approach. One can find a detailed monograph about the Bernstein polynomials in [3,4]. Cárdenas et al. [5] in 2011, defined the Bernstein type operators by $B_n(f \circ \tau^{-1}) \circ \tau$ and showed that its Korovkin set is $\{e_0, \tau, \tau^2\}$ instead of $\{e_0, e_1, e_2\}$. These operators present an interesting byproduct sequence of positive linear operators of polynomial type with nice geometric shape preserving properties, which converge to the identity, which in a certain sense improves B_n in approximating a number of increasing functions, and which, apart from the constant functions, fixes suitable polynomials of a prescribed degree. The notion of convexity with respect to τ plays an important role. Recently, Aral et al. [6] in 2014 defined a similar modification of Szász-Mirakyan type operators obtaining approximation properties of these operators on the interval $[0, \infty)$.

Very recently motivated by the above work, İlarslan et al. [7] introduced a new modification of Lupaş operators [8] using a suitable function ρ, which satisfies the following properties:

(ρ_1) ρ be a continuously differentiable function on $[0, \infty)$,
(ρ_2) $\rho(0) = 0$ and $\inf\limits_{u \in [0,\infty)} \rho'(u) \geq 1$.

The generalized Lupaş operators are defined as

$$\mathcal{L}_m^\rho(f; u) = 2^{-m\rho(u)} \sum_{l=0}^\infty \frac{(m\rho(u))_l}{2^l l!} (f \circ \rho^{-1})\left(\frac{l}{m}\right), \tag{1}$$

for $m \geq 1$, $u \geq 0$, and suitable functions f are defined on $[0, \infty)$. If $\rho(u) = u$ then (1) reduces to the Lupaş operators defined in [8].

İlarslan et al. [7] gave uniform convergence results on a weighted space, where the weight function is $\phi(u) = 1 + \rho^2(u)$ satisfying the conditions (ρ_1) and and (ρ_2) given above, in the sense of Gadjiev's results [9,10]. For the rate of convergence, the authors used a weighted modulus of continuity stated by Holhoş in [11] using the weight function. They obtained a Voronovskaya-type result and monotonicity of the sequence of operators $\mathcal{L}_m^\rho(f;u)$. Moreover, they obtained some quantitative type theorems on weighted spaces.

The purpose of this paper is to define the q-analogue of operators (1) which depend on ρ.

Before proceeding further, let us recall some basic definitions and notations of quantum calculus [12]. For any fixed real number $q > 0$, the q-integer $[l]_q$, for $l \in \mathbb{N}$ (set of natural numbers) are defined as

$$[l]_q := \begin{cases} \frac{(1-q^l)}{(1-q)}, & q \neq 1 \\ l, & q = 1, \end{cases}$$

and the q-factorial by

$$[l]_q! := \begin{cases} [l]_q[l-1]_q \cdots [1]_q, & l \geq 1 \\ 1, & l = 0. \end{cases}$$

The q-Binomial expansion is

$$(u+y)_q^m := (u+y)(u+qy)(u+q^2y) \cdots (u+q^{m-1}y),$$

and the q-binomial coefficients are as follows:

$$\begin{bmatrix} m \\ l \end{bmatrix}_q := \frac{[m]_q!}{[l]_q![m-l]_q!}.$$

The Gauss-formula is defined as:

$$(u+y)_q^m = \sum_{j=0}^{m} \begin{bmatrix} m \\ l \end{bmatrix}_q q^{j(j-1)/2} y^j u^{m-j}.$$

After development of q-calculus, Lupaş [13] introduced the q-Lupaş operator (rational) as follows:

$$L_{m,q}(f;u) = \sum_{l=0}^{m} \frac{f\left(\frac{[l]_q}{[m]_q}\right) \begin{bmatrix} m \\ l \end{bmatrix}_q q^{\frac{l(l-1)}{2}} u^l (1-u)^{m-l}}{\prod_{j=1}^{m}\{(1-u) + q^{j-1}u\}}, \quad (2)$$

and studied its approximation properties.

Similarly, Phillips [14] constructed another q-analogue of Bernstein operators (polynomials) as follows:

$$B_{m,q}(f;u) = \sum_{l=0}^{m} \begin{bmatrix} m \\ l \end{bmatrix}_q u^l \prod_{s=0}^{m-l-1}(1-q^s u) f\left(\frac{[l]_q}{[m]_q}\right), \quad u \in [0,1] \quad (3)$$

where $B_{m,q} : C[0,1] \to C[0,1]$ defined for any $m \in \mathbb{N}$ and any function $f \in C[0,1]$, where $C[0,1]$ denotes the set of all continuous functions on $[0,1]$.

The basis of these operators have been used in Computer Aided Geometric Design (CAGD) to study curves and surfaces. Then it became an active area of research in approximation theory as well as CAGD [15–17]. In the recent past, q-analogues of various operators were investigated by several researchers (see [18–23]).

The q-analogue is a very interesting idea. It can also be used in statistical and biological physics, multi-type directed scalefree percolation, and modeling epidemic spread with awareness and heterogeneous transmission rates in networks.

Persuaded by the above mentioned work, we introduce the q-analogue of operators (1) which depends on a suitable function ρ, as follows:

Definition 1. Let $0 < q < 1$ and $m \in \mathbb{N}$. For $f : [0, \infty) \to \mathbb{R}$, we define q-analogue of generalized Lupaş operators as

$$\mathcal{L}_{m,q}^{\rho}(f; u) = 2^{-[m]_q \rho(u)} \sum_{l=0}^{\infty} \frac{([m]_q \rho(u))_l}{2^l [l]_q!} \left(f \circ \rho^{-1}\right)\left(\frac{[l]_q}{[m]_q}\right), \quad (4)$$

where $([m]_q \rho(u))_l$ is the rising factorial defined as:

$$([m]_q \rho(u))_0 = 1,$$
$$([m]_q \rho(u))_l = ([m]_q \rho(u))([m]_q \rho(u) + 1)([m]_q \rho(u) + 2) \cdots ([m]_q \rho(u) + l - 1), \quad l \geq 0.$$

The operators (4) are linear and positive. For $q = 1$, the operators (4) turn out to be generalized Lupaş operators defined in (1). Next, we prove some auxiliary results for (4).

Lemma 1. Let $\mathcal{L}_{m,q}^{\rho}$ be given by (4). Then for each $u \geq 0$ and $m \in \mathbb{N}$ we have

(i) $\mathcal{L}_{m,q}^{\rho}(1; u) = 1$,
(ii) $\mathcal{L}_{m,q}^{\rho}(\rho; u) = \rho(u)$,
(iii) $\mathcal{L}_{m,q}^{\rho}(\rho^2; u) = \rho^2(u) + \frac{(1+q)}{[m]_q} \rho(u)$,
(iv) $\mathcal{L}_{m,q}^{\rho}(\rho^3; u) = \rho^3(u) q^3 + \frac{(3q^3 + q^2 + 2q)}{[m]_q} \rho^2(u) + \frac{(2q^3 + q^2 + 2q + 1)}{[m]_q^2} \rho(u)$,
(v) $\mathcal{L}_{m,q}^{\rho}(\rho^4; u) = \rho^4(u) q^6 + \frac{(6q^6 + q^5 + 2q^4 + 3q^3)}{[m]_q} \rho^3(u) + \frac{(11q^6 + 3q^5 + 6q^4 + 10q^3 + 3q^2 + 3q)}{[m]_q^2} \rho^2(u)$
$+ \frac{(6q^6 + 2q^5 + 4q^4 + 7q^3 + 3q^2 + 3q + 1)}{[m]_q^3} \rho(u)$.

Proof. By taking into account the recurrence relation $([m]_q \rho(u))_0 = 1, ([m]_q \rho(u))_l = ([m]_q \rho(u))([m]_q \rho(u) + 1)_{l-1}, l \geq 1$, we have

(i)
$$\mathcal{L}_{m,q}^{\rho}(1; u) = 2^{-[m]_q \rho(u)} \sum_{l=0}^{\infty} \frac{([m]_q \rho(u))_l}{2^l [l]_q!} = 1.$$

(ii)
$$\mathcal{L}_{m,q}^{\rho}(\rho; u) = 2^{-[m]_q \rho(u)} \sum_{l=0}^{\infty} \frac{([m]_q \rho(u))_l}{[l]_q! 2^l} \frac{[l]_q}{[m]_q}$$
$$= \frac{2^{-([m]_q \rho(u)+1)} [m]_q \rho(u)}{[m]_q} \sum_{l=0}^{\infty} \frac{([m]_q \rho(u) + 1)_l}{[l]_q! 2^l}$$
$$= \rho(u).$$

(iii)

$$\begin{aligned}\mathcal{L}_{m,q}^{\rho}(\rho^2;u) &= 2^{-[m]_q\rho(u)}\sum_{l=0}^{\infty}\frac{([m]_q\rho(u))_l}{[l]_q!2^l}\frac{[l]_q^2}{[m]_q^2}\\ &= \frac{2^{-[m]_q\rho(u)}}{[m]_q^2}\sum_{l=0}^{\infty}\frac{([m]_qu)_l}{[l]_q!2^l}[l]_q^2\\ &= \frac{2^{-([m]_q\rho(u)+1)}[m]_q\rho(u)}{[m]_q^2}\sum_{l=0}^{\infty}\frac{([m]_q\rho(u)+1)_l}{[l]_q!2^l}[l]_q\\ &= \rho^2(u) + \frac{(1+q)}{[m]_q}\rho(u).\end{aligned}$$

(iv)

$$\mathcal{L}_{m,q}^{\rho}(\rho^3;u) = 2^{-[m]_q\rho(u)}\sum_{l=0}^{\infty}\frac{([m]_q\rho(u))_l}{[l]_q!2^l}\frac{[l]_q^3}{[m]_q^3}$$

Now by using $[l+1]_q = (1+q[l]_q)$ and shifting l to $l+1$, we have

$$\begin{aligned}\mathcal{L}_{m,q}^{\rho}(\rho^3;u) &= \frac{2^{-([m]_q\rho(u)+1)}[m]_q\rho(u)}{[m]_q^3}\sum_{l=0}^{\infty}\frac{([m]_q\rho(u)+1)_l}{[l]_q!2^l}[l+1]_q^2\\ &= \frac{2^{-([m]_q\rho(u)+1)}[m]_q\rho(u)}{[m]_q^3}\sum_{l=0}^{\infty}\frac{([m]_q\rho(u)+1)_l}{[l]_q!2^l}(1+q[l]_q)^2\\ &= \frac{2^{-([m]_q\rho(u)+1)}[m]_q\rho(u)}{[m]_q^3}\sum_{l=0}^{\infty}\frac{([m]_q\rho(u)+1)_l}{[l]_q!2^l},\\ &+ \frac{2^{-([m]_q\rho(u)+1)}[m]_q\rho(u)}{[m]_q^3}\sum_{l=0}^{\infty}\frac{([m]_q\rho(u)+1)_l}{[l]_q!2^l}q^2[l]_q^2\\ &+ \frac{2^{-([m]_q\rho(u)+1)}[m]_q\rho(u)}{[m]_q^3}\sum_{l=0}^{\infty}\frac{([m]_q\rho(u)+1)_l}{[l]_q!2^l}2q[l]_q,\\ &= A + B + C(Say).\end{aligned}$$

Now, let us calculate the values of A, B, and C

$$\begin{aligned}A &= \frac{2^{-([m]_q\rho(u)+1)}[m]_q\rho(u)}{[m]_q^3}\sum_{l=0}^{\infty}\frac{([m]_q\rho(u)+1)_l}{[l]_q!2^l},\\ &= \frac{\rho(u)}{[m]_q^2}.\end{aligned}$$

$$\begin{aligned}B &= \frac{2^{-([m]_q\rho(u)+1)}[m]_q\rho(u)}{[m]_q^3}\sum_{l=0}^{\infty}\frac{([m]_q\rho(u)+1)_l}{[l]_q!2^l}q^2[l]_q^2\\ &= \frac{2^{-([m]_q\rho(u)+2)}([m]_q\rho(u)+1)\rho(u)}{[m]_q^2}q^2\sum_{l=0}^{\infty}\frac{([m]_q\rho(u)+2)_l}{[l]_q!2^l}[l+1]_q\\ &= \frac{2^{-([m]_q\rho(u)+2)}([m]_q\rho(u)+1)\rho(u)}{[m]_q^2}q^2\sum_{l=0}^{\infty}\frac{([m]_q\rho(u)+2)_l}{[l]_q!2^l}(1+q[l]_q)\\ &= \rho^3(u)q^3 + \frac{3\rho^2(u)q^3}{[m]_q} + \frac{2\rho(u)q^3}{[m]_q^2} + \frac{\rho^2(u)q^2}{[m]_q} + \frac{\rho(u)q^2}{[m]_q^2}.\end{aligned}$$

Also,

$$C = \frac{2^{-([m]_q\rho(u)+1)}[m]_q\rho(u)}{[m]_q^3}\sum_{l=0}^{\infty}\frac{([m]_q\rho(u)+1)_l}{[l]_q!2^l}2q[l]_q,$$

$$= \frac{2\rho^2(u)q}{[m]_q} + \frac{2\rho(u)q}{[m]_q^2}.$$

On adding A, B, and C we have,

$$\mathcal{L}_{m,q}^{\rho}(\rho^3;u) = \rho^3(u)q^3 + \frac{(3q^3+q^2+2q)}{[m]_q}\rho^2(u) + \frac{(2q^3+q^2+2q+1)}{[m]_q^2}\rho(u).$$

(v)

$$\mathcal{L}_{m,q}^{\rho}(\rho^4;u) = 2^{-[m]_q\rho(u)}\sum_{l=0}^{\infty}\frac{([m]_q\rho(u))_l}{[l]_q!2^l}\frac{[l]_q^4}{[m]_q^4}.$$

Now, by using $[l+1]_q = (1+q[l]_q)$ and shifting l to $l+1$, we have

$$\mathcal{L}_{m,q}^{\rho}(\rho^4;u) = \frac{2^{-([m]_q\rho(u)+1)}[m]_q\rho(u)}{[m]_q^4}\sum_{l=0}^{\infty}\frac{([m]_q\rho(u)+1)_l}{[l]_q!2^l}[l+1]_q^3$$

$$= \frac{2^{-([m]_q\rho(u)+1)}\rho(u)}{[m]_q^3}\sum_{l=0}^{\infty}\frac{([m]_q\rho(u)+1)_l}{[l]_q!2^l}(1+q[l]_q)^3$$

$$= \frac{2^{-([m]_q\rho(u)+1)}\rho(u)}{[m]_q^3}\sum_{l=0}^{\infty}\frac{([m]_q\rho(u)+1)_l}{[l]_q!2^l}$$

$$+ \frac{2^{-([m]_q\rho(u)+1)}\rho(u)}{[m]_q^3}\sum_{l=0}^{\infty}\frac{([m]_q\rho(u)+1)_l}{[l]_q!2^l}q^3[l]_q^3$$

$$+ \frac{2^{-([m]_q\rho(u)+1)}\rho(u)}{[m]_q^3}\sum_{l=0}^{\infty}\frac{([m]_q\rho(u)+1)_l}{[l]_q!2^l}3q^2[l]_q^2$$

$$+ \frac{2^{-([m]_q\rho(u)+1)}\rho(u)}{[m]_q^3}\sum_{l=0}^{\infty}\frac{([m]_q\rho(u)+1)_l}{[l]_q!2^l}3q[l]_q,$$

$$= D + E + F + G \,(Say).$$

Now, let us calculate the values of D, E, F, and G

$$D = \frac{2^{-([m]_q\rho(u)+1)}\rho(u)}{[m]_q^3}\sum_{l=0}^{\infty}\frac{([m]_q\rho(u)+1)_l}{[l]_q!2^l}$$

$$= \frac{\rho(u)}{[m]_q^3}.$$

$$\begin{aligned}
E &= \frac{2^{-([m]_q \rho(u)+1)} \rho(u)}{[m]_q^3} \sum_{l=0}^{\infty} \frac{([m]_q \rho(u)+1)_l}{[l]_q! 2^l} q^3 [l]_q^3 \\
&= \frac{2^{-([m]_q \rho(u)+2)} \rho(u)([m]_q \rho(u)+1)}{[m]_q^3} q^3 \sum_{l=0}^{\infty} \frac{([m]_q \rho(u)+2)_l}{[l]_q! 2^l} [l+1]_q^2 \\
&= \frac{2^{-([m]_q \rho(u)+2)} \rho(u)([m]_q \rho(u)+1)}{[m]_q^3} q^3 \sum_{l=0}^{\infty} \frac{([m]_q \rho(u)+2)_l}{[l]_q! 2^l} (1+q[l]_q)^2 \\
&= \frac{2^{-([m]_q \rho(u)+2)} \rho(u)([m]_q \rho(u)+1)}{[m]_q^3} q^3 \sum_{l=0}^{\infty} \frac{([m]_q \rho(u)+2)_l}{[l]_q! 2^l} (1+q^2[l]_q^2 + 2q[l]_q) \\
&= \left(\rho^4(u) + \frac{6\rho^3(u)}{[m]_q} + \frac{11\rho^2(u)}{[m]_q^2} + \frac{6\rho(u)}{[m]_q^3} \right) q^6 + \left(\frac{\rho^3(u)}{[m]_q} + \frac{3\rho^2(u)}{[m]_q^2} + \frac{2\rho(u)}{[m]_q^3} \right) q^5 \\
&\quad + \left(\frac{2\rho^3(u)}{[m]_q} + \frac{6\rho^2(u)}{[m]_q^2} + \frac{4\rho(u)}{[m]_q^3} \right) q^4 + \left(\frac{\rho^2(u)}{[m]_q^2} + \frac{\rho(u)}{[m]_q^3} \right) q^3.
\end{aligned}$$

Similarly,

$$\begin{aligned}
F &= \frac{2^{-([m]_q \rho(u)+1)} \rho(u)}{[m]_q^3} \sum_{l=0}^{\infty} \frac{([m]_q \rho(u)+1)_l}{[l]_q! 2^l} 3q^2 [l]_q^2 \\
&= \left(\frac{3\rho^3(u)}{[m]_q} + \frac{9\rho^2(u)}{[m]_q^2} + \frac{6\rho(u)}{[m]_q^3} \right) q^3 + \left(\frac{3\rho^2(u)}{[m]_q^2} + \frac{3\rho(u)}{[m]_q^3} \right) q^2.
\end{aligned}$$

Also,

$$\begin{aligned}
G &= \frac{2^{-([m]_q \rho(u)+1)} \rho(u)}{[m]_q^3} \sum_{l=0}^{\infty} \frac{([m]_q \rho(u)+1)_l}{[l]_q! 2^l} 3q [l]_q \\
&= \left(\frac{3\rho^2(u)}{[m]_q^2} + \frac{3\rho(u)}{[m]_q^3} \right) q.
\end{aligned}$$

On adding D, E, F, and G we have,

$$\begin{aligned}
\mathcal{L}_{m,q}^{\rho}(\rho^4; u) &= \rho^4(u) q^6 + \frac{(6q^6 + q^5 + 2q^4 + 3q^3)}{[m]_q} \rho^3(u) \\
&\quad + \frac{(11q^6 + 3q^5 + 6q^4 + 10q^3 + 3q^2 + 3q)}{[m]_q^2} \rho^2(u) \\
&\quad + \frac{(6q^6 + 2q^5 + 4q^4 + 7q^3 + 3q^2 + 3q + 1)}{[m]_q^3} \rho(u).
\end{aligned}$$

$\square$

Corollary 1. *For $n = 1, 2, 3, 4$ the nth order central moments of $\mathcal{L}_{m,q}^{\rho}$ defined as $\mu_{n,m}^{\rho}(q; u) = \mathcal{L}_{m,q}^{\rho}((\rho(t) - \rho(u))_q^n; u)$, by using linearity of operators (4) and by Lemma 1 we have*

(i) $\mathcal{L}_{m,q}^{\rho}(\rho(t) - \rho(u); u) = \mathcal{L}_{m,q}^{\rho}(\rho(t); u) - \rho(u) \mathcal{L}_{m,q}^{\rho}(1; u) = 0$,

(ii) $\mathcal{L}_{m,q}^{\rho}((\rho(t) - \rho(u))^2; u) = \mathcal{L}_{m,q}^{\rho}(\rho^2(t); u) + \rho^2(u) \mathcal{L}_{m,q}^{\rho}(1; u) - 2\rho(u) \mathcal{L}_{m,q}^{\rho}(\rho(t); u) = \frac{(1+q)}{[m]_q} \rho(u)$,

(iii) $\mathcal{L}_{m,q}^{\rho}((\rho(t) - \rho(u))^3; u) = \mathcal{L}_{m,q}^{\rho}(\rho^3(t); u) - \rho^3(u) \mathcal{L}_{m,q}^{\rho}(1; u) - 3\rho(u) \mathcal{L}_{m,q}^{\rho}(\rho^2(t); u) - 3\rho^2(u) \mathcal{L}_{m,q}^{\rho}(\rho(t); u) = (q^3 - 1)\rho^3(u) + \frac{(3q^3 + q^2 - q - 3)}{[m]_q} \rho^2(u) + \frac{(2q^3 + q^2 + 2q + 1)}{[m]_q^2} \rho(u)$,

(iv) $\mathcal{L}_{m,q}^{\rho}((\rho(t) - \rho(u))^4; u) = \mathcal{L}_{m,q}^{\rho}(\rho^4(t); u) + \rho^4(u) \mathcal{L}_{m,q}^{\rho}(1; u) + 6\rho^2(u) \mathcal{L}_{m,q}^{\rho}(\rho^2(t); u) - 4\rho^3(u) \mathcal{L}_{m,q}^{\rho}(\rho(t); u) - 4\rho(u) \mathcal{L}_{m,q}^{\rho}(\rho^3(t); u) = (q^6 - 4q^3 + 3)\rho^4(u) + \frac{(6q^6 + q^5 + 2q^4 - 9q^3 - 4q^2 - 2q + 6)}{[m]_q} \rho^3(u) + \frac{(11q^6 + 3q^5 + 6q^4 + 2q^3 - q^2 - 5q - 4)}{[m]_q^2} \rho^2(u) + \frac{(6q^6 + 2q^5 + 4q^4 + 7q^3 + 3q^2 + 3q + 1)}{[m]_q^3} \rho(u)$.

Remark 1. *We observe from Lemma 1 and Corollary 1, that for $q = 1$, we get the moments and central moments of generalized Lupaş operators [7].*

2. Weighted Approximation

We start by noting that ρ not only defines a Korovkin-type set $\{1, \rho, \rho^2\}$ but also characterizes growth of the functions that are approximated.

Let $\phi(u) = 1 + \rho^2(u)$ be a weight function satisfying the conditions (ρ_1) and and (ρ_2) given above let $\mathcal{B}_\phi[0, \infty)$ be the weighted space defined by

$$\mathcal{B}_\phi[0, \infty) = \{f : [0, \infty) \to \mathbb{R} \, | \, |f(u)| \leq \mathcal{K}_f \phi(u), u \geq 0\},$$

where $\mathcal{K}_f$ is a constant which depends only on f. $\mathcal{B}_\phi[0, \infty)$ is a normed linear space equipped with the norm

$$\| f \|_\phi = \sup_{u \in [0, \infty)} \frac{|f(u)|}{\phi(u)}.$$

Also, we define the following subspaces of $\mathcal{B}_\phi[0, \infty)$ as

$$\mathcal{C}_\phi[0, \infty) = \{f \in \mathcal{B}_\phi[0, \infty) : \; f \; is \; continuous \; on \; [0, \infty)\},$$

$$\mathcal{C}_\phi^*[0, \infty) = \left\{f \in \mathcal{C}_\phi[0, \infty) : \; \lim_{u \to \infty} \frac{f(u)}{\phi(u)} = \mathcal{K}_f\right\},$$

where $\mathcal{K}_f$ is a constant depending on f and

$$\mathcal{U}_\phi[0, \infty) = \{f \in \mathcal{C}_\phi[0, \infty) : \; \frac{f(u)}{\phi(u)} \; is \; uniformly \; continuous \; on \; [0, \infty)\}.$$

Obviously,

$$\mathcal{C}_\phi^*[0, \infty) \subset \mathcal{U}_\phi[0, \infty) \subset \mathcal{C}_\phi[0, \infty) \subset \mathcal{B}_\phi[0, \infty).$$

For the weighted uniform approximation by linear positive operators acting from $\mathcal{C}_\phi[0, \infty)$ to $\mathcal{B}_\phi[0, \infty)$, we state the following results due to Gadjiev in [9,10].

Lemma 2 ([9])**.** *Let $(\mathcal{A}_m)_{m \geq 1}$ be a sequence of positive linear operators which acts from $\mathcal{C}_\phi[0, \infty)$ to $\mathcal{B}_\phi[0, \infty)$ if and only if the inequality*

$$|\mathcal{A}_m(\phi; u)| \leq \mathcal{K}_m \phi(u), \; u \geq 0,$$

holds, where $\mathcal{K}_m > 0$ is a constant depending on m.

Theorem 1 ([10])**.** *Let $(\mathcal{A}_m)_{m \geq 1}$ be a sequence of positive linear operators, acting from $\mathcal{C}_\phi[0, \infty)$ to $\mathcal{B}_\phi[0, \infty)$ and satisfying*

$$\lim_{m \to \infty} \| \mathcal{L}_m \rho^i - \rho^i \|_\phi = 0, \; i = 0, 1, 2.$$

Then we have

$$\lim_{m \to \infty} \| \mathcal{L}_m(f) - f \|_\phi = 0, \; for \; any \; f \in \mathcal{C}_\phi^*[0, \infty).$$

Remark 2. *It is clear from Lemma 1 and Lemma 2 that the operators $\mathcal{L}_{m,q}^\rho$ act from $\mathcal{C}_\phi[0, \infty)$ to $\mathcal{B}_\phi[0, \infty)$. Also the convergence of these operators are applicable in studying switched linear systems, see: Subspace confinement for switched linear systems, also see in [24,25].*

Theorem 2. Let q_m be a sequence in $(0,1)$, such that $q_m \to 1$ as $m \to \infty$. Then for each function $f \in C^*_\phi[0,\infty)$ we have
$$\lim_{m \to \infty} \| \mathcal{L}^\rho_{m,q_m}(f) - f \|_\phi = 0.$$

Proof. By Lemma 1 (i) and (ii), it is clear that
$$\| \mathcal{L}^\rho_{m,q_m}(1;u) - 1 \|_\phi = 0.$$
$$\| \mathcal{L}^\rho_{m,q_m}(\rho;u) - \rho \|_\phi = 0.$$

and by Lemma 1 (iii), we have
$$\| \mathcal{L}^\rho_{m,q_m}(\rho^2;u) - \rho^2 \|_\phi = \sup_{u \in [0,\infty)} \frac{(1+q)\rho(u)}{[m]_q(1+\rho^2(u))} \leq \frac{1+q}{[m]_q}. \tag{5}$$

Then from Lemma 1 and (5) we get $\lim_{m \to \infty} \| \mathcal{L}^\rho_{m,q_m}(\rho^i) - \rho^i \|_\phi = 0, i = 0,1,2$. Hence, the proof is completed. □

3. Rate of Convergence or Order of Approximation

In this section, we determine the rate of convergence for $\mathcal{L}^\rho_{m,q}$ by weighted modulus of continuity $\omega_\rho(f;\delta)$ which was recently considered by Holhoş [11] as follows:

$$\omega_\rho(f;\delta) = \sup_{u,z \in [0,\infty), |\rho(z) - \rho(u)| \leq \delta} \frac{|f(z) - f(u)|}{\phi(z) + \phi(u)}, \quad \delta > 0, \tag{6}$$

where $f \in \mathcal{C}_\phi[0,\infty)$, with the following properties:

(i) $\omega_\rho(f;0) = 0$,

(ii) $\omega_\rho(f;\delta) \geq 0, \delta \geq 0$ for $f \in \mathcal{C}_\phi[0,\infty)$,

(iii) $\lim_{\delta \to 0} \omega_\rho(f;\delta) = 0$, for each $f \in U_\phi[0,\infty)$.

Theorem 3 ([11]). Let $\mathcal{A}_m : \mathcal{C}_\phi[0,\infty) \to \mathcal{B}_\phi[0,\infty)$ be a sequence of positive linear operators with

$$\| \mathcal{A}_m(\rho^0) - \rho^0 \|_{\phi^0} = a_m, \tag{7}$$
$$\| \mathcal{A}_m(\rho) - \rho \|_{\phi^{\frac{1}{2}}} = b_m, \tag{8}$$
$$\| \mathcal{A}_m(\rho^2) - \rho^2 \|_\phi = c_m, \tag{9}$$
$$\| \mathcal{A}_m(\rho^3) - \rho^3 \|_{\phi^{\frac{3}{2}}} = d_m, \tag{10}$$

where the sequences $a_m, b_m, c_m,$ and d_m converge to zero as $m \to \infty$. Then

$$\| \mathcal{A}_m(f) - f \|_{\phi^{\frac{3}{2}}} \leq (7 + 4a_m + 2c_m)\omega_\rho(f;\delta_m) + \| f \|_\phi a_m, \tag{11}$$

for all $f \in \mathcal{C}_\phi[0,\infty)$, where

$$\delta_m = 2\sqrt{(a_m + 2b_m + c_m)(1 + a_m)} + a_m + 3b_m + 3c_m + d_m.$$

Theorem 4. Let for each $f \in \mathcal{C}_\phi[0,\infty)$ with $0 < q < 1$. Then we have

$$\| \mathcal{L}^\rho_{m,q}(f) - f \|_{\phi^{\frac{3}{2}}} \leq \left(7 + \frac{2(1+q)}{[m]_q}\right)\omega_\rho(f;\delta_{m,q}),$$

where ω_ρ is the weighted modulus of continuity defined in (6) and

$$\delta_{m,q} = 2\sqrt{\frac{(1+q)}{[m]_q} + \frac{3(1+q)}{[m]_q}} + \left((q^3-1) + \frac{(3q^3+q^2+2q)}{[m]_q} + \frac{(2q^3+q^2+2q+1)}{[m]_q^2}\right).$$

Proof. By using Lemma 1, we have

$$\|\mathcal{L}_{m,q}^\rho(\rho^0) - \rho^0\|_{\phi^0} = a_{m,q} = 0,$$

$$\|\mathcal{L}_{m,q}^\rho(\rho) - \rho\|_{\phi^{\frac{1}{2}}} = b_{m,q} = 0,$$

and

$$\|\mathcal{L}_{m,q}^\rho(\rho^2) - \rho^2\|_\phi \leq \frac{(1+q)}{[m]_q} = c_{m,q}.$$

Finally,

$$\|\mathcal{L}_{m,q}^\rho(\rho^3) - \rho^3\|_{\phi^{\frac{3}{2}}} \leq (q^3-1) + \frac{(3q^3+q^2+2q)}{[m]_q} + \frac{(2q^3+q^2+2q+1)}{[m]_q^2} = d_{m,q}.$$

Thus, the sequences $a_{m,q}$, $b_{m,q}$, $c_{m,q}$, and $d_{m,q}$ are calculated. The sequences a_m, b_m, c_m, and d_m converge to zero as $m \to \infty$. Then

$$\|\mathcal{L}_{m,q}^\rho(f) - f\|_{\phi^{\frac{3}{2}}} \leq (7 + 4a_{m,q} + 2c_{m,q})\omega_\rho(f; \delta_{m,q}) + \|f\|_\phi\, a_{m,q}, \tag{12}$$

for all $f \in \mathcal{C}_\phi[0, \infty)$, where

$$\delta_{m,q} = 2\sqrt{(a_{m,q} + 2b_{m,q} + c_{m,q})(1 + a_{m,q})} + a_{m,q} + 3b_{m,q} + 3c_{m,q} + d_{m,q}.$$

Hence, by substituting the values of $a_{m,q}$, $b_{m,q}$, $c_{m,q}$ and $d_{m,q}$ we obtain the desired result. □

Remark 3. *For $\lim\limits_{\delta \to 0} \omega_\rho(f; \delta) = 0$ in Theorem 4, we find*

$$\lim_{m \to \infty} \|\mathcal{L}_{m,q}^\rho(f) - f\|_{\phi^{\frac{3}{2}}} = 0, \quad \text{for } f \in \mathcal{U}_\phi[0, \infty).$$

4. Voronovskaya-Type Theorem

In this section, using a technique developed in [5] by Cardenas-Morales, Garrancho and Raşa, we prove pointwise convergence of $\mathcal{L}_{m,q}^\rho$ by obtaining Voronovskaya-type theorems.

Theorem 5. *Let $f \in \mathcal{C}_\phi[0, \infty)$, $u \in [0, \infty)$ with $0 < q_m < 1$, $q_m \to 1$ as $m \to \infty$. Suppose that $(f \circ \rho^{-1})'$ and $(f \circ \rho^{-1})''$ exist at $\rho(u)$. If $(f \circ \rho^{-1})''$ is bounded on $[0, \infty)$, then we have*

$$\lim_{m \to \infty} [m]_{q_m} \left[\mathcal{L}_{m,q_m}^\rho(f; u) - f(u)\right] = \rho(u) \left(f \circ \rho^{-1}\right)''(\rho(u)).$$

Proof. By using the q-Taylor expansion of $(f \circ \rho^{-1})$ at $\rho(u) \in [0, \infty)$, there exist a point w lying between u and z, then we have

$$\begin{aligned} f(w) = \left(f \circ \rho^{-1}\right)(\rho(w)) &= \left(f \circ \rho^{-1}\right)(\rho(u)) + \left(f \circ \rho^{-1}\right)'(\rho(u))(\rho(w) - \rho(u)) \\ &+ \frac{\left(f \circ \rho^{-1}\right)''(\rho(u))(\rho(w) - \rho(u))^2}{[2]_q} + \lambda_u^q(w)(\rho(w) - \rho(u))^2, \end{aligned} \tag{13}$$

where
$$\lambda_u^q(w) = \frac{(f\circ\rho^{-1})''(\rho(w)) - (f\circ\rho^{-1})''(\rho(u))}{[2]_q}. \tag{14}$$

Therefore, by (14) together with the assumption on f ensures that
$$|\lambda_u^q(w)| \leq \mathcal{K}, \quad \text{for all} \quad w \in [0,\infty)$$

and is convergent to zero as $w \to u$. Now applying the operators (4) to the equality (13), we obtain

$$\begin{aligned}
\left[\mathcal{L}_{m,q_m}^{\rho}(f;u) - f(u)\right] &= \left(f\circ\rho^{-1}\right)'(\rho(u))\mathcal{L}_{m,q_m}^{\rho}\left((\rho(w)-\rho(u));u\right) \\
&+ \frac{(f\circ\rho^{-1})''(\rho(u))\mathcal{L}_{m,q_m}^{\rho}\left((\rho(w)-\rho(y))^2;u\right)}{[2]_q} \\
&+ \mathcal{L}_{m,q_m}^{\rho}\left(\lambda_u^q(w)\left((\rho(w)-\rho(u))^2;u\right)\right).
\end{aligned} \tag{15}$$

By Lemma 1 and Corollary 1, we get
$$\lim_{m\to\infty}[m]_{q_m}\mathcal{L}_{m,q_m}^{\rho}\left((\rho(w)-\rho(u));u\right) = 0, \tag{16}$$

and
$$\lim_{m\to\infty}[m]_{q_m}\mathcal{L}_{m,q_m}^{\rho}\left((\rho(w)-\rho(u))^2;u\right) = [2]_q\rho(u). \tag{17}$$

By estimating the last term on the right hand side of equality (15), we will get the proof.

Since from (14), for every $\epsilon > 0$, $\lim_{w\to u}\lambda_u^{q_m}(w) = 0$. Let $\delta > 0$ such that $|\lambda_u^{q_m}(w)| < \epsilon$ for every $w \geq 0$. Using a Cauchy-Schwartz inequality, we have

$$\begin{aligned}
\lim_{m\to\infty}[m]_{q_m}\mathcal{L}_{m,q_m}^{\rho}\left(|\lambda_u^{q_m}(w)|\,(\rho(w)-\rho(u))^2;u\right) &\leq \epsilon \lim_{m\to\infty}[m]_{q_m}\mathcal{L}_{m,q_m}^{\rho}\left((\rho(w)-\rho(u))^2_{q_m};u\right) \\
&+ \frac{\mathcal{K}}{\delta^2}\lim_{m\to\infty}[m]_{q_m}\mathcal{L}_{m,q_m}^{\rho}\left((\rho(w)-\rho(u))^4_{q_m};u\right).
\end{aligned}$$

Since
$$\lim_{m\to\infty}[m]_{q_m}\mathcal{L}_{m,q_m}^{\rho}\left((\rho(w)-\rho(u))^4_{q_m};u\right) = 0, \tag{18}$$

we obtain
$$\lim_{m\to\infty}[m]_{q_m}\mathcal{L}_{m,q_m}^{\rho}\left(|\lambda_u^{q_m}(w)|\,(\rho(w)-\rho(y))^2_{q_m};y\right) = 0. \tag{19}$$

Thus, by using Equations (16), (17) and (19) to Equation (15) the proof is completed. □

5. Local Approximation

In this section, we present local approximation theorems for the operators $\mathcal{L}_{m,q}^{\rho}$. By $\mathcal{C}_B[0,\infty)$, we denote the space of real-valued continuous and bounded functions f defined on the interval $[0,\infty)$. The norm $\|\cdot\|$ on the space $\mathcal{C}_B[0,\infty)$ is given by

$$\|f\| = \sup_{0\leq u<\infty}|f(x)|.$$

For $\delta > 0$ and $W^2 = \{s \in \mathcal{C}_B[0,\infty): s', s'' \in \mathcal{C}_B[0,\infty)\}$. The $\mathcal{K}$-functional is defined as

$$\mathcal{K}_2(f,\delta) = \inf_{s\in W^2}\{\|f-s\| + \delta\|g''\|\},$$

By Devore and Lorentz ([26], p. 177, Theorem 2.4), there exists an absolute constant $\mathcal{C} > 0$ such that
$$\mathcal{K}(f,\delta) \leq \mathcal{C}\omega_2(f,\sqrt{\delta}). \tag{20}$$

The second order modulus of smoothness is as follows,
$$\omega_2(f,\sqrt{\delta}) = \sup_{0 < h \leq \sqrt{\delta}} \sup_{u \in [0,\infty)} |f(u+2h) - 2f(u+h) + f(u)|$$

where $f \in \mathcal{C}_B[0,\infty)$. The usual modulus of continuity of $f \in \mathcal{C}_B[0,\infty)$ is defined by
$$\omega(f,\delta) = \sup_{0 < h \leq \delta} \sup_{u \in [0,\infty)} |f(u+h) - f(u)|.$$

Theorem 6. *Let $f \in \mathcal{C}_B[0,\infty)$ with $0 < q < 1$. Let ρ be a function satisfying the conditions (ρ_1), (ρ_2), and $\|\rho''\|$ is finite. Then, there exists an absolute constant $\mathcal{C} > 0$ such that*
$$\left|\mathcal{L}^\rho_{m,q}(f;u) - f(u)\right| \leq \mathcal{CK}\left(f, \frac{(1+q)}{[m]_q}\rho(u)\right).$$

Proof. Let $s \in W^2$ and $u, z \in [0,\infty)$. Using Taylor's formula we have
$$s(z) = s(u) + \left(s \circ \rho^{-1}\right)'(\rho(u))(\rho(z) - \rho(u)) + \int_{\rho(u)}^{\rho(z)} (\rho(z) - v)\left(s \circ \rho^{-1}\right)''(v)dv. \tag{21}$$

Using the equality
$$\left(s \circ \rho^{-1}\right)''(\rho(u)) = \frac{s''(u)}{(\rho'(u))^2} - s'(u)\frac{\rho''(u)}{(\rho'(u))^3}. \tag{22}$$

Now, put $v = \rho(y)$ in the last term in equality (21), we get
$$\begin{aligned}\int_{\rho(u)}^{\rho(z)} (\rho(z) - v)\left(s \circ \rho^{-1}\right)''(v)dv &= \int_u^z (\rho(z) - \rho(y))\left[\frac{s''(y)\rho'(y) - s'(y)\rho''(v)}{(\rho'(y))^2}\right]dy \\ &= \int_{\rho(u)}^{\rho(z)} (\rho(z) - v)\frac{s''(\rho^{-1}(v))}{(\rho'(\rho^{-1}(v)))^2}dv \\ &\quad - \int_{\rho(u)}^{\rho(z)} (\rho(z) - v)\frac{s'(\rho^{-1}(v))\rho''(\rho^{-1}(v))}{(\rho'(\rho^{-1}(v)))^3}dv.\end{aligned} \tag{23}$$

By using Lemma 1 and (23) and applying the operator (4) to the both sides of equality (21), we deduce
$$\begin{aligned}\mathcal{L}^\rho_{m,q}(s;u) &= s(u) + \mathcal{L}^\rho_{m,q}\left(\int_{\rho(u)}^{\rho(z)} (\rho(z) - v)\frac{s''(\rho^{-1}(v))}{(\rho'(\rho^{-1}(v)))^2}dv; u\right) \\ &\quad - \mathcal{L}^\rho_{m,q}\left(\int_{\rho(u)}^{\rho(z)} (\rho(z) - v)\frac{s'(\rho^{-1}(v))\rho''(\rho^{-1}(v))}{(\rho'(\rho^{-1}(v)))^3}dv; u\right).\end{aligned}$$

As we know ρ is strictly increasing on $[0,\infty)$ and with condition (ρ_2), we have
$$\left|\mathcal{L}^\rho_{m,q}(s;u) - s(u)\right| \leq \mathcal{M}^\rho_{m,2}(u)\left(\|s''\| + \|s'\|\|\rho''\|\right),$$

where
$$\mathcal{M}^\rho_{m,2}(u) = \mathcal{L}^\rho_{m,q}((\rho(t) - \rho(u))^2; u).$$

For $f \in \mathcal{C}_B[0,\infty)$, we have

$$\begin{aligned}|\mathcal{L}_{m,q}^{\rho}(s;u)| &\leq \|f\circ\rho^{-1}\|2^{-[m]_q\rho(u)}\sum_{l=0}^{\infty}\frac{([m]_q\rho(u))_l}{2^l[l]_q!}\\ &\leq \|f\|\mathcal{L}_{m,q}^{\rho}(1;u) = \|f\|.\end{aligned} \quad (24)$$

Hence we have

$$\begin{aligned}|\mathcal{L}_{m,q}^{\rho}(f;u) - f(u)| &\leq |\mathcal{L}_{m,q}^{\rho}(f-s;u)| + |\mathcal{L}_{m,q}^{\rho}(s;u) - s(u)| + |s(u) - f(u)|\\ &\leq 2\|f-g\| + \frac{(1+q)}{[m]_q}\rho(u)(\|s''\| + \|s'\|\|\rho''\|).\end{aligned}$$

If we choose $\mathcal{C} = \max\{2, \|\rho''\|\}$, then

$$|\mathcal{L}_{m,q}^{\rho}(f;u) - f(u)| \leq \mathcal{C}\left(2\|f-g\| + \frac{(1+q)}{[m]_q}\rho(u)\|s''\|_{W^2}\right).$$

Taking infimum over all $s \in W^2$ we obtain

$$|\mathcal{L}_{m,q}^{\rho}(f;u) - f(u)| \leq \mathcal{C}K\left(f, \frac{(1+q)}{[m]_q}\rho(u)\right).$$

□

Now, we recall local approximation in terms of α order Lipschitz-type maximal functions given in [27]. Let ρ be a function satisfying the conditions (ρ_1), (ρ_2), $0 < \alpha \leq 1$ and $Lip_{\mathcal{M}}(\rho(u); \alpha)$, $\mathcal{M} \geq 0$ is the set of functions f satisfying the inequality

$$|f(z) - f(u)| \leq \mathcal{M}|\rho(z) - \rho(u)|^{\alpha}, u, z \geq 0.$$

Moreover, for a bounded subset $\mathcal{E} \subset [0,\infty)$, we say that the function $f \in \mathcal{C}_B[0,\infty)$ belongs to $Lip_{\mathcal{M}}(\rho(u); \alpha)$, $0 < \alpha \leq 1$ on $\mathcal{E}$ if

$$|f(z) - f(u)| \leq \mathcal{M}_{\alpha,f}|\rho(z) - \rho(u)|^{\alpha}, u \in \mathcal{E} \text{ and } z \geq 0,$$

where $\mathcal{M}_{\alpha,f}$ is a constant depending on α and f.

Theorem 7. *Let ρ be a function satisfying the conditions (ρ_1), (ρ_2). Then for any $f \in Lip_{\mathcal{M}}(\rho(u); \alpha)$, $0 < \alpha \leq 1$ with $0 < q < 1$ and for every $u \in (0,\infty), m \in \mathbb{N}$, we have*

$$|\mathcal{L}_{m,q}^{\rho}(f;u) - f(u)| \leq \mathcal{M}\left(\frac{(1+q)}{[m]_q}\rho(u)\right)^{\frac{\alpha}{2}}, \quad (25)$$

Proof. Assume that $\alpha = 1$. Then, for $f \in Lip_{\mathcal{M}}(\alpha; 1)$ and $u \in (0, \infty)$, we have

$$\begin{aligned}|\mathcal{L}_{m,q}^{\rho}(f;u) - f(u)| &\leq \mathcal{L}_{m,q}^{\rho}(|f(z) - f(u)|; u)\\ &\leq \mathcal{M}\mathcal{L}_{m,q}^{\rho}(|\rho(z) - f(u)|; u).\end{aligned}$$

By applying the Cauchy–Schwartz inequality, we find

$$|\mathcal{L}_{m,q}^\rho(f;u) - f(u)| \le \mathcal{M}[\mathcal{L}_{m,q}^\rho((\rho(t) - \rho(u))^2; u)]^{\frac{1}{2}}$$
$$\le \mathcal{M}\sqrt{\frac{(1+q)\rho(u)}{[m]_q}}.$$

Let us assume that $\alpha \in (0,1)$. Then, for $f \in Lip_\mathcal{M}(\alpha;1)$ and $u \in (0,\infty)$, we have

$$|\mathcal{L}_{m,q}^\rho(f;u) - f(u)| \le \mathcal{L}_{m,q}^\rho(|f(z) - f(u)|; u)$$
$$\le \mathcal{M}\mathcal{L}_{m,q}^\rho(|\rho(z) - f(u)|^\alpha; u).$$

From Hölder's inequality with $p = \frac{1}{\alpha}$ and $q = \frac{1}{1-\alpha}$, for $f \in Lip_\mathcal{M}(\rho(u);\alpha)$, we have

$$|\mathcal{L}_{m,q}^\rho(f;u) - f(u)| \le \mathcal{M}[\mathcal{L}_{m,q}^\rho(|(\rho(t) - \rho(u)|; u)]^\alpha.$$

Finally by the Cauchy–Schwartz inequality, we get

$$|\mathcal{L}_{m,q}^\rho(f;u) - f(u)| \le \mathcal{M}\left(\frac{(1+q)}{[m]_q}\rho(u)\right)^{\frac{\alpha}{2}}.$$

□

A relationship between local smoothness of functions and the local approximation was given by Agratini in [28]. Here we will prove the similar result for operators $\mathcal{L}_{m,q}^\rho$ ($m \in \mathbb{N}$) for functions from $Lip_\mathcal{M}(\rho(u))$ on a bounded subset.

Theorem 8. *Let $\mathcal{E}$ be a bounded subset of $[0,\infty)$ and ρ be a function satisfying the conditions (ρ_1), (ρ_2). Then for any $f \in Lip_\mathcal{M}(\rho(u);\alpha)$, $0 < \alpha \le 1$ on $\mathcal{E}$ $\alpha \in (0,1]$, we have*

$$|\mathcal{L}_{m,q}^\rho(f;u) - f(u)| \le \mathcal{M}_{\alpha,f}\left\{\left[\frac{(1+q)\rho(u)}{[m]_q}\right]^{\frac{\alpha}{2}} + 2[\rho'(u)]^\alpha d^\alpha(u,\mathcal{E})\right\}, u \in [0,\infty), m \in \mathbb{N},$$

where $d(u,\mathcal{E}) = \inf\{\|u - y\| : y \in \mathcal{E}\}$ and $\mathcal{M}_{\alpha,f}$ is a constant depending on α and f.

Proof. Let $\overline{\mathcal{E}}$ be the closure of $\mathcal{E}$ in $[0,\infty)$. Then, there exists a point $u_0 \in \overline{\mathcal{E}}$ such that $d(u,\mathcal{E}) = |u - u_0|$. Using the monotonicity of $\mathcal{L}_{m,q}^\rho$ and the hypothesis of f, we obtain

$$|\mathcal{L}_{m,q}^\rho(f;u) - f(u)| \le \mathcal{L}_{m,q}^\rho(|f(z) - f(u_0)|; u) + \mathcal{L}_{m,q}^\rho(|f(u) - f(u_0)|; u)$$
$$\le \mathcal{M}_{\alpha,f}\left\{\mathcal{L}_{m,q}^\rho(|\rho(z) - \rho(u_0)|^\alpha; u) + |\rho(u) - \rho(u_0)|^\alpha\right\}$$
$$\le \mathcal{M}_{\alpha,f}\left\{\mathcal{L}_{m,q}^\rho(|\rho(z) - \rho(u)|^\alpha; u) + 2|\rho(u) - \rho(u_0)|^\alpha\right\}.$$

By choosing $p = \frac{2}{\alpha}$ and $q = \frac{2}{2-\alpha}$, as well as the fact $|\rho(u) - \rho(u_0)| = \rho'(u)|\rho(u) - \rho(u_0)|$ in the last inequality. Then by using Hölder's inequality we easily conclude

$$|\mathcal{L}_{m,q}^\rho(f;u) - f(u)| \le \mathcal{M}_{\alpha,f}\left\{[\mathcal{L}_{m,q}^\rho((\rho(z) - \rho(u))^2; u)]^{\frac{1}{2}} + 2[\rho'(u)|\rho(u) - \rho(u_0)|]^\alpha\right\}.$$

Hence, by Corollary 1 we get the proof. □

Now, for $f \in \mathcal{C}_B[0,\infty)$, we recall local approximation in terms of α order generalized Lipschitz-type maximal function given by Lenze [29] as

$$\widetilde{\omega}_\alpha^\rho(f;u) = \sup_{z \neq u, z \in (0,\infty)} \frac{|f(z) - f(u)|}{|z - u|^\alpha}, \ u \in [0,\infty) \text{ and } \alpha \in (0,1]. \tag{26}$$

Then we get the next result

Theorem 9. *Let $f \in \mathcal{C}_B[0,\infty)$ and $\alpha \in (0,1]$ with $0 < q < 1$. Then, for all $u \in [0,\infty)$, we have*

$$\left|\mathcal{L}_{m,q}^\rho(f;u) - f(u)\right| \leq \widetilde{\omega}_\alpha^\rho(f;u) \left(\frac{(1+q)}{[m]_q}\rho(u)\right)^{\frac{\alpha}{2}}.$$

Proof. We know that

$$|\mathcal{L}_{m,q}^\rho(f;u) - f(u)| \ \leq \ \mathcal{L}_{m,q}^\rho(|f(t) - f(u)|;u).$$

From Equation (26), we have

$$|\mathcal{L}_{m,q}^\rho(f;u) - f(u)| \ \leq \ \widetilde{\omega}_\alpha^\rho(f;u)\mathcal{L}_{m,q}^\rho(|\rho(z) - \rho(u)|^\alpha;u).$$

From Hölder's inequality with $p = \frac{2}{\alpha}$ and $q = \frac{2}{2-\alpha}$, we have

$$|\mathcal{L}_{m,q}^\rho(f;u) - f(u)| \ \leq \ \widetilde{\omega}_\alpha^\rho(f;u)\left[\mathcal{L}_{m,q}^\rho((\rho(t)-\rho(u))^2;u)\right]^{\frac{\alpha}{2}}$$
$$\leq \ \widetilde{\omega}_\alpha^\rho(f;u)\left(\frac{(1+q)}{[m]_q}\rho(u)\right)^{\frac{\alpha}{2}}.$$

which proves the desired result. □

6. Conclusions

Here, the q-analogue of the generalized Lupaş operators are constructed. We have investigated convergence properties, order of approximation, Voronovskaja-type results and also quantitative estimates for the local approximation. The constructed operators provide better flexibility in approximating functions and rate of convergence which are dependent on the selection of the function ρ and extra parameter q. These operators also possess interesting properties and depending on the selection of q, can obtain better approximation while $q \neq 1$. The basis of these operators can be used to draw curves and surfaces in Computer Aided Geometric Design (CAGD).

Author Contributions: Supervision, A.K. and Z.A.; Writing—original draft, M.Q.; Writing—review and editing, M.M. All authors have read and agreed to the published version of the manuscript.

Funding: This research received no external funding.

Acknowledgments: The authors are thankful to the referees for their valuable comments and suggestions. The first author is also grateful to the Council of Scientific and Industrial Research (CSIR), India, for providing the Senior Research Fellowship (09/1172(0001)/2017-EMR-I).

Conflicts of Interest: We declare that there is no conflict of interest.

References

1. Bernstein, S.N. Démonstation du théorème de Weierstrass fondée sur le calcul de probabilités. *Commun. Soc. Math. Kharkow* **1912**, *13*, 1–2.
2. Weierstrass, K. *Über die Analytische Darstellbarkeit Sogenannter Willkürlicher Functionen Einer Reellen Veränderlichen*; Sitzungsberichte der Königlich Preußischen Akademie der Wissenschaften zu Berlin: Berlin, Gemany, 1885; pp. 633–639, 789–805.

3. Korovkin, P.P. On convergence of linear operators in the space of continuous functions (Russian). *Dokl. Akad. Nauk SSSR (N.S.)* **1953**, *90*, 961–964.
4. Lorentz, G.G. *Bernstein Polynomials*; University of Toronto Press: Toronto, ON, Canada, 1953.
5. Cárdenas-Morales, D.; Garrancho, P.; Rasa, I. Bernstein-type operators which preserve polynomials. *Comput. Math. Appl.* **2011**, *62*, 158–163. [CrossRef]
6. Aral, A.; Inoan, D.; Rasa, I. On the generalized Szász-Mirakyan operators. *Results Math.* **2014**, *65*, 441–452. [CrossRef]
7. İlarslan, H.G.I.; Aral, A.; Başcanbaz-Tunca, G. Generalized Lupaş operators. *AIP Conf. Proc.* **2018**, *1926*, 020019.
8. Lupaş, A. The approximation by some positive linear operators. In *Approximation Theory, Proceedings of the International Dortmund Meeting, IDoMAT 95, Witten, Germany, 13–17 March 1995*; Müller, M.W., Felten, M., Mache, D.H., Eds.; Akademie Verlag: Berlin, Germany, 1995; pp. 201–229.
9. Gadzhiev, A.D. A problem on the convergence of a sequence of positive linear operators on unbounded sets, and theorems that are analogous to P. P. Korovkin's theorem. *Dokl. Akad. Nauk SSSR* **1974**, *218*, 1001–1004. (In Russian)
10. Gadjiev, A.D. Theorems of the type of P. P. Korovkin's theorems. *Math. Zamet* **1976**, *20*, 781–786.
11. Holhoş, A. Quantitative estimates for positive linear operators in weighted spaces. *Gen. Math.* **2008**, *16*, 99–110.
12. Kac, V.; Cheung, P. *Quantum Calculus*; Springer: New York, NY, USA, 2002.
13. Lupaş, A. A q-analogue of the Bernstein operator. In *Seminar on Numerical and Statistical Calculus*; University of Cluj-Napoca: Cluj-Napoca, Rominia, 1987; Volume 9; pp. 85–92.
14. Phillips, G.M. Bernstein polynomials based on the q-integers. *Ann. Numer. Math.* **1997** *4*, 511–518.
15. Khan, K.; Lobiyal, D.K. Bézier curves based on Lupaş (p,q)-analogue of Bernstein functions in CAGD. *J. Comput. Appl. Math.* **2017**, *317*, 458–477. [CrossRef]
16. Khan, K.; Lobiyal, D.K.; Kilicman, A. A de Casteljau Algorithm for Bernstein type Polynomials based on (p,q)-integers. *Appl. Appl. Math.* **2018**, *13*, 997–1017.
17. Khan, K.; Lobiyal, D.K.; Kilicman, A. Bézier curves and surfaces based on modified Bernstein polynomials. *Azerbaijan J. Math.* **2019**, *9*, 3–21.
18. Aral, A.; Doğru, O. Bleimann-Butzer and Hahn operators based on q-integers. *J. Inequalities Appl.* **2008**, *2007*, 079410. [CrossRef]
19. Aral, A.; Gupta, V.; Agarwal, R.P. *Applications of q-Calculus in Operator Theory*; Springer: New York, NY, USA, 2013.
20. Derriennic, M.M. Modified Bernstein polynomials and Jacobi polynomials in q-calculus. *Rend. Circ. Mat. Palermo Ser. II* **2005**, *76*, 269–290.
21. Mahmudov, N.I. Approximation by the q-Szász-Mirakjan Operators. *Abstr. Appl. Anal.* **2012**, *2012*, 754217. [CrossRef]
22. Mishra, V.N.; Patel, P. On generalized integral Bernstein operators based on q-integers. *Appl. Math. Comput.* **2014**, *242*, 931–944. [CrossRef]
23. Mursaleen, M.; Ahasan, M. The Dunkl generalization of Stancu type q-Szász-Mirakjan-Kantrovich operators and some approximation results. *Carpathian J. Math.* **2018**, *34*, 363–370.
24. Shang, Y. Subspace confinement for switched linear systems. *Forum Math.* **2017**, *29*, 693–699. [CrossRef]
25. Wang, N.; Egerstedt, M.; Martin, C. Stability of switched linear systems and the convergence of random products. In Proceedings of the 48h IEEE Conference on Decision and Control (CDC) Held Jointly with 2009 28th Chinese Control Conference, Shanghai, China, 15–18 December 2009; pp. 3721–3726.
26. DeVore, R.A.; Lorentz, G.G. Constructive Approximation, Grundlehren Math. Wiss. In *Fundamental Principales of Mathematical Sciences*; Springer: Berlin, Germany, 1993.
27. Gadzhiev, A.D.; Aral, A. The estimates of approximation by using a new type of weighted modulus of continuity. *Comput. Math. Appl.* **2007**, *54*, 127–135.
28. Agratini, O. On the rate of convergence of a positive approximation process. *Nihonkai Math. J.* **2000**, *11*, 47–56.
29. Lenze, B. On Lipschitz type maximal functions and their smoothness spaces. *Indag. Math.* **1988**, *91*, 53–63. [CrossRef]

© 2020 by the authors. Licensee MDPI, Basel, Switzerland. This article is an open access article distributed under the terms and conditions of the Creative Commons Attribution (CC BY) license (http://creativecommons.org/licenses/by/4.0/).

Article

Some Applications of a New Integral Operator in q-Analog for Multivalent Functions

Qaiser Khan [2], Muhammad Arif [2], Mohsan Raza [3], Gautam Srivastava [4,5], Huo Tang [1,*] and Shafiq ur Rehman [6]

1. School of Mathematics and Computer Sciences, Chifeng University, Chifeng 024000, China
2. Department of Mathematics, Abdul Wali Khan University Mardan, Mardan 23200, Pakistan; qaisermath84@gmail.com (Q.K.); marifmaths@awkum.edu.pk (M.A.)
3. Department of Mathematics, Government College University Faisalabad, Faisalabad 38000, Pakistan; mohsan976@yahoo.com
4. Department of Mathematics and Computer Science, Brandon University, 270 18th Street, Brandon, MB R7A 86A9, Canada; srivastavag@brandonu.ca
5. Research Center for Interneural Computing, China Medical University, Taichung 40402, Taiwan
6. Department of Mathematics, COMSATS University Islamabad, Attock 43600, Pakistan; shafiq@cuiatk.edu.pk
* Correspondence: thth2009@163.com

Received: 15 August 2019; Accepted: 7 November 2019; Published: 3 December 2019

Abstract: This paper introduces a new integral operator in q-analog for multivalent functions. Using as an application of this operator, we study a novel class of multivalent functions and define them. Furthermore, we present many new properties of these functions. These include distortion bounds, sufficiency criteria, extreme points, radius of both starlikeness and convexity, weighted mean and partial sum for this newly defined subclass of multivalent functions are discussed. Various integral operators are obtained by putting particular values to the parameters used in the newly defined operator.

Keywords: p-valent analytic function; Hadamard product; q-integral operator

MSC: Primary 30C45; Secondary 30C50

---

1. Introduction

The study of q-extension of calculus or q-analysis motivated the researchers due to its recent use in different applications. In [1,2], Jackson introduced the theory of q-calculus. We have seen applications of q-analysis in Geometric Function Theory (GFT). They were introduced and applied systematically to the generalized q-hypergeometric functions in [3]. Later, Ismail et al. [4] used the q-differential operator to examine the geometry of starlike function in q-analog. This theory was later extended to the family of q-starlike function with some order by Agrawal and Sahoo [5]. Due to this development in function theory, many researchers were motivated, as we have seen by Srivastava in [6]. They added significant contributions, which has slowly made this research area more attractive to forthcoming researchers. We direct the attention of our readers to [7–12] for more information. Moreover, Kanas et al. [13] used Hadamad product to define the q-extension of the Ruscheweyh operator. They also discussed in detail some intricate applications of this operator.

Mohammad and Darus [14] conducted an elaborate study of this operator. We have also seen similar work by Mahmood and Sokół [15] and Ahmad et al. [16]. Recently, new thoughts by Maslina in [17] were used to create a novel differential operator called generalized q-differential operator with the help of q-hypergeometric functions where the authors conducted an in-depth study of applications of this operator. For further information on the extensions of different operators in q-analog, we direct the readers to [18–22]. The aim of the present article is to introduce a new integral operator in q-analog for multivalent functions using Hadamard product and then study some of its useful applications.

Let $\mathfrak{A}_p$ ($p \in \mathbb{N} = \{1, 2, \ldots\}$) contain multivalent functions of all forms f that can be defined as holomorphic and/or analytic in any given subset $\mathbb{D} = \{z : |z| < 1\}$ that is part of a complex plane $\mathbb{C}$ which also has the series form shown as:

$$f(z) = z^p + \sum_{n=1}^{\infty} a_{n+p} z^{n+p}, \quad (z \in \mathbb{D}). \tag{1}$$

For any two given functions that are analytic in form f and g in $\mathbb{D}$, then we can clearly state that f is subordinate to g, mostly symbolically if it is presented clearly as $f \prec g$ or $f(z) \prec g(z)$, if and only if there exists an analytic function w with the given properties as $w(0) = 0$ and $|w(z)| < 1$ such that $f(z) = g(w(z))$ ($z \in \mathbb{D}$). Moreover, if and only if g can be seen as univalent in $\mathbb{D}$, then we can clearly have:

$$f(z) \prec g(z) \quad (z \in \mathbb{D}) \quad \Longleftrightarrow \quad f(0) = g(0) \quad \text{and} \quad f(\mathbb{D}) \subset g(\mathbb{D}).$$

For analytic functions f of the form Equation (1) and g of the form

$$g(z) = z^p + \sum_{n=1}^{\infty} b_{n+p} z^{n+p}, \quad (z \in \mathbb{D}), \tag{2}$$

the convolution or Hadamard product is defined by

$$(f * g)(z) = z^p + \sum_{n=1}^{\infty} a_{n+p} b_{n+p} z^{n+p}, \quad (z \in \mathbb{D}).$$

For given $q \in (0, 1)$, the derivative in q-analog of f is given by

$$\mathcal{D}_q f(z) = \frac{f(z) - f(qz)}{z(1-q)}, \quad (z \neq 0, q \neq 1). \tag{3}$$

Making use of Equations (1) and (3), we can easily obtain for $n \in \mathbb{N}$ and $z \in \mathbb{D}$

$$\mathcal{D}_q \left\{ \sum_{n=1}^{\infty} a_{n+p} z^{n+p} \right\} = \sum_{n=1}^{\infty} [n+p]_q a_{n+p} z^{n+p-1}, \tag{4}$$

where

$$[n]_q = \frac{1-q^n}{1-q} = 1 + \sum_{k=1}^{n-1} q^k, \quad [0]_q = 0. \tag{5}$$

For $n \in \mathbb{Z}^+ := \mathbb{Z} \setminus \{-1, -2, \ldots\}$, the q-factorial is given as:

$$[n]_q! = \begin{cases} 1, & n = 0, \\ [1]_q [2]_q \cdots [n]_q, & n \in \mathbb{N}. \end{cases}$$

In addition, with $t > 0$, the q-Pochhammer symbol has the form:

$$[t]_{q,n} = ([t]_q)_n = \begin{cases} 1, & n = 0, \\ [t]_q [t+1]_q \cdots [t+n-1]_q, & n \in \mathbb{N}, \end{cases}$$

where $[t]_q$ is given by Equation (5).

For $t > 0$, the gamma function in q-analog is presented as

$$\Gamma_q(t+1) = [t]_q \Gamma_q(t) \text{ and } \Gamma_q(1) = 1.$$

We now consider a function

$$\mathcal{F}_{q,\lambda+p}^{-1}(z) = z^p + \sum_{n=1}^{\infty} \Psi_{n-p} a_{n+p} z^{n+p}, \ (\lambda > -p, \ z \in \mathbb{D}), \tag{6}$$

with

$$\Psi_{n-p} = \frac{[\lambda+p]_{q,n+1-p}}{[n+1-p]_q!}. \tag{7}$$

We can see that the series given in Equation (6) is absolutely convergent in $\mathbb{D}$. Now, we introduce the integral operator $\mathcal{J}_q^{\lambda+p-1} : \mathfrak{A}_p \to \mathfrak{A}_p$ by

$$\mathcal{J}_q^{\lambda+p-1} f(z) = (\mathcal{F}_{q,\lambda+p}^{-1} * f)(z) = z^p + \sum_{n=1}^{\infty} \Psi_{n-p} a_{n+p} z^{n+p} \ (z \in \mathbb{D}), \tag{8}$$

where $\lambda > -p$. We note that

$$\lim_{q \to 1^-} \mathcal{F}_{q,\lambda+p}^{-1}(z) = \frac{z^p}{(1-z)^{\lambda+1}} \text{ and } \lim_{q \to 1^-} \mathcal{J}_q^{\lambda+p-1} f(z) = f(z) * \frac{z^p}{(1-z)^{\lambda+1}}.$$

Various integral operators were obtained by putting particular values to the parameters used in the newly defined operator as given by Equation (7):

(i). Making $p = 1$ in our newly defined operator $\mathcal{J}_q^{\lambda+p-1} f$, we obtain the operator $\mathcal{J}_q^\lambda f$ which was introduced by Arif et al. [20] and is given by

$$\mathcal{J}_q^\lambda f(z) = (\mathcal{F}_{q,\lambda+1}^{-1} * f)(z) = z + \sum_{n=1}^{\infty} \Psi_{n-1} a_{n+1} z^{n+1} \ (z \in \mathbb{D}).$$

(ii). When $q \to 1^-$, the operator defined in Equation (7) leads to the following well-known Noor integral operator for multivalent functions introduced in [23].

$$\mathcal{J}^{\lambda+p-1} f(z) = (\mathcal{F}_{\lambda+p}^{-1} * f)(z) = z^p + \sum_{n=1}^{\infty} \Psi_{n-p} a_{n+p} z^{n+p}, \ (z \in \mathbb{D}).$$

(iii). If we set $p = 1$ along with $q \to 1^-$ in Equation (7), then the operator $\mathcal{J}^{\lambda+p-1} f$ reduced to the following familiar Noor integral operator studied in [24,25].

$$\mathcal{I}^\lambda f(z) = (\mathcal{F}_{\lambda+1}^{-1} * f)(z) = z + \sum_{n=1}^{\infty} \Psi_{n-1} a_{n+1} z^{n+1}, \ (z \in \mathbb{D}).$$

For more details on the q-analog of differential and integral operators, see [17,26,27].

Motivated from the work in [28–35], we now introduce a subfamily $\mathcal{H}_{p,q}^\lambda(\alpha, \mu, \beta)$ of $\mathfrak{A}_p$ by using $\mathcal{J}_q^{\lambda+p-1}$ as follows:

Definition 1. *Let* $f \in \mathfrak{A}_p$. *Then,* $f \in \mathcal{H}_{p,q}^\lambda(\alpha, \mu, \beta)$, *if it satisfies the relation*

$$\left| \frac{z^{1-p} \mathcal{D}_q \mathcal{J}_q^{\lambda+p-1} f(z) - [p]_q}{2\beta \left[z^{1-p} \mathcal{D}_q \mathcal{J}_q^{\lambda+p-1} f(z) - \alpha [p]_q \right] - \left[z^{1-p} \mathcal{D}_q \mathcal{J}_q^{\lambda+p-1} f(z) - [p]_q \right]} \right| < \mu, \tag{9}$$

where $\frac{1}{2} \leqslant \beta < 1, 0 \leqslant \alpha < \frac{1}{2}, 0 < \mu \leqslant 1$ *and* $0 < q < 1$.

By varying the parameters values in the class $\mathcal{H}_{p,q}^{\lambda}(\alpha,\mu,\beta)$, we get many new classes; we list some of them.

(i). For $p=1$, we have $\mathcal{H}_{1,q}^{\lambda}(\alpha,\mu,\beta) \equiv \mathcal{H}_{q}^{\lambda}(\alpha,\mu,\beta)$.

(ii). Taking the limit $q \to 1^-$, we get the class $\mathcal{H}_{p}^{\lambda}(\alpha,\mu,\beta)$.

(iii). Putting $\beta = \frac{1}{2}$, $\mu = 1$ and $\alpha = 0$, we obtain $\mathcal{H}_{p,q}^{\lambda}\left(0,1,\frac{1}{2}\right)$.

(iv). Further, if we put $p=1$ and $q \to 1^-$ in $\mathcal{H}_{p,q}^{\lambda}\left(0,1,\frac{1}{2}\right)$, we have the class $\mathcal{H}_{1}^{\lambda}\left(0,1,\frac{1}{2}\right)$.

Note that we assume throughout our discussion, unless otherwise stated,

$$\frac{1}{2} \leq \beta \leq 1,\ 0 \leq \alpha < 1,\ 0 < \mu \leq 1,\ \lambda > -p,\ 0 < q < 1$$

and all coefficients a_k are positive.

2. The Main Results and Their Consequences

Theorem 1. *If $f \in \mathfrak{A}_p$ has the form of Equation (1) and satisfies the inequality*

$$\sum_{n=1}^{\infty} \Psi_{n-p}[n+p]_q(1+\mu(2\beta-1))|a_{n+p}| \leq 2\mu\beta[p]_q(1-\alpha), \tag{10}$$

then $f \in \mathcal{H}_{p,q}^{\lambda}(\alpha,\mu,\beta)$.

Proof. To show that $f \in \mathcal{H}_{p,q}^{\lambda}(\alpha,\mu,\beta)$, we just need to prove Equation (9). For this, consider

$$L = \left| \frac{\frac{1}{[p]_q}z^{1-p}D_q\mathcal{J}_q^{\lambda+p-1}f(z)-1}{2\beta\left[\frac{1}{[p]_q}z^{1-p}D_q\mathcal{J}_q^{\lambda+p-1}f(z)-\alpha\right]-\left[\frac{1}{[p]_q}z^{1-p}D_q\mathcal{J}_q^{\lambda+p-1}f(z)-1\right]} \right|$$

$$= \left| \frac{z^{1-p}D_q\mathcal{J}_q^{\lambda+p-1}f(z)-[p]_q}{2\beta\left[z^{1-p}D_q\mathcal{J}_q^{\lambda+p-1}f(z)-\alpha[p]_q\right]-\left[z^{1-p}D_q\mathcal{J}_q^{\lambda+p-1}f(z)-[p]_q\right]} \right|.$$

Using Equation (8) with the help of Equations (3) and (4), we can easily obtain

$$L = \left| \frac{\sum_{n=1}^{\infty}\Psi_{n-p}[n+p]_q a_{n+p}z^n}{2\beta[p]_q(1-\alpha) + 2\beta\sum_{n=1}^{\infty}\Psi_{n-p}[n+p]_q a_{n+p}z^n - \sum_{n=1}^{\infty}\Psi_{n-p}[n+p]_q a_{n+p}z^n} \right|$$

$$= \left| \frac{\sum_{n=1}^{\infty}\Psi_{n-p}[n+p]_q a_{n+p}z^n}{2\beta[p]_q(1-\alpha) + (2\beta-1)\sum_{n=1}^{\infty}\Psi_{n-p}[n+p]_q a_{n+p}z^n} \right|$$

$$\leq \frac{\sum_{n=1}^{\infty}\Psi_{n-p}[n+p]_q|a_{n+p}|}{2\beta[p]_q(1-\alpha) - (2\beta-1)\sum_{n=1}^{\infty}\Psi_{n-p}[n+p]_q|a_{n+p}|} < \mu,$$

where we have used the inequality in Equation (10) and this completes the proof. □

Making $\beta = \frac{1}{2}$, $p=1$ along with $q \to 1^-$, we get the following result.

Corollary 1. If $f \in \mathfrak{A}$ and satisfies the inequality

$$\sum_{n=1}^{\infty} \Psi_{n-1}(n+1)|a_{n+1}| \leq \mu(1-\alpha),$$

then $f \in \mathcal{H}^\lambda\left(\alpha, \mu, \frac{1}{2}\right)$.

Theorem 2. If $f \in \mathcal{H}^\lambda_{p,q}(\alpha, \mu, \beta)$ has the form of Equation (1), then

$$r^p - \xi r \leq |f(z)| \leq r^p + \xi r, |z| = r < 1,$$

where

$$\xi = \frac{2\mu\beta [p]_q (1-\alpha)}{(1+\mu(2\beta-1))\Psi_{1-p}[1+p]_q}.$$

Proof. Consider

$$\begin{aligned}
|f(z)| &= \left| z^p + \sum_{n=1}^{\infty} a_{n+p} z^{n+p} \right| \\
&\leq |z|^p + \sum_{n=1}^{\infty} |a_{n+p}| |z|^{n+p} \\
&= r^p + \sum_{n=1}^{\infty} |a_{n+p}| r^{n+p}.
\end{aligned}$$

Since $0 < r < 1$ and $r^{n+p} < r$,

$$|f(z)| \leq r^p + r \sum_{n=1}^{\infty} |a_{n+p}|. \tag{11}$$

Similarly,

$$|f(z)| \geq r^p - r \sum_{n=1}^{\infty} |a_{n+p}|. \tag{12}$$

It can easily be seen that

$$\Psi_{1-p}(1+\mu(2\beta-1))[1+p]_q \sum_{n=1}^{\infty} |a_{n+p}| \leq \sum_{n=1}^{\infty} \Psi_{n-p}(1+\mu(2\beta-1))[n+p]_q |a_{n+p}|.$$

By using the relation in Equation (10), we obtain

$$\Psi_{1-p}(1+\mu(2\beta-1))[1+p]_q \sum_{n=1}^{\infty} |a_{n+p}| \leq 2\mu\beta [p]_q (1-\alpha),$$

which gives

$$\sum_{n=1}^{\infty} |a_{n+p}| \leq \frac{2\mu\beta[p]_q(1-\alpha)}{\Psi_{1-p}(1+\mu(2\beta-1))[1+p]_q}.$$

Now, by using the above relation in Equations (11) and (12), we obtain the result. □

Setting $\beta = \frac{1}{2}$, $p = 1$ along with $q \to 1^-$ in the last theorem, we have

Corollary 2. If $f \in \mathcal{H}^\lambda\left(\alpha, \mu, \frac{1}{2}\right)$, then for $|z| = r < 1$

$$\left(1 - \frac{\mu(1-\alpha)}{2(\lambda+1)}\right) r \leq |f(z)| \leq \left(1 + \frac{\mu(1-\alpha)}{2(\lambda+1)}\right) r.$$

Theorem 3. If $f \in \mathcal{H}^\lambda_{p,q}(\alpha, \mu, \beta)$ has the form of Equation (1), then

$$[p]_q r^{p-1} - \vartheta r \leq |\mathcal{D}_q f(z)| \leq [p]_q r^{p-1} + \vartheta r, |z| = r < 1,$$

where $\vartheta = \frac{2\mu\beta[p]_q(1-\alpha)}{(1+\mu(2\beta-1))\Psi_{1-p}}$.

Proof. By using Equations (3) and (4), we can have

$$\mathcal{D}_q f(z) = [p]_q z^{p-1} + \sum_{n=1}^{\infty} [n+p]_q a_{n+p} z^{n+p-1}.$$

Since $|z|^{p-1} = r^{p-1} < 1$, $r^{n+p-1} \leq r$ and

$$|\mathcal{D}_q f(z)| \leq [p]_q r^{p-1} + r \sum_{n=1}^{\infty} [n+p]_q |a_{n+p}|. \tag{13}$$

Similarly,

$$|\mathcal{D}_q f(z)| \geq [p]_q r^{p-1} - r \sum_{n=1}^{\infty} [n+p]_q |a_{n+p}|. \tag{14}$$

Now, by using Equation (10), we get

$$\Psi_{1-p}(1+\mu(2\beta-1)) \sum_{n=1}^{\infty} [n+p]_q |a_{n+p}| \leq \sum_{n=1}^{\infty} \Psi_{n-p} [n+p]_q (1+\mu(2\beta-1)) |a_{n+p}|.$$

This implies that

$$\sum_{n=1}^{\infty} [n+p]_q |a_{n+p}| \leq \frac{2\mu\beta[p]_q(1-\alpha)}{(1+\mu(2\beta-1))\Psi_{1-p}}.$$

Finally, by using above relation in Equations (13) and (14), we have the result. □

For $q \to 1^-$, we have the following corollary.

Corollary 3. If $f \in \mathcal{H}^\lambda_p(\alpha, \mu, \beta)$, then for $|z| = r < 1$

$$pr^{p-1} - \vartheta r \leq |f'(z)| \leq pr^{p-1} + \vartheta r,$$

where

$$\vartheta = \frac{2\mu\beta(1-\alpha)p}{(1+\mu(2\beta-1))\Psi_{1-p}}.$$

Theorem 4. If $f \in \mathcal{H}^\lambda_{p,q}(\alpha, \mu, \beta)$, then $f \in \mathcal{S}^*_p(\delta)$ for $|z| < r_1$, where

$$r_1 = \left(\frac{(p-\delta)(1+\mu(2\beta-1))\Psi_{n-p}[n+p]_q}{2\mu\beta(1-\alpha)(\delta-p)[p]_q}\right)^{\frac{1}{n}}.$$

Proof. Let $f \in \mathcal{H}_{p,q}^{\lambda}(\alpha, \mu, \beta)$. To show that $f \in \mathcal{S}_p^*(\delta)$, we have to prove that

$$\left| \frac{zf'(z) - pf(z)}{zf'(z) + (p - 2\delta)f(z)} \right| < 1.$$

Using Equation (1), we conclude that

$$\sum_{n=1}^{\infty} \left(\frac{\delta - p}{p - \delta} \right) |a_{n+p}| |z|^n < 1. \tag{15}$$

From Equation (10), it can easily be obtained that

$$\sum_{n=1}^{\infty} \left(\frac{\Psi_{n-p}[n+p]_q(1+\mu(2\beta-1))|}{[p]_q 2\mu\beta(1-\alpha)} \right) |a_{n+p}| < 1.$$

The relation in Equation (15) is true, if the following holds

$$\sum_{n=1}^{\infty} \left(\frac{\delta - p}{p - \delta} \right) |a_{n+p}| |z|^n < \sum_{n=1}^{\infty} \left(\frac{\Psi_{n-p}[n+p]_q(1+\mu(2\beta-1))|}{[p]_q 2\mu\beta(1-\alpha)} \right) |a_{n+p}|,$$

which implies that

$$|z|^n < \frac{(p - \delta)(1 + \mu(2\beta - 1))\Psi_{n-p}[n+p]_q}{2\mu\beta(1-\alpha)(\delta - p)[p]_q}.$$

Therefore,

$$|z| < \left(\frac{(p - \delta)(1 + \mu(2\beta - 1))\Psi_{n-p}[n+p]_q}{2\mu\beta(1-\alpha)(\delta - p)[p]_q} \right)^{\frac{1}{n}} = r_1.$$

Hence, we get the required result. □

Letting $p = 1$ and $q \to 1^-$ in the last theorem, we get the result below.

Corollary 4. *If $f \in \mathcal{H}^{\lambda}(\alpha, \mu, \beta)$, then $f \in \mathcal{S}^*(\delta)$ for $|z| < r_1$, where*

$$r_1 = \left(\frac{(1 - \delta)(1 + \mu(2\beta - 1))[(n+1)\Psi_{n-1}]}{2\mu\beta(1-\alpha)(\delta - 1)} \right)^{\frac{1}{n}}.$$

Theorem 5. *If $f \in \mathcal{H}_{p,q}^{\lambda}(\alpha, \mu, \beta)$, then $f \in \mathcal{C}_p(\delta)$ for $|z| < r_2$, where*

$$r_2 = \left(\frac{p(p - \delta)(1 + \mu(2\beta - 1))\Psi_{n-p}[n+p]_q}{2\mu\beta(1-\alpha)[p]_q(\delta - p)(n+p)} \right)^{\frac{1}{n}}.$$

Proof. Since $f \in \mathcal{C}_p(\delta)$,

$$\left| \frac{zf''(z) - (p - 1)f'(z)}{zf''(z) + (1 - 2\delta + p)f'(z)} \right| < 1.$$

By using Equation (1) and after some simplifications, we get

$$\sum_{n=1}^{\infty} \left(\frac{(\delta - p)(n + p)}{p(p - \delta)} \right) |a_{n+p}| |z|^n < 1. \tag{16}$$

Now, from Equation (10), we can easily obtain that

$$\sum_{n=1}^{\infty} \left(\frac{(1+\mu(2\beta-1))\Psi_{n-p}[n+p]_q}{2\mu\beta(1-\alpha)[p]_q} \right) |a_{n+p}| < 1.$$

The relation in Equation (16) is true if

$$\sum_{n=1}^{\infty} \left(\frac{(\delta-p)(n+p)}{p(p-\delta)} \right) |a_{n+p}| \, |z|^n < \sum_{n=1}^{\infty} \left(\frac{(1+\mu(2\beta-1))\Psi_{n-p}[n+p]_q}{2\mu\beta(1-\alpha)[p]_q} \right) |a_{n+p}|,$$

which gives

$$|z|^n < \left(\frac{p(p-\delta)(1+\mu(2\beta-1))\Psi_{n-p}[n+p]_q}{2\mu\beta(1-\alpha)[p]_q (\delta-p)(n+p)} \right).$$

Hence,

$$|z| < \left(\frac{p(p-\delta)(1+\mu(2\beta-1))\Psi_{n-p}[n+p]_q}{2\mu\beta(1-\alpha)[p]_q (\delta-p)(n+p)} \right)^{\frac{1}{n}} = r_2.$$

Thus, we obtain the required result. □

Substituting $p = 1$ and taking $q \to 1^-$ in the last theorem, we get the corollary below.

Corollary 5. *If $f \in \mathcal{H}^\lambda(\alpha, \mu, \beta)$, then $f \in \mathcal{C}(\delta)$ for $|z| < r_2$, where*

$$r_2 = \left(\frac{(1-\delta)(1+\mu(2\beta-1))\Psi_{n-1}}{2\mu\beta(1-\alpha)(\delta-1)} \right)^{\frac{1}{n}}.$$

Theorem 6. *Let $f_p(z) = z^p$ and*

$$f_k(z) = z^p + \frac{2\mu\beta(1-\alpha)[p]_q}{(1+\mu(2\beta-1))\Psi_{k-2p}[k]_q a_k} z^k, \quad (k \geq n+p). \tag{17}$$

Then, $f \in \mathcal{H}^\lambda_{p,q}(\alpha, \mu, \beta)$, if and only if it can be expressed in the form

$$f(z) = \lambda_p z^p + \sum_{k=n+p}^{\infty} \lambda_k f_k(z), \tag{18}$$

where $\lambda_p \geq 0$, $\lambda_k \geq 0$, $k \geq n+p$ and $\lambda_p + \sum_{k=n+p}^{\infty} \lambda_k = 1$.

Proof. We suppose that f can be written of the form of Equation (18), thus

$$\begin{aligned}
f(z) &= \lambda_p z^p + \sum_{k=n+p}^{\infty} \lambda_k f_k(z), \\
&= \lambda_p z^p + \sum_{k=n+p}^{\infty} \lambda_k \left(z^p + \frac{2\mu\beta(1-\alpha)[p]_q}{(1+\mu(2\beta-1))\Psi_{k-2p}[k]_q a_k} z^k \right) \\
&= z^p + \sum_{k=n+p}^{\infty} \frac{2\mu\beta(1-\alpha)[p]_q \lambda_k}{(1+\mu(2\beta-1))\Psi_{k-2p}[k]_q a_k} z^k.
\end{aligned}$$

This implies that

$$\sum_{k=n+p}^{\infty} \frac{(1+\mu(2\beta-1))\Psi_{k-2p}[k]_q a_k}{2\mu\beta(1-\alpha)[p]_q} \times \frac{2\mu\beta(1-\alpha)[p]_q}{(1+\mu(2\beta-1))\Psi_{k-2p}[k]_q a_k} \lambda_k$$

$$= \sum_{k=n+p}^{\infty} \lambda_k,$$
$$= 1 - \lambda_p \leq 1.$$

Conversely, we suppose that $f_n \in \mathcal{H}^\lambda_{p,q}(\alpha, \mu, \beta)$. Then, by using Equation (10), we have

$$|a_k| \leq \frac{2\mu\beta(1-\alpha)[p]_q}{(1+\mu(2\beta-1))\Psi_{k-2p}[k]_q}, \quad k \geq p+n.$$

By setting

$$\lambda_k = \frac{(1+\mu(2\beta-1))\Psi_{k-2p}[k]_q}{2\mu\beta(1-\alpha)[p]_q} a_k,$$

then

$$\begin{aligned}
f(z) &= z^p + \sum_{k=n+p}^{\infty} a_k z^k \\
&= z^p + \sum_{k=n+p}^{\infty} \frac{2\mu\beta(1-\alpha)[p]_q}{(1+\mu(2\beta-1))\Psi_{k-2p}[k]_q a_k} \lambda_k z^k \\
&= z^p + \sum_{k=n+p}^{\infty} [z^p - f_k(z)] \lambda_k \\
&= \left(1 - \sum_{k=n+p}^{\infty} \lambda_k\right) z^p + \sum_{k=n+p}^{\infty} \lambda_k f_k(z) \\
&= \lambda_p z^p + \sum_{k=n+p}^{\infty} \lambda_k f_k(z).
\end{aligned}$$

This complete the result. □

Putting $p = 1$, $\beta = \frac{1}{2}$ and $q \to 1^-$ in the above theorem, we obtain the upcoming corollary.

Corollary 6. Let $f(z) = z$ and

$$f_k(z) = z + \frac{\mu(1-\alpha)}{k\Psi_{k-2}a_k} z^k, \quad (k \geq n+1). \tag{19}$$

Then, $f \in \mathcal{H}^*_p\left(\alpha, \mu, \frac{1}{2}\right)$, if and only if it can be expressed in the form

$$f(z) = \lambda z + \sum_{k=n+1}^{\infty} \lambda_k f_k(z), \tag{20}$$

where $\lambda \geq 0$, $\lambda_k \geq 0$, $k \geq n+1$ and $\lambda + \sum_{k=n+1}^{\infty} \lambda_k = 1$.

Theorem 7. If $f, g \in \mathcal{H}^\lambda_{p,q}(\alpha, \mu, \beta)$, then $(f * g) \in \mathcal{H}^\lambda_{p,q}(\alpha, \mu, \beta)$, where

$$l \geq \mu \left(\frac{2l\beta(1-\alpha)[p]_q}{\Psi_{n-p}[n+p]_q} - l(2\beta-1) \right). \tag{21}$$

Proof. We have to find largest μ such that

$$\sum_{n=1}^{\infty} \frac{(1+\mu(2\beta-1))\Psi_{n-p}[n+p]_q}{2\mu\beta(1-\alpha)[p]_q} a_{n+p} b_{n+p} \leq 1.$$

Let $f, g \in \mathcal{H}_{p,q}^{\lambda}(\alpha, \mu, \beta)$. Then, using Equation (10), we obtain

$$\sum_{n=1}^{\infty} \frac{(1+l(2\beta-1))\Psi_{n-p}[n+p]_q}{2l\beta(1-\alpha)[p]_q} a_{n+p} \leq 1 \qquad (22)$$

and

$$\sum_{n=1}^{\infty} \frac{(1+l(2\beta-1))\Psi_{n-p}[n+p]_q}{2l\beta(1-\alpha)[p]_q} b_{n+p} \leq 1. \qquad (23)$$

By using Cauchy–Schwarz inequality, we have

$$\sum_{n=1}^{\infty} \frac{(1+l(2\beta-1))\Psi_{n-p}[n+p]_q}{2l\beta(1-\alpha)[p]_q} \sqrt{a_{n+p}b_{n+p}} \leq 1. \qquad (24)$$

Thus, we have to show that

$$\frac{(1+l(2\beta-1))\Psi_{n-p}[n+p]_q}{2l\beta(1-\alpha)[p]_q} a_{n+p}b_{n+p}$$
$$\leq \frac{(1+\mu(2\beta-1))\Psi_{n-p}[n+p]_q}{2\mu\beta(1-\alpha)[p]_q} \sqrt{a_{n+p}b_{n+p}},$$

that is

$$\sqrt{a_{n+p}b_{n+p}} \leq \frac{l(1+\mu(2\beta-1))}{\mu(1+l(2\beta-1))}. \qquad (25)$$

In addition, from Equation (24), we can write

$$\sqrt{a_{n+p}b_{n+p}} \leq \frac{2l\beta(1-\alpha)[p]_q}{(1+l(2\beta-1))\Psi_{n-p}[n+p]_q}. \qquad (26)$$

Consequently, we have to show that

$$\frac{2l\beta(1-\alpha)[p]_q}{(1+l(2\beta-1))\Psi_{n-p}[n+p]_q} \leq \frac{l(1+\mu(2\beta-1))}{\mu(1+l(2\beta-1))}.$$

By simple calculation, we get

$$l \geq \mu \left(\frac{2l\beta(1-\alpha)[p]_q}{\Psi_{n-p}[n+p]_q} - l(2\beta-1) \right),$$

which completes the required result. □

Theorem 8. *Let f_1 and f_2 be in the class $\mathcal{H}_{p,q}^{\lambda}(\alpha, \mu, \beta)$. Then, the weighted mean w_q of f_1 and f_2 is also in the class $\mathcal{H}_{p,q}^{\lambda}(\alpha, \mu, \beta)$.*

Proof. From the definition of weighted mean, we have

$$\begin{aligned} w_q &= \frac{1}{2}[(1-q)f_1(z) + (1+q)f_2(z)] \\ &= \left[(1-q)\left(z^p + \sum_{n=1}^{\infty} a_{n+p}z^{n+p} \right) + (1+q)\left(z^p + \sum_{n=1}^{\infty} b_{n+p}z^{n+p} \right) \right] \\ &= z^p + \sum_{n=1}^{\infty} \frac{1}{2}\left[(1-q)a_{n+p} + (1+q)b_{n+p} \right] z^{n+p}. \end{aligned}$$

Now, using Equation (10), we have

$$\sum_{n=1}^{n} \Psi_{n-p} [n+p]_q (1+\mu(2\beta-1)) a_{n+p} \leq 2\mu\beta [p]_q (1-\alpha)$$

and

$$\sum_{n=1}^{n} \Psi_{n-p} [n+p]_q (1+\mu(1-2\beta)) b_{n+p} \leq 2\mu\beta [p]_q (1-\alpha).$$

Consider

$$\left[\sum_{n=1}^{\infty} \Psi_{n-p} [n+p]_q (1+\mu(2\beta-1))\right] \left[\frac{1}{2}\left[(1-q) a_{n+p} + (1+q) b_{n+p}\right]\right]$$

$$= \frac{1}{2}(1-q) \sum_{n=1}^{\infty} \Psi_{n-p} [n+p]_q (1+\mu(2\beta-1)) a_{n+p}$$

$$+ \frac{1}{2}(1+q) \sum_{n=1}^{\infty} \Psi_{n-p} [n+p]_q (1+\mu(2\beta-1)) b_{n+p}$$

$$\leq \frac{1}{2}(1-q) 2\mu\beta [p]_q (1-\alpha) + \frac{1}{2}(1+q) 2\mu\beta [p]_q (1-\alpha).$$

This shows that $w_q \in \mathcal{H}_{p,q}^{\lambda}(\alpha,\mu,\beta)$. □

3. Applications

The q-calculus has played an important role in the study of almost every branch of mathematics and physics, for example, in the theory of special functions, differential equations, combinatorics, analytic number theory, quantum theory, quantum group, special polynomials, numerical analysis, operator theory and other related theories. Quantum calculus is considered as one of the most active research areas in mathematics and physics. For more details, please refer to [22,36–41].

4. Concluding Remarks and Observations

In this paper, we introduce a new integral operator $\mathcal{J}_q^{\lambda+p-1}$ in q-analog and define the class $\mathcal{H}_{p,q}^{\lambda}(\alpha,\mu,\beta)$ of multivalent functions by using this operator. Several useful properties such as sufficiency criteria, distortion bounds, radius of starlikness and radius of convexity, extreme points, weighted mean and partial sum for this newly defined subclass of multivalent functions are investigated. In addition, we observe that, if we take some suitable parameters p,q,α,μ,β in the results involved, we get the corresponding properties for q-analog of differential and integral operators mentioned in the Introduction.

Author Contributions: Conceptualization, M.A.; Formal analysis, M.R.; Funding acquisition, H.T.; Investigation, Q.K.; Methodology, M.R.; Software, Q.K.; Supervision, M.R., G.S. and H.T.; Visualization, G.S.; Writing—original draft, Q.K.; Writing—review & editing, M.A., S.u.R.

Funding: The research conducted in this article was partly supported by the Natural Science Foundation of the People's Republic of China under Grants 11561001 and 11271045, the Program for Young Talents of Science and Technology in Universities of Inner Mongolia Autonomous Region under Grant NJYT-18-A14, the Natural Science Foundation of Inner Mongolia of the People's Republic of China under Grant 2018MS01026 and the Natural Science Foundation of Chifeng of Inner Mongolia.

Conflicts of Interest: The authors declare no conflict of interest.

References

1. Jackson, F.H. On q-definite integrals. *Q. J. Pure Appl. Math.* **1910**, *41*, 193–203.
2. Jackson, F.H. On q-functions and a certain difference operator. *Earth Environ. Sci. Trans. R. Soc. Edinburgh.* **1909**, *46*, 253–281. [CrossRef]

3. Srivastava, H.M. Univalent functions, fractional calculus, and associated generalized hypergeometric functions. In *Univalent Functions, Fractional Calculus, and Their Applications*; Srivastava, H.M., Owa, S., Eds.; Halsted Press: Chichester, UK; John Wiley and Sons: New York, NY, USA, 1989; pp. 329–354.
4. Ismail, M.E.H.; Merkes, E.; Styer, D. A generalization of starlike functions. *Complex Var. Theory Appl.* **1990**, *14*, 77–84. [CrossRef]
5. Agrawal, S.; Sahoo, S.K. A generalization of starlike functions of order α. *Hokkaido Math. J.* **2017**, *46*, 15–27. [CrossRef]
6. Srivastava, H.M.; Bansal, D. Close-to-convexity of a certain family of q-Mittag-Leffler functions. *J. Nonlinear Var. Anal.* **2017**, *1*, 61–69.
7. Raza, M.; U-Din, M. Close-to-convexity of q-Mittag-Leffler Functions, *C. R. de l'Acad. Bulg. des Sci.* **2018**, *71*, 1581–1591.
8. Ul-Haq, M.; Raza, M.; Arif, M.; Khan, Q.; Tang, H. q-analogue of differential subordinations. *Mathematics* **2019**, *7*, 724. [CrossRef]
9. Shi, L.; Raza, M.; Javed, K.; Hussain, S.; Arif, M. Class of analytic functions defined by q-integral operator in a symmetric region. *Symmetry* **2019**, *11*, 1042. [CrossRef]
10. Mahmood, S.; Jabeen, M.; Malik, S.N.; Srivastava, H.M.; Manzoor, R.; Riaz, S.M.J. Some coefficient inequalities of q-starlike functions associated with conic domain defined by q-derivative. *J. Funct. Spaces* **2018**, *2018*, 8492072. [CrossRef]
11. Srivastava, H.M.; Khan, B.; Khan, N.; Ahmad, Q.Z. Coefficient inequalities for q-starlike functions associated with the Janowski functions. *Hokkaido Math. J.* **2019**, *48*, 407–425. [CrossRef]
12. Srivastava, H.M.; Ahmad, Q.Z.; Khan, N.; Khan, N.; Khan, B. Hankel and Toeplitz determinants for a subclass of q-starlike functions associated with a general conic domain. *Mathematics* **2019**, *7*, 181. [CrossRef]
13. Kanas, S.; Răducanu, D. Some class of analytic functions related to conic domains. *Math. Slovaca* **2014**, *64*, 1183–1196. [CrossRef]
14. Aldweby, H.; Darus, M. Some subordination results on q-analogue of Ruscheweyh differential operator. *Abstr. Appl. Anal.* **2014**, *2014*, 958563. [CrossRef]
15. Mahmmod, S.; Sokół, J. New subclass of analytic functions in conical domain associated with Ruscheweyh q-differential operator. *Results Math.* **2017**, *71*, 1345–1357. [CrossRef]
16. Ahmad, K.; Arif, M.; Liu, J.-L. Convolution properties for a family of analytic functions involving q-analogue of Ruscheweyh differential operator. *Turkish J. Math.* **2019**, *43*, 1712–1720. [CrossRef]
17. Mohammed, A.; Darus, M. A generalized operator involving the q-hypergeometric function. *Matematički Vesnik* **2013**, *65*, 454–465.
18. Arif, M.; Ahmad, B. New subfamily of meromorphic starlike functions in circular domain involving q-differential operator. *Math. Slovaca* **2018**, *68*, 1049–1056. [CrossRef]
19. Arif, M.; Dziok, J.; Raza, M.; Sokół, J. On products of multivalent close-to-star functions. *J. Inequal. Appl.* **2015**, *2015*, 5. [CrossRef]
20. Arif, M.; Haq, M.; Liu, J.-L. A subfamily of univalent functions associated with q-analogue of Noor integral operator. *J. Funct. Spaces* **2018**, *2018*, 3818915. [CrossRef]
21. Arif, M.; Srivastava, H.M.; Umar, S. Some applications of a q-analogue of the Ruscheweyh type operator for multivalent functions. *Revista de la Real Academia de Ciencias Exactas Físicas y Naturales Serie A Matemáticas* **2019**, *113*, 1211–1221. [CrossRef]
22. Shi, L.; Khan, Q.; Srivastava, G.; Liu, J.-L.; Arif, M. A study of multivalent q-starlike functions connected with circular domain. *Mathematics* **2019**, *7*, 670. [CrossRef]
23. Liu, J.-L.; Inayat Noor, K. Some properties of Noor integral operator. *J. Nat. Geom.* **2002**, *21*, 81–90.
24. Noor, K.I. On new classes of integral operators. *J. Natur. Geom.* **1999**, *16*, 71–80.
25. Noor, K.I.; Noor, M.A. On integral operators. *J. Math. Anal. Appl.* **1999**, *238*, 341–352. [CrossRef]
26. Aldawish, I.; Darus, M. Starlikness of q-differential operator involving quantum calculus. *Korean J. Math.* **2014**, *22*, 699–709. [CrossRef]
27. Aldweby, H.; Darus, M. A subclass of harmonic univalent functions associated with q-analogue of Dziok-Srivastava operator. *ISRN Math. Anal.* **2013**, *2013*, 382312. [CrossRef]
28. Seoudy, T.M.; Aouf, M.K. Coefficient estimates of new classes of q-starlike and q-convex functions of complex order. *J. Math. Inequal.* **2016**, *10*, 135–145. [CrossRef]

29. Dziok, J.; Murugusundaramoorthy, G.; Sokoł, J. On certain class of meromorphic functions with positive coefcients. *Acta Math. Sci.* **2012**, *32*, 1–16.
30. Huda, A.; Darus, M. Integral operator defined by q-analogue of Liu-Srivastava operator. *Studia Univ. Babes-Bolyai Ser. Math.* **2013**, *58*, 529–537.
31. Pommerenke, C. On meromorphic starlike functions. *Pacific J. Math.* **1963**, *13*, 221–235. [CrossRef]
32. Uralegaddi, B.A.; Somanatha, C. Certain diferential operators for meromorphic functions. *Houston J. Math.* **1991**, *17*, 279–284.
33. Srivastava, H.M.; Khan, N.; Darus, M.; Rahim, M.T.; Ahmad, Q.Z.; Zeb, Y. Properties of spiral-like close-to-convex functions associated with conic domains. *Mathematics* **2019**, *7*, 706. [CrossRef]
34. Mahmood, S.; Khan, I.; Srivastava, H.M.; Malik, S.N. Inclusion relations for certain families of integral operators associated with conic regions. *J. Inequal. Appl.* **2019**, *2019*, 59. [CrossRef]
35. Srivastava, H.M.; Rafiullah, M.; Arif, M. Some subclasses of close-to-convex mappings associated with conic regions. *Appl. Math. Comput.* **2016**, *285*, 94–102. [CrossRef]
36. Aral, A.; Gupta, V.; Agarwal, R.P. *Applications of q-Calculus in Operator Theory*; Springer Science+Business Media: New York, NY, USA, 2013.
37. Ernst, T. *A Comprehensive Treatment of q-Calculus*; Springer: Berlin, Germany, 2012.
38. Ernst, T. *The History of q-Calculus and a New Method*; U.U.D.M. Report 2000; Department of Mathematics, Upsala University, Springer: Berlin, Germany, 2000; Volume 16.
39. Srivastava, G. Gauging ecliptic sentiment. In Proceedings of the 2018 41st International Conference on Telecommunications and Signal Processing (TSP), Athens, Greece, 4 July 2018; pp. 1–5.
40. Sene, N.; Srivastava, G. Generalized Mittag-Leffler Input Stability of the Fractional Differential Equations. *Symmetry* **2019**, *11*, 608. [CrossRef]
41. Mahmood, S.; Raza, N.; AbuJarad, E.S.; Srivastava, G.; Srivastava, H.M.; Malik, S.N. Geometric Properties of Certain Classes of Analytic Functions Associated with a q-Integral Operator. *Symmetry* **2019**, *11*, 719. [CrossRef]

© 2019 by the authors. Licensee MDPI, Basel, Switzerland. This article is an open access article distributed under the terms and conditions of the Creative Commons Attribution (CC BY) license (http://creativecommons.org/licenses/by/4.0/).

Article

A New Extension of the τ-Gauss Hypergeometric Function and Its Associated Properties

Hari Mohan Srivastava [1,2,3], Asifa Tassaddiq [4,*], Gauhar Rahman [5], Kottakkaran Sooppy Nisar [6] and Ilyas Khan [7]

1. Department of Mathematics and Statistics, University of Victoria, Victoria, BC V8W 3R4, Canada; harimsri@math.uvic.ca
2. Department of Medical Research, China Medical University Hospital, China Medical University, Taichung 40402, Taiwan
3. Department of Mathematics and Informatics, Azerbaijan University, 71 Jeyhun Hajibeyli Street, AZ1007 Baku, Azerbaijan
4. College of Computer and Information Sciences, Majmaah University, Al Majmaah 11952, Saudi Arabia
5. Department of Mathematics, Shaheed Benazir Bhutto University, Sheringal 18000, Upper Dir, Pakistan; gauhar55uom@gmail.com
6. Department of Mathematics, College of Arts and Science, Prince Sattam bin Abdulaziz University, Wadi Aldawaser 11991, Saudi Arabia; ksnisar1@gmail.com or n.sooppy@psau.edu.sa
7. Department of Mathematics, College of Science Al-Zulfi, Majmaah University, Al Majmaah 11952, Saudi Arabia; i.said@mu.edu.sa
* Correspondence: a.tassaddiq@mu.edu.sa

Received: 29 June 2019; Accepted: 16 October 2019; Published: 20 October 2019

Abstract: In this article, we define an extended version of the Pochhammer symbol and then introduce the corresponding extension of the τ-Gauss hypergeometric function. The basic properties of the extended τ-Gauss hypergeometric function, including integral and derivative formulas involving the Mellin transform and the operators of fractional calculus, are derived. We also consider some new and known results as consequences of our proposed extension of the τ-Gauss hypergeometric function.

Keywords: gamma function and its extension; Pochhammer symbol and its extensions; hypergeometric function and its extensions; τ-Gauss hypergeometric function and its extensions; τ-Kummer hypergeometric function; Fox-Wright function

MSC: 33B15; 33B20; 33C05; 33C20; 33C45; 33C60

---

1. Introduction

Throughout this article, we denote the sets of positive integers, negative integers, and complex numbers by $\mathbb{N}$, $\mathbb{Z}^-$, and $\mathbb{C}$, respectively. We also set

$$\mathbb{N}_0 = \mathbb{N} \cup \{0\} \quad \text{and} \quad \mathbb{Z}_0^- = \mathbb{Z}^- \cup \{0\}.$$

During the past few decades, various extensions and generalizations of well-known special functions have been studied by various researchers (see, for example, [1–6]). For example, a two-parameter extension of the gamma function $\Gamma(\xi)$ with the parameters p and v) was defined in [2] by

$$\Gamma_v(\xi;p) = \begin{cases} \sqrt{\dfrac{2p}{\pi}} \displaystyle\int_0^\infty t^{\xi-\frac{3}{2}} e^{-t} K_{v+\frac{1}{2}}\left(\dfrac{p}{t}\right) dt & \left(\min\{\Re(p),\Re(v)\} > 0;\ \xi \in \mathbb{C}\right), \\ \Gamma_p(\xi) & (v = 0;\ \Re(\xi) > 0), \end{cases} \quad (1)$$

where $K_v(z)$ is the modified Bessel function (or the Macdonald function) of order v and $\Gamma_p(\xi)$ was studied in [2,7]. Indeed, if we set $v = 0$ in (1) and make use of the following relationship:

$$K_{\frac{1}{2}}(z) = \sqrt{\frac{\pi}{2z}}\, e^{-z},$$

then this extended gamma function $\Gamma_p(\xi)$ is given by (see [2,7])

$$\Gamma_p(\xi) = \int_0^\infty t^{\xi-1}\, e^{-t-\frac{p}{t}}\, dt \qquad (\Re(p) > 0;\ \Re(\xi) > 0). \tag{2}$$

In the year 2012, Srivastava et al. [8] (see also [9]) defined the incomplete Pochhammar symbols in terms of the incomplete gamma functions. Another generalization of the Pochhammer symbol was defined in [10] by

$$(\xi; p)_\mu = \begin{cases} \dfrac{\Gamma_p(\xi+\mu)}{\Gamma(\xi)} & (\Re(p) > 0;\ \xi, \mu \in \mathbb{C}), \\[2mm] (\xi)_\mu & (p = 0;\ \xi, \mu \in \mathbb{C} \setminus \{0\}). \end{cases} \tag{3}$$

Here, in our present investigation, we first introduce a new extension $(\xi; p, v)_\mu$ of the Pochhammer symbol $(\xi; p)_\mu$ in (3), which is defined by

$$(\xi; p, v)_\mu = \begin{cases} \dfrac{\Gamma_v(\xi+\mu; p)}{\Gamma(\xi)} & (\min\{\Re(p), \Re(v)\} > 0;\ \xi, \mu \in \mathbb{C}), \\[2mm] (\xi; p)_\mu & (v = 0;\ \xi, \mu \in \mathbb{C} \setminus \{0\}), \end{cases} \tag{4}$$

where, as we mentioned above in connection with (3), the generalized Pochhmmer symbol $(\xi; p)_\mu$ was studied by Srivastava et al. [10]. The integral representation of the extended Pochhammer symbol $(\xi; p, v)_\mu$ is given by

$$(\xi; p, v)_\mu = \sqrt{\frac{2p}{\pi}}\, \frac{1}{\Gamma(\xi)} \int_0^\infty t^{\xi+\mu-\frac{3}{2}}\, e^{-t}\, K_{v+\frac{1}{2}}\left(\frac{p}{t}\right) dt, \tag{5}$$

which, in the special case when $v = 0$, yields the following result due to Srivastava et al. [10]):

$$(\xi; p, 0)_\mu = (\xi; p)_\mu = \frac{1}{\Gamma(\xi)} \int_0^\infty t^{\xi+\mu-1}\, e^{-t-\frac{p}{t}}\, dt \tag{6}$$

$$(\Re(p) > 0;\ \Re(\xi+\mu) > 0 \quad \text{when} \quad p = 0).$$

By using the definition (4), we now define an extension of the generalized hypergeometric function ${}_\mathfrak{p}F_\mathfrak{q}$ (with $\mathfrak{p}$ numerator parameters and $\mathfrak{q}$ denominator parameters) as follows:

$${}_\mathfrak{p}F_\mathfrak{q}\left[\begin{array}{c}(\rho_1; p, v), \rho_2, \cdots, \rho_\mathfrak{p}; \\ \sigma_1, \cdots, \sigma_\mathfrak{q};\end{array} z\right] = \sum_{n=0}^\infty \frac{(\rho_1; p, v)_n\, (\rho_2)_n \cdots (\rho_\mathfrak{p})_n}{(\sigma_1)_n \cdots (\sigma_\mathfrak{q})_n}\, \frac{z^n}{n!}, \tag{7}$$

where

$$\rho_j \in \mathbb{C}\ (j = 1, \cdots, \mathfrak{p}) \quad \text{and} \quad \sigma_j \in \mathbb{C} \setminus \mathbb{Z}_0^-\ (j = 1, \cdots, \mathfrak{q}).$$

Another interesting extension of the Pochhammer symbol and the associated hypergeometric functions was recently given by Srivastava et al. in [11].

We next recall that Virchenko et al. [12] studied the following τ-Gauss hypergeometric function $_2R_1^\tau(z)$ defined by (see also [13,14])

$$_2R_1^\tau(z) = {_2R_1}(\delta_1, \delta_2; \delta_3; \tau; z) = \frac{\Gamma(\delta_3)}{\Gamma(\delta_2)} \sum_{n=0}^{\infty} \frac{(\delta_1)_n \, \Gamma(\delta_2 + n\tau)}{\Gamma(\delta_3 + n\tau)} \frac{z^n}{n!} \qquad (8)$$

$$(\tau > 0; \, |z| < 1; \, \Re(\delta_3) > \Re(\delta_2) > 0 \text{ when } |z| = 1),$$

for which they derived an integral representation in the form

$$_2R_1(\delta_1, \delta_2; \delta_3; \tau; z) = \frac{1}{B(\delta_2; \delta_3 - \delta_2)} \int_0^\infty t^{\delta_2 - 1} (1-t)^{\delta_3 - \delta_2 - 1} (1 - zt^\tau)^{-\delta_1} dt \qquad (9)$$

$$(\tau > 0; \, |\arg(1-z)| < \pi; \, \Re(\delta_3) > \Re(\delta_2) > 0)$$

in terms of the classical beta function $B(\alpha, \beta)$ defined by

$$B(\alpha, \beta) := \begin{cases} \int_0^1 t^{\alpha - 1} (1-t)^{\beta = a} dt & (\min\{\Re(\alpha), \Re(\beta)\} > 0), \\ \dfrac{\Gamma(\alpha)\Gamma(\beta)}{\Gamma(\alpha + \beta)} & (\alpha, \beta \in \mathbb{C} \setminus \mathbb{Z}_0^-). \end{cases} \qquad (10)$$

Remark 1. *For $\tau = 1$, (8) and (9) would immediately yield the definition of the Gauss hypergeoemtric function $_2F_1(\delta_1, \delta_2; \delta_3; z)$ and its Eulerian integral representation (see, for details, [15]).*

Remark 2. *The so-called τ-Gauss hypergeometric function in (8) is, in fact, a rather specialized case of the widely-studied Fox-Wright extension $_p\Psi_q$ of the generalized hypergeometric function $_pF_q$ in (7) involving $\mathfrak{p}$ numerator and $\mathfrak{q}$ denominator parameters (see, for example, [16]).*

2. An Extension of the τ-Gauss Hypergeometric Function

In this section, we first introduce the following extension of the τ-Gauss hypergeometric function $_2R_1^\tau(z)$ in terms of the Pochhammer symbol $(\zeta; p, v)_\mu$ defined by (4) for $\delta_1, \delta_2 \in \mathbb{C}$ and $\delta_3 \in \mathbb{C} \setminus \mathbb{Z}_0^-$:

$$_2R_1\big[(\delta_1; p, v), \delta_2; \delta_3; \tau; z\big]$$
$$= \frac{\Gamma(\delta_3)}{\Gamma(\delta_2)} \sum_{n=0}^{\infty} \frac{(\delta_1; p, v)_n \, \Gamma(\delta_2 + n\tau)}{\Gamma(\delta_3 + n\tau)} \frac{z^n}{n!} \qquad (11)$$

$$(p \geqq 0; \, v > 0; \, \tau > 0; \, |z| < 1; \, \Re(\delta_3) > \Re(\delta_2) > 0 \text{ when } |z| = 1 \text{ and } p = 0).$$

Remark 3. *The following are some of the special cases of τ-Gauss hypergeometric functions defined by (11). (i) When $v = 0$, (11) reduces to the following extended τ-Gauss hypergeometric function (see [17]):*

$$_2R_1\big[(\delta_1; p), \delta_2; \delta_3; \tau; z\big] = \frac{\Gamma(\delta_3)}{\Gamma(\delta_2)} \sum_{n=0}^{\infty} \frac{(\delta_1; p)_n \, \Gamma(\delta_2 + n\tau)}{\Gamma(\delta_3 + n\tau)} \frac{z^n}{n!} \qquad (12)$$

$$(p \geqq 0; \, \tau > 0; \, |z| < 1; \, \Re(\delta_3) > \Re(\delta_2) > 0 \text{ when } |z| = 1 \text{ and } p = 0).$$

(ii) When $\tau = 1$, (11) will yield the following extended Gauss hypergeometric function:

$$_2F_1\big[(\delta_1; p, v); \delta_2; \delta_3; z\big] = \sum_{n=0}^{\infty} \frac{(\delta_1; p, v)_n \, (\delta_2)_n}{(\delta_3)_n} \frac{z^n}{n!}.$$

(iii) When $v = 0$ and $\tau = 1$, (11) will reduce to the following extended Gauss hypergeometric function (see [10]):

$$_2F_1\left[(\delta_1;p);\delta_2;\delta_3;z\right] = \sum_{n=0}^{\infty} \frac{(\delta_1;p)_n \, (\delta_2)_n}{(\delta_3)_n} \frac{z^n}{n!}.$$

3. Integral Representations and Derivative Formulas

In this section, we obtain the Eulerian and Laplace-type integral representations and some derivative formulas of the extended τ-Gauss hypergeometric function defined by (11).

Theorem 1. *The following Eulerian representation holds true for* (11):

$$_2R_1\left[(\delta_1;p,v),\delta_2;\delta_3;\tau;z\right]$$
$$= \frac{1}{B(\delta_2,\delta_3-\delta_2)} \int_0^1 t^{\delta_2-1}(1-t)^{\delta_3-\delta_2-1} \, _1F_0[(\delta_1;p,v);\underline{\quad};zt^\tau] \, dt \tag{13}$$

$\left(\Re(p) > 0; \, v > 0; \, \tau > 0; \, |z| < 1; \, \Re(\delta_3) > \Re(\delta_2) > 0\right)$,

where $B(\alpha,\beta)$ denotes the classical beta function defined by (10).

Proof. Using the following well-known identity involving the beta function $B(\alpha,\beta)$:

$$\frac{(\delta_2)_{\tau n}}{(\delta_3)_{\tau n}} = \frac{B(\delta_2+\tau n, \delta_3-\delta_2)}{B(\delta_2;\delta_3-\delta_2)} = \frac{1}{B(\delta_2,\delta_3-\delta_2)} \int_0^1 t^{\delta_2+\tau n-1}(1-t)^{\delta_3-\delta_2-1} \, dt$$

$$\left(\Re(\delta_3) > \Re(\delta_2) > 0\right)$$

in (11) and using the definition (7), we get the desired assertion (13) of Theorem 1. □

Theorem 2. *The following Laplace-type representation holds true for* (11):

$$_2R_1\left[(\delta_1;p,v),\delta_2;\delta_3;\tau;z\right] = \frac{\sqrt{\frac{2p}{\pi}}}{\Gamma(\delta_1)} \int_0^\infty t^{\delta_1-\frac{3}{2}} e^{-t} K_{v+\frac{1}{2}}\left(\frac{p}{t}\right) \, _1\Phi_1^\tau[\delta_2;\delta_3;zt] \, dt \tag{14}$$

$\left(\Re(p) > 0; \, v > 0; \, \tau > 0; \, \Re(z) < 1; \, \Re(\delta_1) > 0\right)$,

where $_1\Phi_1^\tau[\delta_2;\delta_3;zt]$ is the τ-Kummer hypergeometric function defined by

$$_1\Phi_1^\tau(z) = \, _1\Phi_1^\tau[\delta_2;\delta_3;zt] = \frac{\Gamma(\delta_3)}{\Gamma(\delta_2)} \sum_{n=0}^\infty \frac{\Gamma(\delta_2+\tau n)}{\Gamma(\delta_3+\tau n)} \frac{z^n}{n!} \tag{15}$$

$(\tau > 0; \, \delta_2 \in \mathbb{C}; \, \delta_3 \in \mathbb{C} \setminus \mathbb{Z}_0^-)$.

Proof. By first utilizing (5) in (11) and then applying (15), we obtain the assertion (14) of Theorem 2. □

Remark 4. *When $\tau = 1$,* (13) *and* (14) *yield the following special cases:*

$$_2F_1\left[(\delta_1;p,v);\delta_2;\delta_3;z\right]$$
$$= \frac{1}{B(\delta_2,\delta_3-\delta_2)} \int_0^1 t^{\delta_2-1}(1-t)^{\delta_3-\delta_2-1} \, _1F_0[(\delta_1;p,v);\underline{\quad};zt] \, dt \tag{16}$$

and

$$2F_1[(\delta_1; p, v); \delta_2; \delta_3; z] = \frac{\sqrt{\frac{2p}{\pi}}}{\Gamma(\delta_1)} \int_0^\infty t^{\delta_1 - \frac{3}{2}} e^{-t} K_{v+\frac{1}{2}}\left(\frac{p}{t}\right) {}_1F_1[\delta_2; \delta_3; zt] \, dt, \qquad (17)$$

respectively. Similarly, when $v = 0$, our integral representations (13) and (14) reduce to the following known results (see [17]):

$$2R_1[(\delta_1; p), \delta_2; \delta_3; \tau; z]$$
$$= \frac{1}{B(\delta_2, \delta_3 - \delta_2)} \int_0^1 t^{\delta_2 - 1} (1-t)^{\delta_3 - \delta_2 - 1} {}_1F_0[(\delta_1; p); \text{---}; zt^\tau] \, dt$$

and

$$2R_1[(\delta_1; p), \delta_2; \delta_3; \tau; z] = \frac{1}{\Gamma(\delta_1)} \int_0^\infty t^{\delta_1 - 1} e^{-t - \frac{p}{t}} {}_1\Phi_1^\tau[\delta_2; \delta_3; zt] \, dt,$$

respectively. Moreover, when $\tau = 1$ and $v = 0$, (13) and (14) yield the following known results (see [10]):

$$2F_1[(\delta_1; p), \delta_2; \delta_3; z]$$
$$= \frac{1}{B(\delta_2, \delta_3 - \delta_2)} \int_0^1 t^{\delta_2 - 1} (1-t)^{\delta_3 - \delta_2 - 1} {}_1F_0[(\delta_1; p); \text{---}; zt] \, dt$$

and

$$2F_1[(\delta_1; p, v), \delta_2; \delta_3; z] = \frac{1}{\Gamma(\delta_1)} \int_0^\infty t^{\delta_1 - 1} e^{-t - \frac{p}{t}} {}_1F_1[\delta_2; \delta_3; zt] \, dt,$$

respectively.

Theorem 3. *Each of the following derivative formulas holds true for the extended τ-Gauss hypergeometric function defined by* (11):

$$\frac{d^n}{dz^n}\left\{ {}_2R_1[(\delta_1; p, v), \delta_2; \delta_3; \tau; z]\right\}$$
$$= \frac{(\delta_1)_n \, \Gamma(\delta_2 + n\tau)\Gamma(\delta_3)}{\Gamma(\delta_3 + n\tau)\Gamma(\delta_2)} \, {}_2R_1\left[(\delta_1 + n; p, v), \delta_2 + n\tau; \delta_3 + n\tau; \tau; z\right] \qquad (18)$$

and

$$\frac{d^n}{dz^n}\left\{ z^{\delta_3 - 1} \, {}_2R_1[(\delta_1; p, v), \delta_2; \delta_3; \tau; \omega z^\tau]\right\}$$
$$= \frac{z^{\delta_3 - n - 1}\Gamma(\delta_3)}{\Gamma(\delta_3 - n)} \, {}_2R_1\left[(\delta_1; p, v), \delta_2; \delta_3 - n; \tau; \omega z^\tau\right]. \qquad (19)$$

Proof. Upon differentiating both sides of (11) with respect to z, we get

$$\frac{d}{dz}\left\{ {}_2R_1[(\delta_1; p, v), \delta_2; \delta_3; \tau; z]\right\} = \frac{\Gamma(\delta_3)}{\Gamma(\delta_2)} \sum_{n=1}^\infty \frac{(\delta_1; p, v)_n \, \Gamma(\delta_2 + n\tau)}{\Gamma(\delta_3 + n\tau)} \frac{z^{n-1}}{(n-1)!}. \qquad (20)$$

Replacing n by $n+1$ in (20), we have

$$\frac{d}{dz}\left\{ {}_2R_1[(\delta_1; p, v), \delta_2; \delta_3; \tau; z]\right\} = \frac{\Gamma(\delta_3)}{\Gamma(\delta_2)} \sum_{n=0}^\infty \frac{(\delta_1; p, v)_{n+1} \, \Gamma(\delta_2 + (n+1)\tau)}{\Gamma(\delta_3 + (n+1)\tau)} \frac{z^n}{n!},$$

which, after simplification, yields

$$\frac{d}{dz}\left\{ {}_2R_1\big[(\delta_1;p,v),\delta_2;\delta_3;\tau;z\big]\right\}$$
$$= \frac{\delta_1 \Gamma(\delta_3)\Gamma(\delta_2+\tau)}{\Gamma(\delta_3+\tau)\Gamma(\delta_2)}\sum_{n=0}^{\infty}\frac{(\delta_1+1;p,v)_n\,\Gamma(\delta_2+\tau+n\tau)}{\Gamma(\delta_3+\tau+n\tau)}\frac{z^n}{n!}$$
$$= \frac{\delta_1\,\Gamma(\delta_3)\Gamma(\delta_2+\tau)}{\Gamma(\delta_3+\tau)\Gamma(\delta_2)}\,{}_2R_1[(\delta_1+1;p,v),\delta_2+\tau;\delta_3+\tau;\tau;z].$$

By iterating this differentiation process n times, we are led to the desired assertion (18) of Theorem 3.

Similarly, in order to prove the assertion (19) of Theorem 3, we observe that

$$\frac{d^n}{dz^n}\left\{z^{\delta_3-1}\,{}_2R_1\big[(\delta_1;p,v),\delta_2;\delta_3;\tau;\omega z^\tau\big]\right\}$$
$$= \frac{\Gamma(\delta_3)}{\Gamma(\delta_2)}\sum_{m=0}^{\infty}\frac{(\delta_1;p,v)_m\,\Gamma(\delta_2+m\tau)}{\Gamma(\delta_3+m\tau)}\frac{\omega^m}{m!}\frac{d^n}{dz^n}\left\{z^{\delta_3+\tau m-1}\right\}$$
$$= \frac{\Gamma(\delta_3)}{\Gamma(\delta_2)}\sum_{m=0}^{\infty}\frac{(\delta_1;p,v)_m\,\Gamma(\delta_2+m\tau)}{\Gamma(\delta_3+m\tau)}\frac{\omega^m}{m!}$$
$$\cdot \Big[(\delta_3+\tau m-1)(\delta_3+\tau m-2)\cdots(\delta_3+\tau m-n-1)\Big]z^{\delta_3+\tau m-n-1}$$
$$= \frac{z^{\delta_3-n-1}\,\Gamma(\delta_3)}{\Gamma(\delta_2)}\sum_{m=0}^{\infty}\frac{(\delta_1;p,v)_m\,\Gamma(\delta_2+m\tau)}{\Gamma(\delta_3+m\tau)}\frac{(\omega z^\tau)^m}{m!}\frac{\Gamma(\delta_3+m\tau)}{\Gamma(\delta_3+\tau m-n)}$$
$$= \frac{z^{\delta_3-n-1}\Gamma(\delta_3)\Gamma(\delta_3-n)}{\Gamma(\delta_3-n)\Gamma(\delta_2)}\sum_{m=0}^{\infty}\frac{(\delta_1;p,v)_m\,\Gamma(\delta_2+m\tau)}{\Gamma(\delta_3+\tau m-n)}\frac{(\omega z^\tau)^m}{m!},$$

which, in view of (11), gives the derivative formula (19) asserted by Theorem 3. □

4. Application of the Mellin Transform

The well-known Mellin transform of a given integrable function $f(t)$ is defined by

$$\mathfrak{M}\{f(t):t\to s\}=\int_0^\infty t^{s-1}\,f(t)\,dt, \tag{21}$$

provided that the improper integral in (21) exists.

Theorem 4. *The Mellin transform of the extended τ-Gauss hypergeometric function,*

$$_2R_1\big[(\delta_1;p,v),\delta_2;\delta_3;\tau;z\big],$$

is given by

$$\mathfrak{M}\Big\{{}_2R_1\big[(\delta_1;p,v),\delta_2;\delta_3;\tau;z\big]:p\to s\Big\}$$
$$= \frac{2^{s-1}}{\sqrt{\pi}}(\delta_1)_s\,\Gamma\left(\frac{s-v}{2}\right)\Gamma\left(\frac{s+v+1}{2}\right)\,{}_2R_1(\delta_1+s;\delta_2,\delta_3;\tau;z) \tag{22}$$

$$\big(\Re(s-v)>0;\ \Re(\delta_1+s)>-1\big).$$

Proof. Applying the definition (21) of the Mellin transform on both sides of (11), we get

$$\mathfrak{M}\Big\{{}_2R_1\big[(\delta_1;p,v),\delta_2;\delta_3;\tau;z\big]:p\to s\Big\}$$

$$= \int_0^\infty p^{s-1} \left(\frac{\Gamma(\delta_3)}{\Gamma(\delta_2)} \sum_{n=0}^\infty \frac{(\delta_1; p, v)_n \Gamma(\delta_2 + n\tau)}{\Gamma(\delta_3 + n\tau)} \frac{z^n}{n!} \right) dp$$

$$= \frac{\Gamma(\delta_3)}{\Gamma(\delta_2)} \sum_{n=0}^\infty \frac{\Gamma(\delta_2 + n\tau)}{\Gamma(\delta_3 + n\tau)} \frac{z^n}{n!} \frac{1}{\Gamma(\delta_1)} \int_0^\infty p^{s-1} \Gamma_v(\delta_1 + n; p)\, dp. \tag{23}$$

Using the following result given by Chaudhry and Zubair ([2], Eq. 4.105),

$$\int_0^\infty p^{s-1} \Gamma_v(\delta_1 + n; p) dp = \frac{2^{s-1}}{\sqrt{\pi}} \Gamma\left(\frac{s-v}{2}\right) \Gamma\left(\frac{s+v+1}{2}\right) \Gamma(\delta_1 + n + s), \tag{24}$$

in (23), we find that

$$\mathfrak{M}\left\{ {}_2R_1\left[(\delta_1; p, v), \delta_2; \delta_3; \tau; z\right] : p \to s \right\}$$

$$= \frac{2^{s-1}}{\sqrt{\pi}} \Gamma\left(\frac{s-v}{2}\right) \Gamma\left(\frac{s+v+1}{2}\right) \frac{\Gamma(\delta_3)\Gamma(\delta_1 + s)}{\Gamma(\delta_1 + s)\Gamma(\delta_1)\Gamma(\delta_2)}$$

$$\cdot \sum_{n=0}^\infty \frac{\Gamma(\delta_1 + n + s)\Gamma(\delta_2 + n\tau)}{\Gamma(\delta_3 + n\tau)} \frac{z^n}{n!}$$

$$= \frac{2^{s-1}}{\sqrt{\pi}} (\delta_1)_s \Gamma\left(\frac{s-v}{2}\right) \Gamma\left(\frac{s+v+1}{2}\right) \frac{\Gamma(\delta_3)}{\Gamma(\delta_2)}$$

$$\cdot \sum_{n=0}^\infty \frac{(\delta_1 + s)_n \Gamma(\delta_2 + n\tau)}{\Gamma(\delta_3 + n\tau)} \frac{z^n}{n!}, \tag{25}$$

which, in view of (8), yields the Mellin transform formula (22) asserted by Theorem 4. □

5. Use of the Operators of Fractional Calculus

In this section, we recall the operators $\mathfrak{I}_{\rho+}$ and $\mathfrak{D}_{\rho+}$ of the fractional integral and fractional derivatives of order $\mu \in \mathbb{C}$ ($\Re(\mu) > 0$), which are defined by (see [18,19])

$$\left(\mathfrak{I}_{\rho+}^\mu f\right)(x) = \frac{1}{\Gamma(\mu)} \int_0^x \frac{f(t)}{(x-t)^{1-\mu}} dt \qquad (\mu \in \mathbb{C}; \Re(\mu) > 0) \tag{26}$$

and

$$\left(\mathfrak{D}_{\rho+}^\mu f\right)(x) = \frac{d^n}{dx^n} \left\{ \left(\mathfrak{I}_{\rho+}^{n-\mu} f\right)(x) \right\} \qquad (\mu \in \mathbb{C}; \Re(\mu) > 0;\ n = [\Re(\mu)] + 1), \tag{27}$$

respectively.

We now prove the following fractional integral and fractional derivative formulas associated with the extended τ-Gauss hypergeometric function:

$${}_2R_1\left[(\delta_1; p, v), \delta_2; \delta_3; \tau; z\right].$$

Theorem 5. *Let $\rho \in \mathbb{R}_+ = [0, \infty)$, $\delta_1, \delta_2, \delta_3, \omega \in \mathbb{C}$, and $\min\{\Re(mu), \Re(\delta_3), \Re(\tau) > 0\}$. Then the following formulas hold true for $x > \rho$:*

$$\left(\mathfrak{I}_{\rho+}^\mu \left[(t-\rho)^{\delta_3-1} {}_2R_1\left[(\delta_1; p, v), \delta_2; \delta_3; \tau; \omega(t-\rho)^\tau\right]\right]\right)(x)$$

$$= \frac{(x-\rho)^{\delta_3+\mu-1} \Gamma(\delta_3)}{\Gamma(\delta_3 + \mu)} {}_2R_1\left[(\delta_1; p, v), \delta_2; \delta_3 + \mu; \tau; \omega(x-\rho)^\tau\right] \tag{28}$$

and

$$\left(\mathfrak{D}_{\rho+}^{\mu}\left[(t-\rho)^{\delta_3-1}\ {}_2R_1\left[(\delta_1;p,v),\delta_2;\delta_3;\tau;\omega(t-\rho)^{\tau}\right]\right]\right)(x)$$
$$=\frac{(x-\rho)^{\delta_3-\mu-1}\,\Gamma(\delta_3)}{\Gamma(\delta_3-\mu)}\ {}_2R_1\left[(\delta_1;p,v),\delta_2;\delta_3-\mu;\tau;\omega(x-\rho)^{\tau}\right]. \tag{29}$$

Proof. Using the following well-known relation (see [18,19]),

$$\left(\mathfrak{I}_{\rho+}^{\mu}\left[(t-\rho)^{\delta_3-1}\right]\right)(x) = \frac{\Gamma(\delta_3)}{\Gamma(\delta_3+\mu)}(x-\rho)^{\delta_3+\mu-1} \qquad (x>\rho), \tag{30}$$

we have

$$\left(\mathfrak{I}_{\rho+}^{\mu}\left[(t-\rho)^{\delta_3-1}\ {}_2R_1\left[(\delta_1;p,v),\delta_2;\delta_3;\tau;\omega(t-\rho)^{\tau}\right]\right]\right)(x)$$
$$=\left(\mathfrak{I}_{\rho+}^{\mu}\left[\frac{\Gamma(\delta_3)}{\Gamma(\delta_2)}\sum_{n=0}^{\infty}\frac{(\delta_1;p,v)_n\,\Gamma(\delta_2+n\tau)}{\Gamma(\delta_3+n\tau)}\frac{\omega^n}{n!}(t-\rho)^{\delta_3+\tau n-1}\right]\right)$$
$$=\frac{(x-\rho)^{\delta_3+\mu-1}\Gamma(\delta_3)}{\Gamma(\delta_3+\mu)}\ {}_2R_1\left[(\delta_1;p,v),\delta_2;\delta_3+\mu;\tau;\omega(x-\rho)^{\tau}\right],$$

which proves the assertion (28) of Theorem 5.

Next, in view of (27) and (11), we have

$$\left(\mathfrak{D}_{\rho+}^{\mu}\left[(t-\rho)^{\delta_3-1}\ {}_2R_1\left[(\delta_1;p,v),\delta_2;\delta_3;\tau;\omega(t-\rho)^{\tau}\right]\right]\right)(x)$$
$$=\frac{d^n}{dx^n}\left\{\left(\mathfrak{I}_{\rho+}^{n-\mu}\left[(t-\rho)^{\delta_3-1}\ {}_2R_1\left[(\delta_1;p,v),\delta_2;\delta_3;\tau;\omega(t-\rho)^{\tau}\right]\right]\right)(x)\right\}$$
$$=\frac{d^n}{dx^n}\left\{\frac{(x-\rho)^{\delta_3+n-\mu-1}\,\Gamma(\delta_3)}{\Gamma(\delta_3-\mu+n)}\ {}_2R_1\left[(\delta_1;p,v),\delta_2;\delta_3+n-\mu;\tau;\omega(x-\rho)^{\tau}\right]\right\}. \tag{31}$$

Finally, by applying (19) to the equation (31), we are led to the assertion (29) of Theorem 5. □

6. Concluding Remarks

In our present investigation, we have first introduced an extension of the τ-Gauss hypergeometric function in terms of a certain extended Pochhammer symbol. We have then derived its various properties, including (for example) integral representations, derivative formulas, Mellin transform formulas, as well as the fractional integral and fractional derivative formulas. We have observed that by letting $v=0$, the various results derived in this paper will reduce to the corresponding results proved earlier in [17]. Moreover, if we set $\tau=1$, then we get several interesting new or known formulas for the extended Gauss hypergeometric function. Finally, we have observed that, if $v=0$ and $\tau=1$, then we get some new or known results for the extended Gauss hypergeometric function defined and studied by Srivastava et al. [10].

Author Contributions: H.M.S., G.R. and K.S.N.; methodology: H.M.S. and A.T.; formal analysis: A.T. and I.K.; writing—original draft preparation: H.M.S., G.R. and K.S.N.;writing and review and editing: A.T. and I.K.; funding acquisition: A.T. All authors contributed equally.

Funding: This research was funded by Deanship of Scientific Research at Majmaah University grant number (RGP-2019-28).

Acknowledgments: The authors extend their appreciation to the Deanship of Scientific Research at Majmaah University for funding this work under Project Number (RGP-2019-28).

Conflicts of Interest: The authors declare no conflict of interest.

References

1. Arshad, M.; Mubeen, S.; Nisar, K.S.; Rahman, G. Extended Wright-Bessel function and its properties. *Commun. Korean Math. Soc.* **2018**, *33*, 143–155.
2. Chaudhry, M.A.; Zubair, S.M. *On a Class of Incomplete Gamma Functions with Applications*; Chapman and Hall (CRC Press Company): Boca Raton, NJ, USA; London, UK; New York, NY, USA; Washington, DC, USA, 2001.
3. Chaudhry, M.A.; Qadir, A.; Srivastava, H.M.; Paris, R.B. Extended hypergeometric and confluent hypergeometric functions. *Appl. Math. Comput.* **2004**, *159*, 589–602. [CrossRef]
4. Mubeen, S.; Rahman, G.; Nisar, K.S.; Choi, J.; Arshad, M. An extended beta function and its properties. *Far East J. Math. Sci.* **2017**, *102*, 1545–1557. [CrossRef]
5. Özarslan, M.A.; Yilmaz, B. The extended Mittag-Leffler function and its properties. *J. Inequal. Appl.* **2014**, *2014*, 85. [CrossRef]
6. Özergin, E.; Özarslan, M.A.; Altın, A. Extension of gamma, beta and hypergeometric functions. *J. Comput. Appl. Math.* **2011**, *235*, 4601–4610. [CrossRef]
7. Chaudhry, M.A.; Zubair, S.M. Generalized incomplete gamma functions with applications. *J. Comput. Appl. Math.* **1994**, *55*, 99–124. [CrossRef]
8. Srivastava, H.M.; Chaudhry, M.A.; Agarwal, R.P. The incomplete Pochhammer symbols and their applications to hypergeometric and related functions. *Integral Transforms Spec. Funct.* **2012**, *23*, 659–683. [CrossRef]
9. Srivastava, R. Some classes of generating functions associated with a certain family of extended and generalized hypergeometric functions. *Appl. Math. Comput.* **2014**, *243*, 132–137. [CrossRef]
10. Srivastava, H.M.; Çetinkaya, A.; Kıymaz, İ.O. A certain generalized Pochhammer symbol and its applications to hypergeometric functions. *Appl. Math. Comput.* **2014**, *226*, 484–491. [CrossRef]
11. Srivastava, H.M.; Rahman, G.; Nisar, K.S. Some Extensions of the Pochhammer Symbol and the Associated Hypergeometric Functions. *Iran. J. Sci. Tech. Trans. A Sci.* **2019**, *43*, 2601–2606. [CrossRef]
12. Virchenko, N. On some generalizations of the functions of hypergeometric type. *Fract. Calc. Appl. Anal.* **1999**, *2*, 233–244.
13. Galué, L.; Al-Zamel, A.; Kalla, S.L. Further results on generalized hypergeometric functions. *Appl. Math. Comput.* **2013**, *136*, 17–25. [CrossRef]
14. Virchenko, N.; Kalla, S.L.; Al-Zamel, A. Some results on a generalized hypergeometric function. *Integral Transforms Spec. Funct.* **2001**, *12*, 89–100. [CrossRef]
15. Rainville, E.D. *Special Functions*; Macmillan Company: New York, NY, USA, 1960; Reprinted by Chelsea Publishing Company: New York, NY, USA, 1971.
16. Srivastava, H.M.; Karlsson, P.W. *Multiple Gaussian Hypergeometric Series*; Halsted Press (Ellis Horwood Limited): Chichester, UK; John Wiley and Sons: New York, NY, USA; Chichester, UK; Brisbane, Australia; Toronto, ON, Canada, 1985.
17. Parmar, R.K. Extended τ-hypergeometric functions and associated properties. *C. R. Math. Acad. Sci. Paris Sér. I* **2015**, *353*, 421–426. [CrossRef]
18. Kilbas, A.A.; Srivastava, H.M.; Trujillo, J.J. *Theory and Applications of Fractional Differential Equations*; North-Holland Mathematical Studies; Elsevier (North-Holland) Science Publishers: Amsterdam, The Netherlands; London, UK; New York, NY, USA, 2006; Volume 204.
19. Samko, S.G.; Kilbas, A.A.; Marichev, O.I. *Fractional Integrals and Derivatives: Theory and Applications*; Translated from the Russian: Integrals and Derivatives of Fractional Order and Some of Their Applications; ("Nauka i Tekhnika", Minsk, 1987); Gordon and Breach Science Publishers: Reading, UK, 1993.

© 2019 by the authors. Licensee MDPI, Basel, Switzerland. This article is an open access article distributed under the terms and conditions of the Creative Commons Attribution (CC BY) license (http://creativecommons.org/licenses/by/4.0/).

Article

The Second Hankel Determinant Problem for a Class of Bi-Close-to-Convex Functions

Nak Eun Cho [1], Ebrahim Analouei Adegani [2,*], Serap Bulut [3] and Ahmad Motamednezhad [2]

1. Department of Applied Mathematics, College of Natural Sciences, Pukyong National University, Busan 608-737, Korea; necho@pknu.ac.kr
2. Faculty of Mathematical Sciences, Shahrood University of Technology, P.O. Box 316-36155 Shahrood, Iran; a.motamedne@gmail.com
3. Faculty of Aviation and Space Sciences, Kocaeli University, Arslanbey Campus, 41285 Kartepe-Kocaeli, Turkey; serap.bulut@kocaeli.edu.tr
* Correspondence: analoey.ebrahim@gmail.com

Received: 21 September 2019; Accepted: 11 October 2019; Published: 17 October 2019

Abstract: The purpose of the present work is to determine a bound for the functional $H_2(2) = a_2 a_4 - a_3^2$ for functions belonging to the class $\mathcal{C}_\Sigma$ of bi-close-to-convex functions. The main result presented here provides much improved estimation compared with the previous result by means of different proof methods than those used by others.

Keywords: univalent function; second Hankel determinant; subordination; close-to-convex functions; bi-close-to-convex functions

MSC: Primary 30C45; Secondary 30C50

1. Introduction

Let $\mathcal{A}$ be a class of analytic functions in the open unit disk $\mathbb{D} = \{z \in \mathbb{C} : |z| < 1\}$, of the form

$$f(z) = z + \sum_{n=2}^{\infty} a_n z^n \qquad (z \in \mathbb{D}). \tag{1}$$

Let $\mathcal{S}$ be the class of functions $f \in \mathcal{A}$ which are univalent in $\mathbb{D}$. A function $f \in \mathcal{A}$ is said to be *starlike*, if it satisfies the inequality

$$\operatorname{Re}\left(\frac{zf'(z)}{f(z)}\right) > 0 \qquad (z \in \mathbb{D}). \tag{2}$$

We denote by $\mathcal{S}^*$ the class which consists of all functions $f \in \mathcal{A}$ that are starlike. A function $f \in \mathcal{A}$ is said to be *close-to-convex* if there exits a function $g \in \mathcal{S}^*$ such that it satisfies the inequality

$$\operatorname{Re}\left(\frac{zf'(z)}{g(z)}\right) > 0 \qquad (z \in \mathbb{D}). \tag{3}$$

We denote by $\mathcal{C}$ the class which consists of all functions $f \in \mathcal{A}$ that are close-to-convex. We note that $\mathcal{S}^* \subset \mathcal{C} \subset \mathcal{S}$ and that $|a_n| \leq n$ for $f \in \mathcal{S}^*$.

For two functions f and g which are analytic in $\mathbb{D}$, we say that the function f is *subordinate* to g, and write $f(z) \prec g(z)$, if there exists a *Schwarz function* w, that is a function w analytic in $\mathbb{D}$ with $w(0) = 0$ and $|w(z)| < 1$ in $\mathbb{D}$, such that $f(z) = g(w(z))$ for all $z \in \mathbb{D}$. In particular, if the function g is univalent in $\mathbb{D}$, then $f \prec g$ if and only if $f(0) = g(0)$ and $f(\mathbb{D}) \subseteq g(\mathbb{D})$ [1].

In 1976, Noonan and Thomas [2] defined the *q-th Hankel determinant* for integers $n \geq 1$ and $q \geq 1$ by

$$H_q(n) = \begin{vmatrix} a_n & a_{n+1} & \cdots & a_{n+q-1} \\ a_{n+1} & a_{n+2} & \cdots & a_{n+q} \\ \vdots & \vdots & \vdots & \vdots \\ a_{n+q-1} & a_{n+q} & \cdots & a_{n+2q-2} \end{vmatrix} \quad (a_1 = 1).$$

In general, one of the important tools in the theory of univalent functions is the *Hankel determinant*. It is used, for example, in showing that a function of bounded characteristic in $\mathbb{D}$, that is, a function which is a ratio of two bounded analytic functions with its Laurent series around the origin having integral coefficients, is rational [3]. For the use of Hankel determinant in the study of meromorphic functions, see [4]. For detailed information, the readers are encouraged [5,6]. Various properties of these determinants can be found in [7] (Chapter 4). The investigations of Hankel determinants for different classes of analytic functions started in the 1960s. Pommerenke [8] proved that the Hankel determinants of univalent functions satisfy $|H_q(n)| \leq K n^{-(\frac{1}{2}+\beta)q+\frac{3}{2}}$ where $n, q \in \mathbb{N}$, $q \geq 2$, $\beta > 1/4000$ and K depends only on q. Later, Hayman [9] proved that $|H_q(n)| \leq A n^{\frac{1}{2}}$ where $n \in \mathbb{N}$ and A is an absolute constant for areally mean univalent functions. Pommerenke [10] investigated the Hankel determinant of areally mean p-valent functions, univalent functions as well as of starlike functions. For results related to these determinants, see also [11,12].

Note that

$$H_2(1) = \begin{vmatrix} a_1 & a_2 \\ a_2 & a_3 \end{vmatrix} \quad \text{and} \quad H_2(2) = \begin{vmatrix} a_2 & a_3 \\ a_3 & a_4 \end{vmatrix},$$

where the Hankel determinants $H_2(1) = a_3 - a_2^2$ and $H_2(2) = a_2 a_4 - a_3^2$ are well-known as *Fekete-Szegö* and *second Hankel determinant* functionals, respectively. Further, Fekete and Szegö [13] introduced the generalized functional $a_3 - \lambda a_2^2$, where λ is some real number. In recent years, the research on Hankel determinants has focused on the estimation of $|H_2(2)|$. Problems in this field has also been argued by several authors for various classes of univalent functions [14–24].

The Koebe one-quarter theorem [1] ensures that the image of $\mathbb{D}$ under every univalent function $f \in \mathcal{S}$ contains a disk of radius $1/4$. Thus every function $f \in \mathcal{S}$ has an inverse f^{-1}, such that

$$f^{-1}(f(z)) = z \quad (z \in \mathbb{D}), \quad \text{and} \quad f\left(f^{-1}(w)\right) = w \quad \left(|w| < r_0(f); r_0(f) \geq \frac{1}{4}\right),$$

where the inverse f^{-1} has the power series expansion (see [25])

$$f^{-1}(w) = w - a_2 w^2 + \left(2a_2^2 - a_3\right) w^3 - \left(5a_2^3 - 5a_2 a_3 + a_4\right) w^4 + \cdots. \tag{4}$$

A function $f \in \mathcal{A}$ is said to be *bi-univalent* in $\mathbb{D}$ if both f and f^{-1} are univalent in $\mathbb{D}$, in the sense that f^{-1} has a univalent analytic continuation to $\mathbb{D}$. Let Σ denote the class of bi-univalent functions in $\mathbb{D}$. For a brief history of functions in the class Σ and also other different characteristics of these functions and the coefficient problems, see [25–32] and the references therein.

In 2014, Hamidi and Jahangiri [33] defined the class of bi-close-to-convex functions of order α ($0 \leq \alpha < 1$) that this class is denoted by $\mathcal{C}_\Sigma(\alpha)$ and in particular, $\mathcal{C}_\Sigma(0) = \mathcal{C}_\Sigma$.

Definition 1. *A function $f \in \Sigma$ is in the class of bi-close-to-convex functions of order α if the following conditions are satisfied:*

$$\operatorname{Re}\left(\frac{z f'(z)}{g(z)}\right) > \alpha \quad (z \in \mathbb{D}) \tag{5}$$

and

$$\operatorname{Re}\left(\frac{w F'(w)}{G(w)}\right) > \alpha \quad (w \in \mathbb{D}), \tag{6}$$

where the function $F(w) = f^{-1}(w)$ is defined by (4), $g(z) = z + \sum\limits_{n=2}^{\infty} b_n z^n \in \mathcal{S}^*$ and

$$G(w) = w + \sum_{n=2}^{\infty} B_n z^n \in \mathcal{S}^*. \tag{7}$$

Recently, Güney et al. [34] obtained the bound for the second Hankel determinant $H_2(2)$ for the class $\mathcal{C}_\Sigma$ of bi-close-to-convex functions as follows:

Theorem 1. *Let the function f given by (1) be in the class $\mathcal{C}_\Sigma$ and $G(w) = g^{-1}(w)$. Then*

$$|H_2(2)| := \left| a_2 a_4 - a_3^2 \right| \leq \frac{353}{36}.$$

Remark 1. *By means of the subordination, the conditions (5) and (6) are, respectively, equivalent to*

$$\frac{zf'(z)}{g(z)} \prec \frac{1+z}{1-z} \quad \text{and} \quad \frac{wF'(z)}{G(w)} \prec \frac{1+w}{1-w}.$$

The main purpose of this paper is to determine bounds for the functional $H_2(2) = a_2 a_4 - a_3^2$ for functions belonging to the subclass $\mathcal{C}_\Sigma$ of bi-close-to-convex functions, which is a much improved estimation than the previous result given by Güney et al. [34]. We note that our proof method is by means of the subordination and more direct than those used by others and so we get a smaller upper bound and more accurate estimation for the functional $|H_2(2)|$ for functions in the class $\mathcal{C}_\Sigma$.

2. Main Results

Theorem 2. *Let the function f given by (1) be in the class $\mathcal{C}_\Sigma$ and $G(w) = g^{-1}(w)$. Then*

$$|H_2(2)| := \left| a_2 a_4 - a_3^2 \right| \leq \frac{227}{36}.$$

In order to prove our main result, we need the following lemmas.

Lemma 1. [1] *(p. 190) Let u be analytic function in the unit disk $\mathbb{D}$, with $u(0) = 0$, and $|u(z)| < 1$ for all $z \in \mathbb{D}$, with the power series expansion*

$$u(z) = \sum_{n=1}^{\infty} c_n z^n.$$

Then, $|c_n| \leq 1$ for all $n \in \mathbb{N}$. Furthermore, $|c_n| = 1$ for some $n \in \mathbb{N}$ if and only if $u(z) = e^{i\theta} z^n$, $\theta \in \mathbb{R}$.

Lemma 2. [20] *If $\psi(z) = \sum\limits_{n=1}^{\infty} \psi_n z^n$, $z \in \mathbb{D}$, is a Schwarz function with $\psi_1 \in \mathbb{R}$, then*

$$\psi_2 = x\left(1 - \psi_1^2\right),$$
$$\psi_3 = \left(1 - \psi_1^2\right)\left(1 - |x|^2\right) s - \psi_1 \left(1 - \psi_1^2\right) x^2,$$

for some x, s, with $|x| \leq 1$ and $|s| \leq 1$.

Lemma 3. [35] *Let the function $f \in \mathcal{S}^*$ be given by* (1). *Then, for any real number μ,*

$$\left| a_3 - \mu a_2^2 \right| \begin{cases} 3 - 4\mu & \text{if } \mu \leq \frac{1}{2} \\ 1 & \text{if } \frac{1}{2} \leq \mu \leq 1 \\ 4\mu - 3 & \text{if } \mu \geq 1. \end{cases}$$

Lemma 4. [19] *Let the function $f \in \mathcal{S}^*$ be given by* (1). *Then*

$$|H_2(2)| := \left| a_2 a_4 - a_3^2 \right| \leq 1.$$

Equality holds true for the Koebe function $k(z) = \dfrac{z}{(1-z)^2}$.

Lemma 5. [36] *Let the function $f \in \mathcal{S}^*$ be given by* (1). *Then*

$$|a_2 a_3 - a_4| \leq 2.$$

Equality holds true for the Koebe function $k(z) = \dfrac{z}{(1-z)^2}$.

Proof of Theorem 2. As noted in Remark 1, if $f \in \mathcal{C}_\Sigma$, then by definition of subordination, there exist two Schwarz functions u and v, of the form $u(z) = \sum_{n=1}^{\infty} c_n z^n$ and $v(z) = \sum_{n=1}^{\infty} d_n z^n$, $z \in \mathbb{D}$ that we can write

$$\frac{zf'(z)}{g(z)} = \frac{1+u(z)}{1-u(z)} = 1 + 2c_1 z + (2c_2 + 2c_1^2) z^2 + (2c_3 + 4c_1 c_2 + 2c_1^3) z^3 + \cdots$$

and

$$\frac{zF'(z)}{G(z)} = \frac{1+v(w)}{1-v(w)} = 1 + 2d_1 w + (2d_2 + 2d_1^2) w^2 + (2d_3 + 4d_1 d_2 + 2d_1^3) w^3 + \cdots.$$

Equating coefficients in two above relations then gives

$$2a_2 - b_2 = 2c_1, \tag{8}$$

$$3a_3 - b_3 - 2a_2 b_2 + b_2^2 = 2c_2 + 2c_1^2, \tag{9}$$

$$4a_4 - b_4 - 2a_2 b_3 + 2b_2 b_3 - 3a_3 b_2 + 2a_2 b_2^2 - b_2^3 = 2c_3 + 4c_1 c_2 + 2c_1^3, \tag{10}$$

and

$$-2a_2 + b_2 = 2d_1, \tag{11}$$

$$-3a_3 + b_3 - 2a_2 b_2 - b_2^2 + 6a_2^2 = 2d_2 + 2d_1^2, \tag{12}$$

$$-4a_4 + b_4 - 2a_2 b_3 - 3b_2 b_3 - 3a_3 b_2 + 2a_2 b_2^2 + 2b_2^3 - 20a_2^3$$
$$+ 20a_2 a_3 + 6a_2^2 b_2 = 2d_3 + 4d_1 d_2 + 2d_1^3, \tag{13}$$

respectively. From (8) and (11), we get that

$$c_1 = -d_1, \tag{14}$$

Also, according to the proof of [34] (Theorem), it is enough that we set $2c_1$, $2c_2 + 2c_1^2$, $2c_3 + 4c_1c_2 + 2c_1^3$ instead of c_1, c_2, c_3, and $2d_1$, $2d_2 + 2d_1^2$, $2d_3 + 4d_1d_2 + 2d_1^3$ instead of d_1, d_2, d_3 in relations (2.5)–(2.10) in [34], respectively. Thus we can write (2.20) in [34], as given below:

$$\begin{aligned}
|a_2 a_4 - a_3^2| = \Big| &\tfrac{1}{8}(b_2 b_4 - b_3^2) + \tfrac{1}{72}b_3^2 + \tfrac{2}{8}(b_4 - b_2 b_3)c_1 - \tfrac{10}{48}b_2 b_3 c_1 \\
&+ \tfrac{4}{24}\left(b_3 - \tfrac{13}{4}b_2^2\right)c_1^2 \\
&- \tfrac{7}{144}\left(b_3 - \tfrac{19}{14}b_2^2\right)b_2^2 - \tfrac{2}{9}\left(b_3 - \tfrac{19}{16}b_2^2\right)(c_2 - d_2) \\
&- \tfrac{10}{32}b_2 c_1 \left[4c_1^2 - \tfrac{8}{15}(c_2 - d_2) - \tfrac{13}{15}b_2^2\right] \\
&- \tfrac{2}{16}c_1\left[8c_1^3 - \tfrac{4}{3}c_1(c_2 - d_2) - 2(c_3 - d_3) - 4c_1^3 - 4c_1(c_2 + d_2)\right] \\
&+ \tfrac{1}{16}b_2\left[2(c_3 - d_3) + 4c_1^3 + 4c_1(c_2 + d_2)\right] - \tfrac{4}{36}(c_2 - d_2)^2 \Big|.
\end{aligned}$$ (15)

According to Lemma 2 and (14), we find that

$$c_2 - d_2 = \left(1 - c_1^2\right)(x - y) \quad \text{and} \quad c_2 + d_2 = \left(1 - c_1^2\right)(x + y) \quad (16)$$

and

$$c_3 = \left(1 - c_1^2\right)\left(1 - |x|^2\right)s - c_1\left(1 - c_1^2\right)x^2 \quad \text{and} \quad d_3 = \left(1 - d_1^2\right)\left(1 - |y|^2\right)t - d_1\left(1 - d_1^2\right)y^2,$$

where

$$c_3 - d_3 = (1 - c_1^2)\left[(1 - |x|^2)s - (1 - |y|^2)t\right] - c_1(1 - c_1^2)(x^2 + y^2) \quad (17)$$

for some x, y, s, t with $|x| \leq 1$, $|y| \leq 1$, $|s| \leq 1$ and $|t| \leq 1$. Applying (16) and (17) in (15), it follows that

$$\begin{aligned}
|a_2 a_4 - a_3^2| = \Big| &\tfrac{1}{8}(b_2 b_4 - b_3^2) + \tfrac{1}{72}b_3^2 + \tfrac{2}{8}(b_4 - b_2 b_3)c_1 - \tfrac{10}{48}b_2 b_3 c_1 \\
&+ \tfrac{4}{24}\left(b_3 - \tfrac{13}{4}b_2^2\right)c_1^2 - \tfrac{7}{144}\left(b_3 - \tfrac{19}{14}b_2^2\right)b_2^2 \\
&- \tfrac{40}{32}b_2 c_1^3 + \tfrac{26}{96}b_2^3 c_1 - \tfrac{16}{16}c_1^4 + \tfrac{8}{16}c_1^4 + \tfrac{4}{16}b_2 c_1^3 \\
&+ 2\left(1 - c_1^2\right)(x - y)\left[-\tfrac{1}{9}\left(b_3 - \tfrac{19}{16}b_2^2\right) + \tfrac{8}{96}b_2 c_1 + \tfrac{4}{48}c_1^2\right] \\
&+ 2c_1\left(1 - c_1^2\right)(x + y)\left[\tfrac{4}{16}c_1 + \tfrac{2}{16}b_2\right] \\
&+ 2(1 - c_1^2)\left[(1 - |x|^2)s - (1 - |y|^2)t\right]\left[\tfrac{2}{16}c_1 + \tfrac{1}{16}b_2\right] \\
&- 2c_1(1 - c_1^2)(x^2 + y^2)\left[\tfrac{2}{16}c_1 + \tfrac{1}{16}b_2\right] - \tfrac{4}{36}\left(1 - c_1^2\right)^2 (x - y)^2 \Big|.
\end{aligned}$$

Since by Lemma 1, $|c_1| \leq 1$, we assume that $c_1 = c \in [0,1]$. So by utilizing the triangle inequality we have

$$\left|a_2 a_4 - a_3^2\right|$$
$$\leq \frac{1}{8}\left|b_2 b_4 - b_3^2\right| + \frac{1}{72}\left|b_3^2\right| + \frac{2}{8}\left|b_4 - b_2 b_3\right| c + \frac{10}{48}\left|b_3 - \frac{13}{10} b_2^2\right| c |b_2|$$
$$+ \frac{4}{24}\left|b_3 - \frac{13}{4} b_2^2\right| c^2 + \frac{7}{144}\left|b_3 - \frac{19}{14} b_2^2\right| |b_2^2|$$
$$+ \left|-\frac{40}{32} + \frac{4}{16}\right| |b_2| c^3 + \left|-\frac{16}{16} + \frac{8}{16}\right| c^4$$
$$+ 2\left(1 - c^2\right) \left[\frac{1}{9}\left|b_3 - \frac{19}{16} b_2^2\right| + \frac{8}{96}|b_2|c + \frac{4}{48} c^2\right] (|x| + |y|)$$
$$+ 2c\left(1 - c^2\right) \left[\frac{4}{16} c + \frac{2}{16}|b_2|\right] (|x| + |y|)$$
$$+ 2(1 - c^2)\left[\frac{2}{16} c + \frac{1}{16}|b_2|\right]\left[(1 - |x|^2) + (1 - |y|^2)\right]$$
$$+ 2c(1 - c^2)\left[\frac{2}{16} c + \frac{1}{16}|b_2|\right]\left(|x|^2 + |y|^2\right) + \frac{4}{36}\left(1 - c^2\right)^2 \left((|x| + |y|)^2\right)$$
$$= \frac{1}{8}\left|b_2 b_4 - b_3^2\right| + \frac{1}{72}\left|b_3^2\right| + \frac{2}{8}\left|b_4 - b_2 b_3\right| c + \frac{10}{48}\left|b_3 - \frac{13}{10} b_2^2\right| c |b_2|$$
$$+ \frac{4}{24}\left|b_3 - \frac{13}{4} b_2^2\right| c^2 + \frac{7}{144}\left|b_3 - \frac{19}{14} b_2^2\right| |b_2^2| + |b_2| c^3 + \frac{1}{2} c^4$$
$$+ 4(1 - c^2)\left[\frac{2}{16} c + \frac{1}{16}|b_2|\right]$$
$$+ \left(2\left[\frac{1}{9}\left|b_3 - \frac{19}{16} b_2^2\right| + \frac{8}{96}|b_2|c + \frac{4}{48} c^2\right] + 2c\left[\frac{4}{16} c + \frac{2}{16}|b_2|\right]\right) \left(1 - c^2\right) (|x| + |y|)$$
$$+ 2\left(\frac{2}{16} c + \frac{1}{16}|b_2|\right) (c - 1)(1 - c^2)\left(|x|^2 + |y|^2\right) + \frac{4}{36}\left(1 - c^2\right)^2 (|x| + |y|)^2.$$

We now apply Lemmas 3–5 in order to deduce that

$$\left|a_2 a_4 - a_3^2\right|$$
$$\leq \frac{1}{8} + \frac{1}{8} + \frac{2}{4} c + \frac{22}{24} c + \frac{40}{24} c^2 + \frac{17}{36} + 2c^3 + \frac{1}{2} c^4 + \frac{8}{16}(1 - c^2)(c + 1)$$
$$+ \left(2\left[\frac{7}{36} + \frac{16}{96} c + \frac{4}{48} c^2\right] + 2c\left[\frac{4}{16} c + \frac{4}{16}\right]\right) \left(1 - c^2\right) (|x| + |y|)$$
$$+ \frac{4}{16}(c + 1)(c - 1)(1 - c^2)\left(|x|^2 + |y|^2\right)$$
$$+ \frac{4}{36}\left(1 - c^2\right)^2 (|x| + |y|)^2.$$

Now, for $\lambda = |x| \leq 1$ and $\mu = |y| \leq 1$, we obtain

$$\left|a_2 a_4 - a_3^2\right| \leq J_1 + (\lambda + \mu) J_2 + (\lambda^2 + \mu^2) J_3 + (\lambda + \mu)^2 J_4 = L(\lambda, \mu),$$

where

$$J_1 = J_1(c) = \frac{26}{36} + \frac{34}{24}c + \frac{40}{24}c^2 + 2c^3 + \frac{1}{2}c^4 + \frac{8}{16}(1-c^2)(c+1) \geq 0$$

$$J_2 = J_2(c) = \left(\frac{7}{18} + \frac{5}{6}c + \frac{4}{6}c^2\right)\left(1-c^2\right) \geq 0$$

$$J_3 = J_3(c) = -\frac{1}{4}(1-c^2)^2 \leq 0$$

$$J_4 = J_4(c) = \frac{1}{9}\left(1-c^2\right)^2 \geq 0.$$

We now need to maximize the function $L(\lambda, \mu)$ on the closed square $[0,1] \times [0,1]$ for $c \in [0,1]$. With regards to $L(\lambda, \mu) = L(\mu, \lambda)$, it is sufficient to show that there exists the maximum of

$$H(\lambda) = L(\lambda, \lambda) = J_1 + 2\lambda J_2 + 2\lambda^2(J_3 + 2J_4), \tag{18}$$

on $\lambda \in [0,1]$ according to $c \in [0,1]$. We let $c \in [0,1]$. Considering Equation (18) for $0 < \lambda < 1$ and $J_3 + 2J_4 < 0$, we consider for critical point

$$\lambda_0 = \frac{-J_2}{2(J_3 + 2J_4)} = \frac{J_2}{2k} = \frac{18\left(\frac{7}{18} + \frac{5}{6}c + \frac{4}{6}c^2\right)(1-c^2)}{(1-c^2)^2} = \frac{18\left(\frac{7}{18} + \frac{5}{6}c + \frac{4}{6}c^2\right)}{(1-c^2)} > 1$$

for any fixed $c \in [0,1]$, where $k = -(J_3 + 2J_4) > 0$. Therefore, for $\lambda_0 = \frac{J_2}{2k} > 1$, it follows that $k < \frac{J_2}{2} \leq J_2$, and so $J_2 + J_3 + 2J_4 \geq 0$. So,

$$H(0) = J_1 \leq J_1 + 2(J_2 + J_3 + 2J_4) = H(1).$$

Therefore, it follows that

$$\max\{H(\lambda) : \lambda \in [0,1]\} = H(1) = J_1 + 2J_2 + 2J_3 + 4J_4.$$

Therefore, $\max L(\lambda, \mu) = L(1,1)$ on the boundary of the square.
We define the real function W on $(0,1)$ by

$$W(c) = L(1,1) = J_1 + 2J_2 + 2J_3 + 4J_4.$$

Now putting J_1, J_2, J_3 and J_4 in the function W, we have

$$W(c) = -\frac{8}{9}c^4 - \frac{1}{6}c^3 + \frac{11}{6}c^2 + \frac{43}{12}c + \frac{70}{36}.$$

By elementary calculations, we get that $W(c)$ is an increasing function of c. Therefore, we obtain the maximum of $W(c)$ on $c = 1$ and

$$\max W(c) = W(1) = \frac{227}{36}.$$

This completes the proof. □

Example 1. *If we choose the functions*

$$f(z) = z + \frac{z^3}{2}, \quad g(z) = z - \frac{z^3}{3},$$

then will have

$$F(w) = f^{-1}(w) = w - \frac{w^3}{2}, \quad G(w) = g^{-1}(w) = w + \frac{w^3}{3}$$

and so these functions satisfy in Definition 1. Thus function $f \in \Sigma$ is bi-close-to-convex, that is, $f \in \mathcal{C}_\Sigma$ (see for more details, [33]). Therefore, Theorem 2 holds for $f(z) = z + \dfrac{z^3}{2}$.

Remark 2. *The obtained bound for $|a_2 a_4 - a_3^2|$ in Theorem 2 is smaller than and more accurate the estimation given in Theorem 1.*

3. Conclusions

In the present paper, we find a smaller upper bound and more accurate estimation for the functional $|H_2(2)|$ for functions in the class $\mathcal{C}_\Sigma$ with $G(w) = g^{-1}(w)$ which is an improvement of the result obtained by Guney et al. [34]. Obtaining a sharp estimate for $|H_2(2)|$ of the class $\mathcal{C}_\Sigma$ with $G(w) = g^{-1}(w)$ is still an open problem.

Author Contributions: Investigation: N.E.C., E.A.A., S.B. and A.M.

Funding: This research was supported by the Basic Science Research Program through the National Research Foundation of Korea (NRF) funded by the Ministry of Education, Science and Technology (No. 2019R1I1A3A01050861).

Acknowledgments: The authors would like to express their thanks to the referees for their constructive advices and comments that helped to improve this paper.

Conflicts of Interest: The authors declare no conflict of interest.

References

1. Duren, P.L. *Univalent Functions. Grundlehren der mathematischen Wissenschaften*; Springer: New York, NY, USA; Berlin/Heidelberg, Germany; Tokyo, Japan, 1983; Volume 259.
2. Noonan, J.W.; Thomas, D.K. On the second Hankel determinant of areally mean *p*-valent functions. *Trans. Am. Math. Soc.* **1976**, *223*, 337–346.
3. Cantor, D.G. Power series with integral coefficients. *Bull. Am. Math. Soc.* **1963**, *69*, 362–366. [CrossRef]
4. Wilson, R. Determinantal criteria for meromorphic functions. *Proc. Lond. Math. Soc.* **1954**, *4*, 357–374. [CrossRef]
5. Edrei, A. Sur les determinants recurrents et less singularities d'une fonction donee por son developpement de Taylor. *Comput. Math.* **1940**, *7*, 20–88.
6. Polya, G.; Schoenberg, I.J. Remarks on de la Vallee Poussin means and convex conformal maps of the circle. *Pac. J. Math.* **1958**, *8*, 259–334. [CrossRef]
7. Vein, R.; Dale, P. Determinants and Their Applications in Mathematical Physics. In *Applied Mathematical Sciences*; Springer: New York, NY, USA, 1999; Volume 134.
8. Pommerenke, C. On the coefficients and Hankel determinant of univalent functions. *J. Lond. Math. Soc.* **1966**, *41*, 111–112. [CrossRef]
9. Hayman, W.K. On second Hankel determinant of mean univalent functions. *Proc. Lond. Math. Soc.* **1968**, *18*, 77–94. [CrossRef]
10. Pommerenke, C. On the Hankel determinants of univalent functions. *Mathematika* **1967**, *14*, 108–112. [CrossRef]
11. Noor, K.I. On the Hankel determinant problem for strongly close-to-convex functions. *J. Nat. Geom.* **1997**, *11*, 29–34.
12. Noor, K.I. On certain analytic functions related with strongly close-to-convex functions. *Appl. Math. Comput.* **2008**, *197*, 149–157. [CrossRef]
13. Fekete, M.; Szegö, G. Eine Bemerkung Über Ungerade Schlichte Funktionen. *J. Lond. Math. Soc.* **1933**, *8*, 85–89. [CrossRef]
14. Altinkaya, Ş.; Yalçin, S. Upper bound of second Hankel determinant for bi-Bazilevic functions. *Mediterr. J. Math.* **2016**, *13*, 4081–4090. [CrossRef]

15. Çağlar, M.; Deniz, E.; Srivastava, H.M. Second Hankel determinant for certain subclasses of bi-univalent functions. *Turk. J. Math.* **2017**, *41*, 694–706. [CrossRef]
16. Cho, N.E.; Kowalczyk, B.; Kwon, O.S.; Lecko, A.; Sim, Y.J. The bounds of some determinants for starlike functions of order alpha. *Bull. Malays. Math. Sci. Soc.* **2018**, *41*, 523–535. [CrossRef]
17. Cho, N.E.; Kowalczyk, B.; Kwon, O.S.; Lecko, A.; Sim, Y.J. Some coefficient inequalities related to the Hankel determinant for strongly starlike functions of order alpha. *J. Math. Inequal.* **2017**, *11*, 429–439. [CrossRef]
18. Deniz, E.; Çağlar, M.; Orhan, H. Second Hankel determinant for bi-starlike and bi-convex functions of order β. *Appl. Math. Comput.* **2015**, *271*, 301–307. [CrossRef]
19. Janteng, A.; Halim, S.; Darus, M. Coefficient inequality for a function whose derivative has a positive real part. *J. Inequal. Pure Appl. Math.* **2006**, *7*, 1–5.
20. Kanas, S.; Analouei Adegani, E.; Zireh, A. An unified approach to second Hankel determinant of bi-subordinate functions. *Mediterr. J. Math.* **2017**, *14*, 233. [CrossRef]
21. Motamednezhad, A.; Bulboacă, T.; Adegani, E.A.; Dibagar, N. Second Hankel determinant for a subclass of analytic bi-univalent functions defined by subordination. *Turk. J. Math.* **2018**, *42*, 2798–2808. [CrossRef]
22. Orhan, H.; Magesh, N.; Yamini, J. Bounds for the second Hankel determinant of certain bi-univalent functions. *Turk. J. Math.* **2016**, *40*, 679–687. [CrossRef]
23. Tang, H.; Srivastava, H.M.; Sivasubramanian, S.; Gurusamy, P. The Fekete–Szegö functional problems for some classes of m-fold symmetric bi-univalent functions. *J. Math. Inequal.* **2016**, *10*, 1063–1092. [CrossRef]
24. Motamednezhad, A.; Bulut, S.; Analouei Adegani, E. Upper bound of second Hankel determinant for k-bi-subordinate functions. *U.P.B. Sci. Bull. Ser. A* **2019**, *81*, 31–42.
25. Lewin, M. On a coefficient problem for bi-univalent functions. *Proc. Am. Math. Soc.* **1967**, *18*, 63–68.
26. Analouei Adegani, E.; Bulut, S.; Zireh, A. Coefficient estimates for a subclass of analytic bi-univalent functions. *Bull. Korean Math. Soc.* **2018**, *55*, 405–413.
27. Analouei Adegani, E.; Cho, N.E.; Jafari M. Logarithmic coefficients for univalent functions defined by subordination. *Mathematics* **2019**, *7*, 408. [CrossRef]
28. Analouei Adegani, E.; Cho, N.E.; Motamednezhad, A.; Jafari, M. Bi-univalent functions associated with Wright hypergeometric functions. *J. Comput. Anal. Appl.* **2020**, *28*, 261–271.
29. Bulut, S. Coefficient estimates for a new subclass of analytic and bi-univalent functions defined by Hadamard product. *J. Complex Anal.* **2014**, *2014*, 1–7. [CrossRef]
30. Srivastava, H.M.; Gaboury, S.; Ghanim, F. Coefficient estimates for a general subclass of analytic and bi-univalent functions of the Ma-Minda type. *RACSAM* **2018**, *112*, 1157–1168, doi:10.1007/s13398-017-0416-5.
31. Xu, Q.H.; Xiao, H.G.; Srivastava, H.M. A certain general subclass of analytic and bi-univalent functions and associated coefficient estimate problems. *Appl. Math. Comput.* **2012**, *218*, 11461–11465. [CrossRef]
32. Zireh, A.; Analouei Adegani, E.; Bulut, S. Faber polynomial coefficient estimates for a comprehensive subclass of analytic bi-univalent functions defined by subordination. *Bull. Belg. Math. Soc. Simon Stevin* **2016**, *23*, 487–504. [CrossRef]
33. Hamidi, S.G.; Jahangiri, J.M. Faber polynomial coefficient estimates for analytic bi-close-to-convex functions. *C. R. Acad. Sci. Paris Ser. I* **2014**, *352*, 17–20. [CrossRef]
34. Güney, H.Ö.; Murugusundaramoorthy, G.; Srivastava, H.M. The second Hankel determinant for a certain class of bi-close-to-convex functions. *Results Math.* **2019**, *74*, 93. [CrossRef]
35. Keogh, F.R.; Merkes, E.P. A coefficient inequality for certain classes of analytic functions. *Proc. Am. Math. Soc.* **1969**, *20*, 8–12. [CrossRef]
36. Babalola, K.O. On $H_3(1)$ Hankel determinant for some classes of univalent functions. *Inequal. Theory Appl.* **2010**, *6*, 1–7. [CrossRef]

 © 2019 by the authors. Licensee MDPI, Basel, Switzerland. This article is an open access article distributed under the terms and conditions of the Creative Commons Attribution (CC BY) license (http://creativecommons.org/licenses/by/4.0/).

Article

Interesting Explicit Expressions of Determinants and Inverse Matrices for Foeplitz and Loeplitz Matrices

Zhaolin Jiang [1], Weiping Wang [1], Yanpeng Zheng [2,*], Baishuai Zuo [1] and Bei Niu [1]

1. School of Mathematics and Statistics, Linyi University, Linyi 276000, China; jiangzhaolin@lyu.edu.cn or jzh1208@sina.com (Z.J.); weiping227@sina.cn (W.W.); bszuo124@163.com (B.Z.); beiniu@stu.xidian.edu.cn (B.N.)
2. School of Automation and Electrical Engineering, Linyi University, Linyi 276000, China
* Correspondence: zhengyanpeng@lyu.edu.cn

Received: 4 September 2019; Accepted: 6 October 2019; Published: 11 October 2019

Abstract: Foeplitz and Loeplitz matrices are Toeplitz matrices with entries being Fibonacci and Lucas numbers, respectively. In this paper, explicit expressions of determinants and inverse matrices of Foeplitz and Loeplitz matrices are studied. Specifically, the determinant of the $n \times n$ Foeplitz matrix is the $(n+1)$th Fibonacci number, while the inverse matrix of the $n \times n$ Foeplitz matrix is sparse and can be expressed by the nth and the $(n+1)$th Fibonacci number. Similarly, the determinant of the $n \times n$ Loeplitz matrix can be expressed by use of the $(n+1)$th Lucas number, and the inverse matrix of the $n \times n$ ($n > 3$) Loeplitz matrix can be expressed by only seven elements with each element being the explicit expressions of Lucas numbers. Finally, several numerical examples are illustrated to show the effectiveness of our new theoretical results.

Keywords: determinant; inverse; Fibonacci number; Lucas number; Toeplitz matrix; Hankel matrix

MSC: 15A09; 15A15; 15A69; 65F05

1. Introduction

Toeplitz matrices often arise in statistics, econometrics, psychometrics, structural engineering, multichannel filtering, reflection seismology, etc. (see [1,2] and references therein). Furthermore, they have been employed in quite wide fields of applications, especially in the elliptic Dirichlet-periodic boundary value problems [3], solving fractional diffusion equations [4–6], numerical analysis [7], signal processing [7], and system theory [7], etc. Citations of a large number of results have been made in a series of papers and in the monographs of Iohvidov [8] and Heining and Rost [9].

It seems to be an ideal research area and current topic of interest to specify inverses of Toeplitz matrices as well as the special Toeplitz matrices involving famous numbers as entries. Some scholars showed the explicit determinant and inverse of the special matrices involving famous numbers. The authors [10] proposed the invertibility of generalized Lucas skew circulant matrices and provided the determinant and the inverse matrix. Furthermore, the invertibility of generalized Lucas skew left circulant matrices was also discussed. The determinant and the inverse matrix of generalized Lucas skew left circulant matrices were obtained respectively. The determinants and inverses of Tribonacci skew circulant type matrices were discussed in [11]. The authors provided determinants and inverses of circulant matrices with Jacobsthal and Jacobsthal–Lucas numbers in [12]. The explicit determinants of circulant and left circulant matrices including Tribonacci numbers and generalized Lucas numbers were shown based on Tribonacci numbers and generalized Lucas numbers only in [13]. Moreover, four kinds of norms and bounds for the spread of these matrices were discussed respectively. In [14], circulant type matrices with the k-Fibonacci and k-Lucas numbers were considered and the explicit determinant and inverse matrix were presented by constructing the transformation matrices.

Jiang et al. [15] gave the invertibility of circulant type matrices with the sum and product of Fibonacci and Lucas numbers and provided the determinants and the inverses of the these matrices. Jiang and Hong [16] studied exact form determinants of the RSFPLR circulant matrices and the RSLPFL circulant matrices involving Padovan, Perrin, Tribonacci, and the generalized Lucas number by the inverse factorization of a polynomial. It is worthwhile to note that Akbulak and Bozkurt gave the upper and lower bounds for the spectral norms of the Fibonacci and Lucas Toeplitz matrices [17].

In this paper, we will show the explicit determinants and inverses of the Foeplitz matrix and Fankel matrix both involving Fibonacci numbers (see Definitions 1 and 2 below), and the Loeplitz matrix and Lankel matrix both involving Lucas numbers (see Definitions 3 and 4). The main results are obtained by factoring the considered matrices into structured factors, whose determinant and inverse are computed exactly, and then reassembling the factorization. This paper provides a novel characterization of Fibonacci or Lucas numbers as the determinant of Toeplitz matrices containing numbers from the same sequence. In fact, the main contribution of this paper is that Toeplitz matrix, tridiagonal Toeplitz matrices with perturbed corner entries, the Fibonacci number, and the Golden Ratio are connected together.

Here the Fibonacci and Lucas sequences (see, e.g., [18]) are defined by the following recurrence relations, respectively:

$$F_{n+1} = F_n + F_{n-1}(n \geq 1), \quad \text{where} \quad F_0 = 0, F_1 = 1,$$
$$L_{n+1} = L_n + L_{n-1}(n \geq 1), \quad \text{where} \quad L_0 = 2, L_1 = 1,$$
$$F_{-(n+1)} = -F_{-n} + F_{-(n-1)}(n \geq 1), \quad \text{where} \quad F_0 = 0, F_{-1} = 1,$$
$$L_{-(n+1)} = -L_{-n} + L_{-(n-1)}(n \geq 1), \quad \text{where} \quad L_0 = 2, L_{-1} = -1.$$

The following identities are easily attainable

$$F_{-n} = (-1)^{n+1} F_n, \quad L_{-n} = (-1)^n L_n, \tag{1}$$

$$\sum_{i=2}^{n-2} a^i L_{k+i} = \frac{-a^3 L_{k+1} - a^2 L_{k+2} + a^{n-1} L_{n-1+k} + a^n L_{n-2+k}}{a^2 + a - 1}, a \neq \frac{-1 \pm \sqrt{5}}{2}, \tag{2}$$

$$\sum_{i=2}^{n-2} a^i L_{k-i} = \frac{-a^2 L_{k-2} - a^3 L_{k-1} + a^{n-1} L_{k-(n-4)} + a^n L_{k-(n-2)}}{a^2 - a - 1}, a \neq \frac{1 \pm \sqrt{5}}{2}. \tag{3}$$

Definition 1. *An $n \times n$ Foeplitz matrix is defined as a Toeplitz matrix of the form*

$$T_{F,n} = \begin{pmatrix} F_1 & F_2 & \cdots & F_{n-1} & F_n \\ F_{-2} & F_1 & \ddots & \ddots & F_{n-1} \\ \vdots & \ddots & \ddots & \ddots & \vdots \\ F_{-n+1} & \ddots & \ddots & F_1 & F_2 \\ F_{-n} & F_{-n+1} & \cdots & F_{-2} & F_1 \end{pmatrix}_{n \times n}, \tag{4}$$

where $F_1, F_{\pm 2}, \cdots, F_{\pm n}$ are the Fibonacci numbers.

Definition 2. *An $n \times n$ Fankel matrix is defined as a Hankel matrix of the form*

$$H_{F,n} = \begin{pmatrix} F_n & F_{n-1} & \cdots & F_2 & F_1 \\ F_{n-1} & \cdot^{\cdot^{\cdot}} & \cdot^{\cdot^{\cdot}} & F_1 & F_{-2} \\ \vdots & \cdot^{\cdot^{\cdot}} & \cdot^{\cdot^{\cdot}} & \cdot^{\cdot^{\cdot}} & \vdots \\ F_2 & F_1 & \cdot^{\cdot^{\cdot}} & \cdot^{\cdot^{\cdot}} & F_{-n+1} \\ F_1 & F_{-2} & \cdots & F_{-n+1} & F_{-n} \end{pmatrix}_{n \times n}, \tag{5}$$

where F_1, $F_{\pm 2}$, $\cdots$, $F_{\pm n}$ are the Fibonacci numbers.

Definition 3. *An $n \times n$ Loeplitz matrix is defined as a Toeplitz matrix of the form*

$$T_{L,n} = \begin{pmatrix} L_1 & L_2 & \cdots & L_{n-1} & L_n \\ L_{-2} & L_1 & \ddots & \ddots & L_{n-1} \\ \vdots & \ddots & \ddots & \ddots & \vdots \\ L_{-n+1} & \ddots & \ddots & L_1 & L_2 \\ L_{-n} & L_{-n+1} & \cdots & L_{-2} & L_1 \end{pmatrix}_{n \times n}, \qquad (6)$$

where L_1, $L_{\pm 2}$, $\cdots$, $L_{\pm n}$ are the Lucas numbers.

Definition 4. *An $n \times n$ Lankel matrix is defined as a Hankel matrix of the form*

$$H_{L,n} = \begin{pmatrix} L_n & L_{n-1} & \cdots & L_2 & L_1 \\ L_{n-1} & \cdot^{\cdot^{\cdot}} & \cdot^{\cdot^{\cdot}} & L_1 & L_{-2} \\ \vdots & \cdot^{\cdot^{\cdot}} & \cdot^{\cdot^{\cdot}} & \cdot^{\cdot^{\cdot}} & \vdots \\ L_2 & L_1 & \cdot^{\cdot^{\cdot}} & \cdot^{\cdot^{\cdot}} & L_{-n+1} \\ L_1 & L_{-2} & \cdots & L_{-n+1} & L_{-n} \end{pmatrix}_{n \times n}, \qquad (7)$$

where L_1, $L_{\pm 2}$, $\cdots$, $L_{\pm n}$ are the Lucas numbers.

It is easy to check that

$$H_{F,n} = T_{F,n} \hat{I}_n, \qquad (8)$$
$$H_{L,n} = T_{L,n} \hat{I}_n, \qquad (9)$$

where $\hat{I}_n$ is the counteridentity matrix, the square matrix whose elements are all equal to zero except those on the counter-diagonal, which are all equal to 1, which provide us with basic relations between $T_{F,n}$ and $H_{F,n}$, and $T_{L,n}$ and $H_{L,n}$, respectively.

Lemma 1. *([19], Lemma 2.5) Define an $n \times n$ bi-band-Toeplitz matrix by*

$$\mathcal{F}_n(\alpha, \beta) = \begin{pmatrix} \alpha & 0 & \cdots & \cdots & \cdots & 0 \\ \beta & \alpha & \ddots & & & \vdots \\ 0 & \beta & \alpha & \ddots & & \vdots \\ \vdots & \ddots & \ddots & \ddots & \ddots & \vdots \\ \vdots & & \ddots & \beta & \alpha & 0 \\ 0 & \cdots & \cdots & 0 & \beta & \alpha \end{pmatrix}_{n \times n},$$

the inverse of $\mathcal{F}_n(\alpha, \beta)$ can be expressed as

$$\mathcal{F}_n(\alpha, \beta)^{-1} = \begin{pmatrix} \Delta_1 & 0 & \cdots & \cdots & \cdots & \cdots & 0 \\ \Delta_2 & \Delta_1 & \ddots & & & & \vdots \\ \Delta_3 & \Delta_2 & \Delta_1 & \ddots & & & \vdots \\ \vdots & \ddots & \ddots & \ddots & \ddots & & \vdots \\ \Delta_{n-2} & \ddots & \ddots & \ddots & \ddots & \ddots & \vdots \\ \Delta_{n-1} & \Delta_{n-2} & \ddots & \ddots & \Delta_2 & \Delta_1 & 0 \\ \Delta_n & \Delta_{n-1} & \Delta_{n-2} & \cdots & \Delta_3 & \Delta_2 & \Delta_1 \end{pmatrix}_{n \times n},$$

where

$$\Delta_i = \frac{(-\beta)^{i-1}}{\alpha^i}, \; i \geq 1.$$

Remark 1. *This Lemma is a special case of ([19], Lemma 2.5).*

2. The Determinant and Inverse Matrix of Foeplitz, Fankel, Loeplitz, and Lankel Matrices

In this section, we study the determinant and the inverse of Foeplitz, Fankel, Loeplitz, and Lankel matrices by factoring the considered matrices into structured factors, whose determinant and inverse are computed exactly, and then reassembling the factorization. We establish the relationship between the determinant of these matrices and Fibonacci or Lucas numbers.

2.1. Determinant and Inverse Matrix of a Foeplitz Matrix

In this subsection, the determinant and the inverse of the Foeplitz matrix $T_{F,n}$ are studied.

Theorem 1. *Let $T_{F,n}$ be an $n \times n$ Foeplitz matrix defined as in (4). Then $T_{F,n}$ is invertible and*

$$\det T_{F,n} = F_{n+1}, \tag{10}$$

where F_{n+1} is the $(n+1)$th Fibonacci number.

Proof. For $n \leq 3$, it is easy to check that $\det T_{F,1} = 1 = F_2$, $\det T_{F,2} = 2 = F_3$ and $\det T_{F,3} = 3 = F_4$. Therefore, Equation (10) is satisfied. Now, we consider the case $n > 3$. Define two additional nonsingular matrices,

$$A_1 = \begin{pmatrix} 1 & & & & & \\ -F_{-n} & & & & 1 & \\ -F_{-n-1} & & & 1 & -1 & \\ 0 & & 1 & -1 & -1 & \\ \vdots & \iddots & \iddots & \iddots & & \\ 0 & 1 & -1 & -1 & & \end{pmatrix}_{n \times n}, B_1 = \begin{pmatrix} 1 & 0 & \cdots & \cdots & 0 \\ 0 & & & \iddots & 1 \\ \vdots & & \iddots & 1 & 0 \\ \vdots & \iddots & \iddots & \iddots & \vdots \\ 0 & 1 & 0 & \cdots & 0 \end{pmatrix}_{n \times n}.$$

Multiplying $T_{F,n}$ by A_1 from the left, we obtain

$$A_1 T_{F,n} = \begin{pmatrix} F_1 & F_2 & F_3 & \cdots & F_{n-1} & F_n \\ 0 & \alpha_2 & \alpha_3 & \cdots & \alpha_{n-1} & \alpha_n \\ \vdots & \beta_2 & \beta_3 & \cdots & \beta_{n-1} & \beta_n \\ \vdots & 0 & \cdots & 0 & 1 & 0 \\ \vdots & \vdots & \ddots & \ddots & \ddots & \vdots \\ 0 & 0 & 1 & 0 & \cdots & 0 \end{pmatrix}_{n \times n},$$

where

$$\alpha_i = -F_{-n}F_i + F_{-n+i-1}, (i = 2, 3, \cdots, n), \tag{11}$$
$$\beta_i = -F_{-n-1}F_i + F_{-n+i-2}, (i = 2, 3, \cdots, n-1),$$
$$\beta_n = -F_{-n-1}F_n. \tag{12}$$

Then, multiplying $A_1 T_{F,n}$ by B_1 from the right, we have

$$A_1 T_{F,n} B_1 = \begin{pmatrix} F_1 & F_n & F_{n-1} & \cdots & F_3 & F_2 \\ 0 & \alpha_n & \alpha_{n-1} & \cdots & \alpha_3 & \alpha_2 \\ 0 & \beta_n & \beta_{n-1} & \cdots & \beta_3 & \beta_2 \\ 0 & 0 & 1 & 0 & \cdots & 0 \\ \vdots & \vdots & \ddots & \ddots & \ddots & \vdots \\ 0 & 0 & \cdots & 0 & 1 & 0 \end{pmatrix}_{n \times n}, \tag{13}$$

and

$$\begin{aligned} \det(A_1 T_{F,n} B_1) &= \det(A_1) \det(T_{F,n}) \det(B_1) \\ &= F_1[(-1)^{n-1}\beta_2 \alpha_n - (-1)^{n-1}\alpha_2 \beta_n] \\ &= (-1)^{n-1} F_1[(-F_{-n}F_n + F_1)(-F_{-n-1}F_2 + F_{-n}) + F_{-n-1}F_n(-F_{-n}F_2 + F_{-n+1})]. \end{aligned}$$

From the definition of A_1 and B_1, we get

$$\det A_1 = \det B_1 = (-1)^{\frac{(n-1)(n-2)}{2}}.$$

Therefore, we have

$$\begin{aligned} \det T_{F,n} &= (-1)^{n-1} F_1[(-F_{-n}F_n + F_1)(-F_{-n-1}F_2 + F_{-n}) \\ &\quad + F_{-n-1}F_n(-F_{-n}F_2 + F_{-n+1})] \\ &= F_{n+1}. \end{aligned}$$

Since $F_{n+1} \neq 0$, the $n \times n$ Foeplitz matrix is invertible. Thus, the proof is completed. □

Remark 2. *Theorem 1 gives the relationship between the Foeplitz matrix and the Fibonacci number. From the perspective of number theory, the $(n+1)$th Fibonacci number can be represented by the determinant of an $n \times n$ Foeplitz matrix.*

Theorem 2. Let $T_{F,n}$ be an $n \times n$ Foeplitz matrix defined as in (4). The inverse matrix of $T_{F,n}$ is

$$T_{F,n}^{-1} = \begin{pmatrix} \frac{F_n}{F_{n+1}} & -1 & 0 & 0 & 0 & \cdots & \cdots & \cdots & 0 & \frac{(-1)^n}{F_{n+1}} \\ 1 & -1 & -1 & 0 & 0 & \cdots & \cdots & \cdots & 0 & 0 \\ 0 & 1 & -1 & -1 & 0 & & & & & 0 \\ \vdots & \ddots & 1 & -1 & -1 & \ddots & & & & \vdots \\ \vdots & & \ddots & \ddots & \ddots & \ddots & \ddots & & & \vdots \\ \vdots & & & \ddots & \ddots & \ddots & \ddots & \ddots & & \vdots \\ \vdots & & & & \ddots & \ddots & \ddots & \ddots & \ddots & \vdots \\ \vdots & & & & & \ddots & 1 & -1 & -1 & 0 \\ 0 & \cdots & \cdots & \cdots & \cdots & \cdots & 0 & 1 & -1 & -1 \\ -\frac{1}{F_{n+1}} & 0 & \cdots & \cdots & \cdots & \cdots & \cdots & 0 & 1 & \frac{F_n}{F_{n+1}} \end{pmatrix}_{n \times n}, \quad (14)$$

where F_n and F_{n+1} are the nth and $(n+1)$th Fibonacci numbers, respectively.

Proof. For $n = 1$, it is easy to check that

$$T_{F,1} = 1 \text{ and } T_{F,1}^{-1} = \frac{F_1}{F_2}.$$

For $n = 2$, we have

$$T_{F,2} = \begin{pmatrix} 1 & 1 \\ -1 & 1 \end{pmatrix} \text{ and } T_{F,2}^{-1} = \begin{pmatrix} \frac{1}{2} & -\frac{1}{2} \\ \frac{1}{2} & \frac{1}{2} \end{pmatrix},$$

and for $n = 3$, we have

$$T_{F,3} = \begin{pmatrix} 1 & 1 & 2 \\ -1 & 1 & 1 \\ 2 & -1 & 1 \end{pmatrix} \text{ and } T_{F,3}^{-1} = \begin{pmatrix} \frac{2}{3} & -1 & -\frac{1}{3} \\ 1 & -1 & -1 \\ -\frac{1}{3} & 1 & \frac{2}{3} \end{pmatrix},$$

which are in agreement with Equation (14). Now, we consider the case $n \geq 4$. The explicit expression of the inverse of the Foeplitz matrix can be obtained by use of Equation (14). Define addtionally two nonsigular matrices

$$A_2 = \begin{pmatrix} 1 & 0 & \cdots & \cdots & \cdots & \cdots & 0 \\ 0 & 1 & \ddots & & & & \vdots \\ 0 & -\frac{\beta_n}{\alpha_n} & 1 & \ddots & & & \vdots \\ \vdots & \ddots & 0 & \ddots & \ddots & & \vdots \\ \vdots & & \ddots & \ddots & \ddots & \ddots & \vdots \\ \vdots & & & \ddots & \ddots & 1 & 0 \\ 0 & \cdots & \cdots & \cdots & 0 & 0 & 1 \end{pmatrix}_{n \times n}$$

and

$$B_2 = \begin{pmatrix} 1 & -\frac{F_n}{F_1} & \frac{F_n \alpha_{n-1}}{\alpha_n} - F_{n-1} & \cdots & \frac{F_n \alpha_3}{\alpha_n} - F_3 & \frac{F_n \alpha_2}{\alpha_n} - F_2 \\ 0 & 1 & -\frac{\alpha_{n-1}}{\alpha_n} & \cdots & -\frac{\alpha_3}{\alpha_n} & -\frac{\alpha_2}{\alpha_n} \\ \vdots & \ddots & 1 & 0 & \cdots & 0 \\ \vdots & & \ddots & \ddots & \ddots & \vdots \\ \vdots & & & \ddots & 1 & 0 \\ 0 & \cdots & \cdots & \cdots & 0 & 1 \end{pmatrix}_{n \times n},$$

where α_i and β_n are defined as in (11) and (12), respectively.

Multiplying $A_1 T_{F,n} B_1$ by A_2 from the left and by B_2 from the right, we obtain

$$AT_{F,n}B = A_2 A_1 T_{F,n} B_1 B_2$$

$$= \begin{pmatrix} F_1 & 0 & 0 & 0 & \cdots & 0 & 0 \\ 0 & \alpha_n & 0 & 0 & \cdots & 0 & 0 \\ 0 & 0 & \beta_{n-1} - \frac{\beta_n \alpha_{n-1}}{\alpha_n} & \beta_{n-2} - \frac{\beta_n \alpha_{n-2}}{\alpha_n} & \cdots & \beta_3 - \frac{\beta_n \alpha_3}{\alpha_n} & \beta_2 - \frac{\beta_n \alpha_2}{\alpha_n} \\ \vdots & \ddots & 1 & 0 & \cdots & & 0 \\ \vdots & & \ddots & 1 & \ddots & & \vdots \\ \vdots & & & \ddots & \ddots & \ddots & \vdots \\ 0 & \cdots & \cdots & \cdots & 0 & 1 & 0 \end{pmatrix}_{n \times n},$$

where

$$A = A_2 A_1 = \begin{pmatrix} 1 & & & & & 0 \\ -F_n & & & & & 1 \\ \frac{\beta_n F_{-n}}{\alpha_n} - F_{-n-1} & & & 1 & -\frac{\beta_n}{\alpha_n} - 1 \\ 0 & & & 1 & -1 & -1 \\ \vdots & & \ddots & \ddots & \ddots & \\ 0 & 1 & -1 & -1 & & \end{pmatrix}_{n \times n},$$

$$B = B_1 B_2 = \begin{pmatrix} 1 & -\frac{F_n}{F_1} & \frac{F_n \alpha_{n-1}}{\alpha_n} - F_{n-1} & \cdots & \frac{F_n \alpha_3}{\alpha_n} - F_3 & \frac{F_n \alpha_2}{\alpha_n} - F_2 \\ 0 & \cdots & \cdots & \cdots & 0 & 1 \\ \vdots & & & \ddots & 1 & 0 \\ \vdots & & \ddots & \ddots & \ddots & \vdots \\ \vdots & \ddots & 1 & 0 & \cdots & 0 \\ 0 & 1 & -\frac{\alpha_{n-1}}{\alpha_n} & \cdots & -\frac{\alpha_3}{\alpha_n} & -\frac{\alpha_2}{\alpha_n} \end{pmatrix}_{n \times n},$$

with α_i and β_n are defined as in (11) and (12), respectively. In addition, the matrix $AT_{F,n}B$ admits a block partition of the form

$$AT_{F,n}B = N \oplus M, \tag{15}$$

where $N \oplus M$ denotes the direct sum of the matrices N and M, $N = \mathrm{diag}(F_1, \alpha_n)$ is a nonsingular diagonal matrix, and

$$M = \begin{pmatrix} -\frac{\beta_n \alpha_{n-1}}{\alpha_n} + \beta_{n-1} & -\frac{\beta_n \alpha_{n-2}}{\alpha_n} + \beta_{n-2} & \cdots & -\frac{\beta_n \alpha_3}{\alpha_n} + \beta_3 & -\frac{\beta_n \alpha_2}{\alpha_n} + \beta_2 \\ 1 & 0 & \cdots & \cdots & 0 \\ 0 & 1 & \ddots & & \vdots \\ \vdots & \ddots & \ddots & \ddots & \vdots \\ 0 & \cdots & 0 & 1 & 0 \end{pmatrix}_{(n-2)\times(n-2)}$$

From (15), we obtain
$$T_{F,n}^{-1} = B(N^{-1} \oplus M^{-1})A.$$

Based on the defintions of N and M, we have $N^{-1} = \mathrm{diag}(F_1^{-1}, \alpha_n^{-1})$ and

$$M^{-1} = \begin{pmatrix} 0 & 1 & 0 & \cdots & 0 \\ \vdots & \ddots & \ddots & \ddots & \vdots \\ \vdots & & \ddots & \ddots & 0 \\ 0 & \cdots & 0 & 0 & 1 \\ \frac{\alpha_n}{\beta_2\alpha_n - \beta_n\alpha_2} & -\frac{\beta_{n-1}\alpha_n - \beta_n\alpha_{n-1}}{\beta_2\alpha_n - \beta_n\alpha_2} & \cdots & -\frac{\beta_4\alpha_n - \beta_n\alpha_4}{\beta_2\alpha_n - \beta_n\alpha_2} & -\frac{\beta_3\alpha_n - \beta_n\alpha_3}{\beta_2\alpha_n - \beta_n\alpha_2} \end{pmatrix}_{(n-2)\times(n-2)}$$

By direct computation, we have

$$T_{F,n}^{-1} = B(N^{-1} \oplus M^{-1})A = \begin{pmatrix} \frac{F_n}{F_{n+1}} & -1 & 0 & 0 & 0 & \cdots & \cdots & 0 & \frac{(-1)^n}{F_{n+1}} \\ 1 & -1 & -1 & 0 & 0 & \cdots & \cdots & 0 & 0 \\ 0 & 1 & -1 & -1 & 0 & & & & 0 \\ \vdots & \ddots & 1 & -1 & -1 & \ddots & & & \vdots \\ \vdots & & \ddots & \ddots & \ddots & \ddots & & & \vdots \\ \vdots & & & \ddots & \ddots & \ddots & \ddots & & \vdots \\ \vdots & & & & \ddots & 1 & -1 & -1 & 0 \\ 0 & \cdots & \cdots & \cdots & \cdots & 0 & 1 & -1 & -1 \\ -\frac{1}{F_{n+1}} & 0 & \cdots & \cdots & \cdots & \cdots & 0 & 1 & \frac{F_n}{F_{n+1}} \end{pmatrix}_{n\times n}$$

□

Remark 3. *It is well known that if you divide F_n by F_{n+1}, then these ratios get closer and closer to about 0.618, which is known to many people as the Golden Ratio, a number which has fascinated mathematicians, scientists and artists for centuries. Equation (14) can be appreciated in many different ways, and it is easy to see that top-left and bottom-right corner entries of $T_{F,n}^{-1}$ get closer and closer to the Golden Ratio. In fact, Toeplitz matrices, tridiagonal Toeplitz matrices with perturbed corner entries, the Fibonacci number, and the Golden Ratio are all connected by Equation (14).*

2.2. Determinant and Inverse Matrix of a Fankel Matrix

In this subsection, the determinant and the inverse of the Fankel matrix $H_{F,n}$ are studied.

Theorem 3. Let $H_{F,n}$ be an $n \times n$ Fankel matrix defined as in (5). Then $H_{F,n}$ is invertible and

$$\det H_{F,n} = (-1)^{\frac{(n-1)n}{2}} F_{n+1},$$

where F_{n+1} is the $(n+1)$th Fibonacci number.

Proof. From (8), it follows that $\det H_{F,n} = \det \hat{I}_n \det T_{F,n}$. We obtain this conclusion by the fact that $\det \hat{I}_n = (-1)^{\frac{n(n-1)}{2}}$ and Theorem 1. □

Remark 4. *This Theorem gives the relationship between the Fankel matrix and the Fibonacci number. From the standpoint of number theory, the $(n+1)$th Fibonacci number can be expressed as the product of the determinant of an $n \times n$ Fankel matrix and a sign function.*

Theorem 4. Let $H_{F,n}$ be an $n \times n$ Fankel matrix defined as in (5). Then inverse matrix of $H_{F,n}$ is

$$H_{F,n}^{-1} = \begin{pmatrix} -\frac{1}{F_{n+1}} & 0 & 0 & 0 & 0 & \cdots & 0 & 0 & 1 & \frac{F_n}{F_{n+1}} \\ 0 & \cdots & \cdots & \cdots & \cdots & \cdots & 0 & 1 & -1 & -1 \\ \vdots & & & & \cdot\cdot\cdot & 1 & -1 & -1 & 0 \\ \vdots & & & \cdot\cdot\cdot & \cdot\cdot\cdot & \cdot\cdot\cdot & \cdot\cdot\cdot & & \vdots \\ \vdots & & \cdot\cdot\cdot & \cdot\cdot\cdot & \cdot\cdot\cdot & \cdot\cdot\cdot & & & \vdots \\ \vdots & \cdot\cdot\cdot & \cdot\cdot\cdot & \cdot\cdot\cdot & \cdot\cdot\cdot & & & & \vdots \\ 0 & 1 & -1 & -1 & \cdot\cdot\cdot & & & & \vdots \\ 1 & -1 & -1 & 0 & \cdots & \cdots & \cdots & & 0 \\ \frac{F_n}{F_{n+1}} & -1 & 0 & 0 & 0 & \cdots & 0 & 0 & 0 & \frac{(-1)^n}{F_{n+1}} \end{pmatrix}_{n \times n}, \quad (16)$$

where F_n and F_{n+1} are the nth and $(n+1)$th Fibonacci numbers, respectively.

Proof. We obtain this conclusion by formula (8) and Theorem 2. □

Remark 5. *Equation (16) can be appreciated in many different ways, and it is easy to see that bottom-left and top-right corner entries of $H_{F,n}^{-1}$ get closer and closer to the Golden Ratio. In fact, Hankel matrices, sub-tridiagonal Hankel matrices with perturbed corner entries, the Fibonacci number, and the Golden Ratio are all connected by Equation (16).*

2.3. Determinant and Inverse Matrix of a Loeplitz Matrix

In this subsection, the determinant and the inverse of the Loeplitz matrix $T_{L,n}$ are studied.

Theorem 5. Let $T_{L,n}$ be an $n \times n$ Loeplitz matrix defined as in (6). Then $T_{L,n}$ is invertible and

$$\det T_{L,n} = (-1)^{n+1} L_{n+1} - 2^n, \text{ for } n \geq 1, \quad (17)$$

where L_{n+1} is the $(n+1)$th Lucas number.

Proof. For $n \leq 3$, it is easy to check that

$$\det T_{L,1} = 1, \ \det T_{L,2} = -8 \text{ and } \det T_{L,3} = -1.$$

Therefore, Equation (17) is satisfied. Now, we consider the case $n > 3$. Define additional nonsingular matrices,

$$\Delta_1 = \begin{pmatrix} 1 & & & & & & \\ -L_{-n} & & & & & 1 & \\ -L_{-n-1} & & & & 1 & -1 & \\ 0 & & & 1 & -1 & -1 & \\ \vdots & & \iddots & \iddots & \iddots & & \\ 0 & 1 & -1 & -1 & & & \end{pmatrix}_{n \times n}.$$

Multiplying $T_{L,n}$ by Δ_1 from the left, we obtain

$$\Delta_1 T_{L,n} = \begin{pmatrix} L_1 & L_2 & L_3 & \cdots & L_{n-2} & L_{n-1} & L_n \\ 0 & a_2 & a_3 & \cdots & a_{n-2} & a_{n-1} & a_n \\ 0 & b_2 & b_3 & \cdots & b_{n-2} & b_{n-1} & b_n \\ 0 & \cdots & \cdots & 0 & 2 & -1 & 0 \\ \vdots & & \iddots & \iddots & \iddots & \iddots & \vdots \\ \vdots & \iddots & 2 & -1 & \iddots & & \vdots \\ 0 & 2 & -1 & 0 & \cdots & \cdots & 0 \end{pmatrix}_{n \times n},$$

where

$$\begin{cases} a_i = -L_{-n}L_i + L_{-n+i-1}, (i = 2, 3, \cdots, n-1), \\ a_n = -L_{-n}L_n + L_1, \\ b_i = -L_{-n-1}L_i + L_{-n+i-2}, (i = 2, 3, \cdots, n-2), \\ b_{n-1} = -L_{-n-1}L_{n-1} + L_1 - L_{-2}, \\ b_n = -L_{-n-1}L_n + L_2 - L_1. \end{cases} \quad (18)$$

Then, multiplying $\Delta_1 T_{L,n}$ by B_1 from the right, we have

$$\Delta_1 T_{L,n} B_1 = \begin{pmatrix} L_1 & L_n & L_{n-1} & L_{n-2} & \cdots & L_3 & L_2 \\ 0 & a_n & a_{n-1} & a_{n-2} & \cdots & a_3 & a_2 \\ \vdots & b_n & b_{n-1} & b_{n-2} & \cdots & b_3 & b_2 \\ \vdots & 0 & -1 & 2 & 0 & \cdots & 0 \\ \vdots & \vdots & \iddots & \iddots & \iddots & \iddots & \vdots \\ \vdots & \vdots & & \iddots & -1 & 2 & 0 \\ 0 & 0 & \cdots & \cdots & 0 & -1 & 2 \end{pmatrix}_{n \times n}, \quad (19)$$

and

$$\begin{aligned} \det(\Delta_1 T_{L,n} B_1) &= \det(\Delta_1) \det(T_{L,n}) \det(B_1) \\ &= L_1 (a_n \sum_{i=1}^{n-2} 2^{n-2-i} b_{n-i} - b_n \sum_{i=1}^{n-2} 2^{n-2-i} a_{n-i}) \\ &= 2^{n-3}[(-L_{-n-1}L_{n-1} + L_1 - L_{-2})(-L_{-n}L_n + L_1) - (-L_{-n}L_{n-1} + L_{-2}) \\ &\quad \cdot (-L_{-n-1}L_n + L_0)] + \sum_{i=2}^{n-2} 2^{n-2-i}[(-L_{-n-1}L_{n-i} + L_{-i-2})(-L_{-n}L_n + L_1) \\ &\quad - (-L_{-n}L_{n-i} + L_{-i-1})(-L_{-n-1}L_n + L_0)]. \end{aligned}$$

From the definition of Δ_1 and B_1, we get

$$\det \Delta_1 = \det B_1 = (-1)^{\frac{(n-1)(n-2)}{2}}.$$

By formulas (2) and (3), we obtain

$$\det T_{L,n} = (-1)^{n+1} L_{n+1} - 2^n,$$

which completes the proof. □

Remark 6. *This Theorem gives the relationship between the Loeplitz matrix and the Lucas number. From the perspective of number theory, the $(n+1)$th Lucas number can be expressed as the sum of the determinant of $n \times n$ Loeplitz matrix and scalar matrix.*

Theorem 6. *Let $T_{L,n}$ be an $n \times n$ Loeplitz matrix defined as in (6). Then*

$$T_{L,1}^{-1} = 1, \quad T_{L,2}^{-1} = \begin{pmatrix} -\frac{1}{8} & \frac{3}{8} \\ \frac{3}{8} & -\frac{1}{8} \end{pmatrix}, \quad T_{L,3}^{-1} = \begin{pmatrix} 8 & -9 & -5 \\ 15 & -17 & -9 \\ -13 & 15 & 8 \end{pmatrix},$$

and for $n > 3$, $T_{L,n}^{-1}$ is

$$\mathbf{T}_{L,n}^{-1} = \begin{pmatrix} Q_3 & Q_2 & 2^{n-3}Q_1 & \cdots & 2^2 Q_1 & 2Q_1 & Q_1 \\ Q_4 & Q_5 & Q_2 & \ddots & \ddots & 2^2 Q_1 & 2Q_1 \\ 2Q_4 & Q_6 & Q_5 & \ddots & \ddots & \ddots & 2^2 Q_1 \\ \vdots & 2Q_6 & \ddots & \ddots & \ddots & \ddots & \vdots \\ 2^{n-4}Q_4 & \vdots & \ddots & \ddots & Q_5 & Q_2 & 2^{n-3}Q_1 \\ 2^{n-3}Q_4 & 2^{n-4}Q_6 & \cdots & 2Q_6 & Q_6 & Q_5 & Q_2 \\ Q_7 & 2^{n-3}Q_4 & 2^{n-4}Q_4 & \cdots & 2Q_4 & Q_4 & Q_3 \end{pmatrix}_{n \times n}, \quad (20)$$

where

$$Q_1 = \frac{5}{\det T_{L,n}},$$
$$Q_2 = 1 + 2^{n-2} Q_1,$$
$$Q_3 = \frac{\det T_{L,n-1}}{\det T_{L,n}},$$
$$Q_4 = \frac{(-1)^n (L_n + L_{n+2})}{\det T_{L,n}},$$
$$Q_5 = 3 + 2^{n-1} Q_1,$$
$$Q_6 = 5 + 2^n Q_1,$$
$$Q_7 = \frac{2^{n-2}[L_n + (-1)^{n+1} L_{n-1}] + (-1)^n}{\det T_{L,n}},$$
$$\det T_{L,n} = (-1)^{n+1} L_{n+1} - 2^n,$$

and L_j ($j = 1, \pm 2, \cdots, \pm n$) is the jth Lucas number.

Proof. For $n \leq 3$, it is easy to check that

$$T_{L,1}^{-1} = 1, \ T_{L,2}^{-1} = \begin{pmatrix} -\frac{1}{8} & \frac{3}{8} \\ \frac{3}{8} & -\frac{1}{8} \end{pmatrix}, \ T_{L,3}^{-1} = \begin{pmatrix} 8 & -9 & -5 \\ 15 & -17 & -9 \\ -13 & 15 & 8 \end{pmatrix}.$$

Now, we consider the case $n \geq 4$. The explicit expression of the inverse of the Loeplitz matrix can be found by use of Equation (20). Define additionally two nonsingular matrices.

$$\Delta_2 = \begin{pmatrix} 1 & & & & \\ & 1 & & & \\ & -\frac{b_n}{a_n} & 1 & & \\ & & & \ddots & \\ & & & & 1 \end{pmatrix}_{n \times n}$$

and

$$\nabla_2 = \begin{pmatrix} 1 & -\frac{L_n}{L_1} & \tau_{n-1} & \cdots & \tau_3 & \tau_2 \\ 0 & 1 & -\frac{a_{n-1}}{a_n} & \cdots & -\frac{a_3}{a_n} & -\frac{a_2}{a_n} \\ \vdots & 0 & 1 & 0 & \cdots & 0 \\ \vdots & \vdots & \ddots & \ddots & \ddots & \vdots \\ \vdots & \vdots & & \ddots & \ddots & 0 \\ 0 & 0 & \cdots & \cdots & 0 & 1 \end{pmatrix}_{n \times n},$$

where

$$\tau_i = \frac{L_n a_i}{L_1 a_n} - \frac{L_i}{L_1}, (i = 2, 3, \cdots, n-1), \quad (21)$$

with a_i and b_i are defined as in (18).

Multiplying $\Delta_1 T_{L,n} B_1$ by Δ_2 from the left and by ∇_2 from the right, we get

$$\Delta T_{L,n} \nabla = \Delta_2 \Delta_1 T_{L,n} B_1 \nabla_2 = \begin{pmatrix} L_1 & 0 & 0 & 0 & \cdots & 0 & 0 \\ 0 & a_n & 0 & 0 & \cdots & 0 & 0 \\ \vdots & 0 & \gamma_{n-1} & \gamma_{n-2} & \cdots & \gamma_3 & \gamma_2 \\ \vdots & 0 & -1 & 2 & 0 & \cdots & 0 \\ \vdots & \vdots & \ddots & \ddots & \ddots & \ddots & \vdots \\ \vdots & \vdots & & \ddots & \ddots & \ddots & 0 \\ 0 & 0 & \cdots & \cdots & & 0 & -1 & 2 \end{pmatrix}_{n \times n},$$

where

$$\Delta = \Delta_2 \Delta_1 = \begin{pmatrix} 1 & & & & & & 0 \\ -L_{-n} & & & & & & 1 \\ \frac{b_n L_{-n}}{a_n} - L_{-n-1} & & & & 1 & \frac{-b_n}{a_n} - 1 \\ 0 & & & 1 & -1 & -1 \\ \vdots & & \iddots & \iddots & \iddots & \\ 0 & 1 & -1 & -1 & & & \end{pmatrix}_{n \times n},$$

$$\nabla = B_1 \nabla_2 = \begin{pmatrix} 1 & -\frac{L_n}{L_1} & \tau_{n-1} & \cdots & \tau_3 & \tau_2 \\ 0 & \cdots & \cdots & & 0 & 1 \\ \vdots & & & \iddots & 1 & 0 \\ \vdots & & \iddots & \iddots & \iddots & \vdots \\ \vdots & \iddots & 1 & 0 & \cdots & 0 \\ 0 & 1 & -\frac{a_{n-1}}{a_n} & \cdots & -\frac{a_3}{a_n} & -\frac{a_2}{a_n} \end{pmatrix}_{n \times n},$$

$$\gamma_i = -\frac{b_n a_i}{a_n} + b_i, \ (i = 2, 3, \cdots, n-1),$$

with a_i, b_i and τ_i are defined as in (18) and (21), respectively. In addition, the matrix $\Delta T_{L,n} \nabla$ admits a block partition of the form

$$\Delta T_{L,n} \nabla = \mathcal{N} \oplus \mathcal{M}, \tag{22}$$

where $\mathcal{N} \oplus \mathcal{M}$ denotes the direct sum of the matrices $\mathcal{N}$ and $\mathcal{N}$. $\mathcal{N} = \mathrm{diag}(L_1, a_n)$ is a nonsingular diagonal matrix,

$$\mathcal{M} = \begin{pmatrix} \gamma_{n-1} & \gamma_{n-2} & \gamma_{n-3} & \cdots & \gamma_3 & \gamma_2 \\ -1 & 2 & 0 & \cdots & & 0 \\ 0 & -1 & 2 & \ddots & & \vdots \\ \vdots & \ddots & \ddots & \ddots & \ddots & \vdots \\ \vdots & & \ddots & -1 & 2 & 0 \\ 0 & \cdots & & & -1 & 2 \end{pmatrix}_{(n-2) \times (n-2)}.$$

Denote $\ell = \gamma_{n-1} - VC^{-1}U \neq 0$, where $V = (\gamma_{n-2} \ \gamma_{n-3} \ \cdots \ \gamma_3 \ \gamma_2)_{1 \times (n-3)}$,

$$C = \begin{pmatrix} 2 & 0 & \cdots & & 0 \\ -1 & 2 & \ddots & & \vdots \\ 0 & \ddots & \ddots & \ddots & \vdots \\ \vdots & \ddots & -1 & 2 & 0 \\ 0 & \cdots & & -1 & 2 \end{pmatrix}_{(n-3) \times (n-3)}$$

and $U = (-1 \ 0 \ \cdots \ 0)^T_{1 \times (n-3)}$. From (22), we obtain

$$T_{L,n}^{-1} = \nabla (\mathcal{N}^{-1} \oplus \mathcal{M}^{-1}) \Delta.$$

Based on the definitions of $\mathcal{N}$ and $\mathcal{M}$, we have $\mathcal{N}^{-1} = \text{diag}(L_1^{-1}, a_n^{-1})$. By Lemma 1, we get

$$C^{-1} = \begin{pmatrix} \varpi_1 & 0 & \cdots & \cdots & \cdots & \cdots & 0 \\ \varpi_2 & \varpi_1 & \ddots & & & & \vdots \\ \varpi_3 & \varpi_2 & \varpi_1 & \ddots & & & \vdots \\ \vdots & \ddots & \ddots & \ddots & \ddots & & \vdots \\ \varpi_{n-5} & \ddots & & \ddots & \ddots & \ddots & \vdots \\ \varpi_{n-4} & \varpi_{n-5} & \ddots & \ddots & \varpi_2 & \varpi_1 & 0 \\ \varpi_{n-3} & \varpi_{n-4} & \varpi_{n-5} & \cdots & \varpi_3 & \varpi_2 & \varpi_1 \end{pmatrix}_{(n-3)\times(n-3)},$$

where

$$\varpi_i = \frac{1}{2^i}, \ 1 \le i \le n-3.$$

From Lemma 5 in [20], we have

$$\mathcal{M}^{-1} = \begin{pmatrix} \frac{1}{\ell} & -\frac{1}{\ell}VC^{-1} \\ -\frac{1}{\ell}C^{-1}U & C^{-1} + \frac{1}{\ell}C^{-1}UVC^{-1} \end{pmatrix}_{(n-2)\times(n-2)}$$

where

$$VC^{-1} = (\hat{\eta}_1, \hat{\eta}_2, \cdots, \hat{\eta}_{n-3}),$$

$$C^{-1} + \frac{1}{\ell}C^{-1}UVC^{-1} = [m'_{i,j}]_{i,j=1}^{n-3},$$

$$\hat{\eta}_i = \sum_{j=1}^{n-2-i} \gamma_{n-1-j}\varpi_i, \ 1 \le i \le n-3,$$

$$m'_{i,j} = \varpi_{i-j+1} - \frac{\hat{\eta}_j}{\ell}, \ 1 \le j \le i \le n-3,$$

$$m'_{i,j} = -\frac{\hat{\eta}_j}{\ell}, \ 1 \le i < j \le n-3.$$

Therefore, we get

$$\mathcal{N}^{-1} \oplus \mathcal{M}^{-1} = \begin{pmatrix} 1 & 0 & 0 & \cdots & \cdots & \cdots & 0 \\ 0 & \frac{1}{a_n} & 0 & \cdots & \cdots & \cdots & 0 \\ 0 & 0 & \frac{1}{\ell} & -\frac{\hat{\eta}_1}{\ell} & -\frac{\hat{\eta}_2}{\ell} & \cdots & -\frac{\hat{\eta}_{n-3}}{\ell} \\ \vdots & \vdots & \frac{\varpi_1}{\ell} & m'_{1,1} & m'_{1,2} & \cdots & m'_{1,n-3} \\ \vdots & \vdots & \frac{\varpi_2}{\ell} & m'_{2,1} & m'_{2,2} & \cdots & m'_{2,n-3} \\ \vdots & \vdots & \vdots & \vdots & \vdots & \ddots & \vdots \\ 0 & 0 & \frac{\varpi_{n-3}}{\ell} & m'_{n-3,1} & m'_{n-3,2} & \cdots & m'_{n-3,n-3} \end{pmatrix}_{n\times n}.$$

Multiplying $\mathcal{N}^{-1} \oplus \mathcal{M}^{-1}$ by Δ from the right, we obtain

$$(\mathcal{N}^{-1} \oplus \mathcal{M}^{-1})\Delta = \begin{pmatrix} 1 & 0 & \cdots & \cdots & \cdots & \cdots & 0 \\ \frac{-L_{-n}}{a_n} & 0 & \cdots & \cdots & \cdots & 0 & \frac{1}{a_n} \\ c_{1,1} & c_{1,2} & c_{1,3} & c_{1,4} & c_{1,5} & \cdots & c_{1,n} \\ c_{2,1} & c_{2,2} & c_{2,3} & c_{2,4} & c_{2,5} & \cdots & c_{2,n} \\ c_{3,1} & c_{3,2} & c_{3,3} & c_{3,4} & c_{3,5} & \cdots & c_{3,n} \\ \vdots & \vdots & \vdots & \vdots & \vdots & \ddots & \vdots \\ c_{n-2,1} & c_{n-2,2} & c_{n-2,3} & c_{n-2,4} & c_{n-2,5} & \cdots & c_{n-2,n} \end{pmatrix}_{n \times n},$$

where

$$c_{1,1} = \frac{1}{\ell}\left(\frac{b_n L_{-n}}{a_n} - L_{-(n+1)}\right), \quad c_{i,1} = \frac{\omega_{i-1}}{\ell}\left(\frac{b_n L_{-n}}{a_n} - L_{-(n+1)}\right), \quad 2 \leq i \leq n-2,$$

$$c_{1,2} = -\frac{\hat{\eta}_{n-3}}{\ell}, \quad c_{i,2} = m'_{i-1,n-3}, \quad 2 \leq i \leq n-2,$$

$$c_{1,3} = -\frac{\hat{\eta}_{n-4}}{\ell} + \frac{\hat{\eta}_{n-3}}{\ell}, \quad c_{i,3} = m'_{i-1,n-4} - m'_{i-1,n-3}, \quad 2 \leq i \leq n-2,$$

$$c_{1,j} = -\frac{\hat{\eta}_{n-1-j}}{\ell} + \frac{\hat{\eta}_{n-j}}{\ell} + \frac{\hat{\eta}_{n+1-j}}{\ell}, \quad 4 \leq j \leq n-1,$$

$$c_{i,j} = m'_{i-1,n-1-j} - m'_{i-1,n-j} - m'_{i-1,n+1-j}, \quad 2 \leq i \leq n-2, \; 4 \leq i \leq n-1,$$

$$c_{1,n} = \frac{1}{\ell}\left(\frac{-b_n}{a_n} - 1\right) + \frac{\hat{\eta}_1}{\ell}, \quad c_{i,n} = \frac{\omega_{i-1}}{\ell}\left(\frac{-b_n}{a_n} - 1\right) - m'_{i-1,1}, \quad 2 \leq i \leq n-2.$$

By formulas (2) and (3), we have

$$T_{L,n}^{-1} = \nabla(\mathcal{N}^{-1} \oplus \mathcal{M}^{-1})\Delta = \begin{pmatrix} Q_3 & Q_2 & 2^{n-3}Q_1 & \cdots & 2^2 Q_1 & 2Q_1 & Q_1 \\ Q_4 & Q_5 & Q_2 & \ddots & \ddots & 2^2 Q_1 & 2Q_1 \\ 2Q_4 & Q_6 & Q_5 & \ddots & \ddots & \ddots & 2^2 Q_1 \\ \vdots & 2Q_6 & \ddots & \ddots & \ddots & \ddots & \vdots \\ 2^{n-4}Q_4 & \vdots & \ddots & \ddots & Q_5 & Q_2 & 2^{n-3}Q_1 \\ 2^{n-3}Q_4 & 2^{n-4}Q_6 & \cdots & 2Q_6 & Q_6 & Q_5 & Q_2 \\ Q_7 & 2^{n-3}Q_4 & 2^{n-4}Q_4 & \cdots & 2Q_4 & Q_4 & Q_3 \end{pmatrix}_{n \times n},$$

where $Q_i (i = 1, 2, \cdots, 7)$ is the same as in Theorem 6. □

2.4. Determinant and Inverse Matrix of a Lankel Matrix

In this subsection, the determinant and the inverse of the Lankel matrix $H_{L,n}$ are studied.

Theorem 7. Let $H_{L,n}$ be an $n \times n$ Lankel matrix defined as in (7). Then $H_{L,n}$ is invertible and

$$\det H_{L,n} = (-1)^{\frac{n(n-1)}{2}}\left[(-1)^{n+1} L_{n+1} - 2^n\right],$$

where L_{n+1} is the $(n+1)$th Lucas number.

Proof. From formula (9), it follows that $\det H_{L,n} = \det \hat{I}_n \det T_{L,n}$. We obtain the desired conclusion by using $\det \hat{I}_n = (-1)^{\frac{n(n-1)}{2}}$ and Theorem 5. □

Remark 7. *This Theorem gives the relationship between the Lankel matrix and the Lucas number. In terms of number theory, the $(n+1)$th Lucas number can be expressed as the sum of the determinant of $n \times n$ Lankel matrix and scalar matrix.*

Theorem 8. *Let $H_{L,n}$ be an $n \times n$ Lankel matrix defined as in (7). Then*

$$H_{L,1}^{-1} = 1, \quad H_{L,2}^{-1} = \begin{pmatrix} \frac{3}{8} & -\frac{1}{8} \\ -\frac{1}{8} & \frac{3}{8} \end{pmatrix}, \quad H_{L,3}^{-1} = \begin{pmatrix} -13 & 15 & 8 \\ 15 & -17 & -9 \\ 8 & -9 & -5 \end{pmatrix},$$

and for $n > 3$, $H_{L,n}^{-1}$ is

$$H_{L,n}^{-1} = \begin{pmatrix} Q_7 & 2^{n-3}Q_4 & 2^{n-4}Q_4 & \cdots & 2Q_4 & Q_4 & Q_3 \\ 2^{n-3}Q_4 & 2^{n-4}Q_6 & & \cdots & 2Q_6 & Q_6 & Q_5 & Q_2 \\ 2^{n-4}Q_4 & & \vdots & & \ddots & Q_5 & Q_2 & 2^{n-3}Q_1 \\ \vdots & & 2Q_6 & & \ddots & & \ddots & \vdots \\ 2Q_4 & Q_6 & Q_5 & & \ddots & & & 2^2 Q_1 \\ Q_4 & Q_5 & Q_2 & & \ddots & & 2^2 Q_1 & 2Q_1 \\ Q_3 & Q_2 & 2^{n-3}Q_1 & \cdots & & 2^2 Q_1 & 2Q_1 & Q_1 \end{pmatrix}_{n \times n},$$

where $Q_i(i = 1, 2, \cdots, 7)$ is the same as in Theorem 6.

Proof. By formula (9), we have $H_{L,n}^{-1} = \hat{I}_n T_{L,n}^{-1}$. Thus we get the desired conclusion from Theorem 6. □

3. Example

In this section, an example demonstrates the method which was introduced above for the calculation of the determinant and inverse of the Foeplitz matrix and the Loeplitz matrix.

Example 1. *Here we consider an 8×8 Foeplitz matrix:*

$$T_{F,8} = \begin{pmatrix} 1 & 1 & 2 & 3 & 5 & 8 & 13 & 21 \\ -1 & 1 & 1 & 2 & 3 & 5 & 8 & 13 \\ 2 & -1 & 1 & 1 & 2 & 3 & 5 & 8 \\ -3 & 2 & -1 & 1 & 1 & 2 & 3 & 5 \\ 5 & -3 & 2 & -1 & 1 & 1 & 2 & 3 \\ -8 & 5 & -3 & 2 & -1 & 1 & 1 & 2 \\ 13 & -8 & 5 & -3 & 2 & -1 & 1 & 1 \\ -21 & 13 & -8 & 5 & -3 & 2 & -1 & 1 \end{pmatrix}_{8 \times 8}.$$

From formula (10), we obtain

$$\det T_{F,8} = F_9 = 34.$$

As the inverse calculation, if we use the corresponding formulas in Theorems 2, we have $F_8 = 21$, $F_9 = 34$. So we get

$$\mathbf{T}_{F,8}^{-1} = \begin{pmatrix} \frac{21}{34} & -1 & 0 & 0 & 0 & 0 & 0 & \frac{1}{34} \\ 1 & -1 & -1 & 0 & 0 & 0 & 0 & 0 \\ 0 & 1 & -1 & 1 & 0 & 0 & 0 & 0 \\ 0 & 0 & 1 & -1 & -1 & 0 & 0 & 0 \\ 0 & 0 & 0 & 1 & -1 & -1 & 0 & 0 \\ 0 & 0 & 0 & 0 & 1 & -1 & -1 & 0 \\ 0 & 0 & 0 & 0 & 0 & 1 & -1 & -1 \\ -\frac{1}{34} & 0 & 0 & 0 & 0 & 0 & 1 & \frac{21}{34} \end{pmatrix}_{8 \times 8}.$$

Example 2. Here we consider a 5×5 Loeplitz matrix:

$$\mathbf{T}_{L,5} = \begin{pmatrix} 1 & 3 & 4 & 7 & 11 \\ 3 & 1 & 3 & 4 & 7 \\ -4 & 3 & 1 & 3 & 4 \\ 7 & -4 & 3 & 1 & 3 \\ -11 & 7 & -4 & 3 & 1 \end{pmatrix}_{5 \times 5}.$$

From formula (18), we obtain

$$\det \mathbf{T}_{L,5} = -(2L_7 - L_9) - 2^5 = -14.$$

As the inverse calculation, if we use the corresponding formulas in Theorem 6, we have $Q_1 = -\frac{5}{14}$, $Q_2 = -\frac{13}{7}$, $Q_3 = \frac{27}{14}$, $Q_4 = \frac{20}{7}$, $Q_5 = -\frac{19}{7}$, $Q_6 = -\frac{45}{7}$, $Q_7 = -\frac{143}{14}$. So we get

$$\mathbf{T}_{L,5}^{-1} = \begin{pmatrix} \frac{27}{14} & -\frac{13}{7} & -\frac{10}{7} & -\frac{5}{7} & -\frac{5}{14} \\ \frac{20}{7} & -\frac{19}{7} & -\frac{13}{7} & -\frac{10}{7} & -\frac{5}{7} \\ \frac{40}{7} & -\frac{45}{7} & -\frac{19}{7} & -\frac{13}{7} & -\frac{10}{7} \\ \frac{80}{7} & -\frac{90}{7} & -\frac{45}{7} & -\frac{19}{7} & -\frac{13}{7} \\ -\frac{143}{14} & \frac{80}{7} & \frac{40}{7} & \frac{20}{7} & \frac{27}{14} \end{pmatrix}_{5 \times 5}.$$

Author Contributions: Conceptualization, methodology, funding acquisition, Z.J.; writing–original draft preparation, W.W.; investigation, resources, formal analysis, software, B.Z. and B.N.; writing–review and editing, supervision, visualization, Y.Z.

Funding: The research was funded by the National Natural Science Foundation of China (Grant No. 11671187), the Natural Science Foundation of Shandong Province (Grant No. ZR2016AM14) and the PhD Research Foundation of Linyi University (Grant No. LYDX2018BS067), China.

Acknowledgments: The authors are grateful to the anonymous referees for their useful suggestions which improve the contents of this article.

Conflicts of Interest: The authors declare no conflict of interest.

References

1. Mukhexjee, B.N.; Maiti, S.S. On some properties of positive definite Toeplitz matrices and their possible applications. *Linear Algebra Appl.* **1988**, *102*, 211–240. [CrossRef]
2. Basilevsky, A. *Applied Matrix Algebra in the Statistical Sciences*; North Holland: New York, NY, USA, 1983.
3. Bai, Z.Z.; Li, G.Q.; Lu, L.Z. Combinative preconditioners of modified incomplete Cholesky factorization and Sherman-Morrison-Woodbury update for self-adjoint elliptic Dirichlet-periodic boundary value problems. *J. Comput. Math.* **2004**, *22*, 833–856.
4. Chen, H.; Lv, W.; Zhang, T.T. A Kronecker product splitting preconditioner for two-dimensional space-fractional diffusion equations. *J. Comput. Phys.* **2018**, *360*, 1–14. [CrossRef]
5. Chen, H.; Zhang, T.T.; Lv, W. Block preconditioning strategies for time-space fractional diffusion equations. *Appl. Math. Comput.* **2018**, *337*, 41–53. [CrossRef]
6. Chen, H.; Wang X.L.; Li X.L. A note on efficient preconditioner of implicit Runge-Kutta methods with application to fractional diffusion equations. *Appl. Math. Comput.* **2019**, *351*, 116–123. [CrossRef]
7. Grenander, U.; Szego, G. *Toeplitz Forms and Their Applications*; University of California Press: Berkeley, CA, USA, 1958.
8. Iohvidov, I.S. *Hankel and Toeplitz Matrices and Forms (Algebruic Theory)*; Translated by G. Philips and A. Thijse; Birkhäuser: Boston, MA, USA, 1982.
9. Heinig, G.; Rost, K. *Algebraic Methods for Toeplitz-Like Matrices and Operators*; Oper. Theory 13; Birtiuser: Basel, Switzerland, 1984.
10. Zheng, Y.P.; Shon, S. Exact determinants and inverses of generalized Lucas skew circulant type matrices. *Appl. Math. Comput.* **2015**, *270*, 105–113. [CrossRef]
11. Jiang, X.Y.; Hong, K.C. Explicit inverse matrices of Tribonacci skew circulant type matrices. *Appl. Math. Comput.* **2015**, *268*, 93–102. [CrossRef]
12. Bozkurt, D.; Tam, T.Y. Determinants and inverses of circulant matrices with Jacobsthal and Jacobsthal-Lucas Numbers. *Appl. Math. Comput.* **2012**, *219*, 544–551. [CrossRef]
13. Li, J.; Jiang, Z.L.; Lu, F.L. Determinants, norms, and the spread of circulant matrices with Tribonacci and generalized Lucas numbers. *Abstr. Appl. Anal.* **2014**, *2014*, 381829. [CrossRef]
14. Jiang, Z.L.; Gong, Y.P.; Gao, Y. Invertibility and explicit inverses of circulant-type matrices with k-Fibonacci and k-Lucas numbers. *Abstr. Appl. Anal.* **2014**, *2014*, 238953.
15. Jiang, Z.L.; Gong, Y.P.; Gao, Y. Circulant type matrices with the sum and product of Fibonacci and Lucas numbers. *Abstr. Appl. Anal.* **2014**, *2014*, 375251. [CrossRef]
16. Jiang, X.Y.; Hong, K.C. Exact determinants of some special circulant matrices involving four kinds of famous numbers. *Abstr. Appl. Anal.* **2014**, *2014*, 273680. [CrossRef]
17. Akbulak, M.; Bozkurt, D. On the norms of Toeplitz matrices involving Fibonacci and Lucas numbers. *Hacet. J. Math. Stat.* **2008**, *37*, 89–95.
18. Thomas, K. *Fibonacci and Lucas Numbers with Applications*; John Wiley & Sons: New York, NY, USA, 2001.
19. Zuo, B.S.; Jiang, Z.L.; Fu, D.Q. Determinants and inverses of Ppoeplitz and Ppankel matrices. *Special Matrices* **2018**, *6*, 201–215. [CrossRef]
20. Liu, L.; Jiang, Z.L. Explicit form of the inverse matrices of Tribonacci circulant type matrices. *Abstr. Appl. Anal.* **2015**, *2015*, 169726. [CrossRef]

© 2019 by the authors. Licensee MDPI, Basel, Switzerland. This article is an open access article distributed under the terms and conditions of the Creative Commons Attribution (CC BY) license (http://creativecommons.org/licenses/by/4.0/).

Article

A Study of Determinants and Inverses for Periodic Tridiagonal Toeplitz Matrices with Perturbed Corners Involving Mersenne Numbers

Yunlan Wei [1,2], Yanpeng Zheng [1,3,*], Zhaolin Jiang [1,*] and Sugoog Shon [2]

1. School of Mathematics and Statistics, Linyi University, Linyi 276000, China; weiyunlan@lyu.edu.cn
2. College of Information Technology, The University of Suwon, Hwaseong-si 445-743, Korea; sshon@suwon.ac.kr
3. School of Automation and Electrical Engineering, Linyi University, Linyi 276000, China
* Correspondence: zhengyanpeng@lyu.edu.cn (Y.Z.); jiangzhaolin@lyu.edu.cn or jzh1208@sina.com (Z.J.)

Received: 5 July 2019; Accepted: 20 September 2019; Published: 24 September 2019

Abstract: In this paper, we study periodic tridiagonal Toeplitz matrices with perturbed corners. By using some matrix transformations, the Schur complement and matrix decompositions techniques, as well as the Sherman-Morrison-Woodbury formula, we derive explicit determinants and inverses of these matrices. One feature of these formulas is the connection with the famous Mersenne numbers. We also propose two algorithms to illustrate our formulas.

Keywords: determinant; inverse; Mersenne number; periodic tridiagonal Toeplitz matrix; Sherman-Morrison-Woodbury formula

1. Introduction

Mersenne numbers are ubiquitous in combinatorics, group theory, chaos, geometry, physics, etc. [1]. They are generated by the following recurrence [2]:

$$M_{n+1} = 3M_n - 2M_{n-1} \quad \text{where} \quad M_0 = 0,\ M_1 = 1,\ n \geq 1; \tag{1}$$

$$M_{-(n+1)} = \frac{3}{2}M_{-n} - \frac{1}{2}M_{-(n-1)} \quad \text{where} \quad M_0 = 0,\ M_{-1} = -\frac{1}{2},\ n \geq 1. \tag{2}$$

The Binet formula says that the nth Mersenne number $M_n = 2^n - 1$ [3]. One application we would like to mention is that Nussbaumer [4] applied number theoretical transform closely related to Mersenne number to deal with problems of digital filtering and convolution of discrete signals.

In this paper, we study some basic quantities (determinants and inverses) associated with the periodic tridiagonal Toeplitz matrix with perturbed corners of type 1, which is defined as follows

$$\mathbb{A} = \begin{pmatrix} \alpha_1 & 2\hbar & 0 & \cdots & 0 & \gamma_1 \\ 0 & -3\hbar & \ddots & \ddots & & 0 \\ 0 & \hbar & \ddots & \ddots & \ddots & \vdots \\ \vdots & \ddots & \ddots & \ddots & 2\hbar & 0 \\ 0 & & \ddots & \ddots & -3\hbar & 2\hbar \\ \alpha_n & 0 & \cdots & 0 & \hbar & \gamma_n \end{pmatrix}_{n \times n}, \tag{3}$$

where $\alpha_1, \alpha_n, \gamma_1, \gamma_n, \hbar$ are complex numbers with $\hbar \neq 0$. Let $\hat{I}_n$ be the $n \times n$ "reverse unit matrix", which has ones along the secondary diagonal and zeros elsewhere. A matrix of the form $\mathbb{B} := \hat{I}_n \mathbb{A} \hat{I}_n$ is called a periodic tridiagonal Toeplitz matrix with perturbed corners of type 2, we say that $\mathbb{B}$ is induced by $\mathbb{A}$. It is readily seen that $\mathbb{A}$ is a periodic tridiagonal Toeplitz matrix with perturbed corners of type 1 if and only if its transpose $\mathbb{A}^T$ is a periodic tridiagonal Toeplitz matrix with perturbed corners of type 2.

Tridiagonal matrices appear not only in pure linear algebra, but also in many practical applications, such as, parallel computing [5], computer graphics [6], fluid mechanics [7,8], chemistry [9], and partial differential equations [10–15]. Taking linear hyperbolic equation as an example, some scholars have studied some matrices in discretized partial differential equations. Chan and Jin [16] discussed a linear hyperbolic equation considered by Holmgren and Otto [17] in one-dimensional and two-dimensional cases. Here we restate the linear hyperbolic equation in the two-dimensional case,

$$\frac{\partial u(x_1, x_2, t)}{\partial t} + v_1 \frac{\partial u(x_1, x_2, t)}{\partial x_1} + v_2 \frac{\partial u(x_1, x_2, t)}{\partial x_2} = g,$$

where $0 < x_1, x_2 \leq 1$, $t > 0$, $u(x_1, 0, t) = f(x_1 - at)$, , $u(0, x_1, t) = f(x_2 - at)$, $u(x_1, x_2, t) = f(x_1 + x_2), g = (v_1 + v_2 - a)f'$. Here v_1, v_2, and a are positive constants and f is a scalar function with derivative f'. Denote s_1, s_2, k as the two spatial steps and time step respectively. For simplicity, assume that $v_1 = v_2 = v$ and $s_1 = s_2 = s$. The linear hyperbolic equation discretized based on trapezoidal rule in time and center difference in two spaces, respectively. It's coefficient matrix is a tridiagonal matrix with perturbed last row:

$$\wp = \begin{pmatrix} 2 & \oslash & 0 & \cdots & \cdots & \cdots & 0 \\ -\oslash & \ddots & \ddots & \ddots & & & \vdots \\ 0 & \ddots & \ddots & \ddots & \ddots & & \vdots \\ \vdots & \ddots & \ddots & \ddots & \ddots & \ddots & \vdots \\ \vdots & & \ddots & \ddots & \ddots & \ddots & 0 \\ \vdots & & & \ddots & -\oslash & 2 & \oslash \\ 0 & \cdots & \cdots & \cdots & 0 & -2\oslash & 2+2\oslash \end{pmatrix}_{n \times n},$$

where $\oslash = vk/s$. On the other hand, some parallel computing algorithms are also designed for solving tridiagonal systems on graphics processing unit (GPU), which are parallel cyclic reduction [18] and partition methods [19]. Recently, Yang et al. [20] presented a parallel solving method which mixes direct and iterative methods for block-tridiagonal equations on CPU-GPU heterogeneous computing systems, while Myllykoski et al. [21] proposed a generalized graphics processing unit implementation of partial solution variant of the cyclic reduction (PSCR) method to solve certain types of separable block tridiagonal linear systems. Compared to an equivalent CPU implementation that utilizes a single CPU core, PSCR method indicated up to 24-fold speedups.

On the other hand, many studies have been conducted for tridiagonal matrices or periodic tridiagonal matrices, especially for their determinants and inverses [22–30]. Two decades ago, Wittenburg [31] studied the inverse of tridiagonal toeplitz and periodic matrices and applied them to elastostatics and vibration theory. Recently, El-Mikkawy and Atlan [32] proposed a symbolic algorithm based on the Doolittle LU factorization and Jia et al. put forward some algorithms [33–35] based on block diagonalization technique for k-tridiagonal matrix. In 2018, Tim and Emrah [36] used backward continued fractions to derive the LU factorization of periodic tridiagonal matrix and then derived an explicit formula for its inverse. Furthermore, some scholars were attracted by the fact that one could view periodic tridiagonal Toeplitz matrices as a special case of periodic tridiagonal matrices. Shehawey [37] generalized Huang and McColl's [38] work and put forward the inverse formula for periodic tridiagonal Toeplitz matrices.

The rest of the paper is organized as follows: Section 2 describes the detailed derivations of the determinants and inverses of periodic tridiagonal Toeplitz matrices with perturbed corners through matrix transformations, Schur complement and matrix decomposition with the Sherman-Morrison-Woodbury formula [39]. Specifically, the formulas on representation of the determinants and inverses of these typies matrices in the form of products of Mersenne numbers and some initial values. Furthermore, the properties of the periodic tridiagonal Toeplitz matrices with perturbed corners of type 2 can also be obtained. Section 3 presents the numerical results to test the effectiveness of our theoretical results. The final conclusions are given in Section 4.

2. Determinants and Inverses

In this section, we derive explicit formulas for the determinants and inverses of a periodic tridiagonal Toeplitz matrix with perturbed corners. Main effort is made to work out those for periodic tridiagonal Toeplitz matrix with perturbed corners of type 1, since the results for type 2 matrices would follow immediately.

Theorem 1. *Let $\mathbb{A} = (a_{i,j})_{i,j=1}^n$ ($n \geq 3$) be an $n \times n$ periodic tridiagonal Toeplitz matrix with perturbed corners of type* 1. *Then*

$$\det \mathbb{A} = (-\hbar)^{n-2}\{[2M_{n-2}\alpha_1 - 4(M_{n-3}+1)\alpha_n]\hbar + M_{n-1}(\alpha_1\gamma_n - \alpha_n\gamma_1)\}, \tag{4}$$

where M_i ($i = n-3, n-2, n-1$) is the ith Mersenne number.

Proof. Define the circulant matrix

$$\epsilon = (\epsilon_{i,j})_{i,j=1}^n, \tag{5}$$

where

$$\epsilon_{i,j} = \begin{cases} 1, & i = n, j = 1, \\ 1, & j = i+1, \\ 0, & \text{otherwise.} \end{cases}$$

Clearly, ϵ is invertible, and

$$\det \epsilon = (-1)^{n-3}. \tag{6}$$

Multiply $\mathbb{A}$ by ϵ from right and then partition $\mathbb{A}\epsilon$ into four blocks:

$$\mathbb{A}\epsilon = \begin{pmatrix} \gamma_1 & \alpha_1 & 2\hbar & 0 & \cdots & \cdots & \cdots & 0 \\ 0 & 0 & -3\hbar & 2\hbar & 0 & & & \vdots \\ 0 & 0 & \hbar & -3\hbar & 2\hbar & 0 & & \vdots \\ \vdots & \vdots & 0 & \hbar & -3\hbar & 2\hbar & \ddots & \vdots \\ \vdots & \vdots & \vdots & 0 & \ddots & \ddots & \ddots & 0 \\ 0 & \vdots & \vdots & \vdots & & \ddots & \ddots & 2\hbar \\ 2\hbar & 0 & \vdots & \vdots & & & \ddots & -3\hbar \\ \gamma_n & \alpha_n & 0 & 0 & \cdots & \cdots & 0 & \hbar \end{pmatrix}$$

$$= \begin{pmatrix} \mathbb{A}_{11} & \mathbb{A}_{12} \\ \mathbb{A}_{21} & \mathbb{A}_{22} \end{pmatrix}. \tag{7}$$

Since $\mathbb{A}_{22}$ is upper triangular, its determinant is clear which is

$$\det \mathbb{A}_{22} = \hbar^{n-2}. \tag{8}$$

As we assume $\hbar \neq 0$, so $\mathbb{A}_{22}$ is invertible. It is known (see, e.g., ([29], Lemma 2.5)) that $\mathbb{A}_{22}^{-1} = (\ddot{a}_{i,j})_{i,j=1}^n$ where

$$\ddot{a}_{i,j} = \begin{cases} \frac{M_{j-i+1}}{\hbar}, & i \leq j, \\ 0, & i > j, \end{cases}$$

and M_i is the ith Mersenne number.

Next, taking the determinants for both sides of (7) and by (see, e.g., ([40], p. 10)), we get

$$\det(\mathbb{A}\epsilon) = \det \mathbb{A}_{22} \det(\mathbb{A}_{11} - \mathbb{A}_{12}\mathbb{A}_{22}^{-1}\mathbb{A}_{21}). \tag{9}$$

Therefore

$$\det \mathbb{A} = \frac{\det \mathbb{A}_{22} \det(\mathbb{A}_{11} - \mathbb{A}_{12}\mathbb{A}_{22}^{-1}\mathbb{A}_{21})}{\det \epsilon}. \tag{10}$$

To find $\det \mathbb{A}$, we need to evaluate the determinant of $(\mathbb{A}_{11} - \mathbb{A}_{12}\mathbb{A}_{22}^{-1}\mathbb{A}_{21})$. From (7) we have

$$\mathbb{A}_{11} - \mathbb{A}_{12}\mathbb{A}_{22}^{-1}\mathbb{A}_{21} = \begin{pmatrix} \gamma_1 - 2M_{n-2}\gamma_n - 4M_{n-3}\hbar & \alpha_1 - 2M_{n-2}\alpha_n \\ M_{n-1}\gamma_n + 2M_{n-2}\hbar & M_{n-1}\alpha_n \end{pmatrix},$$

and so

$$\det\left(\mathbb{A}_{11} - \mathbb{A}_{12}\mathbb{A}_{22}^{-1}\mathbb{A}_{21}\right) = [4(M_{n-3}+1)\alpha_n - 2M_{n-2}\alpha_1]\hbar - M_{n-1}(\alpha_1\gamma_n - \alpha_n\gamma_1). \tag{11}$$

Finally, applying (6), (8), and (11) to (10), we get the determinant of $\mathbb{A}$, which completes the proof. □

Theorem 2. *Let $\mathbb{A} = (a_{i,j})_{i,j=1}^n (n \geq 3)$ be a nonsingular periodic tridiagonal Toeplitz matrix with perturbed corners of type 1. Then $\mathbb{A}^{-1} = (\breve{a}_{i,j})_{i,j=1}^n$, where*

$$\breve{a}_{i,j} = \begin{cases} \frac{2M_{n-2}\hbar + M_{n-1}\gamma_n}{\psi}, & i=1, j=1, \\ \frac{4M_{n-3}\hbar - \gamma_1 + 2M_{n-2}\gamma_n}{\psi}, & i=1, j=2, \\ \frac{(M_{n-2}+1)\alpha_n}{-\psi}, & i=2, j=1, \\ \frac{2M_{n-3}\alpha_1\hbar + M_{n-2}(\alpha_1\gamma_n - \alpha_n\gamma_1)}{-\psi\hbar}, & i=2, j=2, \\ \frac{3(M_{n-3}+1)\alpha_n}{-\psi}, & i=3, j=1, \\ \frac{(M_{n-3}-1)\alpha_1\hbar + (M_{n-2}+1)\alpha_n\hbar + M_{n-3}(\alpha_1\gamma_n - \alpha_n\gamma_1)}{-\psi\hbar}, & i=3, j=2, \\ 3\breve{a}_{i,j-1} - 2\breve{a}_{i,j-2} + \frac{1}{\hbar}, & i \in \{2,3\}, j=i+1, \\ 3\breve{a}_{i,j-1} - 2\breve{a}_{i,j-2}, & \begin{cases} i \in \{1,2,3\}, i+2 \leq j \leq n; \\ 3 \leq j \leq i \leq n, \end{cases} \\ \frac{3}{2}\breve{a}_{i-1,j} - \frac{1}{2}\breve{a}_{i-2,j}, & \begin{cases} j \in \{1,2\}, 4 \leq i \leq n; \\ 4 \leq i < j \leq n, \end{cases} \end{cases} \tag{12}$$

$$\psi = 2M_{n-2}\alpha_1\hbar - (M_{n-1}+1)\alpha_n\hbar + M_{n-1}(\alpha_1\gamma_n - \alpha_n\gamma_1), \tag{13}$$

and M_i ($i = n-3, n-2, n-1$) is the ith Mersenne number.

Proof. Let $\mathbb{A}^{-1} = (\breve{a}_{i,j})_{i,j=1}^{n}$ and the identity matrix $I_n = (e_{i,j})_{i,j=1}^{n}$, where

$$e_{i,j} = \begin{cases} 1, & i = j, \\ 0, & \text{otherwise.} \end{cases} \tag{14}$$

For a nonsingular $\mathbb{A}$,

$$\mathbb{A}^{-1}\mathbb{A} = \mathbb{A}\mathbb{A}^{-1} = I_n. \tag{15}$$

According to (15), we get

$$e_{i,j} = 2\breve{a}_{i,j-1}\hbar - 3\breve{a}_{i,j}\hbar + \breve{a}_{i,j+1}\hbar, \qquad 1 \le i \le n,\; 2 \le j \le n-1, \tag{16}$$

$$e_{i,j} = \breve{a}_{i-1,j}\hbar - 3\breve{a}_{i,j}\hbar + 2\breve{a}_{i+1,j}\hbar, \qquad 3 \le i \le n-1,\; 1 \le j \le n. \tag{17}$$

Based on (14), we get from (16) that

$$\breve{a}_{i,j} = 3\breve{a}_{i,j-1} - 2\breve{a}_{i,j-2}, \quad \begin{cases} i \in \{1,2,3\}, i+2 \le j \le n; \\ 3 \le j \le i \le n, \end{cases} \tag{18}$$

and $\breve{a}_{i,i+1} = 3\breve{a}_{i,i} - 2\breve{a}_{i,i-1} + \frac{1}{\hbar}$ for $i = 2,3$.

Similarly, from (17), we get that

$$\breve{a}_{i,j} = \frac{3\breve{a}_{i-1,j}}{2} - \frac{\breve{a}_{i-2,j}}{2}, \quad \begin{cases} j \in \{1,2\}, 4 \le i \le n; \\ 4 \le i < j \le n. \end{cases} \tag{19}$$

Therefore, based on the above analysis, we need to determine six initial values, that is, $\breve{a}_{i,j}$ ($i \in \{1,2,3\}, j \in \{1,2\}$), for the recurrence relations (18) and (19) in order to compute the inverse of $\mathbb{A}$. The rest of the proof is devoted to evaluating these particular entries of $\mathbb{A}^{-1}$.

We decompose $\mathbb{A}$ as follows:

$$\mathbb{A} = \hbar \Delta + FG, \tag{20}$$

where $\Delta = 3T_{M,n}^{-1}$, $F = (f_1^T, f_2^T)$, $G = \begin{pmatrix} g_1 \\ g_2 \end{pmatrix}$ with

$$f_1 = \left(\alpha_1 + \frac{2M_n\hbar}{M_{n+1}}, -\hbar, 0, \cdots, 0, \alpha_n - \frac{2\hbar}{M_{n+1}}\right)_{1 \times n},$$

$$f_2 = \left(\gamma_1 - \frac{(M_n+1)\hbar}{M_{n+1}}, 0, \cdots, 0, \gamma_n + \frac{2M_n\hbar}{M_{n+1}}\right)_{1 \times n},$$

$$g_1 = (1, 0, \cdots, 0)_{1 \times n},$$

$$g_2 = (0, \cdots, 0, 1)_{1 \times n},$$

and M_i the ith Mersenne number as before.

It could be verified that $\Delta^{-1} = \frac{1}{3}(t_{ij})_{i,j=1}^{n}$, where

$$t_{ij} = \begin{cases} M_{j-i+1}, & 1 \le i \le j \le n, \\ -2M_{j-i-1}, & 1 \le j < i \le n, \end{cases}$$

and M_{-m} is given in (2) for $m = 1, 2, \ldots$.

Applying the Sherman-Morrison-Woodbury formula (see, e.g., ([39] p. 50)) to (20) gives

$$\mathbb{A}^{-1} = (\hbar\Delta + FG)^{-1} = \frac{1}{\hbar}\Delta^{-1} - \frac{1}{\hbar^2}\Delta^{-1}F(I_n + \frac{1}{\hbar}G\Delta^{-1}F)^{-1}G\Delta^{-1}. \tag{21}$$

Now we compute each component on the right side of (21).
Multiplying respectively Δ^{-1} by G and F from left and right,

$$G\Delta^{-1} = \frac{1}{3}\begin{pmatrix} \eta_1 \\ \eta_2 \end{pmatrix}, \tag{22}$$

$$\Delta^{-1}F = \frac{1}{3}\begin{pmatrix} \xi_1 & \xi_2 \end{pmatrix}, \tag{23}$$

where η_1 and η_2 are row vectors, ξ_1 and ξ_2 are column vectors,

$$\eta_1 = (M_j)_{j=1}^n,$$
$$\eta_2 = (-2M_{j-n-1})_{j=1}^n,$$
$$\xi_1^T = (\xi_{1,1} - 3\hbar, \xi_{2,1}, \xi_{3,1}, \cdots, \xi_{n,1}),$$
$$\xi_{i,1} = M_{n-i+1}\alpha_n - 2M_{-i}\alpha_1, \quad i = 1, 2, \cdots, n,$$
$$\xi_2 = (M_{n-i+1}\gamma_n - 2M_{-i}\gamma_1 + 2M_{n-i}\hbar)_{i=1}^n.$$

Then multiplying (23) by $\frac{G}{\hbar}$ from the left, further adding I_n and computing the inverse of the matrix

$$\left(I_n + \frac{G}{\hbar}\Delta^{-1}F\right)^{-1} = \frac{3\hbar}{h}\begin{pmatrix} -2M_{-n}\gamma_1 + \gamma_n + 3\hbar & -(\gamma_1 + M_n\gamma_n + 2M_{n-1}\hbar) \\ 2M_{-n}\alpha_1 - \alpha_n & \alpha_1 + M_n\alpha_n \end{pmatrix},$$

where $h = M_{n+1}[M_{1-n}(\alpha_1\gamma_n - \alpha_n\gamma_1) + M_{2-n}\alpha_1\hbar - \alpha_n\hbar]$. Multiplying the pervious formula $\left(I_n + \frac{G}{\hbar}\Delta^{-1}F\right)^{-1}$ by $\Delta^{-1}F$ from the left and by $G\Delta^{-1}$ from the right, respectively, yields

$$\Delta^{-1}F\left(I_n + \frac{1}{\hbar}G\Delta^{-1}F\right)^{-1}G\Delta^{-1} = (k_{ij})_{i,j=1}^n, \tag{24}$$

where

$$k_{1j} = \frac{\theta_j'\hbar^3 + (\theta_j''\gamma_1 + \theta_j'''\gamma_n)\hbar^2}{M_{n+1}\psi} + \frac{M_j\hbar}{3}, \qquad 1 \le j \le n,$$

$$k_{ij} = \frac{(\alpha_1\eta_{ij}' + \alpha_n\eta_{ij}'')\hbar^2 + (\alpha_1\gamma_n - \alpha_n\gamma_1)\eta_{ij}'''\hbar}{3M_{n+1}\psi}, \qquad 2 \le i \le n, 1 \le j \le n,$$

$$\psi = 2M_{n-2}\alpha_1\hbar - (M_{n-1} + 1)\alpha_n\hbar + M_{n-1}(\alpha_1\gamma_n - \alpha_n\gamma_1),$$

$$\theta_j' = 3M_j(M_{n-1} + 1) - M_{n-1}M_{n-j+1}(M_j + 1), \qquad 1 \le j \le n,$$

$$\theta_j'' = M_n M_j - M_{n-j+1}(M_{j-1} + 1), \qquad 1 \le j \le n,$$

$$\theta_j''' = M_j(M_{n-1} + 1) - M_n M_{n-j+1}(M_{j-1} + 1), \qquad 1 \le j \le n,$$

$$\eta_{ij}' = 2M_n M_{n-i}M_j - 3M_i M_j(M_{n-i} + 1) + M_n M_{i-1}M_{n-j+1}(M_{j-i+1} + 1), \quad 2 \le i \le n, 1 \le j \le n,$$

$$\eta_{ij}'' = M_{i-1}M_{n+j-1}(M_{n+j-1} + 1) - M_{n-i+2}M_j(M_{n-1} + 1), \qquad 2 \le i \le n, 1 \le j \le n,$$

$$\eta_{ij}''' = M_{n+1}[M_{n-i}M_j + M_{i-1}M_{n-j+1}(M_{j-i} + 1)], \qquad 2 \le i \le n, 1 \le j \le n.$$

From (21) and (24), we have

$$(\check{a}_{i,j})_{i,j=1}^n = \frac{1}{\hbar}\Delta^{-1} - \frac{1}{\hbar^2}(k_{ij})_{i,j=1}^n, \tag{25}$$

where

$$\check{a}_{i,j} = \frac{M_{j-i+1}}{3\hbar} - \frac{k_{i,j}}{\hbar^2}, \qquad 1 \le i \le j \le n, \tag{26}$$

$$\check{a}_{i,j} = -\frac{2M_{j-i-1}}{3\hbar} - \frac{k_{i,j}}{\hbar^2}, \qquad 1 \le j < i \le n. \tag{27}$$

By (26) we compute,

$$\check{a}_{1,1} = \frac{2M_{n-2}\hbar + M_{n-1}\gamma_n}{\psi},$$

$$\check{a}_{1,2} = \frac{4M_{n-3}\hbar - \gamma_1 + 2M_{n-2}\gamma_n}{\psi},$$

$$\check{a}_{2,2} = \frac{2M_{n-3}\alpha_1\hbar + M_{n-2}(\alpha_1\gamma_n - \alpha_n\gamma_1)}{-\psi\hbar}.$$

By (27) we compute,

$$\check{a}_{2,1} = \frac{(M_{n-2}+1)\alpha_n}{-\psi},$$

$$\check{a}_{3,1} = \frac{3(M_{n-3}+1)\alpha_n}{-\psi},$$

$$\check{a}_{3,2} = \frac{(M_{n-3}-1)\alpha_1\hbar + (M_{n-2}+1)\alpha_n\hbar + M_{n-3}(\alpha_1\gamma_n - \alpha_n\gamma_1)}{-\psi\hbar}.$$

This completes the proof. □

Remark 1. *Formulas* (26) *and* (27) *would give an analytic formula for* $\mathbb{A}^{-1}$. *However, there is a big advantage of* (12) *from computational consideration as we shall see from Section* 3.

The next two theorems are parallel results for type 1 matrices.

Theorem 3. *Let* $\mathbb{A}$ *be a periodic tridiagonal Toeplitz matrix with perturbed corners of type* 1 *and* $\mathbb{B}$ *be a periodic tridiagonal Toeplitz matrix with perturbed corners of type* 2, *which is induced by* $\mathbb{A}$. *Then*

$$\det \mathbb{B} = (-\hbar)^{n-2}\{[2M_{n-2}\alpha_1 - 4(M_{n-3}+1)\alpha_n]\hbar + M_{n-1}(\alpha_1\gamma_n - \alpha_n\gamma_1)\}.$$

Proof. Since $\det \mathbb{B} = \det \hat{I}_n \det \mathbb{A} \det \hat{I}_n$, we obtain this conclusion by using Theorem 1 and $\det \hat{I}_n = (-1)^{\frac{n(n-1)}{2}}$. □

Theorem 4. *Let* $\mathbb{A}$ *be a periodic tridiagonal Toeplitz matrix with perturbed corners of type* 1 *and* $\mathbb{B}$ *be a periodic tridiagonal Toeplitz matrix with perturbed corners of type* 2, *which is induced by* $\mathbb{A}$. *Then*

$$\mathbb{B}^{-1} = (\check{a}_{n+1-i,n+1-j})_{i,j=1}^n,$$

where $\check{a}_{i,j}$ is the same as (12).

Proof. It follows immediately from $\mathbb{B}^{-1} = \hat{I}_n^{-1}\mathbb{A}^{-1}\hat{I}_n^{-1} = \hat{I}_n\mathbb{A}^{-1}\hat{I}_n$ and Theorem 2. □

3. Algorithms

In this section, we give two algorithms for finding the determinant and inverse of a periodic tridiagonal Toeplitz matrix with perturbed corners of type 1, which is called $\mathbb{A}$. Besides, we analyze these algorithms to illustrate our theoretical results.

Firstly, based on Theorem 1, we give an algorithm for computing determinant of $\mathbb{A}$ as following:

Based on Algorithm 1, we make a comparison of the total number operations for determinant of $\mathbb{A}$ between LU decomposition and Algorithm 1 in Table 1. Specifically, we get that the total number operation for the determinant of $\mathbb{A}$ is $2n + 11$, which can be reduced to $O(logn)$ (see, [41] pp. 226–227).

Table 1. Comparison of the total number operations for determinant of $\mathbb{A}$.

Algorithms	Number Operations
LU decomposition algorithm	$13n - 15$
Algorithm 1	$2n + 11$

Algorithm 1: The determinant of a periodic tridiagonal Toeplitz matrix with perturbed corners of type 1

Step 1: Input $\alpha_1, \alpha_n, \gamma_1, \gamma_n, \hbar$, order n and generate Mersenne numbers
M_i ($i = n - 3$, $n - 2$, $n - 1$) by (1).
Step 2: Calculate and output the determinant of $\mathbb{A}$ by (4).

Next, based on Theorem 2, we give an algorithm for computing inverse of $\mathbb{A}$ as following:

Algorithm 2: The inverse of a periodic tridiagonal Toeplitz matrix with perturbed corners of type 1

Step 1: Input $\alpha_1, \alpha_n, \gamma_1, \gamma_n, \hbar$, order n and generate Mersenne numbers
M_i ($i = n - 3$, $n - 2$, $n - 1$) by (1).
Step 2: Calculate ψ by (13) and six initial values $\breve{a}_{1,1}, \breve{a}_{1,2}, \breve{a}_{2,1}, \breve{a}_{2,2}, \breve{a}_{3,1}, \breve{a}_{3,2}$ by (12).
Step 3: Calculate the remaining elements of the inverse:

$$\breve{a}_{2,3} = 3\breve{a}_{2,2} - 2\breve{a}_{2,1} + \frac{1}{\hbar},$$

$$\breve{a}_{3,4} = 3\breve{a}_{3,2} - 2\breve{a}_{3,1} + \frac{1}{\hbar},$$

$$\breve{a}_{i,j} = 3\breve{a}_{i,j-1} - 2\breve{a}_{i,j-2}, \ i \in \{1,2,3\}, i+2 \leq j \leq n,$$

$$\breve{a}_{i,j} = 3\breve{a}_{i,j-1} - 2\breve{a}_{i,j-2}, \ i \in \{1,2,3\}, 3 \leq j \leq i \leq n,$$

$$\breve{a}_{i,j} = \frac{3}{2}\breve{a}_{i-1,j} - \frac{1}{2}\breve{a}_{i-2,j}, \ j \in \{1,2\}, 4 \leq i \leq n,$$

$$\breve{a}_{i,j} = \frac{3}{2}\breve{a}_{i-1,j} - \frac{1}{2}\breve{a}_{i-2,j}, \ 4 \leq i < j \leq n.$$

Step 4: Output the inverse $\mathbb{A}^{-1} = (\breve{a}_{i,j})_{i,j=1}^n$.

To test the effectiveness of Algorithm 2, we compare the total number of operations for the inverse of $\mathbb{A}$ between LU decomposition and Algorithm 2 in Table 2. The total number operation of LU decomposition is $\frac{5n^3}{6} + 3n^2 + \frac{91n}{6} - 21$, whereas that of Algorithm 2 is $\frac{7n^2}{2} - \frac{3n}{2} + 30$.

Table 2. Comparison of the total number operations for inverse of $\mathbb{A}$.

Algorithms	Number Operations
LU decomposition algorithm	$\frac{5n^3}{6} + 3n^2 + \frac{91n}{6} - 21$
Algorithm 2	$\frac{7n^2}{2} - \frac{3n}{2} + 30$

4. Discussion

In this paper, explicit determinants and inverses of periodic tridiagonal Toeplitz matrices with perturbed corners are represented by the famous Mersenne numbers. This helps to reduce the total number of operations during the calculation process. Some recent research related to our present work can be found in [42–48]. Among them, Qi et al. presented some closed formulas for the Horadam polynomials in terms of a tridiagonal determinant and derived closed formulas for the generalized Fibonacci polynomials, the Lucas polynomials, the Pell-Lucas polynomials, and the Chebyshev polynomials of the first kind in terms of tridiagonal determinants.

5. Conclusions

Mersenne numbers are remarkably wide-spread in many diverse areas of the mathematical, biological, physical, chemical, engineering, and statistical sciences. In this paper, we present explicit formulas for the determinants and inverses of periodic tridiagonal Toeplitz matrices with perturbed corners. The representation of the determinant in the form of products of the Mersenne numbers and some initial values from matrix transformations and Schur complement. For the inverse, our main approaches include the use of matrix decomposition with the Sherman-Morrison-Woodbury formula. Especially, the inverse is just determined by six initial values. To test our method's effectiveness, we propose two algorithms for finding the determinant and inverse of periodic tridiagonal Toeplitz matrices with perturbed corners and compare the total number of operations for the two basic quantities between different algorithms. After comparison, we draw a conclusion that our algorithms are superior to LU decomposition to some extent.

Author Contributions: Conceptualization, methodology, funding acquisition, Z.J. and Y.Z.; investigation, resources, formal analysis, software, Y.Z. and Y.W.; writing—original draft preparation, Y.W.; writing—review and editing, supervision, visualization, S.S.

Funding: The research was funded by Natural Science Foundation of Shandong Province (Grant No. ZR2016AM14), National Natural Science Foundation of China (Grant No.11671187) and the PhD Research Foundation of Linyi University (Grant No.LYDX2018BS067).

Acknowledgments: The authors are grateful to the anonymous referees for their useful suggestions which improve the contents of this article.

Conflicts of Interest: The authors declare no conflict of interest.

References

1. Krizek, M.; Luca, F.; Somer, L. *17 Lectures on Fermat Numbers: From Number Theory to Geometry*; Springer Science & Business Media: New York, NY, USA, 2013.
2. Robinson, R.M. Mersenne and Fermat numbers. *Proc. Am. Math. Soc.* **1954**, *5*, 842–846. [CrossRef]
3. Sloane, N.J.A. Mersenne Primes (of Form $2^p - 1$ Where p Is a Prime). 1964. Available online: https://oeis.org/A000668 (accessed on 24 August 2019).
4. Nussbaumer, H.J. *Fast Fourier Transform and Convolution Algorithms*; Springer Science & Business Media: New York, NY, USA, 2012.
5. Li, J.M.; Zheng, Z.G.; Tian, Q.; Zhang, G.Y.; Zheng, F.Y.; Pan, Y.Y. Research on tridiagonal matrix solver design based on a combination of processors. *Comput. Electr. Eng.* **2017**, *62*, 1–16. [CrossRef]
6. Bender, J.; Müller, M.; Otaduy, M.; Matthias, T.; Miles, M. A survey on position-based simulation methods in computer graphics. *Comput. Graph. Forum.* **2014**, *33*, 228–251. [CrossRef]

7. Vanka, S.P. 2012 Freeman scholar lecture: computational fluid dynamics on graphics processing units. *J. Fluids Eng.* **2013**, *135*, 061401. [CrossRef]
8. Hosamani, S.M.; Kulkarni, B.B.; Boli, R.G.; Gadag, V.M. QSPR analysis of certain graph theocratical matrices and their corresponding energy. *Appl. Math. Nonlinear Sci.* **2017**, *2*, 131–150. [CrossRef]
9. Jantschi, L. The eigenproblem translated for alignment of molecules. *Symmetry* **2019**, *11*, 1027. [CrossRef]
10. Fischer, C.; Usmani, R. Properties of some tridiagonal matrices and their application to boundary value problems. *SIAM J. Numer. Anal.* **1969**, *6*, 127–142. [CrossRef]
11. Wang, J.; Meng, F.W. Interval oscillation criteria for second order partial differential systems with delays. *J. Comput. Appl. Math.* **2008**, *212*, 397–405. [CrossRef]
12. Feng, Q.H.; Meng, F.W. Explicit solutions for space-time fractional partial differential equations in mathematical physics by a new generalized fractional Jacobi elliptic equation-based sub-equation method. *Optik* **2016**, *127*, 7450–7458. [CrossRef]
13. Shao, J.; Zheng, Z.W.; Meng, F.W. Oscillation criteria for fractional differential equations with mixed nonlinearities. *Adv. Differ. Equ-ny.* **2013**, *2013*, 323. [CrossRef]
14. Sun, Y.G.; Meng, F.W. Interval criteria for oscillation of second-order differential equations with mixed nonlinearities. *Appl. Math. Comput.* **2008**, *198*, 375–381. [CrossRef]
15. Xu, R.; Meng, F.W. Some new weakly singular integral inequalities and their applications to fractional differential equations. *J. Inequal. Appl.* **2016**, *2016*, 78. [CrossRef]
16. Chan, R.H.; Jin, X.Q. Circulant and skew-circulant preconditioners for skew-Hermitian type Toeplitz systems. *BIT* **1991**, *31*, 632–646. [CrossRef]
17. Holmgren, S.; Otto, K. *Iterative Solution Methods and Preconditioners for Non-Symmetric Non-Diagonally Dominant Block-Tridiagonal Systems of Equations*; Uppsala Univ.: Uppsala, Sweden, 1989.
18. Hockney, R.W.; Jesshope, C.R. *Parallel Computers*; Adam Hilger: Bristol, UK, 1981.
19. Wang, H.H. A parallel method for tridiagonal equations. *ACM Trans. Math. Softw.* **1981**, *7*, 170–183. [CrossRef]
20. Yang, W.D.; Li, K.L.; Li, K.Q. A parallel solving method for block-tridiagonal equations on CPU-GPU heterogeneous computing systems. *J. Supercomput.* **2017**, *73*, 1760–1781. [CrossRef]
21. Myllykoski, M.; Rossi, T.; Toivanen, J. On solving separable block tridiagonal linear systems using a GPU implementation of radix-4 PSCR method. *J. Parallel Distrib. Comput.* **2018**, *115*, 56–66. [CrossRef]
22. Jiang, X.Y.; Hong, K. Skew cyclic displacements and inversions of two innovative patterned matrices. *Appl. Math. Comput.* **2017**, *308*, 174–184. [CrossRef]
23. Jiang, X.Y.; Hong, K.; Fu, Z.W. Skew cyclic displacements and decompositions of inverse matrix for an innovative structure matrix. *J. Nonlinear Sci. Appl.* **2017**, *10*, 4058–4070. [CrossRef]
24. Zheng, Y.P.; Shon, S.; Kim, J. Cyclic displacements and decompositions of inverse matrices for CUPL Toeplitz matrices. *J. Math. Anal. Appl.* **2017**, *445*, 727–741. [CrossRef]
25. Jiang, Z.L.; Wang, D.D. Explicit group inverse of an innovative patterned matrix. *Appl. Math. Comput.* **2016**, *274*, 220–228. [CrossRef]
26. Jiang, Z.L.; Chen, X.T.; Wang, J.M. The explicit inverses of CUPL-Toeplitz and CUPL-Hankel matrices. *E. Asian J. Appl. Math.* **2017**, *7*, 38–54. [CrossRef]
27. Da Fonseca, C.M.; Yılmaz, F. Some comments on *k*-tridiagonal matrices: Determinant, spectra and inversion. *Appl. Math. Comput.* **2015**, *270*, 644–647. [CrossRef]
28. El-Mikkawy, M. A new computational algorithm for solving periodic tri-diagonal linear systems. *Appl. Math. Comput.* **2005**, *161*, 691–696. [CrossRef]
29. Zuo, B.S.; Jiang, Z.L.; Fu, D.Q. Determinants and inverses of Ppoeplitz and Ppankel matrices. *Special Matrices* **2018**, *6*, 201–215. [CrossRef]
30. Klymchuk, T. Regularizing algorithm for mixed matrix pencils. *Appl. Math. Nonlinear Sci.* **2017**, *2*, 123–130. [CrossRef]
31. Wittenburg, J. Inverses of tridiagonal Toeplitz and periodic matrices with applications to mechanics. *J. Appl. Maths. Mechs.* **1998**, *62*, 575–587. [CrossRef]
32. El-Mikkawy, M.; Atlan, F. A new recursive algorithm for inverting general *k*-tridiagonal matrices. *Appl. Math. Lett.* **2015**, *44*, 34–39. [CrossRef]
33. Jia, J.T.; Sogabe, T.; El-Mikkawy, M. Inversion of *k*-tridiagonal matrices with Toeplitz structure. *Comput. Math. Appl.* **2013**, *65*, 116–125. [CrossRef]

34. Jia, J.T.; Li, S.M. Symbolic algorithms for the inverses of general *k*-tridiagonal matrices. *Comput. Math. Appl.* **2015**, *70*, 3032–3042. [CrossRef]
35. Jia, J.T.; Li, S.M. On the inverse and determinant of general bordered tridiagonal matrices. *Comput. Math. Appl.* **2015**, *69*, 503–509. [CrossRef]
36. Tim, H.; Emrah, K. An analytical approach: Explicit inverses of periodic tridiagonal matrices. *J. Comput. Appl. Math.* **2018**, *335*, 207–226.
37. El-Shehawey, M.; El-Shreef,G.; ShAl-Henawy, A. Analytical inversion of general periodic tridiagonal matrices. *J. Math. Anal. Appl.* **2008**, *345*, 123–134. [CrossRef]
38. Huang, Y.; McColl, W.F. Analytical inversion of general tridiagonal matrices. *J. Phys. A-Math. Gen.* **1997**, *30*, 7919. [CrossRef]
39. Golub, G.H.; Van Loan, C.F. *Matrix Computations*, 3rd ed.; The John Hopkins University Press: Baltimore, MA, USA, 1996.
40. Zhang, F.Z. *The Schur Complement and Its Applications*; Springer Science & Business Media: New York, NY, UAS, 2006.
41. Rosen, K.H. *Discrete Mathematics and Its Applications*; McGraw-Hill: New York, NY, USA, 2011.
42. Zheng, D.Y. Matrix methods for determinants of Pascal-like matrices. *Linear Algebra Appl.* **2019**, *577*, 94–113. [CrossRef]
43. Moghaddamfar, A.R.; Salehy, S.N. Determinant Representations of Sequences: A Survey. *Spec. Matrices* **2014**, *2*. [CrossRef]
44. Cobeli, C.; Zaharescu, A. Promenade around Pascal Triangle-Number Motives. *Bull. Math. Soc. Sci. Math. Roumanie* **2013**, *56*, 73–98.
45. Mirashe, N.; Moghaddamfar, A.R.; Mozafari, S.H. The determinants of matrices constructed by subdiagonal, main diagonal and superdiagonal. *Lobachevskii J. Math.* **2010**, *31*, 295–306. [CrossRef]
46. Qi, F.; Kızılateş, C.; Du, W.S. A closed formula for the Horadam polynomials in terms of a tridiagonal determinant. *Symmetry* **2019**, *11*, 782. [CrossRef]
47. Qi, F.; Guo, B.N. Some determinantal expressions and recurrence relations of the Bernoulli polynomials. *Mathematics* **2016**, *4*, 65. [CrossRef]
48. Sharma, D.; Sen, M. Inverse eigenvalue problems for two special acyclic matrices. *Mathematics* **2016**, *4*, 12. [CrossRef]

© 2019 by the authors. Licensee MDPI, Basel, Switzerland. This article is an open access article distributed under the terms and conditions of the Creative Commons Attribution (CC BY) license (http://creativecommons.org/licenses/by/4.0/).

Article

Upper Bound of the Third Hankel Determinant for a Subclass of Close-to-Convex Functions Associated with the Lemniscate of Bernoulli

Hari M. Srivastava [1,2], Qazi Zahoor Ahmad [3], Maslina Darus [4], Nazar Khan [3,*], Bilal Khan [3], Naveed Zaman [3] and Hasrat Hussain Shah [5]

[1] Department of Mathematics and Statistics, University of Victoria, Victoria, BC V8W 3R4, Canada; harimsri@math.uvic.ca
[2] Department of Medical Research, China Medical University Hospital, China Medical University, Taichung 40402, Taiwan
[3] Department of Mathematics, Abbottabad University of Science and Technology, Abbottabad 22010, Pakistan; zahoorqazi5@gmail.com (Q.Z.A.); bilalmaths789@gmail.com (B.K.); zamannaveed162@gmail.com (N.Z.)
[4] School of Mathematical Sciences, Faculty of Sciences and Technology, Universiti Kebangsaan Malaysia, Bangi 43600, Selangor, Malaysia; maslina@ukm.edu.my
[5] Department of Mathematical Sciecnes, Balochistan University of Information Technology, Engineering and Management Sciences, Quetta 87300, Pakistan; hasrat@mail.ustc.edu.cn
* Correspondence: nazarmaths@gmail.com

Received: 26 June 2019; Accepted: 5 September 2019; Published: 14 September 2019

Abstract: In this paper, our aim is to define a new subclass of close-to-convex functions in the open unit disk $\mathbb{U}$ that are related with the right half of the lemniscate of Bernoulli. For this function class, we obtain the upper bound of the third Hankel determinant. Various other related results are also considered.

Keywords: analytic functions; close-to-convex functions; subordination; lemniscate of Bernoulli; Hankel determinant

MSC: primary 05A30, 30C45; secondary 11B65, 47B38

---

1. Introduction

By $\mathcal{H}(\mathbb{U})$ we denote the class of functions which are analytic in the open unit disk

$$\mathbb{U} = \{z : z \in \mathbb{C} \quad \text{and} \quad |z| < 1\},$$

where $\mathbb{C}$ is the set of complex numbers. We also let $\mathcal{A}$ be the class of analytic functions having the following form:

$$f(z) = z + \sum_{n=2}^{\infty} a_n z^n \qquad (\forall\, z \in \mathbb{U}), \tag{1}$$

and which are normalized by the following conditions:

$$f(0) = 0 \quad \text{and} \quad f'(0) = 1.$$

We denote by $\mathcal{S}$ the class of functions in $\mathcal{A}$, which are univalent in $\mathbb{U}$.

A function $f \in \mathcal{A}$ is called starlike in $\mathbb{U}$ if it satisfies the following inequality:

$$\Re\left(\frac{zf'(z)}{f(z)}\right) > 0 \qquad (\forall\, z \in \mathbb{U}).$$

The class of all such functions is denoted by $\mathcal{S}^*$. For $f \in \mathcal{S}^*$, one can find that (see [1]):

$$|a_n| \leq n \quad \text{for} \quad n = 2, 3, \ldots \tag{2}$$

Next, by $\mathcal{K}$, we denote the class of close-to-convex functions in $\mathbb{U}$ that satisfy the following inequality:

$$\Re\left(\frac{zf'(z)}{g(z)}\right) > 0 \quad (\forall z \in \mathbb{U}),$$

for some $g \in \mathcal{S}^*$.

An example of a function, which is close-to-convex in $\mathbb{U}$, is given by:

$$F(z) = \frac{z - e^{2i\alpha} \cos \alpha z^2}{(1 - e^{i\alpha} z)^2} \quad (0 < \alpha < \pi)$$

which maps $\mathbb{U}$ onto the complex z-plane excluding a vertical slit (see [2] where some interesting properties of this function are obtained).

Moreover, by $\mathcal{SL}^*$, we denote the class of functions $f \in \mathcal{A}$ that satisfy the following inequality:

$$\left|\left(\frac{zf'(z)}{f(z)}\right)^2 - 1\right| < 1 \quad (\forall z \in \mathbb{U}).$$

Thus a function $f \in \mathcal{SL}^*$ is such that $\frac{zf'(z)}{f(z)}$ lies in the region bounded by the right half of the lemniscate of Bernoulli given by the following relation:

$$\left|w^2 - 1\right| < 1,$$

where

$$w = \frac{zf'(z)}{f(z)}.$$

The above defined class was introduced by Sokół et al. (see [3]) and studied by the many authors (see, for example, [4–6]).

Next, if two functions f and g are analytic in $\mathbb{U}$, we say that the function f is subordinate to the function g and write:

$$f \prec g \quad \text{or} \quad f(z) \prec g(z),$$

if there exists a Schwarz function $w(z)$ that is analytic in $\mathbb{U}$ with:

$$w(0) = 0 \quad \text{and} \quad |w(z)| < 1,$$

such that:

$$f(z) = g(w(z)).$$

Furthermore, if the function g is univalent in $\mathbb{U}$, then we have the following equivalence (see, for example, [7]; see also [8]):

$$f(z) \prec g(z) \quad (z \in \mathbb{U}) \rightleftarrows f(0) = g(0) \quad \text{and} \quad f(\mathbb{U}) \subset g(\mathbb{U}).$$

We next denote by $\mathcal{P}$ the class of analytic functions p which are normalized by $p(0) = 1$ and have the following form:

$$p(z) = 1 + \sum_{n=1}^{\infty} p_n z^n, \tag{3}$$

such that:
$$\Re(p(z)) > 0 \qquad (\forall z \in \mathbb{U}).$$

In recent years, several interesting subclasses of analytic and multivalent functions have been introduced and investigated (see, for example, [9–16]). Motivated and inspired by recent and ongoing research, we introduce and investigate here a new subclass of close-to-convex functions in $\mathbb{U}$ which are associated with the lemniscate of Bernoulli by using some techniques similar to those that were used earlier by Sokół and Stankiewicz (see [3]).

Definition 1. *A function f of the form of Equation (1) is said to be in the class $\mathcal{KL}^*$ if and only if:*

$$\left|\left(\frac{zf'(z)}{g(z)}\right)^2 - 1\right| < 1 \qquad (4)$$

for some $g \in \mathcal{S}^*$. Equivalently, we have:

$$\frac{zf'(z)}{g(z)} \prec \sqrt{1+z} \qquad (\forall z \in \mathbb{U})$$

for some $g \in \mathcal{S}^*$.

Thus, clearly, a function $f \in \mathcal{KL}^*$ is such that $\frac{zf'(z)}{g(z)}$ lies in the region bounded by the right half of the lemniscate of Bernoulli given by the following relation:

$$\left|w^2 - 1\right| < 1.$$

A closer look at the above series development of f suggests that many properties of the function f may be affected (or implied) by the size of its coefficients. The coefficient problem has been reformulated in the more special manner of estimating $|a_n|$, that is, the modulus of the nth coefficient. In 1916, Bieberbach conjectured that the nth coefficient of a univalent function is less or equal to that of the Koebe function.

Closely related to the Bieberbach conjecture is the problem of finding sharp estimates for the coefficients of odd univalent functions, which has the most general form of the square-root transformation of a function $f \in \mathcal{S}$:

$$l(z) = \sqrt{f(z^2)} = z + c_3 z^3 + c_5 z^5 \ldots$$

For odd univalent functions, Littlewood and Parley in 1932 proved that, for each postive integer n, the modulus $|c_{2n+1}|$ is less than an absolute constant M. For $M = 1$, the bound becomes the Littlewood–Parley conjecture.

Let $n \geqq 0$ and $q \geqq 1$. Then the qth Hankel determinant is defined as follows:

$$H_q(n) = \begin{vmatrix} a_n & a_{n+1} & \cdot & \cdot & a_{n+q-1} \\ a_{n+1} & \cdot & & & \cdot \\ \cdot & \cdot & & & \cdot \\ \cdot & \cdot & & & \cdot \\ \cdot & \cdot & & & \cdot \\ a_{n+q-1} & \cdot & \cdot & \cdot & a_{n+2(q-1)} \end{vmatrix}$$

The Hankel determinant plays a vital role in the theory of singularities [17] and is useful in the study of power series with integer coefficients (see [18–20]). Noteworthy, several authors obtained the sharp upper bounds on $H_2(2)$ (see, for example, [5,21–29]) for various classes of functions. It is a well-known fact for the Fekete-Szegö functional that:

$$\left|a_3 - a_2^2\right| = H_2(1).$$

This functional is further generalized as follows:

$$\left|a_3 - \mu a_2^2\right|$$

for some real or complex number μ. Fekete and Szegö gave sharp estimates of $\left|a_3 - \mu a_2^2\right|$ for μ real and $f \in \mathcal{S}$, the class of normalized univalent functions in $\mathbb{N}$. It is also known that the functional $\left|a_2 a_4 - a_3^2\right|$ is equivalent to $H_2(2)$. Babalola [30] studied the Hankel determinant $H_3(1)$ for some subclasses of analytic functions. In the present investigation, our focus is on the Hankel determinant $H_3(1)$ for the above-defined function class $\mathcal{KL}^*$.

2. A Set of Lemmas

Lemma 1. *(see [31]) Let:*

$$p(z) = 1 + p_1 z + p_2 z^2 + \cdots$$

be in the class $\mathcal{P}$ of functions with positive real part in $\mathbb{U}$. Then, for any number v:

$$\left|p_2 - v p_1^2\right| \begin{cases} -4v + 2 & (v \leq 0) \\ 2 & (0 \leq v \leq 1) \\ 4v - 2 & (v \geq 1). \end{cases} \quad (5)$$

When $v < 0$ or $v > 1$, the equality holds true in Equation (5) if and only if:

$$p(z) = \frac{1+z}{1-z}$$

or one of its rotations. If $0 < v < 1$, then the equality holds true in Equation (5) if and only if:

$$p(z) = \frac{1+z^2}{1-z^2}$$

or one of its rotations. If $v = 0$, the equality holds true in Equation (5) if and only if:

$$p(z) = \left(\frac{1+\rho}{2}\right)\frac{1+z}{1-z} + \left(\frac{1-\rho}{2}\right)\frac{1-z}{1+z} \quad (0 \leq \rho \leq 1)$$

or one of its rotations. If $v = 1$, then the equality in Equation (5) holds true if $p(z)$ is a reciprocal of one of the functions such that the equality holds true in the case when $v = 0$.

Lemma 2. *[32,33] Let:*

$$p(z) = 1 + p_1 z + p_2 z^2 + \cdots$$

be in the class $\mathcal{P}$ of functions with positive real part in $\mathbb{U}$. Then:

$$2p_2 = p_1^2 + x\left(4 - p_1^2\right)$$

for some x ($|x| \leqq 1$) and:
$$4p_3 = p_1^3 + 2\left(4 - p_1^2\right)p_1 x - \left(4 - p_1^2\right)p_1 x^2 + 2\left(4 - p_1^2\right)\left(1 - |x|^2\right)z$$
for some z ($|z| \leqq 1$).

Lemma 3. *[1] Let:*
$$p(z) = 1 + p_1 z + p_2 z^2 + \cdots$$
be in the class $\mathcal{P}$ of functions with positive real part in $\mathbb{U}$. Then:
$$|p_k| \leqq 2 \qquad (k \in \mathbb{N}).$$

The inequality is sharp.

3. Main Results and Their Demonstrations

In this section, we will prove our main results.

Theorem 1. *Let $f \in \mathcal{KL}^*$ and be of the form of Equation (1). Then:*
$$\left|a_3 - \mu a_2^2\right| \leqq \begin{cases} \frac{1}{48}(62 - 75\mu) & \left(\mu < \frac{38}{75}\right) \\ \frac{1}{2} & \left(\frac{38}{75} \leqq \mu \leqq \frac{86}{75}\right) \\ \frac{1}{48}(75\mu - 62) & \left(\mu > \frac{86}{75}\right). \end{cases}$$

It is asserted also that:
$$\left|a_3 - \mu a_2^2\right| + \frac{1}{3}\left(3\mu - \frac{38}{25}\right)|a_2|^2 \leqq \frac{1}{2} \qquad \left(\frac{38}{75} < \mu \leqq \frac{62}{75}\right)$$
and:
$$\left|a_3 - \mu a_2^2\right| + \frac{1}{3}\left(\frac{86}{25} - 3\mu\right)|a_2|^2 \leqq \frac{1}{2} \qquad \left(\frac{62}{75} < \mu \leqq \frac{86}{75}\right).$$

Proof. If $f \in \mathcal{KL}^*$, then it follows from definition that:
$$\frac{zf'(z)}{g(z)} \prec \phi(z) \qquad (\text{for some } g \in \mathcal{S}^*), \tag{6}$$
where:
$$\phi(z) = (1 + z)^{\frac{1}{2}}.$$

Define a function $p(z)$ by:
$$p(z) = \frac{1 + w(z)}{1 - w(z)} = 1 + p_1 z + p_2 z^2 + \cdots.$$

It is clear that $p(z) \in \mathcal{P}$. This implies that:
$$w(z) = \frac{p(z) - 1}{p(z) + 1}.$$

In addition, from Equation (6), we have:

$$\frac{zf'(z)}{g(z)} \prec \phi(z)$$

with:

$$\phi(w(z)) = \left(\frac{2p(z)}{p(z)+1}\right)^{\frac{1}{2}}.$$

We now have:

$$\left(\frac{2p(z)}{p(z)+1}\right)^{\frac{1}{2}} = 1 + \frac{1}{4}p_1 z + \left[\frac{1}{4}p_2 - \frac{5}{32}p_1^2\right]z^2 + \left[\frac{1}{4}p_3 - \frac{5}{16}p_1 p_2 + \frac{13}{128}p_1^3\right]z^3$$
$$+ \left[\frac{1}{4}p_4 - \frac{5}{16}p_1 p_3 + \frac{39}{128}p_2 p_1^2 - \frac{5}{32}p_2^2 - \frac{141}{2048}p_1^4\right]z^4 + \cdots.$$

Similarly, we get:

$$\frac{zf'(z)}{g(z)} = 1 + [2a_2 - b_2]z + \left[3a_3 - 2a_2 b_2 - b_3 + b_2^2\right]z^2$$
$$+ \left[4a_4 - 2a_2 b_3 - 3a_3 b_2 + 2b_2 b_3 + 2a_2 b_2^2 - b_4 - b_2^3\right]z^3 + \cdots.$$

Therefore, upon comparing the corresponding coefficients and by using Equation (2), we find that:

$$a_2 = \frac{5}{8}p_1, \tag{7}$$

$$a_3 = \frac{1}{4}p_2 + \frac{19}{96}p_1^2 \tag{8}$$

$$a_4 = \frac{7}{48}p_3 + \frac{9}{64}p_1 p_2 + \frac{91}{1536}p_1^3. \tag{9}$$

We thus obtain:

$$\left|a_3 - \mu a_2^2\right| = \frac{1}{4}\left|p_2 - \frac{1}{48}(75\mu - 38)p_1^2\right|. \tag{10}$$

Finally, by applying Lemma 1 in conjunction with Equation (10), we obtain the result asserted by Theorem 1. □

Theorem 2. *Let* $f \in \mathcal{KL}^*$ *and be of the form of Equation* (1). *Then:*

$$\left|a_2 a_4 - a_3^2\right| \leq \frac{9105}{36416}. \tag{11}$$

Proof. Making use of Equations (7)–(9), we have:

$$a_2 a_4 - a_3^2 = \left(\frac{35}{384}p_1 p_3 + \frac{45}{512}p_1^2 p_2 + \frac{455}{12288}p_1^4\right) - \left(\frac{1}{4}p_2 + \frac{19}{96}p_1^2\right)^2$$
$$= \frac{35}{384}p_1 p_3 - \frac{1}{16}p_2^2 - \frac{17}{1536}p_1^2 p_2 - \frac{79}{36864}p_1^4$$
$$= \frac{1}{36864}\left(3360 p_1 p_3 - 2304 p_2^2 - 408 p_1^2 p_2 - 79 p_1^4\right).$$

With the value of p_2 and p_3 from Lemma 2, using triangular inequality and replacing $|x| < 1$ by ρ and p_1 by p, we have:

$$\left|a_2 a_4 - a_3^2\right| = \frac{1}{36864}\left[19p^4 + 1680p\left(4-p^2\right) + 324\left(4-p^2\right)p^2\rho\right.$$
$$\left. + \rho^2\left(4-p^2\right)\left(264p^2 - 1680p + 2304\right)\right]$$
$$= F(p,\rho). \qquad (12)$$

Differentiating Equation (12) with respect to ρ, we have:

$$\frac{\partial F}{\partial \rho} = \frac{1}{36864}\left[324\left(4-p^2\right)p^2 + 2\rho\left(4-p^2\right)\left(264p^2 - 1680p + 2304\right)\right].$$

It is clear that:

$$\frac{\partial F(p,\rho)}{\partial \rho} > 0,$$

which shows that $F(p,\rho)$ is an increasing function on the closed interval $[0,1]$. This implies that the maximum value occurs at $\rho = 1$, that is:

$$\max\{F(p,\rho)\} = F(p,1) = G(p).$$

We now have:

$$G(p) = \frac{1}{36864}\left[-569p^4 + 48p^2 + 9216\right]. \qquad (13)$$

Differentiating Equation (13) with respect to p, we have:

$$G'(p) = \frac{1}{36864}\left[-2276p^3 + 96p\right]$$

Differentiating the above equation again with respect to p, we have:

$$G''(p) = \frac{1}{36864}\left[-6828p^2 + 96\right] < 0.$$

For $p = 0$, this shows that the maximum value of $G(p)$ occurs at $p = 0$. Hence we obtain:

$$\left|a_2 a_4 - a_3^2\right| \leq \frac{9105}{36416},$$

which completes the proof of Theorem 2. □

Theorem 3. *Let $f \in \mathcal{KL}^*$ and of the form of Equation (1). Then:*

$$|a_2 a_3 - a_4| \leq \frac{7}{24}.$$

Proof. We make use of Equations (7)–(9), along with Lemma 2. Since $p_1 \leq 2$, by Lemma 3, let $p_1 = p$ and assume without restriction that $p \in [0,2]$. Then, taking the absolute value and applying the triangle inequality with $\rho = |x|$, we obtain:

$$|a_2 a_3 - a_4| \leq \frac{1}{1536}\left\{55p^3 + 100p\rho\left(4-p^2\right) + 112\left(4-p^2\right)\right.$$
$$\left. + 56\rho^2(p-2)\left(4-p^2\right)\right\}$$
$$=: F(\rho).$$

Differentiating $F(\rho)$ with respect to ρ, we have:

$$F'(\rho) = \frac{1}{1536}\left\{100\rho\left(4-p^2\right) + 112\rho\left(p-2\right)\left(4-p^2\right)\right\}.$$

For $0 < \rho < 1$ and fixed $p \in (0, 2)$, it can easily be seen that:

$$\frac{\partial F}{\partial \rho} < 0.$$

This shows that $F_1(p, \rho)$ is a decreasing function of ρ, which contradicts our assumption. Therefore, we have:

$$\max F(p, \rho) = F(p, 0) = G(p).$$

This implies that:

$$G'(p) = \frac{1}{1536}\left\{165p^2 - 224p\right\}$$

and:

$$G''(p) = \frac{1}{1536}\left\{330p - 224\right\} < 0$$

for $p = 0$. Thus, clearly, $p = 0$ is the point of maximum. Hence we get the required result asserted by Theorem 3. □

To prove Theorem 4, we need Lemma 4.

Lemma 4. *If a function f of the form of Equation (1) is in the class $\mathcal{KL}^*$, then:*

$$|a_2| \leqq \frac{5}{4}, \quad |a_3| \leqq \frac{31}{24}, \quad |a_4| \leqq \frac{85}{64} \quad \text{and} \quad |a_5| \leqq \frac{859}{640}.$$

These estimates are sharp.

Proof. The proof of Lemma 4 is similar to that of a known result which was proved by Sokół (see [6]). Therefore, we here choose to omit the details involved in the proof of Lemma 4. □

Theorem 4. *Let $f \in \mathcal{KL}^*$ and be of the form of Equation (1). Then:*

$$|H_3(1)| \leqq \frac{1509169}{1092480}.$$

Proof. Since:

$$|H_3(1)| \leqq |a_3|\left|\left(a_2a_4 - a_3^2\right)\right| + |a_4|\left|(a_2a_3 - a_4)\right| + |a_5|\left|\left(a_1a_3 - a_2^2\right)\right|. \quad (14)$$

By Theorem 2, we have:

$$\left|a_2a_4 - a_3^2\right| \leqq \frac{9105}{36416}. \quad (15)$$

In addition, by Theorem 3, we get:

$$|a_2a_3 - a_4| \leqq \frac{7}{24}. \quad (16)$$

Now, using the fact that $a_1 = 1$, as well as Theorem 1 with $\mu = 1$, Lemma 4, Equations (15) and (16) in conjunction with Equation (14), we have the required result asserted by Theorem 4. □

4. Conclusions

Using the concept of the principle of subordination, we have introduced a new subclass of close-to-convex functions in $\mathbb{U}$, associated with the limniscate of Bernoulli. We have then derived the upper bound on $H_3(1)$ for this subclass of close-to-convex functions in $\mathbb{U}$, which is associated with the limniscate of Bernoulli. Our main results are stated and proved as Theorems 1–4. These general results are motivated essentially by the earlier works which are pointed out in this presentation.

Author Contributions: Conceptualization, Q.Z.A. and N.K.; methodology, N.K.; software, B.K.; validation, H.M.S.; formal analysis, H.M.S.; Writing—Original draft preparation, H.M.S.; Writing—Review and editing, H.M.S.; supervision, H.M.S. H.H.S. revised the article as per suggestions from Referees.

Funding: The third author is partially supported by UKM grant: GUP-2017-064.

Conflicts of Interest: The authors declare no conflict of interest.

References

1. Duren, P.L. *Univalent Functions*; Grundlehren der Mathematischen Wissenschaften, Band 259; Springer: New York, NY, USA; Berlin/Heidelberg, Germany; Tokyo, Japan, 1983.
2. Goodman, A.W.; Saff, E.B. Functions that are convex in one direction. *Proc. Am. Math. Soc.* **1979**, *73*, 183–187. [CrossRef]
3. Sokół, J.; Stankiewicz, J. Radius of convexity of some subclasses of strongly starlike functions. *Zeszyty Nauk. Politech. Rzeszowskiej Mat.* **1996**, *19*, 101–105.
4. Ali, R.M.; Cho, N.E.; Ravichandran, V.; Kumar, S.S. Differential subordination for functions associated with the lemniscate of Bernoulli. *Taiwan J. Math.* **2012**, *16*, 1017–1026. [CrossRef]
5. Raza, M.; Malik, S.N. Upper bound of the third Hankel determinant for a class of analytic functions related with lemniscate of Bernoulli. *J. Inequal. Appl.* **2013**, *2013*, 412. [CrossRef]
6. Sokół, J. Coefficient estimates in a class of strongly starlike functions. *Kyungpook Math. J.* **2009**, *49*, 349–353. [CrossRef]
7. Miller, S.S.; Mocanu, P.T. Differential subordination and univalent functions. *Mich. Math. J.* **1981**, *28*, 157–171. [CrossRef]
8. Miller, S.S.; Mocanu, P.T. *Differential Subordination: Theory and Applications*; Series on Monographs and Textbooks in Pure and Applied Mathematics, No. 225; Marcel Dekker Incorporated: New York, NY, USA; Basel, Switzerland, 2000.
9. Srivastava, H.M.; Tahir, M.; Khan, B.; Ahmad, Q.Z.; Khan, N. Some general classes of q-starlike functions associated with the Janowski functions. *Symmetry* **2019**, *11*, 292. [CrossRef]
10. Aldweby, H.; Darus, M. On Fekete-Szegö problems for certain subclasses defined by q-derivative. *J. Funct. Spaces* **2017**, *2017*, 7156738. [CrossRef]
11. Noor, K.I.; Khan, N.; Darus, M.; Ahmad, Q.Z.; Khan, B. Some properties of analytic functions associated with conic type regions. *Intern. J. Anal. Appl.* **2018**, *16*, 689–701.
12. Hussain, S.; Khan, S.; Zaighum, M.A.; Darus, M. Applications of a q-Sălăgean type operator on multivalent functions. *J. Inequal. Appl.* **2018**, *2018*, 301. [CrossRef]
13. Kanas, S.; Srivastava, H.M. Linear operators associated with k-uniformly convex functions. *Integral Transforms Spec. Funct.* **2000**, *9*, 121–132. [CrossRef]
14. Srivastava, H.M.; Eker, S.S. Some applications of a subordination theorem for a class of analytic functions. *Appl. Math. Lett.* **2008**, *21*, 394–399. [CrossRef]
15. Rasheed, A.; Hussain, S.; Zaighum, M.A.; Darus, M. Class of analytic function related with uniformly convex and Janowski's functions. *J. Funct. Spaces* **2018**, *2018*, 4679857. [CrossRef]
16. Srivastava, H.M.; Owa, S. *Current Topics in Analytic Function Theory*; World Scientific Publishing: Hackensack, NJ, USA, 1992.
17. Dienes, P. *The Taylor Series: An Introduction to the Theory of Functions of a Complex Variable*; New York-Dover Publishing Company: Mineola, NY, USA, 1957.
18. Cantor, D.G. Power series with integral coefficients. *Bull. Am. Math. Soc.* **1963**, *69*, 362–366. [CrossRef]
19. Edrei, A. Sur les determinants recurrents et less singularities d'une fonction donee por son developpement de Taylor. *Comput. Math.* **1940**, *7*, 20–88.

20. Pólya, G.; Schoenberg, I.J. Remarks on de la Vallée Poussin means and convex conformal maps of the circle. *Pac. J. Math.* **1958**, *8*, 259–334. [CrossRef]
21. Mahmood, S.; Srivastava, H.M.; Khan, N.; Ahmad, Q.Z.; Khan, B.; Ali, I. Upper bound of the third Hankel determinant for a subclass of q-starlike functions. *Symmetry* **2019**, *11*, 347. [CrossRef]
22. Janteng, A.; Abdulhalirn, S.; Darus, M. Coefficient inequality for a function whose derivative has positive real part. *J. Inequal. Pure Appl. Math.* **2006**, *50*, 1–5.
23. Mishra, A.K.; Gochhayat, P. Second Hankel determinant for a class of analytic functions defined by fractional derivative. *Int. J. Math. Math. Sci.* **2008**, *2008*, 153280. [CrossRef]
24. Singh, G.; Singh, G. On the second Hankel determinant for a new subclass of analytic functions. *J. Math. Sci. Appl.* **2014**, *2*, 1–3.
25. Srivastava, H.M.; Ahmad, Q.Z.; Khan, N.; Khan, N.; Khan, B. Hankel and Toeplitz determinants for a subclass of q-starlike functions associated with a general conic domain. *Mathematics* **2019**, *7*, 181. [CrossRef]
26. Srivastava, H.M.; Khan, B.; Khan, N.; Ahmad, Q.Z. Coefficient inequalities for q-starlike functions associated with the Janowski functions. *Hokkaido Math. J.* **2019**, *48*, 407–425. [CrossRef]
27. Shi, L.; Srivastava, H.M.; Arif, M.; Hussain, S.; Khan, H. An investigation of the third hankel determinant problem for certain subfamilies of univalent functions involving the exponential function. *Symmetry* **2019**, *11*, 598. [CrossRef]
28. Güney, H.Ö; Murugusundaramoorthy, G.; Srivastava, H.M. The second hankel determinant for a certain class of bi-close-to-convex function. *Results Math.* **2019**, *74*, 93. [CrossRef]
29. Sun, Y.; Wang, Z.-G.; Rasila, A. On Hankel determinants for subclasses of analytic functions and close-to-convex harmonic mapping. *arXiv* **2017**, arXiv:1703.09485.
30. Babalola, K.O. On $H_3(1)$ Hankel determinant for some classes of univalent functions. *Inequal. Theory Appl.* **2007**, *6*, 1–7.
31. Ma, W.C.; Minda, D. A unified treatment of some special classes of univalent functions. In *Proceedings of the Conference on Complex Analysis*; Li, Z., Ren, F., Yang, L., Zhang, S., Eds.; International Press: Cambridge, UK, 1994; pp. 157–169.
32. Libera, R.J.; Zlotkiewicz, E.J. Early coefficient of the inverse of a regular convex function. *Proc. Am. Math. Soc.* **1982**, *85*, 225–230. [CrossRef]
33. Libera, R.J.; Zlotkiewicz, E.J. Coefficient bounds for the inverse of a function with derivative in $\mathcal{P}$. *Proc. Am. Math. Soc.* **1983**, *87*, 251–257. [CrossRef]

© 2019 by the authors. Licensee MDPI, Basel, Switzerland. This article is an open access article distributed under the terms and conditions of the Creative Commons Attribution (CC BY) license (http://creativecommons.org/licenses/by/4.0/).

Article

Transformation of Some Lambert Series and Cotangent Sums

Namhoon Kim

Department of Mathematics Education, Hongik University, 94 Wausan-ro, Mapo-gu, Seoul 04066, Korea; nkim@hongik.ac.kr

Received: 29 July 2019; Accepted: 8 September 2019; Published: 11 September 2019

Abstract: By considering a contour integral of a cotangent sum, we give a simple derivation of a transformation formula of the series $A(\tau, s) = \sum_{n=1}^{\infty} \sigma_{s-1}(n) e^{2\pi i n \tau}$ for complex s under the action of the modular group on τ in the upper half plane. Some special cases directly give expressions of generalized Dedekind sums as cotangent sums.

Keywords: Lambert series; cotangent sum; modular transformation; Dedekind sum

MSC: 11F99; 11F20

1. Introduction

For $s \in \mathbb{C}$, let $\sigma_s(n) = \sum_{d \mid n} d^s$ be the sum of positive divisors function. For τ in the upper half plane $\mathfrak{H}$, consider

$$A(\tau, s) = \sum_{n=1}^{\infty} \sigma_{s-1}(n) e^{2\pi i n \tau} = \sum_{n=1}^{\infty} n^{s-1} \frac{e^{2\pi i n \tau}}{1 - e^{2\pi i n \tau}}. \qquad (1)$$

$A(\tau, s)$ is an entire function of s for every $\tau \in \mathfrak{H}$ and a Lambert series in $q = e^{2\pi i \tau}$ for every $s \in \mathbb{C}$. The study of transformation of $A(\tau, s)$ under the action of the modular group

$$\tau \mapsto \frac{a\tau + b}{c\tau + d} \qquad \left(\begin{pmatrix} a & b \\ c & d \end{pmatrix} \in \mathrm{SL}(2, \mathbb{Z})\right) \qquad (2)$$

has been a classical subject. Since $A(\tau, s)$ is manifestly invariant under translation $\tau \mapsto \tau + b$ for $b \in \mathbb{Z}$, one only needs to consider transformations (2) with $c > 0$. The main result of this article is the following transformation formula of $A(\tau, s)$. For $\tau \in \mathfrak{H}, s \in \mathbb{C}$ and $\begin{pmatrix} a & b \\ c & d \end{pmatrix} \in \mathrm{SL}(2, \mathbb{Z})$ with $c > 0$,

$$(c\tau + d)^{-s} A\left(\frac{a\tau + b}{c\tau + d}, s\right) = A(\tau, s) + \frac{1}{2}\left(1 - (c\tau + d)^{-s}\right) \zeta(1 - s)$$
$$+ \frac{c^{s-1} \sec \frac{\pi s}{2}}{8(c\tau + d)^{s/2}} \int_C (-z)^{s-1} \left(c + \sum_{j=0}^{c-1} \cot \pi \left(i\sqrt{c\tau + d}\, z - \frac{jd}{c}\right) \cot \pi \left(\frac{iz}{\sqrt{c\tau + d}} - \frac{j}{c}\right)\right) dz \qquad (3)$$

where C is the Hankel contour that encloses the nonnegative real axis in the clockwise direction but not any other poles of the integrand (see Definition 1 and Theorem 1).

Many previously known transformation formulas can be derived as special cases of (3) by considering some particular values of s (see Corollary 1). When s is an even positive integer, $A(\tau, s)$ appears in the Fourier series of the holomorphic Eisenstein series of weight s and satisfies a simple transformation law [1,2] (Corollary 1 (i) and (ii)). The case $s = 0$ is closely related to the Dedekind eta-function

$$\eta(\tau) = e^{\pi i \tau / 12} \prod_{n=1}^{\infty} \left(1 - e^{2\pi i n \tau}\right) \qquad (\tau \in \mathfrak{H})$$

as a branch of $\log \eta$ on $\mathfrak{H}$ is given by

$$\log \eta(\tau) = \frac{\pi i \tau}{12} - A(\tau, 0). \qquad (4)$$

We can thus derive the transformation law of $\log \eta(\tau)$ by setting $s = 0$ in (3) (Corollary 1 (iii)). Among many other proofs of the eta-transformation formula, we mention the work of Siegel [3], who gave a simple proof of it under $\tau \mapsto -\frac{1}{\tau}$ by considering a certain contour integral of a product of cotangent functions. This method was generalized by Rademacher to the full modular group in [4].

The transformation property of $A(\tau, 0)$ also has many applications; for example, it is used in Rademacher's derivation of an analytic formula of the partition function [5], which improved the result of Hardy and Ramanujan [6]. The formula also brings out the notion of the Dedekind sum, which has interesting arithmetic properties [7]. In Corollary 1 (iii), we obtain the cotangent sum representation of the Dedekind sum directly from (3).

The transformation formula of $A(\tau, s)$ when s is an even negative integer, which led to the idea of generalized Dedekind sums, was found by Apostol [8]. A computational error in [8] was corrected by Mikolás [9] and Iseki [10]. We obtain this formula in Corollary 1 (iv), also with cotangent sum expressions for the generalized Dedekind sums.

The general transformation property of $A(\tau, s)$ for complex s was first studied by Lewittes [11] in connection with certain generalized Eisenstein series. Berndt derived a transformation formula involving an integral expression in [12], which is further generalized in [13]. We note that the formula (3) involves an integral expression different from the formula in [12]. Studying the behavior of an integral of a more general cotangent sum may be an interesting topic for further investigation.

Throughout this work, logarithms and powers are taken with the principal argument.

2. Transformation Formula

Definition 1. *Let $c, d \in \mathbb{Z}$ with $(c, d) = 1$ and $c > 0$. For $\tau \in \mathfrak{H}$ and $s \in \mathbb{C}$, we denote $\rho = \sqrt{c\tau + d}$ and define*

$$I_{c,d}(\tau, s) = \frac{c^{s-1}}{8\rho^s} \int_C (-z)^{s-1} \left(c + \sum_{j=0}^{c-1} \cot \pi \left(i\rho z - \frac{jd}{c} \right) \cot \pi \left(\frac{iz}{\rho} - \frac{j}{c} \right) \right) dz \qquad (5)$$

where C is the Hankel contour that encloses the nonnegative real axis $[0, \infty)$ in the clockwise direction excluding any other poles of the integrand.

Consider the integrand in (5). Since

$$\cot \pi z = -i \left(1 + 2 \frac{e^{2\pi i z}}{1 - e^{2\pi i z}} \right), \qquad (6)$$

$\cot \pi z \to -i$ exponentially as $\operatorname{Im} z \to \infty$. Since both $i\rho$ and i/ρ are in the upper half plane, the integrand in (5) decays exponentially as $\operatorname{Re} z \to \infty$ and thus (5) is an entire function in s.

We also note that for $s \in \mathbb{Z}$, the integral (5) along C reduces to the integral around the origin, and $I_{c,d}(\tau, s) = 0$ whenever s is an odd integer as the integrand becomes even.

We write the following standard argument as a lemma.

Lemma 1. *For $\operatorname{Re} s < 0$,*

$$I_{c,d}(\tau, s) = \frac{\pi i c^{s-1}}{4\rho^s} \sum_{z \in \mathbb{C} \setminus [0, \infty)} \operatorname{Res} \left((-z)^{s-1} \sum_{j=0}^{c-1} \cot \pi \left(i\rho z - \frac{jd}{c} \right) \cot \pi \left(\frac{iz}{\rho} - \frac{j}{c} \right) \right) \qquad (7)$$

where the sum denotes the sum of all residues in $\mathbb{C} \setminus [0, \infty)$.

Proof. Let $K_N = C_N + S_N$ be the keyhole contour in $\mathbb{C}\setminus[0,\infty)$, where C_N is the part of C in Definition 1 in the region $|z| \leq N$, and S_N traverses along the circle $|z| = N$.

We can choose a sequence $N \to \infty$ such that each K_N stays well away from the poles of the integrand of (5). The integral over C_N is $2\pi i$ times the sum of the residues inside K_N minus the integral along S_N.

The lemma follows since the integral along S_N vanishes as $N \to \infty$ for $\operatorname{Re} s < 0$. Under this assumption, the sum of the residues in $\mathbb{C}\setminus[0,\infty)$ is absolutely convergent, and the term $(-z)^{s-1}c$ of the integrand can be ignored as it is holomorphic in the slit plane. □

We now restate and prove (3) using the notation introduced above.

Theorem 1. For $\tau \in \mathfrak{H}$ and $s \in \mathbb{C}$, let $A(\tau, s)$ be the series given by (1). For $\gamma = \left(\begin{smallmatrix} a & b \\ c & d \end{smallmatrix}\right) \in \operatorname{SL}(2, \mathbb{Z})$ with $c > 0$, we have

$$(c\tau + d)^{-s} A(\gamma\tau, s) = A(\tau, s) + \frac{1}{2}\left(1 - (c\tau+d)^{-s}\right)\zeta(1-s) + \left(\sec\frac{\pi s}{2}\right) I_{c,d}(\tau, s) \tag{8}$$

where $\gamma\tau = \frac{a\tau+b}{c\tau+d}$, $I_{c,d}(\tau, s)$ is given by (5) and $\zeta(s)$ is the Riemann zeta-function.

Proof. Since both sides of (8) are entire in s, it suffices to prove the equality for $\operatorname{Re} s < 0$. We use Lemma 1 to compute $I_{c,d}(\tau, s)$.

For each $0 \leq j < c$, the factor $\cot\pi(i\rho z - jd/c)$ of (7) has poles at $z = (k + jd/c)/(i\rho)$ for $k \in \mathbb{Z}$ with residue $1/(\pi i\rho)$. Hence the contribution to $I_{c,d}(\tau, s)$ from the residues of (7) at these poles, excluding $z = 0$, is given by

$$\frac{ic^{s-1}}{4\rho^s} \sideset{}{'}\sum_{k\in\mathbb{Z}, 0\leq j<c} \frac{1}{i\rho}\left(-\left(k+\frac{jd}{c}\right)\frac{1}{i\rho}\right)^{s-1} \cot\pi\left(\left(k+\frac{jd}{c}\right)\frac{1}{\rho^2} - \frac{j}{c}\right) \tag{9}$$

where the prime in the sum indicates that the term with $(k, j) = (0, 0)$ is omitted.

Since $(c, d) = 1$, every integer can be uniquely written in the form $kc + jd$ for $k \in \mathbb{Z}$ and $0 \leq j < c$. Let $-n = kc + jd$. Since $ad \equiv 1 \bmod c$, we have $na \equiv -j \bmod c$, and (9) equals

$$\frac{ic^{s-1}}{4\rho^s} \sum_{n\neq 0} \frac{1}{i\rho}\left(\frac{n}{i\rho c}\right)^{s-1} \cot\pi\left(-\frac{n}{\rho^2 c} + \frac{na}{c}\right). \tag{10}$$

We now use the fact that $\cot z$ is odd. For n positive,

$$\frac{1}{i\rho}\left(\frac{n}{i\rho c}\right)^{s-1} - \frac{1}{i\rho}\left(-\frac{n}{i\rho c}\right)^{s-1} = \left(\frac{n}{c}\right)^{s-1}\left(e^{-\frac{\pi is}{2}} + e^{\frac{\pi is}{2}}\right)\rho^{-s} = 2\left(\cos\frac{\pi s}{2}\right)\left(\frac{n}{c}\right)^{s-1}\rho^{-s},$$

and by (6), (10) becomes

$$\frac{1}{2\rho^{2s}}\cos\frac{\pi s}{2}\left(\zeta(1-s) + 2\sum_{n>0} n^{s-1} \frac{e^{2\pi i\left(-\frac{n}{\rho^2 c} + \frac{na}{c}\right)}}{1 - e^{2\pi i\left(-\frac{n}{\rho^2 c} + \frac{na}{c}\right)}}\right). \tag{11}$$

Furthermore, since

$$-\frac{n}{\rho^2 c} + \frac{na}{c} = \frac{n(a\tau + \frac{ad-1}{c})}{c\tau + d} = n\frac{a\tau + b}{c\tau + d},$$

the sum in (11) is in fact

$$\sum_{n>0} n^{s-1} \frac{e^{2\pi in\frac{a\tau+b}{c\tau+d}}}{1 - e^{2\pi in\frac{a\tau+b}{c\tau+d}}} = A\left(\frac{a\tau+b}{c\tau+d}, s\right).$$

We now treat the other poles in a similar fashion. For $0 \leq j < c$, the factor $\cot \pi(iz/\rho - j/c)$ of (7) has poles at $z = (k+j/c)(\rho/i)$ for $k \in \mathbb{Z}$ with residue $\rho/(\pi i)$. Summing over the residues of (7) at these poles, excluding $z = 0$, we obtain

$$\frac{ic^{s-1}}{4\rho^s} \sum_{k \in \mathbb{Z}, 0 \leq j < c}{}' \frac{\rho}{i}\left(-\left(k+\frac{j}{c}\right)\frac{\rho}{i}\right)^{s-1} \cot \pi \left(\left(k+\frac{j}{c}\right)\rho^2 - \frac{jd}{c}\right). \tag{12}$$

Let $n = kc + j$, and (12) equals

$$\frac{ic^{s-1}}{4\rho^s} \sum_{n \neq 0} \frac{\rho}{i}\left(-\frac{n\rho}{ic}\right)^{s-1} \cot \pi \left(\frac{n\rho^2}{c} - \frac{nd}{c}\right). \tag{13}$$

For n positive,

$$\frac{\rho}{i}\left(-\frac{n\rho}{ic}\right)^{s-1} - \frac{\rho}{i}\left(\frac{n\rho}{ic}\right)^{s-1} = -2\left(\cos \frac{\pi s}{2}\right)\left(\frac{n}{c}\right)^{s-1} \rho^s,$$

and thus (13) is

$$-\frac{1}{2} \cos \frac{\pi s}{2} \left(\zeta(1-s) + 2 \sum_{n > 0} n^{s-1} \frac{e^{2\pi i \left(\frac{n\rho^2}{c} - \frac{nd}{c}\right)}}{1 - e^{2\pi i \left(\frac{n\rho^2}{c} - \frac{nd}{c}\right)}} \right), \tag{14}$$

and since $n\rho^2/c - nd/c = n(c\tau+d)/c - nd/c = n\tau$, the sum in (14) is

$$\sum_{n > 0} n^{s-1} \frac{e^{2\pi i n \tau}}{1 - e^{2\pi i n \tau}} = A(\tau, s).$$

Finally, adding (11) with (14), we obtain

$$I_{c,d}(\tau, s) = \frac{1}{2\rho^{2s}} \cos \frac{\pi s}{2} \left(\zeta(1-s) + 2A(\gamma \tau, s) \right) - \frac{1}{2} \cos \frac{\pi s}{2} \left(\zeta(1-s) + 2A(\tau, s) \right)$$

and the theorem follows by dividing both sides by $\cos \frac{\pi s}{2}$. □

3. Discussion of Some Special Cases

Corollary 1. *For $\tau \in \mathfrak{H}$ and $\gamma = \begin{pmatrix} a & b \\ c & d \end{pmatrix} \in SL(2, \mathbb{Z})$ with $c > 0$, we have the following transformation formulas for $A(\tau, s)$ for s an even integer.*

(i) For any integer $m \geq 2$,

$$(c\tau + d)^{-2m} A(\gamma \tau, 2m) = A(\tau, 2m) + \frac{1}{2}\left(1 - (c\tau + d)^{-2m}\right) \zeta(1 - 2m)$$

and thus, if we define $\mathbb{G}_{2m}(\tau) = \frac{1}{2}\zeta(1 - 2m) + A(\tau, 2m)$, it is a modular form of weight $2m$.

(ii)

$$(c\tau + d)^{-2} A(\gamma \tau, 2) = A(\tau, 2) - \frac{1}{24}\left(1 - (c\tau + d)^{-2}\right) + \frac{ic}{4\pi(c\tau + d)},$$

and thus, if we let $\mathbb{G}_2(\tau) = -\frac{1}{24} + A(\tau, 2)$, then

$$\mathbb{G}_2(\gamma \tau) = (c\tau + d)^2 \mathbb{G}_2(\tau) + \frac{ic}{4\pi}(c\tau + d).$$

(iii)
$$A(\gamma\tau, 0) = A(\tau, 0) - \frac{1}{2}\log(c\tau + d) + \frac{\pi i}{4} - \frac{\pi i}{12c}\left(\frac{1}{c\tau + d} + c\tau + d\right) + \pi i\, \mathfrak{s}(d, c)$$

where the Dedekind sum $\mathfrak{s}(d, c)$ satisfies

$$\mathfrak{s}(d, c) = \frac{1}{4c}\sum_{j=1}^{c-1} \cot\pi\left(\frac{jd}{c}\right)\cot\pi\left(\frac{j}{c}\right),$$

and

$$\log\eta(\tau + b) = \frac{\pi i b}{12} + \log\eta(\tau) \qquad (b \in \mathbb{Z})$$

$$\log\eta(\gamma\tau) = \log\eta(\tau) + \frac{1}{2}\log(c\tau + d) - \frac{\pi i}{4} + \frac{\pi i(a+d)}{12c} - \pi i\, \mathfrak{s}(d, c).$$

(iv) For any integer $m > 0$,

$$(c\tau + d)^{2m} A(\gamma\tau, -2m) = A(\tau, -2m) + \frac{1}{2}\left(1 - (c\tau + d)^{2m}\right)\zeta(1 + 2m)$$

$$+ \frac{i}{\pi c^{2m+1}} \sum_{k=0}^{m+1} \zeta(2k)\zeta(2m + 2 - 2k)(c\tau + d)^{2k-1}$$

$$+ \frac{\pi^{2m+1} i}{4c^{2m+1}} \sum_{k=0}^{2m}\sum_{j=1}^{c-1} \cot^{(k)}\pi\left(\frac{jd}{c}\right)\cot^{(2m-k)}\pi\left(\frac{j}{c}\right)(c\tau + d)^k$$

where $\cot^{(n)} z = \frac{1}{n!}\left(\frac{d}{dz}\right)^n \cot z$.

Proof. (i) and (ii) follow from (8) since

$$(-1)^m I_{c,d}(\tau, 2m) = 0 \qquad (m \geq 2)$$

$$-I_{c,d}(\tau, 2) = -\frac{\pi i c}{4\rho^2}\operatorname*{Res}_{z=0}\left(z\cot\pi(i\rho z)\cot\pi\left(\frac{iz}{\rho}\right)\right) = \frac{ic}{4\pi\rho^2}.$$

Let us show (iii). Since $\lim_{s\to 0} \frac{1}{2}\left(1 - (c\tau + d)^{-s}\right)\zeta(1-s) = -\frac{1}{2}\log(c\tau + d)$, we have

$$A(\gamma\tau, 0) = A(\tau, 0) - \frac{1}{2}\log(c\tau + d) + I_{c,d}(\tau, 0)$$

where

$$I_{c,d}(\tau, 0) = \frac{1}{8c}\int_C (-z)^{-1}\left(c + \sum_{j=0}^{c-1}\cot\pi\left(i\rho z - \frac{jd}{c}\right)\cot\pi\left(\frac{iz}{\rho} - \frac{j}{c}\right)\right)dz$$

$$= \frac{\pi i}{4} + \frac{\pi i}{4c}\operatorname*{Res}_{z=0}\left(\frac{1}{z}\sum_{j=0}^{c-1}\cot\pi\left(i\rho z - \frac{jd}{c}\right)\cot\pi\left(\frac{iz}{\rho} - \frac{j}{c}\right)\right). \tag{15}$$

The $j = 0$ term in (15) gives

$$\frac{\pi i}{4c}\operatorname*{Res}_{z=0}\left(\frac{1}{z}\cot\pi(i\rho z)\cot\pi\left(\frac{iz}{\rho}\right)\right) = -\frac{\pi i}{12c}\left(\frac{1}{c\tau + d} + c\tau + d\right),$$

and the sum over $0 < j < c$ in (15) becomes

$$\frac{\pi i}{4c} \sum_{j=1}^{c-1} \cot \pi \left(\frac{jd}{c}\right) \cot \pi \left(\frac{j}{c}\right) = \pi i\, \mathsf{s}(d,c),$$

proving the claim for $A(\tau, 0)$ in (iii). The transformation formula for $\log \eta(\tau)$ now follows from (4) and the equality

$$\frac{\pi i}{12}\left(\frac{a\tau + b}{c\tau + d} - \tau\right) + \frac{\pi i}{12c}\left(\frac{1}{c\tau + d} + c\tau + d\right) = \frac{\pi i(a+d)}{12c}.$$

For (iv), let $s = -2m$ in (8) for a positive integer m. Then,

$$(c\tau + d)^{2m} A(\gamma\tau, -2m) = A(\tau, -2m) + \frac{1}{2}\left(1 - (c\tau + d)^{2m}\right)\zeta(1 + 2m) + (-1)^m I_{c,d}(\tau, -2m)$$

where

$$(-1)^m I_{c,d}(\tau, -2m) = \frac{\pi i \rho^{2m}(-1)^m}{4c^{2m+1}} \operatorname*{Res}_{z=0} \left(\frac{1}{z^{2m+1}} \sum_{j=0}^{c-1} \cot \pi \left(i\rho z - \frac{jd}{c}\right) \cot \pi \left(\frac{iz}{\rho} - \frac{j}{c}\right)\right). \tag{16}$$

From the Laurent expansions

$$\cot \pi(i\rho z) = -\frac{2}{\pi} \sum_{k \geq 0} \zeta(2k)(i\rho z)^{2k-1}$$

$$\cot \pi \left(\frac{iz}{\rho}\right) = -\frac{2}{\pi} \sum_{k \geq 0} \zeta(2k) \left(\frac{iz}{\rho}\right)^{2k-1},$$

the $j = 0$ term in (16) is

$$\frac{\pi i \rho^{2m}(-1)^m}{4c^{2m+1}} \frac{4}{\pi^2} \sum_{k_1 + k_2 = m+1} \zeta(2k_1)\zeta(2k_2)(i\rho)^{2k_1 - 1} \left(\frac{i}{\rho}\right)^{2k_2 - 1}$$

$$= \frac{i}{\pi c^{2m+1}} \sum_{k_1 + k_2 = m+1} \zeta(2k_1)\zeta(2k_2) \rho^{2(m+k_1-k_2)}$$

$$= \frac{i}{\pi c^{2m+1}} \sum_{k=0}^{m+1} \zeta(2k)\zeta(2m+2-2k)(c\tau + d)^{2k-1},$$

and from the Taylor expansions for $0 < j < c$,

$$\cot \pi \left(i\rho z - \frac{jd}{c}\right) = -\sum_{k \geq 0} (-\pi i \rho z)^k \cot^{(k)} \pi \left(\frac{jd}{c}\right)$$

$$\cot \pi \left(\frac{iz}{\rho} - \frac{j}{c}\right) = -\sum_{k \geq 0} \left(-\frac{\pi i z}{\rho}\right)^k \cot^{(k)} \pi \left(\frac{j}{c}\right),$$

the sum over $0 < j < c$ in (16) is

$$\frac{\pi i \rho^{2m}(-1)^m}{4c^{2m+1}} \sum_{k_1+k_2=2m} \sum_{j=1}^{c-1} \cot^{(k_1)} \pi\left(\frac{jd}{c}\right) \cot^{(k_2)} \pi\left(\frac{j}{c}\right) (-\pi i \rho)^{k_1} \left(-\frac{\pi i}{\rho}\right)^{k_2}$$

$$= \frac{\pi^{2m+1} i}{4c^{2m+1}} \sum_{k_1+k_2=2m} \sum_{j=1}^{c-1} \cot^{(k_1)} \pi\left(\frac{jd}{c}\right) \cot^{(k_2)} \pi\left(\frac{j}{c}\right) \rho^{2m+k_1-k_2}$$

$$= \frac{\pi^{2m+1} i}{4c^{2m+1}} \sum_{k=0}^{2m} \sum_{j=1}^{c-1} \cot^{(k)} \pi\left(\frac{jd}{c}\right) \cot^{(2m-k)} \pi\left(\frac{j}{c}\right) (c\tau+d)^k,$$

as claimed. □

For $x \in \mathbb{R}$, we define the periodic Bernoulli functions $\widetilde{B}_n(x)$, $n \geq 0$, by the following identity as formal power series in t:

$$\sum_{n=0}^{\infty} \widetilde{B}_n(x) \frac{t^n}{n!} = \begin{cases} \frac{te^{\{x\}t}}{e^t-1} & \text{if } x \in \mathbb{R}\setminus\mathbb{Z} \\ \frac{t}{e^t-1} + \frac{t}{2} & \text{if } x \in \mathbb{Z}, \end{cases} \quad (17)$$

where $\{x\} = x - \lfloor x \rfloor$ is the fractional part of x. Under the condition of Corollary 1 (iv), the series $A(\tau, -2m)$ for any integer $m > 0$ is known to satisfy the formula ([8–10,12])

$$(c\tau+d)^{2m} A(\gamma\tau, -2m) = A(\tau, -2m) + \frac{1}{2}\left(1 - (c\tau+d)^{2m}\right) \zeta(1+2m)$$
$$+ \frac{(2\pi i)^{2m+1}}{2(2m+2)!} \sum_{k=0}^{2m+2} \binom{2m+2}{k} \sum_{j=0}^{c-1} \widetilde{B}_k(j/c) \widetilde{B}_{2m+2-k}(jd/c) (-(c\tau+d))^{k-1}, \quad (18)$$

while the sums

$$\sum_{j=0}^{c-1} \widetilde{B}_k(j/c) \widetilde{B}_{2m+2-k}(jd/c)$$

which appear in (18) are regarded as generalized Dedekind sums.

Remark 1. *The notation $\overline{B}_n(x)$ is often used for $\widetilde{B}_n(x)$ in (17), but it is also used for*

$$\overline{B}_n(x) = B_n(\{x\}) \quad (n \geq 0) \quad (19)$$

where $B_n(x)$ denotes the Bernoulli polynomials. With (19), $\widetilde{B}_n(x)$ and $\overline{B}_n(x)$ only differ by $\frac{1}{2}$ when $n = 1$ and $x \in \mathbb{Z}$. For $m > 0$ and $0 \leq k \leq 2m+2$, one can see that the equality

$$\sum_{j=0}^{c-1} \overline{B}_k(j\alpha/c) \overline{B}_{2m+2-k}(j\beta/c) = \sum_{j=0}^{c-1} \widetilde{B}_k(j\alpha/c) \widetilde{B}_{2m+2-k}(j\beta/c) \quad (\alpha, \beta, c \in \mathbb{Z}; c > 0) \quad (20)$$

holds as $\overline{B}_{2m+1}(x) = \widetilde{B}_{2m+1}(x)$ is a periodic odd function, vanishing at 0 and $\frac{1}{2}$. Therefore, (18) can be stated in the same way in either notation.

In Corollary 1, we have already obtained the cotangent sum representations for the Dedekind sum and its generalizations in (18). On the other hand, we should also be able to obtain Corollary 1 (iv) by assuming (18) and expanding it in discrete Fourier series. We now present this alternative derivation of Corollary 1 (iv).

We first give a proof of the following lemma (cf. [14]), using the generating function (17).

Lemma 2. For $j, c \in \mathbb{Z}$ with $c > 0$ and $n \geq 0$,

$$\widetilde{B}_n(j/c) = \left(\frac{B_n}{c^n}\right)_{n \neq 1} + \left(n! \left(\frac{i}{2c}\right)^n \sum_{k=1}^{c-1} \cot^{(n-1)} \pi \left(\frac{k}{c}\right) e^{2\pi i j k/c}\right)_{n \geq 1}$$

where the first term is not present for $n = 1$ (so that it is only present for even n) and the second term only for $n \geq 1$, and $\cot^{(n)} z = \frac{1}{n!} \left(\frac{d}{dz}\right)^n \cot z$.

Proof. Let $F_x(t)$ denote the right-hand side of (17). For $\mu = e^{2\pi i/c}$, we let

$$F_{j/c}(t) = \sum_{k=0}^{c-1} a_k(t) \mu^{jk} \qquad (j \in \mathbb{Z}). \tag{21}$$

By multiplying both sides of (21) by μ^{-jl} and summing over j mod c,

$$c a_l(t) = \sum_{j=0}^{c-1} F_{j/c}(t) \mu^{-jl} = \sum_{j=0}^{c-1} \frac{t e^{(j/c)t}}{e^t - 1} \mu^{-jl} + \frac{t}{2} \qquad (0 \leq l < c). \tag{22}$$

Now, as formal power series in t and x,

$$\sum_{j=0}^{c-1} \frac{t e^{(j/c)t}}{e^t - 1} x^j = \frac{t}{e^t - 1} \sum_{j=0}^{c-1} (e^{t/c} x)^j = \frac{t}{e^t - 1} \frac{1 - e^t x^c}{1 - e^{t/c} x}. \tag{23}$$

By putting $x = \mu^{-l}$ in (23), we obtain from (22)

$$a_l(t) = \frac{t/c}{e^{t/c} \mu^{-l} - 1} + \frac{t}{2c} = \frac{(t/c) e^{2\pi i l/c - t/c}}{1 - e^{2\pi i l/c - t/c}} + \frac{t}{2c}.$$

For $l = 0$, we have

$$a_0(t) = \frac{t/c}{e^{t/c} - 1} + \frac{t}{2c} = \sum_{n \geq 0, n \neq 1} \frac{B_n}{c^n} \frac{t^n}{n!}.$$

For $l \neq 0$,

$$a_l(t) = \frac{(t/c) e^{2\pi i l/c - t/c}}{1 - e^{2\pi i l/c - t/c}} + \frac{t}{2c} = \frac{it}{2c} \cot\left(\frac{\pi l}{c} + \frac{it}{2c}\right) = \sum_{n \geq 0} t^{n+1} \left(\frac{i}{2c}\right)^{n+1} \cot^{(n)} \pi \left(\frac{l}{c}\right).$$

The lemma follows by taking the coefficient of $\frac{t^n}{n!}$ in (21). □

Proof of Corollary 1 (iv). We consider the following sum in (18),

$$\frac{(2\pi i)^{2m+1}}{2(2m+2)!} \sum_{k=0}^{2m+2} \binom{2m+2}{k} \sum_{j=0}^{c-1} \widetilde{B}_k(j/c) \widetilde{B}_{2m+2-k}(jd/c)(-(c\tau+d))^{k-1}. \tag{24}$$

We expand $\widetilde{B}_k(j/c)$ and $\widetilde{B}_{2m+2-k}(jd/c)$ in (24) by Lemma 2 and sum over j. The first cross terms give, for $k = 2l$,

$$\frac{(2\pi i)^{2m+1}}{2(2m+2)!} \sum_{l=0}^{m+1} \binom{2m+2}{2l} \sum_{j=0}^{c-1} \frac{B_{2l}}{c^{2l}} \frac{B_{2m+2-2l}}{c^{2m+2-2l}} (-(c\tau+d))^{2l-1}. \tag{25}$$

Since $\zeta(2n) = (-1)^{n+1} \frac{(2\pi)^{2n}}{2(2n)!} B_{2n}$ and as the identical sum over j gives a factor of c, (25) simplifies to

$$\frac{i}{\pi c^{2m+1}} \sum_{l=0}^{m+1} \zeta(2l)\zeta(2m+2-2l)(c\tau+d)^{2l-1}. \tag{26}$$

The rest of surviving cross terms give

$$\frac{(2\pi i)^{2m+1}}{2} \sum_{k=1}^{2m+1} \sum_{j=0}^{c-1} \left(\left(\frac{i}{2c}\right)^k \sum_{\alpha=1}^{c-1} \cot^{(k-1)} \pi\left(\frac{\alpha}{c}\right) e^{2\pi i j \alpha/c} \right)$$
$$\times \left(\left(\frac{i}{2c}\right)^{2m+2-k} \sum_{\beta=1}^{c-1} \cot^{(2m+1-k)} \pi\left(\frac{\beta}{c}\right) e^{2\pi i j d \beta/c} \right) (-(c\tau+d))^{k-1}$$

which, since

$$\sum_{j=0}^{c-1} e^{2\pi i j \alpha/c} e^{2\pi i j d \beta/c} = \begin{cases} c & \text{if } \alpha + d\beta \equiv 0 \bmod c \\ 0 & \text{otherwise,} \end{cases}$$

equals

$$\frac{\pi^{2m+1} i}{4 c^{2m+1}} \sum_{k=1}^{2m+1} \sum_{\beta=1}^{c-1} \cot^{(k-1)} \pi\left(\frac{d\beta}{c}\right) \cot^{(2m+1-k)} \pi\left(\frac{\beta}{c}\right) (c\tau+d)^{k-1}. \tag{27}$$

Hence (24) equals to the sum of (26) and (27), which proves equivalence of (18) and Corollary 1 (iv). □

Funding: This work was supported by the 2018 Hongik University Research Fund.

Acknowledgments: The author wishes to thank Jeong Seog Ryu, Hi-joon Chae and Joongul Lee for helpful comments.

Conflicts of Interest: The author declares no conflict of interest.

References

1. Apostol, T.M. *Modular Functions and Dirichlet Series in Number Theory*, 2nd ed.; Graduate Texts in Mathematics; Springer: New York, NY, USA, 1990; Volume 41. [CrossRef]
2. Bruinier, J.H.; van der Geer, G.; Harder, G.; Zagier, D. *The 1-2-3 of Modular Forms*; Universitext; Lectures from the Summer School on Modular Forms and their Applications held in Nordfjordeid, June 2004; Ranestad, K., Ed.; Springer: Berlin, Germany, 2008. [CrossRef]
3. Siegel, C.L. A simple proof of $\eta(-1/\tau) = \eta(\tau)\sqrt{\tau/i}$. *Mathematika* **1954**, *1*, 4. [CrossRef]
4. Rademacher, H. On the transformation of $\log \eta(\tau)$. *J. Indian Math. Soc.* **1955**, *19*, 25–30.
5. Rademacher, H. On the expansion of the partition function in a series. *Ann. Math.* **1943**, *44*, 416–422. [CrossRef]
6. Hardy, G.H.; Ramanujan, S. Asymptotic formulæ in combinatory analysis [Proc. London Math. Soc. (2) **17** (1918), 75–115]. In *Collected Papers of Srinivasa Ramanujan*; AMS Chelsea Publ.: Providence, RI, USA, 2000; pp. 276–309.
7. Rademacher, H.; Grosswald, E. *Dedekind Sums*; The Carus Mathematical Monographs, No. 16; The Mathematical Association of America: Washington, DC, USA, 1972.
8. Apostol, T.M. Generalized Dedekind sums and transformation formulae of certain Lambert series. *Duke Math. J.* **1950**, *17*, 147–157. [CrossRef]
9. Mikolás, M. Über gewisse Lambertsche Reihen. I. Verallgemeinerung der Modulfunktion $\eta(\tau)$ und ihrer Dedekindschen Transformationsformel. *Math. Z.* **1957**, *68*, 100–110. [CrossRef]
10. Iseki, S. The transformation formula for the Dedekind modular function and related functional equations. *Duke Math. J.* **1957**, *24*, 653–662. [CrossRef]
11. Lewittes, J. Analytic continuation of Eisenstein series. *Trans. Am. Math. Soc.* **1972**, *171*, 469–490. [CrossRef]
12. Berndt, B.C. Generalized Dedekind eta-functions and generalized Dedekind sums. *Trans. Am. Math. Soc.* **1973**, *178*, 495–508. [CrossRef]

13. Berndt, B.C. Generalized Eisenstein series and modified Dedekind sums. *J. Reine Angew. Math.* **1974**, *272*, 182–193. [CrossRef]
14. Beck, M. Dedekind cotangent sums. *Acta Arith.* **2003**, *109*, 109–130. [CrossRef]

© 2019 by the author. Licensee MDPI, Basel, Switzerland. This article is an open access article distributed under the terms and conditions of the Creative Commons Attribution (CC BY) license (http://creativecommons.org/licenses/by/4.0/).

Article
Euler Sums and Integral Connections

Anthony Sofo [1,*] and Amrik Singh Nimbran [2]

1. College of Engineering and Science, Victoria University, P. O. Box 14428, Melbourne City, Victoria 8001, Australia
2. B3-304, Palm Grove Heights, Ardee City, Gurugram, Haryana 122003, India
* Correspondence: anthony.sofo@vu.edu.au

Received: 13 August 2019; Accepted: 4 September 2019; Published: 9 September 2019

Abstract: In this paper, we present some Euler-like sums involving partial sums of the harmonic and odd harmonic series. First, we give a brief historical account of Euler's work on the subject followed by notations used in the body of the paper. After discussing some alternating Euler sums, we investigate the connection of integrals of inverse trigonometric and hyperbolic type functions to generate many new Euler sum identities. We also give some new identities for Catalan's constant, Apery's constant and a fast converging identity for the famous $\zeta(2)$ constant.

Keywords: Euler sums; Catalan's constant; Trigamma function; integral representation; closed form; ArcTan and ArcTanh functions; partial fractions

1. Historical Background and Preliminaries

1.1. Euler's Work

We begin by touching on the historical background of Euler sums. The 20th century British mathematician G. H. Hardy remarked in *A Mathematician's Apology* (1940): " 'Immortality' may be a silly word, but probably a mathematician has the best chance of whatever it may mean." Leonhard Euler (1707–1783), the versatile and prolific Swiss mathematician, achieved immortality through his pioneering work on infinite series and products which began when he was 22. He gained instant fame in youth by solving the Basel problem (suggested to him by Johann Bernoulli who had tried and failed) which asked for the exact sum of the series of reciprocals of squares of natural numbers. He generalized the problem and introduced the famous *Euler zeta function*: $\sum_{n=1}^{\infty} \frac{1}{n^m}$, $m \geq 2$. Euler extended the concept of *factorial* to positive rational numbers. His investigation (motivated by Christian Goldbach, the secretary of the Petersburg Academy where Euler worked) laid the foundation for another famous function, the *gamma function*, the digamma function (the logarithmic derivative of the gamma function) and the *polygamma functions* formed by repeated differentiation of the digamma function. Nicholas Oresme (1323–1382) had showed that the sum $\sum_{k=1}^{\infty} \frac{1}{k}$ diverges. Euler investigated its partial sums (called *harmonic numbers*), $\sum_{k=1}^{n} \frac{1}{k}$, detected their link with logarithms, and introduced his ubiquitous constant γ in the process: $\gamma = \lim_{n \to \infty} \{H_n - \log(n)\}$, where n is any positive integer. He also gave the formula: $H_n = \int_0^1 \frac{1-x^n}{1-x} dx$ which can be established by writing the integrand as a geometric progression and then integrating it term by term.

Euler sums have their origin in the correspondence between Goldbach and Euler after Euler's departure from Russia in 1741. In his letter of 6 December 1742, Goldbach informed Euler that he had calculated the sums of the following series ([1] p. 741):

$$1 - \frac{1}{2}\left(1 \mp \frac{1}{2}\right) + \frac{1}{3}\left(1 \mp \frac{1}{2} + \frac{1}{3}\right) - \frac{1}{4}\left(1 \mp \frac{1}{2} + \frac{1}{3} \mp \frac{1}{4}\right) + \cdots,$$

$$1 - \frac{1}{2}\left(1 - \frac{1}{2^2}\right) + \frac{1}{3}\left(1 - \frac{1}{2^2} + \frac{1}{3^2}\right) - \frac{1}{4}\left(1 - \frac{1}{2^2} + \frac{1}{3^2} - \frac{1}{4^2}\right) + \cdots,$$

$$\frac{1}{2^2}(1) - \frac{1}{3^2}\left(1 - \frac{1}{2}\right) + \frac{1}{4^2}\left(1 - \frac{1}{2} + \frac{1}{3}\right) - \frac{1}{5^2}\left(1 - \frac{1}{2} + \frac{1}{3} - \frac{1}{4}\right) + \cdots$$

However, in his next letter (24th December), he modified his claims stating that they arose out of an error which led to serendipitous discovery of Euler sums:

> when I recently reconsidered the supposed sums of the two series mentioned at the end of my last letter, I perceived at once that they had arisen by a mere writing mistake. However, of this indeed the proverb says "If he had not erred, he should have achieved less"; for on that occasion I came upon the summations of some other series which otherwise I should hardly have looked for, much less discovered.

Goldbach then ([1] p. 747) proposed to compute the sum:

$$1 + \frac{1}{2^n}\left(1 + \frac{1}{2^m}\right) + \frac{1}{3^n}\left(1 + \frac{1}{2^m} + \frac{1}{3^m}\right) + \cdots$$

where m, n are not equal even integers. He further gave the sum $\frac{\pi^4}{72}$ for series with $m = 1, n = 3$. This led Euler to investigate this sum. In his reply of 5 January 1743 to Goldbach, he recorded various sums including ([1] p. 752, G, I):

$$1 - \frac{1}{2}\left(1 + \frac{1}{2}\right) + \frac{1}{3}\left(1 + \frac{1}{2} + \frac{1}{3}\right) - \frac{1}{4}\left(1 + \frac{1}{2} + \frac{1}{3} + \frac{1}{4}\right) + \cdots = \frac{\pi^2}{12} - \frac{\log^2(2)}{2},$$

and

$$1 - \frac{1}{2}\left(1 - \frac{1}{2}\right) + \frac{1}{3}\left(1 - \frac{1}{2} + \frac{1}{3}\right) - \frac{1}{4}\left(1 - \frac{1}{2} + \frac{1}{3} - \frac{1}{4}\right) + \cdots = \frac{\pi^2}{12} + \frac{\log^2(2)}{2},$$

which appeared in §108 (page 509) of his *Continuatio Fragmentorum ex Adversariis mathematicis depormtorum* (Opera Postuma 1, 1862, 487–518) now numbered E819.

Their correspondence culminated in a 47-page paper [2] by Euler (1776) using a cumbersome notation:

$$\int \frac{1}{z^m}\left(\frac{1}{y^n}\right) = 1 + \frac{1}{2^m}\left(1 + \frac{1}{2^n}\right) + \frac{1}{3^m}\left(1 + \frac{1}{2^n} + \frac{1}{3^n}\right) + \cdots$$

In the paper mentioned, he employed three methods (which he called *Prima Methodus, Secunda Methodus* and *Tertia Methodus*) to discover formulas representing these sums in terms of zeta values. First, he multiplied the involved series to obtain this formula (§4, p.144):

$$\int \frac{1}{z^m}\left(\frac{1}{y^n}\right) + \int \frac{1}{z^n}\left(\frac{1}{y^m}\right) = \int \frac{1}{z^m} \int \frac{1}{z^n} + \int \frac{1}{z^{m+n}}.$$

After investigating cases with combined exponents of $z, y = 2, 3, \ldots, 10$, he considered the sum with $n = 1$ in the notation given earlier. He recorded a general formula (§22, p. 165) which, for $m = 2, 3, 4, \ldots$, can be written as:

$$EU(m) = \sum_{n=1}^{\infty} \frac{H_n}{n^m} = \frac{m+2}{2}\zeta(m+1) - \frac{1}{2}\sum_{k=1}^{m-2}\zeta(m-k)\zeta(k+1). \tag{1}$$

1.2. Post-Euler Development

Nielsen (1906) built on and supplemented Euler's work by supplying proof of the general Formula (1) using the method of partial fractions in ([3] pp. 47–49). Ramanujan noted a few Euler sums in his notebooks sometime in the 1910s. Sitaramachandrarao, an Indian mathematician, obtained identities for various alternating Euler sums in 1987 while discussing a formula of Ramanujan [4].

Georghiou and Philippou [5] established Euler's formula as well as:

$$\sum_{n=1}^{\infty} \frac{H_n}{n^{2s+1}} = \frac{1}{2}\sum_{r=2}^{2s}(-1)^r \zeta(r)\zeta(2s+2-r), \quad s \geq 1.$$

They also gave this formula:

$$\sum_{n=1}^{\infty} \frac{H_n^{(2)}}{n^{2s+1}} = \zeta(2)\zeta(2s+1) - \frac{(s+2)(2s+1)}{2}\zeta(2s+3)$$
$$+ 2\sum_{r=2}^{s+1}(r-1)\zeta(2r-1)\zeta(2s+4-2r), \quad s \geq 1,$$

where $H_n^{(m)} = \sum_{\ell=1}^{n} \ell^{-m}$, $m \geq 1, n \geq 1$. They remarked at the end that it was still an open question to give a closed form of $\sum_{n=1}^{\infty} \frac{H_n^{(p)}}{n^q}$ for any integers $p \geq 1$ and $q \geq 2$ in terms of the zeta function. Surprisingly, they did not mention Euler or his work even once.

The publication of De Doelder's paper [6] in 1991 and three papers submitted during July, August and October 1993 by the Borweins and their co-researchers [7–9] produced a revival of interest in these sums among a number of mathematicians. Crandall and Buhler [10] established various series expansions of $\sum_{n=2}^{\infty} \frac{1}{n^s} \sum_{m=1}^{n-1} \frac{1}{m^r}$. Various approaches have been employed to get such sums. De Doelder summed some series by evaluating integrals and using the digamma function. A quadratic sum $\sum_{k=1}^{\infty} \frac{\{\psi(k) - \psi(1)\}^2}{k^2} = \sum_{n=1}^{\infty} \left(\frac{H_n}{n+1}\right)^2 = \frac{11\pi^4}{360}$ was derived by De Doelder and an associated sum conjectured by Enrico Au-Yeung (an under-graduate student at Waterloo, ([7] p. 17) and ([8] p. 1191)) on the basis of his computations was established by means of generating functions and Parseval's identity for Fourier series by the Borweins. In fact, the associated sum $\sum_{n=1}^{\infty} \left(\frac{H_n}{n}\right)^2 = \frac{17\pi^4}{360}$ had been computed by Sandham [11] in 1948, but it remained generally unknown. Chu [12] made use of the hypergeometric method for deriving some Euler sums. Jung, Cho and Choi [13] use integrals from Lewin's book to evaluate different Euler sums. In particular, they show that $\sum_{n=1}^{\infty} \frac{H_n}{(n+1)^3} = \frac{\zeta(4)}{4}$ can be obtained from the integral (7.65) recorded in ([14] p. 204). Flajolet and Salvy [15] used contour integration. Freitas [16] transformed De Doelder's sum into a double integral and then evaluated the integral directly. Experimental evaluation of Euler sums involving heavy use of *Maple* and *Mathematica* led to the discovery of many new sums; some of these were established later. Euler sums have been a popular topic for engaging many mathematicians for last many years.

1.3. Notations and Representations of Harmonic Numbers

We use the following notations throughout our paper:

1. $H_n = \sum_{k=1}^{n} \frac{1}{k}$; $H'_n = \sum_{k=1}^{n} \frac{(-1)^{k+1}}{k}$.

2. $H_n^p = \left(\sum_{k=1}^{n} \frac{1}{k}\right)^p$; $H_n^{(p)} = \sum_{k=1}^{n} \frac{1}{k^p}$.

3. $h_n = \sum_{k=1}^{n} \frac{1}{2k-1} = H_{2n} - \frac{1}{2}H_n$; $h'_n = \sum_{k=1}^{n} \frac{(-1)^{k+1}}{2k-1}$.

$H_n^{(p)}$, the generalized harmonic number of order p, is sometime denoted by $\zeta_n(p)$ (Louis Comtet's *Advanced Combinatorics*, 1974, p. 217, [7b]). The alternating harmonic numbers, H'_n, have been termed *skew-harmonic numbers*. We call h_n the *odd harmonic numbers* and follow Berndt's notation ([17] p. 249, (8.2)) in preference to the symbol O_n used by Borweins [8].

Taking the log of the two sides and then differentiating this product representation

$$\frac{1}{\Gamma(x)} = xe^{\gamma x} \prod_{n=1}^{\infty} \left(1 + \frac{x}{n}\right) e^{-x/n}$$

yields

$$\psi(x) = -\gamma - \frac{1}{x} + \sum_{n=1}^{\infty}\left(\frac{1}{n} - \frac{1}{x+n}\right) = -\gamma + \sum_{n=1}^{\infty}\left(\frac{1}{n} - \frac{1}{x+n-1}\right),$$

leading to the telescoping sum:

$$\psi(1+x) - \psi(x) = \sum_{n=1}^{\infty}\left(\frac{1}{x+n-1} - \frac{1}{x+n}\right) = \frac{1}{x}$$

and for integral arguments $n \geq 1$, $\psi(n+1) - \psi(1) = H_n$ with $\psi(1) = -\gamma$.

We also have:

$$h_n = \frac{1}{2}\left[\psi\left(n + \frac{1}{2}\right) - \psi\left(\frac{1}{2}\right)\right].$$

Choi ([18] Equation (3.7)) defined the harmonic numbers in terms of log-sine functions

$$H_n = -4n \int_0^{\frac{\pi}{2}} \log(\sin x) \sin x (\cos x)^{2n-1} \, dx \tag{2}$$

$$= -4n \int_0^{\frac{\pi}{2}} \log(\cos x) \cos x (\sin x)^{2n-1} \, dx.$$

An unusual, but intriguing representation has recently been given by Ciaurri et al. ([19] Equation (10)) as follows:

$$H_n = \pi \int_0^1 \left(x - \frac{1}{2}\right) \left(\frac{\cos\left(\frac{(4n+1)\pi x}{2}\right) - \cos\left(\frac{\pi x}{2}\right)}{\sin\left(\frac{\pi x}{2}\right)}\right) dx. \tag{3}$$

We may introduce here the Lerch transcendent,

$$\Phi(z, t, a) = \sum_{m=0}^{\infty} \frac{z^m}{(m+a)^t},$$

which is defined for $|z| < 1$ and $\Re(a) > 0$ and satisfies the recurrence

$$\Phi(z, t, a) = z\, \Phi(z, t, a+1) + a^{-t}.$$

The Lerch transcendent generalizes the Hurwitz zeta function at $z = 1$,

$$\Phi(1, t, a) = \sum_{m=0}^{\infty} \frac{1}{(m+a)^t}$$

and the polylogarithm, (see [14]), or de-Jonquière's function, when $a = 1$,

$$\mathrm{Li}_t(z) := \sum_{m=1}^{\infty} \frac{z^m}{m^t}, \ t \in \mathbb{C} \text{ when } |z| < 1;\ \Re(t) > 1 \text{ when } |z| = 1.$$

The polylogarithm of negative integer order arises in the sums of the form

$$\sum_{j \geq 1} j^n z^j = \mathrm{Li}_{-n}(z) = \frac{1}{(1-z)^{n+1}} \sum_{i=0}^{n-1} \left\langle \begin{array}{c} n \\ i \end{array} \right\rangle z^{n-i},$$

where the Eulerian number $\left\langle \begin{array}{c} n \\ i \end{array} \right\rangle = \sum_{j=0}^{i+1} (-1)^j \binom{n+1}{j} (i-j+1)^n$, see [20]. The polygamma function of order k for $z \neq -1, -2, \ldots$

$$\psi^{(k)}(z) = \frac{d^k}{dz^k}\{\psi(z)\} = (-1)^{k+1} k! \sum_{r=0}^{\infty} \frac{1}{(r+z)^{k+1}}$$

and has the recurrence

$$\psi^{(k)}(z+1) = \psi^{(k)}(z) + \frac{(-1)^k k!}{z^{k+1}}.$$

The connection between the polygamma function and harmonic numbers is given by

$$H_z^{(\alpha+1)} = \zeta(\alpha+1) + \frac{(-1)^\alpha}{\alpha!} \psi^{(\alpha)}(z+1),\ z \neq -1, -2, -3, \ldots \tag{4}$$

and the multiplication formula for a positive integer m is (Abramowitz and Stegun, *Handbook of Mathematical Functions*, p. 260, 6.4.8)

$$\psi^{(k)}(mz) = \delta_{k,0} \log(m) + \frac{1}{m^{k+1}} \sum_{j=0}^{m-1} \psi^{(k)}\left(z + \frac{j}{m}\right), \tag{5}$$

where $\delta_{k,r}$ is the Kronecker delta.

We define the alternating zeta function (or Dirichlet eta function) $\eta(z)$ as

$$\eta(z) := \sum_{n=1}^{\infty} \frac{(-1)^{n+1}}{n^z} = \left(1 - 2^{1-z}\right) \zeta(z), \tag{6}$$

where $\eta(1) = \log 2$. If we put

$$S(p, q) := \sum_{n=1}^{\infty} \frac{(-1)^{n+1} H_n^{(p)}}{n^q},$$

in the case where p and q are both positive integers and weight $(p+q)$ is an odd integer, Flajolet and Salvy [15] gave the identity:

$$2S(p,q) = \{1-(-1)^p\}\zeta(p)\eta(q) + \eta(p+q)$$
$$+ 2(-1)^p \sum_{i+2k=q} \binom{p+i-1}{p-1} \zeta(p+i)\eta(2k)$$
$$- 2 \sum_{j+2k=p} \binom{q+j-1}{q-1} (-1)^j \eta(q+j)\eta(2k), \qquad (7)$$

where $\eta(0) = \frac{1}{2}$, $\eta(1) = \log 2$, $\zeta(1) = 0$, and $\zeta(0) = -\frac{1}{2}$ in accordance with the analytic continuation of the Riemann zeta function. For odd weight $(p+q)$, we have from [9] and ([15] Th.3.1):

$$BW(p,q) = \sum_{n=1}^{\infty} \frac{H_n^{(p)}}{n^q} = (-1)^p \sum_{j=1}^{\left[\frac{q}{2}\right]} \binom{p+q-2j-1}{p-1} \zeta(p+q-2j)\zeta(2j)$$
$$+ \frac{1-(-1)^p}{2}\zeta(p)\zeta(q) + (-1)^p \sum_{j=1}^{\left[\frac{p}{2}\right]} \binom{p+q-2j-1}{q-1} \zeta(p+q-2j)\zeta(2j)$$
$$+ \frac{(-1)^p}{2} \left\{ (-1)^p - \binom{p+q-1}{p} - \binom{p+q-1}{q} \right\} \zeta(p+q), \qquad (8)$$

where $[x]$ is the largest integer contained in x. It appears that some isolated cases of $BW(p,q)$, for even weight $(p+q)$, can be expressed in zeta terms, but, in general, almost certainly, for even weight $(p+q)$, and $p \geq 2$, no general closed form expression exists for $BW(p,q)$. Some examples with even weight are

$$\sum_{n=1}^{\infty} \frac{H_n^{(2)}}{n^4} = \zeta^2(3) - \frac{1}{3}\zeta(6), \quad \sum_{n=1}^{\infty} \frac{H_n^{(4)}}{n^4} = \frac{13}{12}\zeta(8).$$

There are also the two general forms of identities,

$$2\sum_{n=1}^{\infty} \frac{H_n^{(2p+1)}}{n^{2p+1}} = \zeta^2(2p+1) + \zeta(4p+2),$$

and

$$2\sum_{n=1}^{\infty} \frac{H_n^{(2p)}}{n^{2p}} = \left(1 - \frac{(4p)!B_{2p}^2}{2((2p)!)^2 B_{4p}}\right) \zeta(4p)$$

for $p \in \mathbb{N}$ and where B_p are the signed Bernoulli numbers of the first kind.

Sofo [21], furthermore, developed the half integer Euler sums. For positive integers m, p and odd weight $m + p$,

$$W(m,p) = \sum_{n=1}^{\infty} \frac{H_{\frac{n}{2}}^{(m)}}{n^p} = (-1)^p \sum_{r=1}^{m} 2^{m-1} \binom{m+p-1-r}{m-r} \times$$
$$[BW(r, m+p-r) - S(r, m+p-r)]$$
$$+ (-1)^{p+1} \sum_{r=2}^{m} \frac{1}{2^{p-r}} \binom{m+p-1-r}{m-r} \zeta(r)\zeta(m+p-r)$$
$$+ (-1)^{p+1} \sum_{k=2}^{p-1} \frac{(-1)^k}{2^{p-k}} \binom{m+p-1-k}{p-k} \zeta(k)\zeta(m+p-k). \qquad (9)$$

In addition,
$$\sum_{n\geq 1} \frac{H_n^{(m)}}{n^p} = 2^{p-1} \sum_{n=1}^{\infty} \frac{H_{\frac{n}{2}}^{(m)}}{n^p} \left\{1-(-1)^{n+1}\right\}.$$

We obtain the alternating Euler identity at half integer value,
$$\sum_{n\geq 1} \frac{(-1)^{n+1} H_{\frac{n}{2}}^{(m)}}{n^p} = W(m,p) - 2^{1-p} BW(m,p). \tag{10}$$

Some other Euler identities are given in [22–37].

1.4. Cauchy Product and Relevant Generating Functions

The multiplication of series by Cauchy's rule (Ch.VI of A. L. Cauchy's *Cours d'Analyse de l'École Royale Polytechnique*, 1821) is given by:
$$\left(\sum_{k=0}^{\infty} a_k x^k\right)\left(\sum_{k=0}^{\infty} b_k x^k\right) = \sum_{k=0}^{\infty} c_k x^k$$

with $c_k = \sum_{j=0}^{k} a_j b_{k-j}$. Cauchy showed (pp. 147-8, Th.6) that if the two convergent series on LHS are composed *only of positive terms*, i.e., they are *absolutely* convergent, then so is their product. Abel's theorem (1826) states that if all three series (including conditionally convergent) converge to sums, say s_a, s_b, s_c respectively, then $s_a \cdot s_b = s_c$ necessarily.

A function $G(x)$ represented by the power series: $G(x) := \sum_{n=0}^{\infty} a_n x^n$ is known as *ordinary generating function* for the sequence $\{a_n\}$.

The Cauchy product of $(1-x)^{-1}$ and $\log(1-x)$ yields the generating function ([38] p. 54, 1.514.6):
$$-\frac{\log(1-x)}{1-x} = \sum_{n=1}^{\infty} H_n x^n, \tag{11}$$

while the Cauchy product of $(1-x)^{-1}$ and $\log(1+x)$ yields:
$$\frac{\log(1+x)}{1-x} = \sum_{n=1}^{\infty} H'_n x^n, \tag{12}$$

both series converging for $|x|<1$.

1.5. Few Sums for Further Use

The Cauchy product of the numerical series $\sum_{k=0}^{\infty} \frac{(-1)^{k+1}}{k+1}$ with itself gives:
$$2\sum_{n=1}^{\infty} \frac{(-1)^{n+1}}{n+1}\left(1+\frac{1}{2}+\cdots+\frac{1}{n}\right) = 2\sum_{n=1}^{\infty}(-1)^{n+1}\frac{H_{n+1}}{n+1} - 2\sum_{n=1}^{\infty}\frac{(-1)^{n+1}}{(n+1)^2}.$$

Furthermore, if $u_1, u_2, u_3, \ldots$ is a decreasing sequence of positive numbers with limit 0, then the series $u_1 - \frac{1}{2}(u_1+u_2) + \frac{1}{3}(u_1+u_2+u_3) - \ldots$ is convergent. Hence, the series on LHS converges (with sum $= 2(-\log 2)^2$) and so does the first series on RHS because the second series too is convergent with sum $= -2(1-\frac{\pi^2}{12})$. We thus get:
$$\sum_{n=1}^{\infty} (-1)^{n+1} \frac{H_n}{n+1} = \frac{\log^2(2)}{2}.$$

From this, we deduce by a shift of the index:

$$\sum_{n=1}^{\infty}(-1)^{n+1}\frac{H_n}{n} = \frac{\pi^2}{12} - \frac{\log^2(2)}{2},$$

leading to

$$\sum_{n=1}^{\infty}(-1)^{n+1}\frac{H_n}{n(n+1)} = \frac{\pi^2}{12} - \log^2(2).$$

The Cauchy product of $\arctan x = \sum_{n=0}^{\infty}\frac{(-1)^n}{2n+1}x^n$ with itself is the series ([39] p. 162, Ex. 13):

$$(\arctan x)^2 = \sum_{n=1}^{\infty}(-1)^{n+1}\frac{h_n}{n}x^{2n}. \tag{13}$$

The series are absolutely convergent when $|x| < 1$ and convergent when $x = 1$. ([40] p. 81, Ex. 2) Therefore,

$$\sum_{n=1}^{\infty}(-1)^{n+1}\frac{h_n}{n} = \frac{\pi^2}{16}. \tag{14}$$

It occurs in ([41] pp. 322–323).

The software *WolframAlpha* gives:

$$\int_0^1 (\arctan x)^2\, dx = \frac{\pi}{16}(\pi + 4\log 2) - G,$$

where G is Catalan's constant defined by $G = \sum_{n=1}^{\infty}\frac{(-1)^{n-1}}{(2n-1)^2} = 0.9159655\ldots$

Thus, we have:

$$\sum_{n=1}^{\infty}(-1)^{n+1}\frac{h_n}{n(2n+1)} = \sum_{n=1}^{\infty}(-1)^{n+1}\left[\frac{h_n}{n} - \frac{2h_n}{2n+1}\right] = \frac{\pi^2}{16} + \frac{\pi\log 2}{4} - G. \tag{15}$$

Putting the value from Equation (14) in Equation (15) leads to:

$$\sum_{n=1}^{\infty}(-1)^{n+1}\frac{h_n}{2n+1} = \frac{G}{2} - \frac{\pi\log 2}{8}. \tag{16}$$

WolframAlpha gives:

$$\int_0^1 \frac{(\arctan x)^2}{x}\, dx = \frac{1}{8}[4\pi G - 7\zeta(3)]$$

yielding ([4] p.14, (4.16)):

$$\sum_{n=1}^{\infty}(-1)^{n+1}\frac{h_n}{n^2} = \pi G - \frac{7\zeta(3)}{4}. \tag{17}$$

2. Euler-Like Alternating Sums

2.1. A New Sum

It is easy to see that

$$H_{2n}^2 = \left(\frac{1}{2}H_n + h_n\right)^2 = \frac{1}{4}H_n^2 + H_n h_n + h_n^2.$$

Therefore,

$$\sum_{n=1}^{\infty}(-1)^{n+1}\frac{H_{2n}^2}{n} = \sum_{n=1}^{\infty}\frac{(-1)^{n+1}H_n^2}{4n} + \sum_{n=1}^{\infty}\frac{(-1)^{n+1}H_n h_n}{n} + \sum_{n=1}^{\infty}\frac{(-1)^{n+1}h_n^2}{n}. \quad (18)$$

We discovered this sum independently:

$$\sum_{n=1}^{\infty}\frac{(-1)^{n+1}H_n^2}{n} = \frac{3\zeta(3)}{4} - \frac{\pi^2 \log 2}{12} + \frac{\log^3(2)}{3}, \quad (19)$$

which on shifting the index yields

$$\sum_{n=1}^{\infty}\frac{(-1)^{n+1}H_n^2}{n+1} = -\frac{\zeta(3)}{4} + \frac{\pi^2 \log 2}{12} - \frac{\log^3(2)}{3}$$

so that

$$\sum_{n=1}^{\infty}\frac{(-1)^{n+1}H_n^2}{n(n+1)} = \zeta(3) - \frac{\pi^2 \log 2}{6} + \frac{2\log^3(2)}{3},$$

and we have an identity for Apery's constant,

$$\sum_{n=1}^{\infty}\frac{(-1)^{n+1}(2n+1)H_n^2}{n(n+1)} = \frac{\zeta(3)}{2}.$$

Later on, we found the first two formulas recorded in ([12] Equations (4.4b) and (4.4c)), borrowed from De Doelder ([6] Equation (12)) who recorded the first sum with a typographical error.

We may also record here a related sum

$$\sum_{n=1}^{\infty}\frac{(-1)^{n+1}H_n^{(2)}}{n} = \zeta(3) - \frac{\pi^2 \log 2}{12}, \quad (20)$$

obtained via the formula of Flajolet and Salvy noted above.

We find this sum in ([6], Equation (21)):

$$\sum_{n=1}^{\infty}\frac{(-1)^{n+1}h_n^2}{n} = \frac{7\zeta(3)}{16}. \quad (21)$$

This sum occurs in ([12], Equation (4.3b)):

$$\sum_{n=1}^{\infty}(-1)^{n+1}\frac{H_{n-1}+H_n}{n}h_n = \frac{7\zeta(3)}{8} - \frac{\pi^2 \log 2}{8},$$

that is,

$$\sum_{n=1}^{\infty}(-1)^{n+1}\frac{2H_n h_n}{n} - \sum_{n=1}^{\infty}(-1)^{n+1}\frac{h_n}{n^2} = \frac{7\zeta(3)}{8} - \frac{\pi^2 \log 2}{8}.$$

We thus obtain using Equation (17):

$$\sum_{n=1}^{\infty}\frac{(-1)^{n+1}H_n h_n}{n} = -\frac{7\zeta(3)}{16} - \frac{\pi^2 \log 2}{16} + \frac{\pi G}{2}. \quad (22)$$

Using the values from Equations (19)–(22) in Equation (18), we deduce this *new* sum:

$$\sum_{n=1}^{\infty}(-1)^{n+1}\frac{H_{2n}^2}{2n} = \frac{3\zeta(3)}{32} - \frac{\pi^2 \log 2}{24} + \frac{\log^3(2)}{24} + \frac{\pi G}{4}. \quad (23)$$

2.2. Double Sums for Catalan's Constant

We discovered this nice sum:

$$(1)1 - \left(1 - \frac{1}{3}\right)\frac{1}{2} + \left(1 - \frac{1}{3} + \frac{1}{5}\right)\frac{1}{3} - \cdots = \sum_{n=1}^{\infty}(-1)^{n+1}\frac{h'_n}{n} = G. \tag{24}$$

$$\sum_{n=1}^{\infty}(-1)^{n+1}\frac{h'_n}{n} = \sum_{n=1}^{\infty}(-1)^{n+1}\frac{1}{n}\int_0^1 \frac{1-(-1)^n x^{2n}}{1+x^2}dx$$

$$= \sum_{n=1}^{\infty}(-1)^{n+1}\frac{1}{n}\int_0^1 \frac{1}{1+x^2}dx$$

$$- \int_0^1 \left(\sum_{n=1}^{\infty}(-1)^{2n+1}\frac{x^{2n}}{n}\right)\frac{1}{1+x^2}dx$$

$$= \log(2)\frac{\pi}{4} - \int_0^1 \frac{\log(1-x^2)}{1+x^2}dx$$

$$= \frac{\pi}{4}\log 2 - \left[\frac{\pi}{4}\log(2) - G\right] = G,$$

as

$$\int_0^1 \frac{\log(1-x^2)}{1+x^2}dx = \int_0^1 \frac{\log(1+x)}{1+x^2}dx + \int_0^1 \frac{\log(1-x)}{1+x^2}dx$$

and ([38] p. 556, (4.291.8), (4.291.10)):

$$\int_0^1 \frac{\log(1+x)}{1+x^2}dx = \frac{\pi}{8}\log 2; \quad \int_0^1 \frac{\log(1-x)}{1+x^2}dx = \frac{\pi}{8}\log 2 - G. \square$$

We also found this sum by shifting the index in Equation (16):

$$\sum_{n=1}^{\infty}(-1)^{n+1}\frac{h_n}{2n-1} = \frac{G}{2} + \frac{\pi}{8}\log 2, \tag{25}$$

for which Chu derives a wrong value $\frac{\pi^2}{24} + \frac{\pi \log(2)}{8}$ in ([12] Equation (4.5c)).

The combination of Equations (16) and (25) yields:

$$\sum_{n=1}^{\infty}\frac{(-1)^{n+1}h_n}{4n^2-1} = \frac{\pi \log 2}{8}, \tag{26}$$

and a new double sum for Catalan's constant:

$$\sum_{n=1}^{\infty}\frac{(-1)^{n+1}n\, h_n}{4n^2-1} = \frac{G}{4}. \tag{27}$$

We discovered this related sum:

$$\sum_{n=1}^{\infty}(-1)^{n+1}\frac{h'_n}{2n-1} = \frac{3\pi^2}{32}, \tag{28}$$

$$\sum_{n=1}^{\infty}(-1)^{n+1}\frac{h'_n}{2n-1} = \sum_{n=1}^{\infty}(-1)^{n+1}\frac{1}{2n-1}\int_0^1 \frac{1-(-1)^n x^{2n}}{1+x^2}\,dx$$
$$= \sum_{n=1}^{\infty}(-1)^{n+1}\frac{1}{2n-1}\int_0^1 \frac{1}{1+x^2}\,dx$$
$$\quad -\int_0^1 \left(\sum_{n=1}^{\infty}(-1)^{2n+1}\frac{x^{2n}}{2n-1}\right)\frac{1}{1+x^2}\,dx$$
$$= \left(\frac{\pi}{4}\right)^2 + \int_0^1 \frac{x\arctan h(x)}{1+x^2}\,dx$$
$$= \frac{\pi^2}{16} + \frac{\pi^2}{32} = \frac{3\pi^2}{32}.$$

3. More Alternating Sums

3.1. Miscellaneous Double Sums

We derived these sums earlier:

$$\sum_{n=1}^{\infty}(-1)^{n+1}\frac{H_n}{n+1} = \frac{\log^2 2}{2}; \quad \sum_{n=1}^{\infty}(-1)^{n+1}\frac{H_n}{n} = \frac{\pi^2}{12} - \frac{\log^2(2)}{2}.$$

Furthermore,

$$\sum_{n=1}^{\infty}(-1)^{n+1}\frac{H_n}{(n+1)^2} = \frac{\zeta(3)}{8}; \quad \sum_{n=1}^{\infty}(-1)^{n+1}\frac{H_n}{n^2} = \frac{5\zeta(3)}{8}.$$

Since $H_{2n} = h_n + \frac{H_n}{2}$, we deduce:

$$\sum_{n=1}^{\infty}(-1)^{n+1}\frac{h_n}{n+1} = \frac{\pi}{16}(8-\pi) - \log 2; \quad \sum_{n=1}^{\infty}(-1)^{n+1}\frac{h_n}{n} = \frac{\pi^2}{16}$$

and

$$\sum_{n=1}^{\infty}(-1)^{n+1}\frac{h_n}{(n+1)^2} = \frac{7\zeta(3)}{4} - \frac{\pi^2}{12} - \pi G + \pi - 2\log 2.$$

We also have these:

$$\sum_{n=1}^{\infty}(-1)^{n+1}\frac{H_{2n}}{n^3} = -5\operatorname{Li}_4\left(\frac{1}{2}\right) + \frac{13\pi^4}{192} - \frac{35\zeta(3)\log 2}{8} + \frac{5\pi^2\log^2(2)}{24} - \frac{5\log^4(2)}{24}.$$

$$\sum_{n=1}^{\infty}(-1)^{n+1}\frac{H_n}{n^3} = -2\operatorname{Li}_4\left(\frac{1}{2}\right) + \frac{11\pi^4}{360} - \frac{7\zeta(3)\log 2}{4} + \frac{\pi^2\log^2(2)}{12} - \frac{\log^4(2)}{12}.$$

Its terms (except the one with π^4) are $2/5$ times of the previous one. See the next sum with the same terms (except the one with π^4):

$$\sum_{n=1}^{\infty}(-1)^{n+1}\frac{H_n^2}{n^2} = -2\operatorname{Li}_4\left(\frac{1}{2}\right) + \frac{41\pi^4}{1440} - \frac{7\zeta(3)\log 2}{4} + \frac{\pi^2\log^2(2)}{12} - \frac{\log^4(2)}{4}.$$

This sum is simpler:

$$\sum_{n=1}^{\infty}(-1)^n\frac{H_n^{(3)}}{n} = \int_0^1 \frac{\operatorname{Li}_3(-x)}{x(1+x)}\,dx = -\frac{19\pi^4}{1440} + \frac{3\zeta(3)\log 2}{4},$$

$$\sum_{n=1}^{\infty}(-1)^{n+1}\frac{h_n}{n^3}=-4\operatorname{Li}_4\left(\frac{1}{2}\right)+\frac{151\pi^4}{2880}-\frac{7\zeta(3)\log 2}{2}+\frac{\pi^2\log^2(2)}{6}-\frac{\log^4(2)}{6},$$

$$\sum_{n=1}^{\infty}(-1)^{n+1}\frac{H_{2n}}{(n+1)^3}=5\operatorname{Li}_4\left(\frac{1}{2}\right)-\frac{181\pi^4}{2880}+\frac{35\zeta(3)\log 2}{8}-\frac{5\pi^2\log^2(2)}{24}$$
$$+\frac{5\log^4 2}{24}-\frac{3\zeta(3)}{4}-\frac{\pi^2}{6}+2\pi-4\log 2,$$

$$\sum_{n=1}^{\infty}(-1)^{n+1}\frac{H_n}{(n+1)^3}=2\operatorname{Li}_4\left(\frac{1}{2}\right)-\frac{15\pi^4}{720}+\frac{7\zeta(3)\log 2}{4}-\frac{\pi^2\log^2(2)}{12}+\frac{\log^4(2)}{12},$$

$$\sum_{n=1}^{\infty}(-1)^{n+1}\frac{h_n}{(n+1)^3}=4\operatorname{Li}_4\left(\frac{1}{2}\right)-\frac{151\pi^4}{2880}+\frac{7\zeta(3)\log 2}{2}-\frac{\pi^2\log^2(2)}{6}$$
$$+\frac{\log^4(2)}{6}-\frac{3\zeta(3)}{4}-\frac{\pi^2}{6}+2\pi-4\log 2.$$

3.2. Sums Involving Harmonic Numbers with Multiple Arguments

For multiple arguments, we know that, for $p \in \mathbb{N}$, $h_{pn} = H_{2pn} - \frac{1}{2}H_{pn}$. From [42], we obtain

$$\sum_{n\geq 1}\frac{(-1)^{n+1}H_{pn}}{n}=\frac{1+p^2}{4p}\zeta(2)-\frac{1}{2}\sum_{j=0}^{p-1}\log^2\left(2\sin\left(\frac{(2j+1)\pi}{2p}\right)\right),$$

$$\sum_{n\geq 1}\frac{(-1)^{n+1}H_{pn}}{n(n+1)}=\sum_{n\geq 1}\frac{(-1)^{n+1}H_{pn}}{n}-\sum_{n\geq 1}\frac{(-1)^{n+1}H_{pn}}{n+1}=\frac{p}{2}\zeta(2)$$
$$+H_{p-1}\ln 2-\sum_{j=0}^{p-1}\log^2\left(2\sin\left(\frac{(2j+1)\pi}{2p}\right)\right)-\frac{1}{2}\sum_{j=1}^{p-1}\frac{1}{j}\left(H_{-\frac{j}{2p}}-H_{-\left(\frac{p+j}{2p}\right)}\right),$$

from which we ascertain

$$\sum_{n\geq 1}\frac{(-1)^{n+1}h_{pn}}{n(n+1)}=\frac{3p}{4}\zeta(2)+\frac{1}{2}\left(H_{2p-1}-H_{p-1}\right)\log 2$$
$$+\frac{1}{2}\sum_{j=0}^{p-1}\log^2\left(2\sin\left(\frac{(2j+1)\pi}{2p}\right)\right)-\sum_{j=0}^{2p-1}\log^2\left(2\sin\left(\frac{(2j+1)\pi}{4p}\right)\right)$$
$$+\frac{1}{4}\sum_{j=1}^{p-1}\frac{1}{j}\left(H_{-\frac{j}{2p}}-H_{-\left(\frac{p+j}{2p}\right)}\right)-\frac{1}{2}\sum_{j=1}^{2p-1}\frac{1}{j}\left(H_{-\frac{j}{4p}}-H_{-\left(\frac{p+j}{4p}\right)}\right).$$

For the non-alternating case, we have from [43],

$$\sum_{n\geq 1}\frac{H_{pn}}{n(n+1)}=\frac{1}{p}\zeta(2)-\sum_{j=1}^{p-1}\frac{1}{j}H_{-\frac{j}{p}}$$

and, therefore, after some simplification

$$\sum_{n\geq 1}\frac{h_{pn}}{n(n+1)}=\frac{1}{2}\sum_{j=1}^{p-1}\frac{1}{j}H_{-\frac{j}{p}}-\sum_{j=1}^{2p-1}\frac{1}{j}H_{-\frac{j}{2p}}=-\sum_{j=1}^{p}\frac{1}{(2j-1)}H_{-\frac{(2j-1)}{2p}}.$$

A simple calculation gives us

$$\sum_{n\geq 1}\frac{h_{2n}}{n(2n+1)}=\log(2)+\frac{1}{2}\pi-\frac{3}{4}\zeta(2).$$

3.3. Few Relations between Pairs of Sums

We discovered these relations:

$$\sum_{n=1}^{\infty}\frac{(-1)^{n+1}h_n^2}{2n-1}+\sum_{n=1}^{\infty}\frac{(-1)^{n+1}h_n}{(2n+1)^2}=\frac{\pi^3}{48}+\frac{\pi\log^2(2)}{16},$$

and

$$\sum_{n=1}^{\infty}\frac{(-1)^{n+1}h_n^2}{2n+1}+\sum_{n=1}^{\infty}\frac{(-1)^{n+1}h_n}{(2n+1)^2}=\frac{\pi^3}{96}-\frac{\pi\log^2(2)}{16}.$$

Thus,

$$\sum_{n=1}^{\infty}\frac{(-1)^{n+1}h_n^2}{2n-1}-\sum_{n=1}^{\infty}\frac{(-1)^{n+1}h_n^2}{2n+1}=\frac{\pi^3}{96}+\frac{\pi\log^2(2)}{8},$$

leading to the sum:

$$\sum_{n=1}^{\infty}\frac{(-1)^{n+1}h_n^2}{4n^2-1}=\frac{\pi^3}{192}+\frac{\pi\log^2(2)}{16}.$$

This relation was deduced from ([12] Equation (4.6c)):

$$\sum_{n=1}^{\infty}\frac{(-1)^{n+1}h_n^2}{2n-1}-\sum_{n=1}^{\infty}\frac{(-1)^{n+1}h_n}{(2n-1)^2}=-\frac{\pi^3}{96}+\frac{\pi\log^2(2)}{16}.$$

We thus have:

$$\sum_{n=1}^{\infty}\frac{(-1)^{n+1}h_n^2}{2n-1}+\sum_{n=1}^{\infty}\frac{(-1)^{n+1}h_n}{(2n+1)^2}=\frac{\pi^3}{48}+\frac{\pi\log^2(2)}{16},$$

$$\sum_{n=1}^{\infty}\frac{(-1)^{n+1}h_n}{(2n-1)^2}+\sum_{n=1}^{\infty}\frac{(-1)^{n+1}h_n}{(2n+1)^2}=\frac{\pi^3}{32}=\sum_{n=1}^{\infty}\frac{(-1)^{n+1}}{(2n-1)^3},$$

$$\sum_{n=1}^{\infty}(-1)^{n+1}\frac{h_n}{n(n+1)}=\frac{\pi^2}{8}-\frac{\pi}{2}+\log 2,$$

$$\sum_{n=1}^{\infty}(-1)^{n+1}\frac{h_n}{n^2(n+1)^2}=-\frac{\pi^2}{3}+2\pi-4\log 2.$$

The two preceding sums have similar terms yielding a nice sum for $\zeta(2)$:

$$\sum_{n=1}^{\infty}(-1)^{n+1}\left(\frac{2n+1}{n(n+1)}\right)^2 h_n=\frac{\pi^2}{6}, \tag{29}$$

$$\sum_{n=1}^{\infty}(-1)^{n+1}\frac{h_n}{n^3(n+1)^3}=\frac{151\pi^4}{1440}-7\zeta(3)\log 2+\frac{\pi^2\log^2(2)}{3}$$
$$-\frac{\log^4 2}{3}+\frac{3\zeta(3)}{4}+\frac{7\pi^2}{6}-8\pi+16\log 2-8\text{Li}_4\left(\frac{1}{2}\right).$$

4. Integrals and Euler Sums

In this section, we will explore the connection of integrals involving inverse trigonometric and hyperbolic functions with Euler sums, thereby producing a number of interesting and new identities.

Theorem 1. *Let $a \in \mathbb{R}^+ \cup \{0\}, \delta \in \mathbb{R}^+ \setminus \{0\}, p \in \mathbb{N} \cup \{0\}$ and $q \in \mathbb{R}^+$; then,*

$$I(a,\delta,p,q) = \int_0^1 x^a (\tanh^{-1}(\delta x^q))^2 \log^p(x) \, dx$$

$$= (-1)^p \, p! \sum_{n \geq 1} \frac{\delta^{2n} h_n}{n(2qn+a+1)^{p+1}},$$

where

$$h_n = H_{2n} - \frac{1}{2}H_n = \log(2) + \frac{1}{2}H_{n-\frac{1}{2}}$$

and H_n, H_{2n} are harmonic numbers with unitary argument and double argument.

Proof. We have that

$$(\tanh^{-1}(\delta x^q))^2 = \sum_{n \geq 1} \frac{\delta^{2n} h_n x^{2qn}}{n},$$

therefore

$$I(a,\delta,p,q) = \sum_{n \geq 1} \frac{\delta^{2n} h_n}{n} \int_0^1 x^{2qn+a} \log^p(x) \, dx,$$

and, integrating by parts, yields

$$I(a,\delta,p,q) = (-1)^p \, p! \sum_{n \geq 1} \frac{\delta^{2n} h_n}{n(2qn+a+1)^{p+1}}.$$

□

Certain particular interesting cases can be explicitly represented and we detail some cases in the next few corollaries.

Corollary 1. *Let $a = -1, \delta = \pm 1, p \in \mathbb{N} \cup \{0\}$ and $q \in \mathbb{R}^+$; then,*

$$I(-1,\pm 1, p, q) = \int_0^1 \frac{1}{x} (\tanh^{-1}(\pm x^q))^2 \log^p(x) \, dx$$

$$= \frac{(-1)^p \, p!}{(2q)^{p+1}} \sum_{n \geq 1} \frac{h_n}{n^{p+2}}.$$

For $p = 0$,

$$I(-1,\pm 1, 0, q) = \int_0^1 \frac{1}{x} (\tanh^{-1}(\pm x^q))^2 \, dx$$

$$= \frac{1}{(2q)^1} \sum_{n \geq 1} \frac{h_n}{n^2} = \frac{7}{8q}\zeta(3).$$

For $p = 1$,

$$I(-1, \pm 1, 1, q) = \int_0^1 \frac{1}{x} (\tanh^{-1}(\pm x^q))^2 \log(x)\, dx$$

$$= -\frac{1}{(2q)^2} \sum_{n \geq 1} \frac{h_n}{n^3} = \frac{1}{4q^2}\left(4L(3) - \frac{35}{8}\zeta(4)\right),$$

where

$$L(3) = \frac{11}{4}\zeta(4) - \frac{7}{4}\zeta(3)\log 2 + \frac{1}{2}\zeta(2)\log^2(2) - \frac{1}{12}\log^4(2) - 2\mathrm{Li}_4\left(\frac{1}{2}\right). \tag{30}$$

For p even, let $p = 2(m-1)$, $m \in \mathbb{N}$

$$I(-1, \pm 1, 2(m-1), q) = \int_0^1 \frac{1}{x}(\tanh^{-1}(\pm x^q))^2 \log^{2m-2}(x)\, dx$$

$$= -\frac{(2m-2)!}{(2q)^{2m+1}} \sum_{n \geq 1} \frac{h_n}{n^{2m}} = \frac{(2m-2)!}{(2q)^{2m+1}} Q(m),$$

where, from paper [44],

$$Q(m) = \sum_{n \geq 1} \frac{h_n}{n^{2m}} = \frac{2^{2m+1}-1}{4}\zeta(2m+1) \tag{31}$$

$$-\frac{1}{2}\sum_{j=1}^{m-1}\left(2^{2m-2j+1} - 1\right)\zeta(2j)\zeta(2m-2j+1).$$

For

$$I\left(-1, \pm\frac{1}{3}, 0, 1\right) = \int_0^1 \frac{1}{x}\left(\tanh^{-1}\left(\pm\frac{1}{3}x\right)\right)^2 dx$$

$$= \frac{1}{2}\sum_{n \geq 1} \frac{h_n}{3^{2n} n^2}$$

$$= \frac{1}{8}\mathrm{Li}_3\left(\frac{1}{4}\right) + \frac{1}{4}\mathrm{Li}_2\left(\frac{1}{4}\right)\log 2 - \frac{1}{12}\log\left(\frac{27}{16}\right)\log^2(2),$$

where we have the identity

$$\sum_{n \geq 1} \frac{H_n}{3^{2n} n^2} = \mathrm{Li}_3\left(\frac{1}{9}\right) + \zeta(3) - \mathrm{Li}_3\left(\frac{8}{9}\right)\log(2) - \mathrm{Li}_2\left(\frac{8}{9}\right)\log\left(\frac{9}{8}\right) - \log^2\left(\frac{9}{8}\right)\log 3.$$

Corollary 2. *Let* $a = 2q - 1, \delta = \pm 1, p \in \mathbb{N} \cup \{0\}$ *and* $q \in \mathbb{R}^+$; *then,*

$$I(2q - 1, \pm 1, p, q) = \int_0^1 \frac{1}{x}(\tanh^{-1}(\pm x^q))^2 \log^p(x)\, dx$$

$$= \frac{(-1)^p p!}{(2q)^{p+1}} \sum_{n \geq 1} \frac{h_n}{n(n+1)^{p+1}}$$

$$= \frac{(-1)^p p!}{(2q)^{p+1}} \sum_{n \geq 1} h_n \left(\frac{1}{n(n+1)} - \sum_{j=2}^{p+1} \frac{1}{(n+1)^j}\right).$$

For $p = 2, q = 1$

$$I(1, \pm 1, 2, 1) = \int_0^1 x (\tanh^{-1}(\pm x))^2 \log^2(x) \, dx$$

$$= \frac{1}{4} \sum_{n \geq 1} \frac{h_n}{n(n+1)^3}$$

$$= \frac{1}{4} \sum_{n \geq 1} h_n \left(\frac{1}{n(n+1)} - \sum_{j=2}^{3} \frac{1}{(n+1)^j} \right)$$

$$= L(3) + \frac{7}{2} \log(2) - \frac{3}{4} \zeta(2) - \frac{11}{16} \zeta(3) - \frac{35}{32} \zeta(4),$$

where we have evaluated

$$\sum_{n \geq 1} \frac{h_n}{n(n+1)} = 2 \log 2, \quad \sum_{n \geq 1} \frac{h_n}{(n+1)^2} = \zeta(2) - 4 \log(2) + \frac{7}{4} \zeta(3),$$

$$\sum_{n \geq 1} \frac{h_n}{(n+1)^3} = 2\zeta(2) - 4L(3) - 8 \log 2 + \zeta(3) + \frac{35}{8} \zeta(4).$$

For $\delta = \pm \frac{1}{3}$,

$$I\left(1, \pm \frac{1}{3}, 0, 1\right) = \int_0^1 x \left\{ \tanh^{-1}\left(\pm \frac{1}{3} x\right) \right\}^2 dx = \frac{1}{2} \sum_{n \geq 1} \frac{h_n}{9^n n(n+1)}$$

$$= 15 \log 2 - 9 \log 3 - \log^2(2),$$

where

$$\sum_{n \geq 1} \frac{H_n}{9^n n(n+1)} = \operatorname{Li}_2\left(\frac{1}{9}\right) - 4 \log^2\left(\frac{9}{8}\right).$$

We can decompose the two sums into:

$$\sum_{n=1}^{\infty} \frac{h_n}{9^n n} = \frac{1}{4} \log^2(2); \quad \sum_{n=1}^{\infty} \frac{h_n}{9^{n+1}(n+1)} = \frac{1}{4} \log^2(2) - \frac{10}{3} \log 2 + 2 \log 3,$$

and

$$\sum_{n=1}^{\infty} \frac{H_n}{9^n n} = \operatorname{Li}_2\left(\frac{1}{9}\right) + \frac{1}{2} \log^2\left(\frac{9}{8}\right); \quad \sum_{n=1}^{\infty} \frac{H_n}{9^{n+1}(n+1)} = \frac{1}{2} \log^2\left(\frac{9}{8}\right).$$

Moreover,

$$\sum_{n=1}^{\infty} \frac{H_{2n}}{9^{2n+1}(2n+1)} = \frac{1}{4} \left[\log^2\left(\frac{9}{8}\right) - \log^2\left(\frac{10}{9}\right) \right] = \frac{1}{4} \log\left(\frac{81}{80}\right) \log\left(\frac{5}{4}\right).$$

An interesting connection with these given four sums is the alternating identity

$$\sum_{n=1}^{\infty} \frac{(-1)^{n+1} H_n}{3^n (n+1)} = \frac{3}{2} \log^2\left(\frac{4}{3}\right),$$

from which, by extrapolation, we are able to obtain the very fast converging alternating identity

$$\zeta(2) = 2 \sum_{n=1}^{\infty} \frac{(-1)^{n+1} h_n}{3^{n-1} n}.$$

Corollary 3. For $a = -1 - q$

$$I(-q-1, \pm 1, p, q) = \int_0^1 x^{-1-q} (\tanh^{-1}(\pm x^q))^2 \log^p(x) \, dx$$

$$= \frac{(-1)^p \, p!}{(2q)^{p+1}} \sum_{n \geq 1} \frac{h_n}{n(2n-1)^{p+1}}$$

$$= \frac{(-1)^p \, p!}{q^{p+1}} \sum_{n \geq 1} h_n \left(\frac{(-1)^p}{n(2n-1)} + 2 \sum_{j=2}^{p+1} \frac{(-1)^{p+1+j}}{(2n-1)^j} \right)$$

$$= \frac{p!}{q^{p+1}} \left(\zeta(2) + 2 \sum_{j=2}^{p+1} (-1)^{1+j} V(j) \right),$$

where, from paper [44], we have evaluated, for j even,

$$V(j) = \sum_{n \geq 1} \frac{h_n}{(2n-1)^j} = \frac{1}{4} (\zeta(j+1) + \eta(j+1)) + \frac{1}{2} (\zeta(j) + \eta(j)) \log 2$$

$$- \frac{1}{2^{j+1}} \sum_{k=1}^{\frac{j}{2}-1} \left(2^{2k} - 1 \right) \zeta(2k) \zeta(j - 2k + 1). \qquad (32)$$

For the case

$$I(-2, \pm 1, 2, 1) = \int_0^1 x^{-2} (\tanh^{-1}(\pm x))^2 \log^2(x) \, dx = 2L(3) - \frac{5}{16} \zeta(4)$$

$$+ \frac{7}{2} \zeta(3) \log 2 - \frac{7}{4} \zeta(3) - 3\zeta(2) \log 2 + 2\zeta(2),$$

where we have the results

$$\sum_{n \geq 1} \frac{h_n}{n(2n-1)} = \zeta(2), \ \sum_{n \geq 1} \frac{h_n}{(2n-1)^2} = \frac{3}{4} \zeta(2) \log 2 + \frac{7}{16} \zeta(3),$$

$$\sum_{n \geq 1} \frac{h_n}{(2n-1)^3} = \frac{21}{2} L(3) + \frac{7}{8} \zeta(3) \log 2 - \frac{5}{64} \zeta(4).$$

For

$$I\left(-2, \pm \frac{1}{2}, 0, 1\right) = \int_0^1 x^{-2} \left\{ \tanh^{-1}\left(\pm \frac{1}{2} x\right) \right\}^2 dx = \sum_{n \geq 1} \frac{h_n}{2^{2n} n (2n-1)}$$

$$= \frac{1}{2} \zeta(2) - \frac{1}{2} \text{Li}_2 \left(\frac{1}{3} \right) - \frac{1}{8} \log \left(\frac{243}{16} \right) \log 3,$$

where we have the identity

$$\sum_{n \geq 1} \frac{H_{2n}}{2^{2n} n (2n-1)} = \frac{5}{4} \zeta(2) + \frac{3}{2} \log \left(\frac{3}{2} \right) - \frac{1}{2} \log 2 - \frac{3}{2} \text{Li}_2 \left(\frac{2}{3} \right) - \frac{3}{2} \log \left(\frac{3}{2} \right) \log 3.$$

Higher powers of the $\tanh^{-1}(x)$ function yield more Euler sum identities, and we have the following corollary.

Corollary 4.

$$\int_0^1 x^a (\tanh^{-1}(x))^3 \, dx = \sum_{n\geq 1} \frac{h_n}{n} \sum_{j\geq 1} \frac{1}{(2j-1)} \int_0^1 x^{2n+2j+a-1} dx$$

$$= \sum_{n\geq 1} \sum_{j\geq 1} \frac{h_n}{n(2j-1)(2n+2j+a)}$$

$$= \sum_{n\geq 1} \frac{h_n \left(2\log(2) + H_{n+\frac{a}{2}}\right)}{n(2n+a+1)}.$$

Various cases follow, for $a = -1$,

$$\int_0^1 \frac{1}{x} (\tanh^{-1}(x))^3 \, dx = \sum_{n\geq 1} \frac{h_n \left(2\log(2) + H_{n-\frac{1}{2}}\right)}{4n^2} = \sum_{n\geq 1} \frac{h_n^2}{4n^2} = \frac{45}{32} \zeta(4).$$

For $a = 1$,

$$\int_0^1 x (\tanh^{-1}(x))^3 \, dx = \frac{1}{4} \sum_{n\geq 1} \frac{h_n \left(2\log(2) + H_{n+\frac{1}{2}}\right)}{n(n+1)}$$

$$= \frac{1}{4} \sum_{n\geq 1} \left(\frac{2h_n}{n(n+1)(2n+1)} + \frac{h_n^2}{n(n+1)} \right) = \frac{3}{4} \zeta(2).$$

From here, we have the identities

$$\sum_{n\geq 1} \frac{h_n}{n(n+1)(2n+1)} = \zeta(2) - 2\log 2, \quad \sum_{n\geq 1} \frac{h_n^2}{n(n+1)} = \frac{1}{2}\zeta(2) + 2\log 2,$$

and we have the new identities

$$\zeta(2) = \frac{2}{3} \sum_{n\geq 1} \frac{h_n}{n(n+1)} \left(h_n + \frac{1}{2n+1} \right),$$

$$\sum_{n\geq 1} \frac{h_n H_{\frac{n}{2}}}{n(n+1)} = \frac{21}{16} \zeta(3) - \frac{9}{2} \log^2(2) + 2G + \frac{1}{2} \zeta(2).$$

For $a = -2$,

$$\int_0^1 \frac{1}{x^2} (\tanh^{-1}(x))^3 \, dx = \frac{1}{2} \sum_{n\geq 1} \frac{h_n (2\log(2) + H_{n-1})}{n(2n-1)}$$

$$= \sum_{n\geq 1} \left(\frac{h_n \log 2}{n(2n-1)} + \frac{h_n H_n}{2n(2n-1)} - \frac{h_n}{2n^2(2n-1)} \right) = \frac{3}{2} \zeta(3).$$

From here, we have the identities

$$\sum_{n\geq 1} \frac{h_n}{n(2n-1)} = \zeta(2), \quad \sum_{n\geq 1} \frac{h_n}{n^2} = \frac{7}{4} \zeta(3),$$

$$\sum_{n\geq 1} \frac{h_n H_n}{n(2n-1)} = \frac{5}{4} \zeta(3) - 2\zeta(2) \log 2 + 2\zeta(2).$$

Consider the quartic,

$$\int_0^1 \frac{1}{x}(\tanh^{-1}(x))^4\,dx = \frac{1}{2}\sum_{n\geq 1}\sum_{j\geq 1}\frac{h_n h_j}{nj(n+j)} = \frac{93}{32}\zeta(5).$$

$$\int_0^1 \frac{1}{\sqrt{x}}(\tanh^{-1}(\tfrac{1}{2}x^{\frac{1}{4}}))^4\,dx = \frac{1}{2}\sum_{n\geq 1}\sum_{j\geq 1}\frac{h_n h_j}{2^{2n+2j-1}nj(n+j+1)}$$

$$= 36\zeta(3) + 22\log^3(3) - \frac{3}{8}\log^4(3) - 48\log(2)\log^2(3) + 48\log(3)\mathrm{Li}_2\left(-\frac{1}{3}\right) + 48\mathrm{Li}_3\left(-\frac{1}{3}\right).$$

For powers of the arctan x function, we shall find some further alternating harmonic Euler sums.

Theorem 2. *Let $a \in \mathbb{R}^+ \cup \{0\}, \delta \in \mathbb{R}^+ \setminus \{0\}, p \in \mathbb{N} \cup \{0\}$ and $q \in \mathbb{R}^+$; then,*

$$J(a,\delta,p,q) = \int_0^1 x^a \arctan^2(\delta x^q)\log^p(x)\,dx$$

$$= (-1)^p p!\sum_{n\geq 1}\frac{(-1)^{n+1}\delta^{2n}h_n}{n(2qn+a+1)^{p+1}},$$

where

$$h_n = H_{2n} - \frac{1}{2}H_n = \log 2 + \frac{1}{2}H_{n-\frac{1}{2}}$$

and H_n, H_{2n} are harmonic numbers with unitary argument and double argument.

Proof. We have that

$$\arctan^2(\delta x^q) = \sum_{n\geq 1}\frac{(-1)^{n+1}\delta^{2n}h_n x^{2qn}}{n};$$

therefore,

$$J(a,\delta,p,q) = \sum_{n\geq 1}\frac{(-1)^{n+1}\delta^{2n}h_n}{n}\int_0^1 x^{2qn+a}\log^p(x)\,dx$$

$$= (-1)^p p!\sum_{n\geq 1}\frac{(-1)^{n+1}\delta^{2n}h_n}{n(2qn+a+1)^{p+1}}.$$

□

Certain particular interesting cases can be explicitly represented and we detail some cases in the next few corollaries.

Corollary 5. *Let $a = -1, \delta = \pm 1, p \in \mathbb{N} \cup \{0\}$ and $q \in \mathbb{R}^+$; then,*

$$J(-1,\pm 1,p,q) = \int_0^1 \frac{1}{x}\arctan^2(\pm x^q)\log^p(x)\,dx$$

$$= \frac{(-1)^p p!}{(2q)^{p+1}}\sum_{n\geq 1}\frac{(-1)^{n+1}h_n}{n^{p+2}}.$$

For $p = 1$, and using the results of the previous sections,

$$J(-1, \pm 1, 1, q) = \int_0^1 \frac{1}{x} \arctan^2(\pm x^q) \log(x) \, dx = \frac{1}{(2q)^2} \sum_{n \geq 1} \frac{(-1)^{n+1} h_n}{n^3}$$

$$= \frac{1}{(2q)^2} \left\{ \frac{151}{32} \zeta(4) - \frac{7}{2} \zeta(3) \log 2 + \zeta(2) \log^2(2) - \frac{1}{6} \log^4(2) - 4\text{Li}_4\left(\frac{1}{2}\right) \right\}.$$

Corollary 6. Let $a = 2q - 1, \delta = \pm 1, p \in \mathbb{N} \cup \{0\}$ and $q \in \mathbb{R}^+$; then,

$$J(2q - 1, \pm 1, p, q) = \int_0^1 x^{2q-1} \arctan^2(\pm x^q) \log^p(x) \, dx$$

$$= \frac{(-1)^p p!}{(2q)^{p+1}} \sum_{n \geq 1} \frac{(-1)^{n+1} h_n}{n(n+1)^{p+1}}.$$

$$= \frac{(-1)^p p!}{(2q)^{p+1}} \sum_{n \geq 1} (-1)^{n+1} h_n \left(\frac{1}{n} - \sum_{j=1}^{p+1} \frac{1}{(n+1)^j} \right).$$

For $p = 2$, and using the results of the previous sections,

$$J(2q - 1, \pm 1, 2, q) = \int_0^1 x^{2q-1} \arctan^2(\pm x^q) \log^2(x) \, dx$$

$$= \frac{2}{(2q)^{p+1}} \sum_{n \geq 1} (-1)^{n+1} h_n \left(\frac{1}{n} - \sum_{j=1}^{3} \frac{1}{(n+1)^j} \right)$$

$$= \frac{1}{(2q)^2} \left(\begin{array}{c} \pi G + \frac{151}{32} \zeta(4) - \frac{7}{2} \zeta(3) \log 2 + \zeta(2) \log^2(2) - \frac{1}{6} \log^4(2) \\ -4\text{Li}_4(\frac{1}{2}) - \frac{3\pi}{2} + \frac{9}{4} \zeta(2) + 7 \log 2 - \zeta(3) \end{array} \right).$$

Corollary 7. Let $a = -q - 1, \delta = \pm 1, p \in \mathbb{N} \cup \{0\}$ and $q \in \mathbb{R}^+$; then,

$$J(-q - 1, \pm 1, p, q) = \int_0^1 x^{-q-1} \arctan^2(\pm x^q) \log^p(x) \, dx$$

$$= \frac{(-1)^p p!}{q^{p+1}} \sum_{n \geq 1} \frac{(-1)^{n+1} h_n}{n(2n-1)^{p+1}}.$$

$$= \frac{(-1)^p p!}{q^{p+1}} \sum_{n \geq 1} (-1)^{n+1} h_n \left(\frac{(-1)^{p+1}}{n} + 2 \sum_{j=1}^{p+1} \frac{(-1)^{j+p+1}}{(2n-1)^j} \right).$$

For $p = 0$,

$$J(2q - 1, \pm 1, 0, q) = \int_0^1 x^{-q-1} \arctan^2(\pm x^q) \, dx$$

$$= \frac{1}{q} \sum_{n \geq 1} (-1)^{n+1} h_n \left(-\frac{1}{n} + \frac{2}{(2n-1)} \right)$$

$$= \frac{1}{q} \left(\frac{\pi}{4} \log 2 + G - \frac{3}{8} \zeta(2) \right).$$

Higher powers of the arctan (x) function yield more Euler sum identities, and we have the following corollary.

Corollary 8.

$$\int_0^1 x^a \arctan^3(x)\, dx = \sum_{n\geq 1} \frac{(-1)^{n+1} h_n}{n} \int_0^1 x^{2n+a} \arctan(x)\, dx$$

$$= \sum_{n\geq 1} \frac{(-1)^{n+1} h_n}{4n} \left(\frac{\pi + H_{\frac{2n+a-2}{4}} - H_{\frac{2n+a}{4}}}{(2n+a+1)} \right).$$

Various cases follow, for $a = 0$, and using the double argument of the polygamma function

$$\int_0^1 \arctan^3(x)\, dx = \sum_{n\geq 1} \frac{(-1)^{n+1} h_n}{4n} \left(\frac{\pi + H_{\frac{2n-2}{4}} - H_{\frac{2n}{4}}}{2n+1} \right)$$

$$= \sum_{n\geq 1} \frac{(-1)^{n+1} h_n}{4n(2n+1)} \left(\pi - 2\log(2) + 2H_n - 2H_{\frac{n}{2}} \right)$$

$$= \frac{63}{64}\zeta(3) - \frac{3}{4}\pi G + \frac{\pi^3}{64} + \frac{18}{32}\zeta(2)\log 2.$$

Manipulating this integral identity gives us the new Euler sums,

$$\sum_{n\geq 1} \frac{(-1)^{n+1} h_n H_{\frac{n}{2}}}{n} = \pi G - \frac{35}{32}\zeta(3) - \frac{9}{8}\zeta(2)\log 2,$$

$$\sum_{n\geq 1} \frac{(-1)^{n+1} h_n \left(H_{\frac{n}{2}} - H_n\right)}{2n+1} = \frac{21}{32}\zeta(3) - \frac{1}{4}\pi G + \frac{3}{4}\pi \log^2(2) - \frac{1}{2}G\log 2,$$

$$\sum_{n\geq 1} \frac{h_{2n}}{n(2n+1)} = \log 2 + \frac{\pi}{2} - \frac{3}{4}\zeta(2).$$

For $a = -1$,

$$\int_0^1 \frac{1}{x} \arctan^3(x)\, dx = \sum_{n\geq 1} \frac{(-1)^{n+1} h_n}{8n^2} \left(\pi + H_{\frac{n}{2}-\frac{3}{4}} - H_{\frac{n}{2}-\frac{1}{4}} \right)$$

$$= \frac{9}{8}\zeta(2) G + \frac{1}{1024} \left(\psi^{(3)}\left(\frac{3}{4}\right) - \psi^{(3)}\left(\frac{1}{4}\right) \right)$$

$$= \frac{9}{8}\zeta(2) G - \frac{3}{2}\beta(4),$$

where the Dirichlet function $\beta(4) = \sum_{j\geq 1} \frac{(-1)^{j+1}}{(2j-1)^4}$. Here, we note that

$$\psi^{(3)}\left(\frac{3}{4}\right) - \psi^{(3)}\left(\frac{1}{4}\right) = 8\pi^4 - 768\beta(4) - \left(8\pi^4 + 768\beta(4)\right) = -1536\beta(4);$$

moreover, using the integral identity, we obtain the new Euler sum

$$\sum_{n\geq 1}\frac{(-1)^{n+1}h_n}{n^2}\left(H_{\frac{n}{2}+\frac{1}{4}}-H_{\frac{n}{2}-\frac{1}{4}}\right) = 3\zeta(2)G - 12\beta(4) + \frac{7}{4}\pi\zeta(3) + 4\pi G$$
$$-7\zeta(3) + 8G - 3\zeta(2) - 2\pi\log 2.$$

For the quartic

$$\int_0^1 \frac{1}{x}\arctan^4(x)\,dx = \sum_{n\geq 1}\sum_{j\geq 1}\frac{(-1)^{n+j}h_nh_j}{jn(j+n)}$$
$$= \frac{1}{16}\pi^3 G + \frac{93}{32}\zeta(5) + \frac{3}{2}\pi\beta(4),$$

and

$$\int_0^1 x\arctan^6(x)\,dx = \sum_{n\geq 1}\sum_{j\geq 1}\sum_{k\geq 1}\frac{(-1)^{n+j+k+1}h_nh_jh_k}{2jkn(j+k+n+1)}$$
$$= \frac{15}{32}\pi^3 G + \frac{20925}{1024}\zeta(5) - \frac{45}{4}\pi\beta(4) - \frac{3}{1024}\pi^5$$
$$+ \frac{945}{4096}\zeta(6) - \frac{675}{256}\zeta(4)\log 2 + \frac{405}{256}\zeta(2)\zeta(3).$$

5. Conclusions

In the first three sections of the paper, we treated miscellaneous Euler sums, particularly, the alternating sums. We developed many new Euler type identities. In particular, we have developed some new identities for the Catalan constant, Apery's constant and Euler's famous $\zeta(2)$ constant. In the fourth section of this paper, we have demonstrated and explored the connection of integrals involving trigonometric and hyperbolic functions with Euler sums. We have evaluated particular integrals related to Euler sums, many of which are not amenable to a mathematical software package. Integrals dealing with powers of arctangent and hyperbolic functions will be further developed in a forthcoming paper.

Author Contributions: Conceptualization, A.S. and A.S.N.; methodology, A.S and A.S.N.; software, A.S.N.; validation, A.S.; formal analysis, A.S. and A.S.N.; investigation, A.S. and A.S.N.; writing–original draft preparation, A.S.N.; writing–review and editing, A.S.; visualization, A.S.N.; supervision, A.S.; project administration, A.S. and A.S.N.

Funding: This research received no external funding.

Conflicts of Interest: The authors declare no conflict of interest.

References

1. Lemmermeyer, F.; Mattmüller, M. (Eds.) *Correspondence of Leonhard Euler with Christian Goldbach: Part I*; Springer: Basel, Switzerland, 2015.
2. Euler, L. Meditationes circa singulare serierum genus. *Novi Comment. Acad. Sci. Petropolitanae* **1776**, *20*, 140–186. Available online: http://eulerarchive.maa.org/ (accessed on 16 July 2019).
3. Nielsen, N. *Handbuch der Theorie der Gammafunktion*; Reprinted by Chelsea Publishing Company, Bronx, New York, 1965; Druck und Verlag von B. G. Teubner: Leipzig, Germany, 1906.
4. Sitaramachandrarao, R. A Formula of S. Ramanujan. *J. Number Theory* **1987**, *25*, 1–19. [CrossRef]
5. Georghiou, C.; Philippou, A.N. Harmonic Sums and the Zeta Function. *Fibonacci Quart.* **1983**, *21*, 29–36.
6. Doelder, P.J.D. On some series containing $\psi(x) - \psi(y)$ and $(\psi(x) - \psi(y))^2$ for certain values of x and y. *J. Comput. Appl. Math.* **1991**, *37*, 125–141. [CrossRef]

7. Bailey, D.H.; Borwein, J.M.; Girgensohn, R. Experimental Evaluation of Euler Sums. *Exp. Math.* **1994**, *3*, 17–30. [CrossRef]
8. Borwein, D.; Borwein, J.M. On an intriguing integral and some series related to $\zeta(4)$. *Proc. Am. Math. Soc.* **1995**, *123*, 1191–1198.
9. Borwein, D.; Borwein, J.M.; Girgensohn, R. Explicit Evaluation of Euler Sums. *Proc. Edinburgh Math. Soc.* **1995**, *38*, 277–294. [CrossRef]
10. Crandall, R.E.; Buhler, J.P. On the evaluation of Euler sums. *Exp. Math.* **1994**, *3*, 275–285. [CrossRef]
11. Sandham, H.F. Advanced Problem 4305. *Am. Math. Mon.* **1948**, *55*, 431. [CrossRef]
12. Chu, W. Hypergeometric series and the Riemann zeta function. *Acta Arith.* **1997**, LXXXII.2, 104–118. [CrossRef]
13. Jung, M.; Cho, Y.J.; Choi, J. Euler Sums Evaluatable From Integrals. *Commun. Korean Math. Soc.* **2004**, *19*, 545–555. [CrossRef]
14. Lewin, L. *Polylogarithms and Associated Functions*; Title of Original 1958 Edition: Dilogarithms and Associated Functions; Elsevier: North Holland, NY, USA, 1981.
15. Flajolet, P.; Salvy, B. Euler Sums and Contour Integral Representations. *Exp. Math.* **1998**, *7*, 15–35. [CrossRef]
16. Freitas, P. Integrals of Polylogarithmic Functions, Recurrence Relations, and Associated Euler Sums. *Math. Comp.* **2005**, *74*, 1425–1440. [CrossRef]
17. Berndt, B.C. *Ramanujan's Notebooks Part I*; Springer: New York, NY, USA, 1985.
18. Choi, J. Log-sine and log-cosine integrals. *Honam Math. J.* **2013**, *35*, 137–146. [CrossRef]
19. Ciaurri, O.; Navas, L.M.; Ruiz, F.J.; Varona, J.L. A simple computation of $\zeta(2k)$ by using Bernoulli polynomials and a telescoping series. *arXiv* **2013**, arXiv:1209.5030.v2.
20. Shang, Y. A class of combinatorial functions for Eulerian numbers. *Matematika* **2012**, *28*, 151–154.
21. Sofo, A. General order Euler sums with rational argument. *Integral Transform. Spec. Funct.* **2019**. [CrossRef]
22. Choi, J.; Srivastava, H.M. Series involving the Zeta functions and a family of generalized Goldbach-Euler series. *Am. Math. Mon.* **2014**, *121*, 229–236. [CrossRef]
23. Choi, J.; Srivastava, H.M. Some applications of the gamma and polygamma functions involving convolutions of the Rayleigh functions, multiple Euler sums and log-sine integrals. *Math. Nachr.* **2009**, *282*, 1709–1723. [CrossRef]
24. Chu, W.; Esposito, F.L. Evaluation of Apéry-like series through multisection method. *J. Class. Anal.* **2018**, *12*, 55–77.
25. Ryotaro, H. On Euler's formulae for double zeta values. *Kyushu J. Math.* **2018**, *72*, 15–24.
26. Li, A.; Qin, H. The representation of the generalized linear Euler sums with parameters. *Integral Transform. Spec. Funct.* **2019**, *30*, 656–681. [CrossRef]
27. Liu, H.; Wang, W. Gauss's theorem and harmonic number summation formulae with certain mathematical constants. *J. Differ. Equ. Appl.* **2019**, *25*, 313–330. [CrossRef]
28. Mező, I. A family of polylog-trigonometric integrals. *Ramanujan J.* **2018**, *46*, 161–171. [CrossRef]
29. Petojević, A.; Srivastava, H.M. Computation of Euler's type sums of the products of Bernoulli numbers. *Appl. Math. Lett.* **2009**, *22*, 796–801. [CrossRef]
30. Sofo, A. Shifted harmonic sums of order two. *Commun. Korean Math. Soc.* **2014**, *29*, 239–255. [CrossRef]
31. Sofo, A. New classes of harmonic number identities. *J. Integer Seq.* **2012**, *15*, 12.
32. Sofo, A.; Cvijović, D. Extensions of Euler harmonic sums. *Appl. Anal. Discret. Math.* **2012**, *6*, 317–328. [CrossRef]
33. Sofo, A.; Srivastava, H.M. A family of shifted harmonic sums. *Ramanujan J.* **2015**, *37*, 89–108. [CrossRef]
34. Teo, L.P. Alternating double Euler sums, hypergeometric identities and a theorem of Zagier. *J. Math. Anal. Appl.* **2018**, *462*, 777–800. [CrossRef]
35. Wang, W.; Lyu, Y. Euler sums and Stirling sums. *J. Number Theory* **2018**, *185*, 160–193. [CrossRef]
36. Wei, C.; Wang, X. Whipple-type $_3F_2$ series and summation formulae involving generalized harmonic numbers. *Int. J. Number Theory* **2018**, *14*, 2385–2407. [CrossRef]
37. Xu, C.; Cai, Y. On harmonic numbers and nonlinear Euler sums. *J. Math. Anal. Appl.* **2018**, *466*, 1009–1042. [CrossRef]
38. Gradshteyn, I.S.; Ryzhik, I.M. *Table of Integrals, Series, and Products*; Academic Press: New York, NY, USA, 2007.
39. Bromwich, T.J. *An Introduction to the Theory of Infinite Series*; Macmillan: London, UK, 1908.
40. Ferrar, W.L. *A Text-Book of Convergence*; Oxford University Press: Oxford, UK, 1938.

41. Knopp, K. *Theory and Application of Infinite Series*; Blackie & Son: New York, NY, USA, 1951.
42. Sofo, A. Evaluation of integrals with hypergeometric and logarithmic functions. *Open Math.* **2018**, *16*, 63–74.
43. Sofo, A. General order Euler sums with multiple argument. *J. Number Theory* **2018**, *189*, 255–271. [CrossRef]
44. Nimbran, A.S.; Sofo, A. New Interesting Euler Sums. *J. Class Anal.* **2019**, Accepted for publication in August.

© 2019 by the authors. Licensee MDPI, Basel, Switzerland. This article is an open access article distributed under the terms and conditions of the Creative Commons Attribution (CC BY) license (http://creativecommons.org/licenses/by/4.0/).

Article

Appell-Type Functions and Chebyshev Polynomials

Pierpaolo Natalini [1] and Paolo Emilio Ricci [2],*

[1] Dipartimento di Matematica e Fisica, Università degli Studi Roma Tre, Largo San Leonardo Murialdo, 1, 00146 Roma, Italy
[2] Section of Mathematics, International Telematic University UniNettuno, Corso Vittorio Emanuele II, 39, 00186 Roma, Italy
* Correspondence: paoloemilioricci@gmail.com

Received: 13 June 2019; Accepted: 26 July 2019; Published: 30 July 2019

Abstract: In a recent article we noted that the first and second kind Cebyshev polynomials can be used to separate the real from the imaginary part of the Appell polynomials. The purpose of this article is to show that the same classic polynomials can also be used to separate the even part from the odd part of the Appell polynomials and of the Appell–Bessel functions.

Keywords: Bessel functions; Appell–Bessel functions; generating functions; Chebyshev polynomials

MSC: AMS 2010 Mathematics Subject Classifications: 33C99–11B83–12E10

1. Introduction

In a recent article [1] we noted that the first and second type Chebyshev polynomials can be used to separate the real from the imaginary part of the Appell polynomials. This is just one of the countless applications of these classic polynomials in Function theory. The purpose of this article is to highlight another of their applications, which is, in some way, analogous to the previous one. In fact everybody knows that the even and the odd part of a function $F(x)$ are derived, in a trivial form, by the equations $\frac{1}{2}[F(x) + F(-x)]$ and $\frac{1}{2}[F(x) - F(-x)]$.

However, in the case of more complicated expressions of the F function, for example in relation to the Appell-type functions, the result is not so obvious, and it will be shown here that it can be obtained using another property of the first and second kind Chebyshev polynomials.

The article is organized as follows. In the first section we use a formula, similar to that of Euler, to separate the even part from the odd part of a binomial power containing hyperbolic functions, showing a connection with the Chebyshev polynomials considered outside their orthogonality interval. Then the results are applied to the case of the Appell polynomials and, in the following sections, to the case of the first kind Bessel functions and to the recently introduced Appell–Bessel functions [2].

It is noteworthy that the study of Appell's polynomials, and of their various extensions, has been considered in both earlier [3] and recent times [4–6]. In these articles have been shown applications to difference equations, expansions in polynomial series and has been analyzed the relative internal structure.

2. Recalling the Chebyshev Polynomials

It is well known that the first kind Chebyshev polynomias $T_n(x)$ can be defined outside the interval $[-1, 1]$, by using their expression in terms of hyperbolic functions [7].

Putting $x = \cosh\theta$, and $f(\theta) = \cosh\theta + \sinh\theta$, we can write:

$$[f(\theta)]^n = (\cosh\theta + \sinh\theta)^n = \sum_{h=0}^{[n/2]} \binom{n}{2h} [\cosh\theta]^{n-2h}([\cosh\theta]^2 - 1)^h +$$

$$+ \sinh\theta \sum_{h=0}^{[(n-1)/2]} \binom{n}{2h+1} [\cosh\theta]^{n-2h-1}([\cosh\theta]^2 - 1)^h = \quad (1)$$

$$= \mathcal{E}[f(\theta)]^n + \mathcal{O}[f(\theta)]^n,$$

where $\mathcal{E}[f(\theta)]^n$ and $\mathcal{O}[f(\theta)]^n$ denote the even and odd part of the $[f(\theta)]^n$ function. Therefore, using the explicit expression of Chebyshev polynomias, we find

$$[f(\theta)]^n = (\cosh\theta + \sinh\theta)^n = T_n(\cosh\theta) + \sinh\theta\, U_{n-1}(\cosh\theta), \quad (2)$$

and

$$\mathcal{E}[f(\theta)]^n = T_n(\cosh\theta), \qquad \mathcal{O}[f(\theta)]^n = \sinh\theta\, U_{n-1}(\cosh\theta). \quad (3)$$

Equations (2) and (3) can be interpreted as an Euler-type formula, owing the analogy with the classical one:

$$[\exp(i\theta)]^n = (\cos\theta + i\sin\theta)^n = \sum_{h=0}^{[n/2]} \binom{n}{2h} [\cos\theta]^{n-2h}([\cos\theta]^2 - 1)^h +$$

$$+ i\sin\theta \sum_{h=0}^{[(n-1)/2]} \binom{n}{2h+1} [\cos\theta]^{n-2h-1}([\cos\theta]^2 - 1)^h = \quad (4)$$

$$= \Re[(\exp(i\theta))^n] + i\Im[(\exp(i\theta))^n].$$

Consequences of the Euler-Type Formula

Using the expansions proven in [8] (but in the case of hyperbolic functions), we find:

Theorem 1. *The Taylor expansions hold:*

$$e^{x\tau}\cosh(y\tau) = \sum_{n=0}^{\infty} \tilde{C}_n(x,y) \frac{\tau^n}{n!},$$

$$e^{x\tau}\sinh(y\tau) = \sum_{n=0}^{\infty} \tilde{S}_n(x,y) \frac{\tau^n}{n!}, \quad (5)$$

where

$$\tilde{C}_n(x,y) = \sum_{j=0}^{[n/2]} \binom{h}{2j} x^{n-2j} y^{2j}$$

$$\tilde{S}_n(x,y) = \sum_{j=0}^{[(n-1)/2]} \binom{h}{2j+1} x^{n-2j-1} y^{2j+1}. \quad (6)$$

Proof. The result follows by using the product of series:
I. Cauchy product involving an even function.

$$\sum_{k=0}^{\infty} c_k \frac{\tau^k}{k!} \sum_{k=0}^{\infty} d_k \frac{\tau^{2k}}{(2k)!} = \sum_{k=0}^{\infty} \left[\sum_{h=0}^{\left[\frac{k}{2}\right]} \binom{k}{2h} c_{k-2h} d_h \right] \frac{\tau^k}{k!}. \tag{7}$$

II. Cauchy product involving an odd function.

$$\sum_{k=0}^{\infty} a_k \frac{\tau^k}{k!} \sum_{k=0}^{\infty} b_k \frac{\tau^{2k+1}}{(2k+1)!} = \sum_{k=0}^{\infty} \left[\sum_{h=0}^{\left[\frac{k-1}{2}\right]} \binom{k}{2h+1} a_{k-2h-1} b_h \right] \frac{\tau^k}{k!}. \tag{8}$$

As a consequence of Equations (5) and (6) we find:

$$e^{x\tau}[\cosh(y\tau) + \sinh(y\tau)] = e^{(x+y)\tau} = \sum_{n=0}^{\infty} [\tilde{C}_n(x,y) + \tilde{S}_n(x,y)] \frac{\tau^n}{n!}, \tag{9}$$

and putting $x = \cosh\theta$, $y = \sinh\theta$

$$\exp[(\cosh\theta + \sinh\theta)\tau] = \sum_{n=0}^{\infty} (\cosh\theta + \sinh\theta)^n \frac{\tau^n}{n!} =$$

$$= \sum_{n=0}^{\infty} [\tilde{C}_n(\cosh\theta, \sinh\theta) + \tilde{S}_n(\cosh\theta, \sinh\theta)] \frac{\tau^n}{n!}. \tag{10}$$

Therefore, we conclude that

$$\tilde{C}_n(\cosh\theta, \sinh\theta) = T_n(\cosh\theta)$$
$$\tilde{S}_n(\cosh\theta, \sinh\theta) = \sinh\theta\, U_{n-1}(\cosh\theta), \tag{11}$$

and furthermore:

$$\tilde{C}_n(x, \sqrt{x^2-1}) = T_n(x), \quad \tilde{S}_n(x, \sqrt{x^2-1}) = \sqrt{x^2-1}\, U_{n-1}(x). \tag{12}$$

□

3. The Even and Odd Part of Appell Polynomials

In this section we show how to represent the even and odd part of Appell polynomials. Consider the Appell polynomials [9–11], defined by the generating function [12]

$$A(t)\, e^{xt} = \sum_{n=0}^{\infty} \alpha_n(x) \frac{t^n}{n!}, \tag{13}$$

where

$$A(t) = \sum_{n=0}^{\infty} a_k \frac{t^k}{k!}. \tag{14}$$

Putting $y := \sqrt{x^2-1}$, we have

$$A(t)\, e^{(x+y)t} = \sum_{n=0}^{\infty} A_n(x,y) \frac{t^n}{n!}, \tag{15}$$

where

$$A_n(x,y) = A_n(x, \sqrt{x^2-1}) = \alpha_n(x + \sqrt{x^2-1}). \tag{16}$$

By using the Cauchy product we find:

$$A_n(x,y) = \sum_{k=0}^{n} \binom{n}{k} a_{n-k} (x+y)^k = \mathcal{E}[A_n(x,y)] + \mathcal{O}[A_n(x,y)], \tag{17}$$

and putting $x = \cosh\theta$, $y = \sqrt{x^2-1} = \sinh\theta$,

$$A_n(x,y) = \sum_{k=0}^{n} \binom{n}{k} a_{n-k} (\cosh\theta + \sinh\theta)^k = \sum_{k=0}^{n} \binom{n}{k} a_{n-k} [\tilde{C}_k(x,y) + \tilde{S}_k(x,y)] =$$

$$= \sum_{k=0}^{n} \binom{n}{k} a_{n-k} [T_k(x) + y\, U_{k-1}(x)].$$

Therefore, by recalling (2) and (3), we conclude that

$$\mathcal{E}[A_n(x, \sqrt{x^2-1})] = \sum_{k=0}^{n} \binom{n}{k} a_{n-k} T_k(x),$$

$$\mathcal{O}[A_n(x, \sqrt{x^2-1})] = \sqrt{x^2-1} \sum_{k=0}^{n} \binom{n}{k} a_{n-k} U_{k-1}(x). \tag{18}$$

Remark 1. *Note that Equation (18) can be applied in general, since, when $|x| > 1$, the position $x + \sqrt{x^2-1} = u$ is equivalent to $x = \frac{u^2-1}{2u}$, and $\sqrt{x^2-1} = \frac{u^2-1}{2u}$, so that Equation (13) becomes:*

$$A(t)\, e^{ut} = \sum_{n=0}^{\infty} \alpha_n(u)\, \frac{t^n}{n!}. \tag{19}$$

Therefore, we can conclude with the theorem.

Theorem 2. *The even and odd part of the Appell polynomials $\alpha_n(u)$ defined by the generating function (19) can be represented, in terms of the first and second kind Chebyshev polynomials, by the equations:*

$$\mathcal{E}[\alpha_n(u)] = \sum_{k=0}^{n} \binom{n}{k} a_{n-k} T_k\left(\frac{u^2+1}{2u}\right),$$

$$\mathcal{O}[\alpha_n(u)] = \frac{u^2-1}{2u} \sum_{k=0}^{n} \binom{n}{k} a_{n-k} U_{k-1}\left(\frac{u^2+1}{2u}\right). \tag{20}$$

4. 1st Kind Bessel Functions

We consider here the first kind of Bessel functions with integer order [13], defined by the generating function [12]:

$$e^{\frac{x}{2}\left(t-\frac{1}{t}\right)} = \sum_{n=-\infty}^{\infty} J_n(x)\, t^n. \tag{21}$$

Putting, for shortness:

$$\tilde{J}_0(x) := \tfrac{1}{2} J_0(x), \qquad \tilde{J}_k(x) := J_k(x), \quad (k \geq 1). \tag{22}$$

Equation (21) writes:

$$e^{\frac{x}{2}\left(t-\frac{1}{t}\right)} = \sum_{n=0}^{\infty} \tilde{J}_n(x) \, t^n + \sum_{n=0}^{\infty} (-1)^n \, \tilde{J}_n(x) \, t^{-n}. \tag{23}$$

Note that, using the notation

$$e^{\frac{x}{2}\left(t-\frac{1}{t}\right)} = \mathcal{E}\left[e^{\frac{x}{2}\left(t-\frac{1}{t}\right)}\right] + \mathcal{O}\left[e^{\frac{x}{2}\left(t-\frac{1}{t}\right)}\right], \tag{24}$$

the even and odd part must be understood with respect to both t and x, and we find:

$$\mathcal{E}\left[e^{\frac{x}{2}\left(t-\frac{1}{t}\right)}\right] = \sum_{n=0}^{\infty} \tilde{J}_{2n}(x) \left(t^{2n} + t^{-2n}\right),$$

$$\mathcal{O}\left[e^{\frac{x}{2}\left(t-\frac{1}{t}\right)}\right] = \sum_{n=0}^{\infty} J_{2n+1}(x) \left(t^{2n+1} - t^{-(2n+1)}\right). \tag{25}$$

Representation by Chebyshev Polynomials

Inverting the equation $\tau = \frac{1}{2}\left(t - \frac{1}{t}\right)$, we have:

$$t = \tau + \sqrt{\tau^2 + 1}. \tag{26}$$

Theorem 3. *The generating function (21) can be represented in terms of Chebyshev polynomials by:*

$$e^{x\tau} = 2 \sum_{n=0}^{\infty} \tilde{J}_{2n}(x) \, T_{2n}(\tau) + 2\sqrt{\tau^2 + 1} \sum_{n=0}^{\infty} J_{2n+1}(x) \, U_{2n}(\tau), \tag{27}$$

so that the even and odd part of the Bessel functions defined by Equation (21) are given by

$$\mathcal{E}\left[\sum_{k=-\infty}^{\infty} J_k(x) \, t^k\right] = 2 \sum_{k=0}^{\infty} \tilde{J}_{2k}(x) \, T_{2k}(\tau),$$

$$\mathcal{O}\left[\sum_{k=-\infty}^{\infty} J_k(x) \, t^k\right] = 2\sqrt{\tau^2 + 1} \sum_{k=0}^{\infty} J_{2k+1} \, U_{2k}(\tau). \tag{28}$$

Proof. Equation (21) writes:

$$e^{x\tau} = \sum_{n=0}^{\infty} \tilde{J}_{2n}(x) \left(t^{2n} + t^{-2n}\right) + \sum_{n=0}^{\infty} J_{2n+1}(x) \left(t^{2n+1} - t^{-(2n+1)}\right), \tag{29}$$

so that

$$\left(t^{2n} + t^{-2n}\right) = 2\, T_{2n}(\tau), \quad \left(t^{2n+1} - t^{-(2n+1)}\right) = 2\sqrt{\tau^2 + 1}\, U_{2n}(\tau), \tag{30}$$

and the result is proven. □

5. Appel–Bessel Functions

Several mixed-type (or hybrid) functions have been recently considered. The starting point of these type of special functions can be found in [14,15]. In this section we consider the Appel–Bessel functions introduced in [2].

Definition 1. *The Appel–Bessel functions [2] are defined by generating function:*

$$G(x,t) = A\left[\tfrac{x}{2}\left(t-\tfrac{1}{t}\right)\right] \exp\left[\tfrac{x}{2}\left(t-\tfrac{1}{t}\right)\right] = \sum_{k=-\infty}^{\infty} [{}_A J_k(x)]\, t^k, \tag{31}$$

where

$$A(\tau) = \sum_{k=0}^{\infty} a_k \frac{\tau^k}{k!}, \quad (a_0 \neq 0). \tag{32}$$

Since

$$G(x,t) = \sum_{k=-\infty}^{\infty} [{}_A J_k(x)]\, t^k = \sum_{k=-\infty}^{\infty} [{}_A J_{-k}(x)]\, t^{-k}, \tag{33}$$

we find

$$[{}_A J_{-k}(x)] = (-1)^k\, [{}_A J_k(x)]. \tag{34}$$

Furthermore, from

$$G(x,-t) = G(-x,t) = \sum_{k=-\infty}^{\infty} [{}_A J_k(x)]\,(-1)^k\, t^k = \sum_{k=-\infty}^{\infty} [{}_A J_k(-x)]\, t^k, \tag{35}$$

we find

$$[{}_A J_k(-x)] = (-1)^k\, [{}_A J_k(x)]. \tag{36}$$

That is, the same symmetry properties of the ordinary 1st kind Bessel functions still hold for the Appell–Bessel functions.

5.1. *Representation of the Appell–Bessel Functions*

Even in this case we put, for shortness:

$$[{}_A \tilde{J}_0(x)] := \tfrac{1}{2}[{}_A J_0(x)], \quad [{}_A \tilde{J}_k(x)] := [{}_A J_k(x)], \quad (k \geq 1). \tag{37}$$

Theorem 4. *The generating function (31) can be represented in terms of Chebyshev polynomials by:*

$$A(x\,\tau)\, e^{x\tau} = 2\sum_{n=0}^{\infty} [{}_A \tilde{J}_{2n}(x)]\, T_{2n}(\tau) + 2\sqrt{\tau^2+1} \sum_{n=0}^{\infty} [{}_A J_{2n+1}(x)]\, U_{2n}(\tau), \tag{38}$$

so that the even and odd part of the Appell-Bessel functions are given by

$$\mathcal{E}\left[\sum_{k=-\infty}^{\infty} [{}_A J_k(x)]\, t^k\right] = 2\sum_{k=0}^{\infty} [{}_A \tilde{J}_{2k}(x)]\, T_{2k}(\tau),$$

$$\mathcal{O}\left[\sum_{k=-\infty}^{\infty} [{}_A J_k(x)]\, t^k\right] = 2\sqrt{\tau^2+1} \sum_{k=0}^{\infty} [{}_A J_{2k+1}]\, U_{2k}(\tau). \tag{39}$$

Proof. Using the symmetry properties (34) and (36), the same technique applied in Section 3 gives the result in the present case. □

5.2. Connection with the Appel–Bessel Functions

Theorem 5. *The following equation holds:*

$$\sum_{k=0}^{\infty} \sum_{h=0}^{k} \binom{k}{h} a_{k-h} \frac{(x\tau)^k}{k!} = \qquad (40)$$

$$= 2 \sum_{n=0}^{\infty} [_A \tilde{J}_{2n}(x)] T_{2n}(\tau) + 2\sqrt{\tau^2+1} \sum_{n=0}^{\infty} [_A J_{2n+1}(x)] U_{2n}(\tau).$$

Proof. By using the Cauchy product we find:

$$A(x\tau) e^{x\tau} = \sum_{k=0}^{\infty} a_k \frac{(x\tau)^k}{k!} \sum_{k=0}^{\infty} \frac{(x\tau)^k}{k!} = \sum_{k=0}^{\infty} \sum_{h=0}^{k} \binom{k}{h} a_{k-h} \frac{(x\tau)^k}{k!}. \qquad (41)$$

Therefore, the result follows by comparing Equations (38) and (41). □

6. Conclusions

It has been shown that the first and second kind Chebyshev polynomials play an important role in separating the even part from the odd part of several polynomials and special functions, which include the Appell polynomials, the first kind Bessel functions and the recently introduced Appell–Bessel functions [2]. This is another remarkable property of Chebyshev's classic polynomials within Function theory, which seems to be the counterpart of another, highlighted in [1], which showed its role in separating the real from the imaginary part of Appell's polynomials.

Author Contributions: Review and editing, P.N.; investigation, original draft preparation, P.E.R.

Funding: This research received no external funding.

Conflicts of Interest: The authors declare that they have not received funds from any institution and that they have no conflict of interest.

References

1. Srivastava, H.M.; Ricci, P.E.; Natalini, P. A Family of Complex Appell Polynomial Sets. *Rev. Real Acad. Cienc. Exactas Fís. Natur. Ser. A Mat.* **2019**, *113*, 2359–2371. [CrossRef]
2. Khan, S.; Naikoo, S.A. Certain discrete Bessel convolutions of the Appell polynomials. *Miskolc Math. Notes* **2019**, *20*, 271–279.
3. Srivastava, H.M. Some characterizations of Appell and q-Appell polynomials. *Ann. Mat. Pura Appl.* **1982**, *130*, 321–329. [CrossRef]
4. Pintér, Á.; Srivastava, H.M. Addition theorems for the Appell polynomials and the associated classes of polynomial expansions. *Aequ. Math.* **2013**, *85*, 483–495. [CrossRef]
5. Srivastava, H.M.; Özarslan, M.A.; Yaşar, B.Y. Difference equations for a class of twice-iterated Δ_h-Appell sequences of polynomials. *Rev. Real Acad. Cienc. Exactas Fís. Natur. Ser. A Mat.* **2019**, *113*, 1851–1871. [CrossRef]
6. Verde-Star, L.; Srivastava, H.M. Some binomial formulas of the generalized Appell form. *J. Math. Anal. Appl.* **2002**, *274*, 755–771. [CrossRef]
7. Rivlin, T.J. *The Chebyshev Polynomials*; J. Wiley: New York, NY, USA, 1990.
8. Masjed-Jamei, M.; Beyki, M.R.; Koepf, W. A New Type of Euler Polynomials and Numbers. *Mediterr. J. Math.* **2018**, *15*, 138. [CrossRef]
9. Appell, P. Sur une classe de polynômes. *Ann. Sci. Ec. Norm. Sup.* **1880**, *9*, 119–144. [CrossRef]
10. Appell, P.; Kampé de Fériet, J. *Fonctions Hypergéométriques et Hypersphériques*. *Polynômes d'Hermite*; Gauthier-Villars: Paris, France, 1926.
11. Sheffer, I.M. Some properties of polynomials sets of zero type. *Duke Math. J.* **1939**, *5*, 590–622. [CrossRef]

12. Srivastava, H.M.; Manocha, H.L. *A Treatise on Generating Functions*; Ellis Horwood Limited: Chichester, UK; John Wiley and Sons: New York, NY, USA; Chichester, UK; Brisbane, Australia; Toronto, ON, Canada, 1984.
13. Watson, G.N. *A Treatise on the Theory of Bessel Functions*, 2nd ed.; Cambridge Univ. Press: Cambridge, UK, 1966.
14. Dattoli, G. Hermite-Bessel and Laguerre-Bessel functions: A by-product of the monomiality principle. In *Advanced Special Functions and Applications*; Cocolicchio, D., Dattoli, G., Srivastava, H.M., Eds.; Aracne Editrice: Rome, Italy, 2000; pp. 147–164.
15. Dattoli, G.; Ricci, P.E.; Srivastava, H.M. Advanced Special Functions and Related Topics in Probability and in Differential Equations. In Proceedings of the Melfi School on Advanced Topics in Mathematics and Physics, Melfi, Italy, 24–29 June 2001.

 © 2019 by the authors. Licensee MDPI, Basel, Switzerland. This article is an open access article distributed under the terms and conditions of the Creative Commons Attribution (CC BY) license (http://creativecommons.org/licenses/by/4.0/).

Article

A Study of Multivalent q-starlike Functions Connected with Circular Domain

Lei Shi [1], Qaiser Khan [2], Gautam Srivastava [3,4], Jin-Lin Liu [5] and Muhammad Arif [2,*]

1. School of Mathematics and Statistics, Anyang Normal University, Anyang 455002, China
2. Department of Mathematics, Abdul Wali Khan University Mardan, Mardan 23200, KP, Pakistan
3. Department of Mathematics and Computer Science, Brandon University, 270 18th Street, Brandon, MB R7A 6A9, Canada
4. Research Center for Interneural Computing, China Medical University, Taichung 40402, Taiwan
5. Department of Mathematics, Yangzhou University, Yangzhou 225002, China
* Correspondence: marifmaths@awkum.edu.pk

Received: 18 June 2019; Accepted: 24 July 2019; Published: 27 July 2019

Abstract: Starlike functions have gained popularity both in literature and in usage over the past decade. In this paper, our aim is to examine some useful problems dealing with q-starlike functions. These include the convolution problem, sufficiency criteria, coefficient estimates, and Fekete–Szegö type inequalities for a new subfamily of analytic and multivalent functions associated with circular domain. In addition, we also define and study a Bernardi integral operator in its q-extension for multivalent functions. Furthermore, we will show that the class defined in this paper, along with the obtained results, generalizes many known works available in the literature.

Keywords: multivalent functions; q-Ruschweyh differential operator; q-starlike functions; circular domain; q-Bernardi integral operator

1. Introduction

The study of q-extension of calculus and q-analysis has attracted and motivated many researchers because of its applications in different parts of mathematical sciences. Jackson was one of the main contributors among all mathematicians who initiated and established the theory of q-calculus [1,2]. As an interesting sequel to [3], in which the q-derivative operator was used for the first time for studying the geometry of q-starlike functions, a firm footing of the usage of the q-calculus in the context of Geometric Function Theory was provided and the basic (or q-) hypergeometric functions were first used in Geometric Function Theory in a book chapter by Srivastava (see, for details, [4] (pp. 347 et seq.)). The theory of q-starlike functions was later extended to various families of q-starlike functions by Agrawal and Sahoo in [5] (see also the recent investigations on this subject by Srivastava et al. [6–11]). Motivated by these q-developments in Geometric Function Theory, many authors added their contributions in this direction which has made this research area much more attractive in works like [4,12].

In 2014, Kanas and Răducanu [13] used the familiar Hadamad products to define a q-extension of the Ruscheweyh operator and discussed important applications of this operator in detail. Moreover, the extensive study of this q-Ruscheweyh operator was further made by Mohammad and Darus [14] and Mahmood and Sokół in [15]. Recently, a new idea was presented by Darus [16] that introduced a new differential operator called a generalized q-differential operator, with the help of q-hypergeometric functions where they studied some useful applications of this operator. For the recent extensions of different operators in q-analogue, see the work in [17–19]. The operator defined in [13] was extended further for multivalent functions by Arif et al. in [20] where they investigated its important applications.

The aim of this paper is to define a family of multivalent q-starlike functions associated with circular domains and to study some of its useful properties.

Background

Let $\mathfrak{A}_p$ $(p \in \mathbb{N} = \{0,1,2,\ldots\})$ contain all multivalent functions say f that are holomorphic or analytic in a subset $\mathbb{D} = \{z : |z| < 1\}$ of a complex plane $\mathbb{C}$ and having the series form:

$$f(z) = z^p + \sum_{l=1}^{\infty} a_{l+p} z^{l+p}, \ (z \in \mathbb{D}). \tag{1}$$

For two analytic functions f and g in $\mathbb{D}$, then f is subordinate to g, symbolically presented as $f \prec g$ or $f(z) \prec g(z)$, if we can find an analytic function w with the properties $w(0) = 0$ & $|w(z)| < 1$ such that $f(z) = g(w(z))$. Also, if g is univalent in $\mathbb{D}$, then we have

$$f(z) \prec g(z) \iff f(0) = g(0) \text{ and } f(\mathbb{D}) \subset g(\mathbb{D}).$$

For given $q \in (0,1)$, the derivative in q-analogue of f is given by

$$\mathcal{D}_q f(z) = \frac{f(z) - f(qz)}{z(1-q)}, \ (z \neq 0, q \neq 1). \tag{2}$$

Making (1) and (2), we easily get that for $n \in \mathbb{N}$ and $z \in \mathbb{D}$:

$$\mathcal{D}_q \left\{ \sum_{n=1}^{\infty} a_{n+p} z^{n+p} \right\} = \sum_{n=1}^{\infty} [n+p]_q a_{n+p} z^{n+p-1}, \tag{3}$$

where

$$[n]_q = \frac{1-q^n}{1-q} = 1 + \sum_{l=1}^{n-1} q^l, \ [0,q] = 0.$$

For $n \in \mathbb{Z}^* := \mathbb{Z} \setminus \{-1,-2,\ldots\}$, the q-number shift factorial is given as

$$[n]_q! = \begin{cases} 1, & n = 0, \\ [1]_q [2]_q \cdots [n]_q, & n \in \mathbb{N}. \end{cases}$$

Also, with $x > 0$, the q-analogue of the Pochhammer symbol has the form

$$[x,q]_{qn} = \begin{cases} 1, & n = 0, \\ [x,q][x+1,q] \cdots [x+n-1,q], & n \in \mathbb{N}, \end{cases}$$

and, for $x > 0$, the Gamma function in q-analogue is presented as

$$\Gamma_q(x+1) = [x,q] \Gamma_q(t) \text{ and } \Gamma_q(1) = 1.$$

We now consider a function

$$\Phi_p(q, \mu+1; z) = z^p + \sum_{n=2}^{\infty} \Lambda_{n+p} z^{n+p}, \ (\mu > -1, z \in \mathbb{D}), \tag{4}$$

with

$$\Lambda_{n+p} = \frac{[\mu+1, q]_{n+p}}{[n+p]_q!}. \tag{5}$$

The series defined in (4) converges absolutely in $\mathbb{D}$. Using $\Phi_p(q, \mu; z)$ with $\mu > -1$ and idea of convolution, Arif et al. [20] established a differential operator $\mathcal{L}_q^{\mu+p-1} : \mathfrak{A}_p \to \mathfrak{A}_p$ by

$$\mathcal{L}_q^{\mu+p-1} f(z) = \Phi_p(q, \mu; z) * f(z) = z^p + \sum_{n=2}^{\infty} \Lambda_{n+p} \, a_{n+p} z^{n+p}, \quad (z \in \mathbb{D}). \tag{6}$$

We also note that

$$\lim_{q \to 1^-} \Phi_p(q, \mu; z) = \frac{z^p}{(1-z)^{\mu+1}} \quad \text{and} \quad \lim_{q \to 1^-} \mathcal{L}_q^{\mu+p-1} f(z) = f(z) * \frac{z^p}{(1-z)^{\mu+1}}.$$

Now, when $q \to 1^-$, the operator defined in (6) becomes the familiar differential operator investigated in [21] and further, setting $p = 1$, we get the most familiar operator known as Ruscheweyh operator [12] (see also [22,23]). Also, for different types of operators in q-analogue, see the works [16,17,19,24–26].

Motivated from the work studied in [3,18,27–29], we establish a family $\mathcal{S}_p^*(q, \mu, A, B)$ using the operator $\mathcal{L}_q^{\mu+p-1}$ as follows:

Definition 1. *Suppose that $q \in (0,1)$ and $-1 \leqq B < A \leqq 1$. Then, $f \in \mathfrak{A}_p$ belongs to the set $\mathcal{S}_p^*(q, \mu, A, B)$, if it satisfies*

$$\frac{z D_q \mathcal{L}_q^{\mu+p-1} f(z)}{[p, q] \mathcal{L}_q^{\mu+p-1} f(z)} \prec \frac{1 + Az}{1 + Bz}, \tag{7}$$

where the function $\frac{1+Az}{1+Bz}$ is known as Janowski function studied in [30].

Alternatively,

$$f \in \mathcal{S}_p^*(q, \mu, A, B) \iff \left| \frac{\frac{z D_q \mathcal{L}_q^{\mu+p-1} f(z)}{[p,q] \mathcal{L}_q^{\mu+p-1} f(z)} - 1}{A - B \frac{z D_q \mathcal{L}_q^{\mu+p-1} f(z)}{[p,q] \mathcal{L}_q^{\mu+p-1} f(z)}} \right| < 1. \tag{8}$$

Note: We will assume throughout our discussion, unless otherwise stated,

$$-1 \leqq B < A \leqq 1, \ q \in (0,1), p \in \mathbb{N}, \text{ and } \mu > -1.$$

2. A Set of Lemmas

Lemma 1. *[31] Let $h(z) = 1 + \sum_{n=1}^{\infty} d_n z^n \prec K(z) = 1 + \sum_{n=1}^{\infty} k_n z^n$ in $\mathbb{D}$. If $K(z)$ is convex univalent in $\mathbb{D}$, then,*

$$|d_n| \leqq |k_1|, \text{ for } n \geqq 1.$$

Lemma 2. *Let $\mathcal{W}$ contain all functions w that are analytic in $\mathbb{D}$, which satisfies $w(0) = 0 \ \& \ |w(z)| < 1$ if the function $w \in \mathcal{W}$, given by*

$$w(z) = \sum_{k=1}^{\infty} w_k z^k \ (z \in \mathbb{D}).$$

Then, for $\lambda \in \mathbb{C}$, we have

$$\left| w_2 - \lambda w_1^2 \right| \leqq \max\{1; |\lambda|\}, \tag{9}$$

and

$$\left| w_3 + \frac{1}{4} w_1 w_2 + \frac{1}{16} w_1^3 \right| \leqq 1. \tag{10}$$

These results are the best possible.

For the first and second part, see references [32,33], respectively.

3. Main Results and Their Consequences

Theorem 1. *Let $f \in \mathfrak{A}_p$ have the series form (1) and satisfy the inequality given by*

$$\sum_{n=1}^{\infty} \wedge_{n+p} \left([n+p,q](1-B) - [p,q](1-A) \right) |a_{n+p}| \leqq [p,q](A-B). \tag{11}$$

Then, $f \in \mathcal{S}_p^*(q, \mu, A, B)$.

Proof. To show $f \in \mathcal{S}_p^*(q, \mu, A, B)$, we just need to show the relation (8). For this, we consider

$$\left| \frac{\frac{zD_q \mathcal{L}_q^{\mu+p-1} f(z)}{[p,q]\mathcal{L}_q^{\mu+p-1} f(z)} - 1}{A - B \frac{zD_q \mathcal{L}_q^{\mu+p-1} f(z)}{[p,q]\mathcal{L}_q^{\mu+p-1} f(z)}} \right| = \left| \frac{zD_q \mathcal{L}_q^{\mu+p-1} f(z) - [p,q]\mathcal{L}_q^{\mu+p-1} f(z)}{A[p,q]\mathcal{L}_q^{\mu+p-1} f(z) - BzD_q \mathcal{L}_q^{\mu+p-1} f(z)} \right|.$$

Using (6), and with the help of (11) and (3), we have

$$= \left| \frac{[p,q]z^p + \sum_{n=1}^{\infty} \wedge_{n+p} a_{n+p}[n+p,q]z^{n+p} - [p,q]\left(z^p + \sum_{n=1}^{\infty} \wedge_{n+p} a_{n+p} z^{n+p}\right)}{A[p,q]\left(z^p + \sum_{n=1}^{\infty} \wedge_{n+p} a_{n+p} z^{n+p}\right) - B\left([p,q]z^p + \sum_{n=1}^{\infty} \wedge_{n+p} a_{n+p}[n+p,q]z^{n+p}\right)} \right|$$

$$= \left| \frac{\sum_{n=1}^{\infty} \wedge_{n+p} a_{n+p}([n+p,q] - [p,q])z^{n+p}}{(A-B)[p,q]z^p + \sum_{n=1}^{\infty} \wedge_{n+p} a_{n+p}(A[p,q] - B[n+p,q])z^{n+p}} \right|$$

$$\leqq \frac{\sum_{n=1}^{\infty} \wedge_{n+p} |a_{n+p}|([n+p,q] - [p,q])|z|^{n+p}}{(A-B)[p,q]|z|^p - \sum_{n=1}^{\infty} \wedge_{n+p} |a_{n+p}|(A[p,q] - B[n+p,q])|z|^{n+p}}$$

$$\leqq \frac{\sum_{n=1}^{\infty} \wedge_{n+p} |a_{n+p}|([n+p,q] - [p,q])}{(A-B)[p,q] - \sum_{n=1}^{\infty} \wedge_{n+p} |a_{n+p}|(A[p,q] - B[n+p,q])} < 1,$$

where we have used the inequality (11) and this completes the proof. □

Varying the parameters μ, b, A, and B in the last Theorem, we get the following known results discussed earlier in [34].

Corollary 1. *Let $f \in \mathfrak{A}$ be given by (1) and satisfy the inequality*

$$\sum_{n=2}^{\infty} ([n,q](1-B) - 1 + A) |a_n| \leqq A - B.$$

Then, the function $f \in \mathcal{S}_q^*[A, B]$.

By choosing $q \to 1^-$ in the last corollary, we get the known result proved by Ahuja [22] and, furthermore, for $A = 1 - \alpha$ and $B = -1$, we obtain the result for the family $\mathcal{S}^*(\xi)$ which was proved by Silverman [35].

Theorem 2. *Let $f \in \mathcal{S}_p^*(q, \mu, A, B)$ be of the form (1). Then,*

$$|a_{p+1}| \leqq \frac{\psi_1(A-B)}{\wedge_{1+p}}, \tag{12}$$

and for $n \geqq 2$,

$$|a_{n+p}| \leqq \frac{(A-B)\psi_n}{\wedge_{n+p}} \prod_{t=1}^{n-1} \left(1 + \frac{[p,q](A-B)}{([p+t,q] - [p,q])} \right), \tag{13}$$

where
$$\psi_n := \psi_n(p,q) = \frac{[p,q]}{([n+p,q]-[p,q])}. \tag{14}$$

Proof. If $f \in \mathcal{S}_p^*(q,\mu,A,B)$, then by definition we have
$$\frac{zD_q \mathcal{L}_q^{\mu+p-1} f(z)}{[p,q]\mathcal{L}_q^{\mu+p-1} f(z)} = \frac{1+Aw(z)}{1+Bw(z)}. \tag{15}$$

Let us put
$$p(z) = 1 + \sum_{n=1}^{\infty} d_n z^n = \frac{1+Aw(z)}{1+Bw(z)}.$$

Then, by Lemma 1, we get
$$|d_n| \leqq A - B. \tag{16}$$

Now, from (15) and (6), we can write
$$z^p + \sum_{n=1}^{\infty} \frac{[n+p,q]}{[p,q]} \wedge_{n+p} a_{n+p} z^{n+p} = \left(1 + \sum_{n=1}^{\infty} d_n z^n\right)\left(z^p + \sum_{n=1}^{\infty} \wedge_{n+p} a_{n+p} z^{n+p}\right). \tag{17}$$

Equating coefficients of z^{n+p} on both sides,
$$\wedge_{n+p}([n+p,q]-[p,q])a_{n+p} = [p,q]\wedge_{n+p-1} a_{n+p-1} d_1 + \cdots + [p,q]\wedge_{1+p} a_{1+p} d_{n-1}.$$

Taking absolute on both sides and then using (16), we have
$$\wedge_{n+p}([n+p,q]-[p,q])|a_{n+p}| \leqq [p,q](A-B)\left(1 + \sum_{k=1}^{n-1} \wedge_{k+p}|a_{k+p}|\right),$$

and this further implies
$$|a_{n+p}| \leqq \frac{(A-B)\psi_n}{\wedge_{n+p}}\left(1 + \sum_{k=1}^{n-1} \wedge_{k+p}|a_{k+p}|\right), \tag{18}$$

where ψ_n is given by (14). So, for $n=1$, we have from (18)
$$|a_{p+1}| \leqq \frac{(A-B)\psi_1}{\wedge_{1+p}},$$

and this shows that (12) holds for $n=1$. To prove (13), we apply mathematical induction. Therefore, for $n=2$, we have from (12):
$$|a_{p+2}| \leqq \frac{(A-B)\psi_2}{\wedge_{2+p}}\left(1 + \wedge_{1+p}|a_{1+p}|\right),$$

using (12), we have
$$|a_{p+2}| \leqq \frac{(A-B)\psi_2}{\wedge_{2+p}}\left(1 + (A-B)\psi_1\right),$$

which clearly shows that (13) holds for $n=2$. Let us assume that (13) is true for $n \leqq m-1$, that is,
$$|a_{m-1+p}| \leqq \frac{(A-B)\psi_{m-1}}{\wedge_{m+p-1}} \prod_{t=1}^{m-2}(1+(A-B)\psi_t).$$

Consider

$$|a_{m+p}| \leq \frac{(A-B)\psi_m}{\wedge_{m+p}}\left(1+\sum_{k=1}^{m-1}\wedge_{k+p}|a_{k+p}|\right)$$

$$= \frac{(A-B)\psi_m}{\wedge_{m+p}}\left\{1+(A-B)\psi_1+\ldots+(A-B)\psi_{m-1}\prod_{t=1}^{m-2}(1+(A-B)\psi_t)\right\}$$

$$= \frac{(A-B)\psi_m}{\wedge_{m+p}}\prod_{t=1}^{m-1}\left(1+\frac{[p,q](A-B)}{([p+t,q]-[p,q])}\right),$$

this implies that the given result is true for $n=m$. Hence, using mathematical induction, we achieve the inequality (13). □

Theorem 3. *Let $f \in \mathcal{S}_p^*(q,\mu,A,B)$, and be given by (1). Then, for $\lambda \in \mathbb{C}$*

$$\left|a_{p+2}-\lambda a_{p+1}^2\right| \leq \frac{(A-B)\psi_2}{\Lambda_{p+2}}\{1;\ |v|\},$$

where v is given by

$$v = (B-(A-B)\psi_1) + \frac{\Lambda_{p+2}\psi_1^2}{\Lambda_{p+1}^2\psi_2}(A-B)\lambda. \tag{19}$$

Proof. Let $f \in \mathcal{S}_p^*(q,\mu,A,B)$, and consider the right-hand side of (15), we have

$$\frac{1+Aw(z)}{1+Bw(z)} = \left(1+A\sum_{k=1}^{\infty}w_k z^k\right)\left(1+B\sum_{k=1}^{\infty}w_k z^k\right)^{-1},$$

where

$$w(z) = \sum_{k=1}^{\infty}w_k z^k,$$

and after simple computations, we can rewrite

$$\frac{1+Aw(z)}{1+Bw(z)} = 1+(A-B)w_1 z+(A-B)\left\{w_2-Bw_1^2\right\}z^2+\ldots. \tag{20}$$

Now, for the left hand side of (15), we have

$$\frac{zD_q\mathcal{L}_q^{\mu+p-1}f(z)}{[p,q]\mathcal{L}_q^{\mu+p-1}f(z)} = \left(1+\sum_{n=1}^{\infty}\frac{[n+p,q]}{[p,q]}\Lambda_{n+p}a_{n+p}z^n\right)\left(1+\sum_{n=1}^{\infty}\Lambda_{n+p}a_{n+p}z^n\right)^{-1}$$

$$= 1+\frac{\Lambda_{1+p}}{\psi_1}a_{1+p}z+\left(\frac{\Lambda_{2+p}a_{2+p}}{\psi_2}-\frac{\Lambda_{1+p}^2 a_{1+p}^2}{\psi_1}\right)z^2+\ldots. \tag{21}$$

From (20) and (21), we have

$$a_{p+1} = \frac{\psi_1}{\Lambda_{p+1}}(A-B)w_1, \tag{22}$$

$$a_{p+2} = \frac{(A-B)\psi_2}{\Lambda_{p+2}}\left\{w_2+((A-B)\psi_1-B)w_1^2\right\}. \tag{23}$$

Now, consider

$$\left|a_{p+2} - \lambda a_{p+1}^2\right| = \left|\frac{(A-B)\psi_2}{\Lambda_{p+2}}\left\{w_2 + ((A-B)\psi_1 - B)w_1^2\right\} - \lambda \frac{\psi_1^2}{\Lambda_{p+1}^2}(A-B)^2 w_1^2\right|$$

$$= \frac{(A-B)\psi_2}{\Lambda_{p+2}}\left|w_2 - \left\{(B-(A-B)\psi_1) + \frac{\Lambda_{p+2}\psi_1^2}{\Lambda_{p+1}^2 \psi_2}(A-B)\lambda\right\}w_1^2\right|,$$

using Lemma 2, we have

$$\left|a_{p+2} - \lambda a_{p+1}^2\right| \leqq \frac{(A-B)\psi_2}{\Lambda_{p+2}}\{1;\ |v|\},$$

where v is given by

$$v = (B-(A-B)\psi_1) + \frac{\Lambda_{p+2}\psi_1^2}{\Lambda_{p+1}^2 \psi_2}(A-B)\lambda.$$

This completes the proof. □

Theorem 4. *Let $f \in \mathcal{S}_p^*(q,\mu,A,B)$ and be given by* (1). *Then,*

$$\left|a_{p+3} - \frac{q+2}{q^2+q+1}\frac{\Lambda_{1+p}\Lambda_{2+p}}{\Lambda_{3+p}}a_{p+2}a_{p+1} + \frac{1}{[3,q]}\frac{\Lambda_{1+p}^3}{\Lambda_{3+p}}a_{p+1}^3\right| \leqq (A-B)\left\{\frac{4(2B-1)^2+1}{8\Lambda_{3+p}}\right\}\psi_3,$$

where ψ_n and Λ_{n+p} are defined by (14) *and* (5), *respectively.*

Proof. From the relations (20) and (21), we have

$$\left(a_{p+3} - \frac{q+2}{q^2+q+1}\frac{\Lambda_{1+p}\Lambda_{2+p}}{\Lambda_{3+p}}a_{p+2}a_{p+1} + \frac{1}{[3,q]}\frac{\Lambda_{1+p}^3}{\Lambda_{3+p}}a_{p+1}^3\right) = \frac{(A-B)\psi_3}{\Lambda_{3+p}}\left\{w_3 - 2Bw_1w_2 + B^2w_1^3\right\},$$

equivalently, we have

$$\left|\left(a_{p+3} - \frac{q+2}{q^2+q+1}\frac{\Lambda_{1+p}\Lambda_{2+p}}{\Lambda_{3+p}}a_{p+2}a_{p+1} + \frac{1}{[3,q]}\frac{\Lambda_{1+p}^3}{\Lambda_{3+p}}a_{p+1}^3\right)\right|$$

$$= \frac{(A-B)\psi_3}{\Lambda_{3+p}}\left|\left(w_3 + \frac{1}{4}w_1w_2 + \frac{1}{16}w_1^3\right) - \frac{16B^2-1}{16}\left(w_2 - w_1^2\right) + \frac{16B^2-32B-5}{16}w_2\right|$$

$$\leqq \frac{(A-B)\psi_3}{\Lambda_{3+p}}\left\{1 + \frac{16B^2-1}{16} + \frac{16B^2-32B-5}{16}\right\}$$

$$\leqq \frac{(A-B)\psi_3}{\Lambda_{3+p}}\left\{\frac{16B^2-16B+5}{8}\right\},$$

where we have used (9) and (10). This completes the proof. □

Theorem 5. *Let $f \in \mathfrak{A}_p$ be given by* (1). *Then, the function f is in the class $\mathcal{S}_p^*(q,\mu,A,B)$, if and only if*

$$\frac{e^{i\theta}(B-[p,q]A)}{z}\left[\mathcal{L}_q^{\mu+p-1}f(z)*\left(\frac{(N+1)z^p-qLz^{p+1}}{(1-z)(1-qz)}\right)\right] \neq 0, \tag{24}$$

for all

$$N = N_\theta = \frac{([p,q]-1)e^{-i\theta}}{([p,q]A-B)},$$

$$L = L_\theta = \frac{(e^{-i\theta}+[p,q]A)}{([p,q]A-B)},$$
(25)

and also for $N = 0, L = 1$.

Proof. Since the function $f \in \mathcal{S}_p^*(q, \mu, A, B)$ is analytic in $\mathbb{D}$, it implies that $\mathcal{L}_q^{\mu+p-1} f(z) \neq 0$ for all $z \in \mathbb{D}^* = \mathbb{D}\setminus\{0\}$—that is

$$\frac{e^{i\theta}(B-[p,q]A)}{z}\mathcal{L}_q^{\mu+p-1}f(z) \neq 0 \quad (z \in \mathbb{D}),$$

and this is equivalent to (24) for $N = 0$ and $L = 1$. From (7), according to the definition of the subordination, there exists an analytic function w with the property that $w(0) = 0$ and $|w(z)| < 1$ such that

$$\frac{zD_q\mathcal{L}_q^{\mu+p-1}f(z)}{[p,q]\mathcal{L}_q^{\mu+p-1}f(z)} = \frac{1+A\omega(z)}{1+B\omega(z)} \quad (z \in \mathbb{D}),$$

which is equivalent for $z \in \mathbb{D}, 0 \leqq \theta < 2\pi$

$$\frac{zD_q\mathcal{L}_q^{\mu+p-1}f(z)}{[p,q]\mathcal{L}_q^{\mu+p-1}f(z)} \neq \frac{1+Ae^{i\theta}}{1+Be^{i\theta}},$$
(26)

and further written in a more simplified form

$$\left(1+Be^{i\theta}\right)zD_q\mathcal{L}_q^{\mu+p-1}f(z) - [p,q]\left(1+Ae^{i\theta}\right)\mathcal{L}_q^{\mu+p-1}f(z) \neq 0.$$
(27)

Now, using the following convolution properties in (27)

$$\mathcal{L}_q^{\mu+p-1}f(z) * \frac{z^p}{(1-z)} = \mathcal{L}_q^{\mu+p-1}f(z) \quad \text{and} \quad \mathcal{L}_q^{\mu+p-1}f(z) * \frac{z^p}{(1-z)(1-qz)} = zD_q\mathcal{L}_q^{\mu+p-1}f(z),$$

then, simple computation gives

$$\frac{1}{z}\left[\mathcal{L}_q^{\mu+p-1}f(z) * \left(\frac{(1+Be^{i\theta})z^p}{(1-z)(1-qz)} - \frac{[p,q](1+Ae^{i\theta})z^p}{(1-z)}\right)\right] \neq 0,$$

or equivalently

$$\frac{(B-[p,q]A)e^{i\theta}}{z}\left[\mathcal{L}_q^{\mu+p-1}f(z) * \left(\frac{(N+1)z^p - Lqz^{p+1}}{(1-z)(1-qz)}\right)\right] \neq 0,$$

which is the required direct part.

Assume that (11) holds true for $L_\theta - 1 = N_\theta = 0$, it follows that

$$\frac{e^{i\theta}(B-[p,q]A)}{z}\mathcal{L}_q^{\mu+p-1}f(z) \neq 0, \text{ for all } z \in \mathbb{D}.$$

Thus, the function $h(z) = \frac{zD_q \mathcal{L}_q^{\mu+p-1} f(z)}{[p,q] \mathcal{L}_q^{\mu+p-1} f(z)}$ is analytic in $\mathbb{D}$ and $h(0) = 1$. Since we have shown that (27) and (11) are equivalent, therefore we have

$$\frac{zD_q \mathcal{L}_q^{\mu+p-1} f(z)}{[p,q] \mathcal{L}_q^{\mu+p-1} f(z)} \neq \frac{1 + Ae^{i\theta}}{1 + Be^{i\theta}} \quad (z \in \mathbb{D}). \tag{28}$$

Suppose that

$$H(z) = \frac{1 + Az}{1 + Bz}, \quad z \in \mathbb{D}.$$

Now, from relation (28) it is clear that $H(\partial \mathbb{D}) \cap h(\mathbb{D}) = \phi$. Therefore, the simply connected domain $h(\mathbb{D})$ is contained in a connected component of $\mathbb{C} \setminus H(\partial \mathbb{D})$. The univalence of the function h, together with the fact that $H(0) = h(0) = 1$, shows that $h \prec H$, which shows that $f \in \mathcal{S}_p^*(q, \mu, A, B)$. □

We now define an integral operator for the function $f \in \mathfrak{A}_p$ as follows:

Definition 2. *Let $f \in \mathfrak{A}_p$. Then, $\mathcal{L} : \mathfrak{A}_p \to \mathfrak{A}_p$ is called the q-analogue of Benardi integral operator for multivalent functions defined by $\mathcal{L}(f) = F_{\eta,p}$ with $\eta > -p$, where $F_{\eta,p}$ is given by*

$$F_{\eta,p}(z) = \frac{[\eta + p, q]}{z^\eta} \int_0^z t^{\eta - 1} f(t) d_q t, \tag{29}$$

$$= z^p + \sum_{n=1}^{\infty} \frac{[\eta + p, q]}{[\eta + p + n, q]} a_{n+p} z^{n+p}, \quad (z \in \mathbb{D}). \tag{30}$$

We easily obtain that the series defined in (30) converges absolutely in $\mathbb{D}$. Now, if $q \to 1$, then the operator $F_{\eta,p}$ reduces to the integral operator studied in [29] and further by taking $p = 1$, we obtain the q-Bernardi integral operator introduced in [36]. If $q \to 1$ and $p = 1$, we obtain the familiar Bernardi integral operator [37].

Theorem 6. *If f is of the form (1), it belongs to the family $\mathcal{S}_p^*(q, \mu, A, B)$ and*

$$F_{\eta,p}(z) = z^p + \sum_{n=1}^{\infty} b_{n+p} z^{n+p}, \tag{31}$$

where $F_{\eta,p}$ is the integral operator given by (29), then

$$|b_{p+1}| \leq \frac{[\eta + p, q]}{[\eta + p + 1, q]} \frac{\psi_1 (A - B)}{\wedge_{1+p}},$$

and for $n \geq 2$

$$|b_{p+n}| \leq \frac{[\eta + p, q]}{[\eta + p + n, q]} \frac{(A - B) \psi_n}{\wedge_{n+p}} \prod_{t=1}^{n-1} \left(1 + \frac{[p, q](A - B)}{([p + t, q] - [p, q])}\right),$$

where ψ_n and $\wedge_{n+p}$ are defined by (14) and (5), respectively.

Proof. The proof follows easily by using (30) and Theorem 2. □

Theorem 7. *Let $f \in \mathcal{S}_p^*(q, \mu, A, B)$ and be given by (1). In addition, if $F_{\eta,p}$ is the integral operator is defined by (29) and is of the form (31), then for $\sigma \in \mathbb{C}$*

$$\left| b_{p+2} - \sigma b_{p+1}^2 \right| \leq \frac{[\eta + p, q]}{[\eta + p + 2, q]} \frac{(A - B) \psi_2}{\wedge_{p+2}} \{1; |v|\},$$

where

$$v = (B - (A - B)\psi_1) + \frac{\Lambda_{p+2}\psi_1^2}{\Lambda_{p+1}^2 \psi_2}(A - B)\frac{[\eta + p, q][\eta + p + 2, q]}{[\eta + p + 1, q]^2}\sigma. \qquad (32)$$

Proof. From (30) and (31), we easily have

$$\begin{aligned} b_{p+1} &= \frac{[\eta + p, q]}{[\eta + p + 1, q]} a_{p+1}, \\ b_{p+2} &= \frac{[\eta + p, q]}{[\eta + p + 2, q]} a_{p+2}. \end{aligned}$$

Now,

$$\left| b_{p+2} - \sigma b_{p+1}^2 \right| = \frac{[\eta + p, q]}{[\eta + p + 2, q]} \left| a_{p+2} - \sigma \frac{[\eta + p, q][\eta + p + 2, q]}{([\eta + p + 1, q])^2} a_{p+1}^2 \right|.$$

By using (22) and (23), we have

$$\left| b_{p+2} - \sigma b_{p+1}^2 \right| = \frac{[\eta + p, q]}{[\eta + p + 2, q]} \frac{(A - B)}{\Lambda_{p+2}} \left| w_2 - v w_1^2 \right|,$$

where v is given by (32). Applying (9), we get

$$\left| b_{p+2} - \sigma b_{p+1}^2 \right| \leq \frac{[\eta + p, q]}{[\eta + p + 2, q]} \frac{(A - B)}{\Lambda_{p+2}} \{1, |v|\}.$$

Hence, we have the required result. □

4. Future Work

The idea presented in this paper can easily be implemented to define some more subfamilies of analytic and univalent functions connected with different image domains [38–40].

5. Conclusions

In this article, we have defined a new class of multivalent q-starlike functions by using multivalent q-Ruscheweyh differential operator. We studied some interesting problems, which are helpful to study the geometry of the image domain, and also used some of the achieved results to find the growth of Hankel determinant. The idea of this determinant is applied in the theory of singularities [39] and in the study of power series with integral coefficients. For deep insight, the reader is invited to read [38–44]. Further, we have generalized the Bernardi integral operator and defined the multivalent q-Bernardi integral operator. Some useful properties of this class of multivalent functions have been studied.

Author Contributions: The authors have equally contributed to accomplish this research work.

Funding: This article is supported financially by the Anyang Normal University, Anyang 455002, Henan, China.

Conflicts of Interest: The authors agree with the contents of the manuscript, and there are no conflicts of interest among the authors.

References

1. Jackson, F.H. On q-functions and a certain difference operator. *Earth Environ. Sci. Trans. R. Soc. Edinburgh* **1909**, *46*, 253–281. [CrossRef]
2. Jackson, F.H. On q-definite integrals. *Q. J. Pure Appl. Math.* **1910**, *41*, 193–203.
3. Ismail, M.E.H.; Merkes, E.; Styer, D. A generalization of starlike functions. *Complex Var. Theory Appl.* **1990**, *14*, 77–84. [CrossRef]

4. Srivastava, H.M. Univalent functions, fractional calculus, and associated generalized hypergeometric functions. In *Univalent Functions, Fractional Calculus, and Their Applications*; Srivastava, H.M., Owa, S., Eds.; Halsted Press: Chichester, UK; John Wiley and Sons: New York, NY, USA, 1989; pp. 329–354.
5. Agrawal, S.; Sahoo, S.K. A generalization of starlike functions of order α. *Hokkaido Math. J.* **2017**, *46*, 15–27. [CrossRef]
6. Mahmood, S.; Ahmad, Q.Z.; Srivastava, H.M.; Khan, N.; Khan, B.; Tahir, M. A certain subclass of meromorphically q-starlike functions associated with the Janowski functions. *J. Inequal. Appl.* **2019**, *2019*, 88. [CrossRef]
7. Mahmood, S.; Jabeen, M.; Malik, S.N.; Srivastava, H.M.; Manzoor, R.; Riaz, S.M.J. Some coefficient inequalities of q-starlike functions associated with conic domain defined by q-derivative. *J. Funct. Spaces* **2018**, *2018*, 8492072. [CrossRef]
8. Mahmood, S.; Srivastava, H.M.; Khan, N.; Ahmad, Q.Z.; Khan, B.; Ali, I. Upper bound of the third Hankel determinant for a subclass of q-starlike functions. *Symmetry* **2019**, *11*, 347. [CrossRef]
9. Srivastava, H.M.; Ahmad, Q.Z.; Khan, N.; Khan, B. Hankel and Toeplitz determinants for a subclass of q-starlike functions associated with a general conic domain. *Mathematics* **2019**, *7*, 181. [CrossRef]
10. Srivastava, H.M.; Khan, B.; Khan, N.; Ahmad, Q.Z. Coeffcient inequalities for q-starlike functions associated with the Janowski functions. *Hokkaido Math. J.* **2019**, *48*, 407–425. [CrossRef]
11. Srivastava, H.M.; Tahir, M.; Khan, B.; Ahmad, Q.Z.; Khan, N. Some general classes of q-starlike functions associated with the Janowski functions. *Symmetry* **2019**, *11*, 292. [CrossRef]
12. Ruscheweyh, S. New criteria for univalent functions. *Proc. Am. Math. Soc.* **1975**, *49*, 109–115. [CrossRef]
13. Kanas, S.; Răducanu, D. Some class of analytic functions related to conic domains. *Math. Slovaca* **2014**, *64*, 1183–1196. [CrossRef]
14. Aldweby, H.; Darus, M. Some subordination results on q-analogue of Ruscheweyh differential operator. *Abstr. Appl. Anal.* **2014**, *2014*, 958563. [CrossRef]
15. Mahmood, S.; Sokół, J. New subclass of analytic functions in conical domain associated with Ruscheweyh q-differential operator. *Results Math.* **2017**, *71*, 1345–1357. [CrossRef]
16. Mohammed, A.; Darus, M. A generalized operator involving the q-hypergeometric function. *Matematički Vesnik* **2013**, *65*, 454–465.
17. Ahmad, B.; Arif, M. New subfamily of meromorphic convex functions in circular domain involving q-operator. *Int. J. Anal. Appl.* **2018**, *16*, 75–82.
18. Arif, M.; Dziok, J.; Raza, M.; Sokół, J. On products of multivalent close-to-star functions. *J. Inequal. Appl.* **2015**, *2015*, 5. [CrossRef]
19. Arif, M.; Haq, M.; Liu, J.-L. A subfamily of univalent functions associated with q-analogue of Noor integral operator. *J. Funct. Spaces* **2018**, *2018*, 3818915. [CrossRef]
20. Arif, M.; Srivastava, H.M.; Umar, S. Some applications of a q-analogue of the Ruscheweyh type operator for multivalent functions. *Rev. Real Acad. Cienc. Exactas Fís. Natur. Ser. A Mat. (RACSAM)* **2019**, *113*, 1211–1221. [CrossRef]
21. Goel, R.M.; Sohi, N.S. A new criterion for p-valent functions. *Proc. Am. Math. Soc.* **1980**, *78*, 353–357.
22. Ahuja, O.P. Families of analytic functions related to Ruscheweyh derivatives and subordinate to convex functions. *Yokohama Math. J.* **1993**, *41*, 39–50.
23. Noor, K.I.; Arif, M. On some applications of Ruscheweyh derivative. *Comput. Math. Appl.* **2011**, *62*, 4726–4732. [CrossRef]
24. Aldweby, H.; Darus, M. A subclass of harmonic univalent functions associated with q-analogue of Dziok-Srivastava operator. *ISRN Math. Anal.* **2013**, *2013*, 382312. [CrossRef]
25. Aldawish, I.; Darus, M. Starlikeness of q-differential operator involving quantum calculus. *Korean J. Math.* **2014**, *22*, 699–709. [CrossRef]
26. Arif, M.; Ahmad, B. New subfamily of meromorphic starlike functions in circular domain involving q-differential operator. *Math. Slovaca* **2018**, *68*, 1049–1056. [CrossRef]
27. Mahmood, S.; Arif, M.; Malik, S.N. Janowski type close-to-convex functions associated with conic regions. *J. Inequal. Appl.* **2017**, *2017*, 259. [CrossRef]
28. Seoudy, T.M.; Aouf, M.K. Coefficient estimates of new classes of q-starlike and q-convex functions of complex order. *J. Math. Inequal.* **2016**, *10*, 135–145. [CrossRef]

29. Wang, Z.G.; Raza, M.; Ayaz, M.; Arif, M. On certain multivalent functions involving the generalized Srivastava-Attiya operator. *J. Nonlinear Sci. Appl.* **2016**, *9*, 6067–6076. [CrossRef]
30. Janowski, W. Some extremal problems for certain families of analytic functions. *Annales Polonici Mathematici* **1973**, *28*, 297–326. [CrossRef]
31. Rogosinski, W. On the coefficients of subordinate functions. *Proc. Lond. Math. Soc.* **1943**, *48*, 48–82. [CrossRef]
32. Keogh, F.R.; Merkes, E.P. A coefficient inequality for certain classes of analytic functions. *Proc. Am. Math. Soc.* **1969**, *20*, 8–12. [CrossRef]
33. Sokół, J.; Thomas, D.K. Cefficient estimates in a class of strongly starlike functions. *Kyungpook Math. J.* **2009**, *49*, 349–353. [CrossRef]
34. Seoudy, T.M.; Aouf, M.K. Convolution properties for certain classes of analytic functions defined by q-derivative operator. *Abstr. Appl. Anal.* **2014**, *2014*, 846719. [CrossRef]
35. Silverman, H.; Silvia, E.M.; Telage, D. Convolution conditions for convexity starlikeness and spiral-likness. *Mathematiche Zeitschrift* **1978**, *162*, 125–130. [CrossRef]
36. Noor, K.I.; Riaz, S.; Noor, M.A. On q-Bernardi integral operator. *TWMS J. Pure Appl. Math.* **2017**, *8*, 3–11.
37. Bernardi, S.D. Convex and starlike univalent functions. *Trans. Am. Math. Soc.* **1969**, *135*, 429–446. [CrossRef]
38. Cantor, D.G. Power series with integral coefficients. *Bull. Am. Math. Soc.* **1963**, *69*, 362–366. [CrossRef]
39. Dienes, P. *The Taylor Series*; Dover: New York, NY, USA, 1957.
40. Edrei, A. Sur les determinants recurrents et less singularities d'une fonction donee por son developpement de Taylor. *Comput. Math.* **1940**, *7*, 20–88.
41. Polya, G.; Schoenberg, I.J. Remarks on de la Vallee Poussin means and convex conformal maps of the circle. *Pacific J. Math.* **1958**, *8*, 259–334. [CrossRef]
42. Mahmood, S.; Srivastava, G.; Srivastava, H.M.; Abujarad, E.S.; Arif, M.; Ghani, F. Sufficiency Criterion for A Subfamily of Meromorphic Multivalent Functions of Reciprocal Order with Respect to Symmetric Points. *Symmetry* **2019**, *11*, 764. [CrossRef]
43. Mahmood, S.; Raza, N.; AbuJarad, E.S.; Srivastava, G.; Srivastava, H.M.; Malik, S.N. Geometric Properties of Certain Classes of Analytic Functions Associated with a q-Integral Operator. *Symmetry* **2019**, *11*, 719. [CrossRef]
44. Sene, N.; Srivastava, G. Generalized Mittag-Leffler Input Stability of the Fractional Differential Equations. *Symmetry* **2019**, *11*, 608. [CrossRef]

© 2019 by the authors. Licensee MDPI, Basel, Switzerland. This article is an open access article distributed under the terms and conditions of the Creative Commons Attribution (CC BY) license (http://creativecommons.org/licenses/by/4.0/).

Article

A Novel Integral Equation for the Riemann Zeta Function and Large t-Asymptotics

Konstantinos Kalimeris [1],* and Athanassios S. Fokas [1,2],*

[1] Department of Applied Mathematics and Theoretical Physics, University of Cambridge, Cambridge CB3 0WA, UK
[2] Viterbi School of Engineering, University of Southern California, Los Angeles, CA 90089-2560, USA
* Correspondence: kk364@cam.ac.uk (K.K.); tf227@cam.ac.uk (A.S.F.)

Received: 7 June 2019; Accepted: 18 July 2019; Published: 20 July 2019

Abstract: Based on the new approach to Lindelöf hypothesis recently introduced by one of the authors, we first derive a novel integral equation for the square of the absolute value of the Riemann zeta function. Then, we introduce the machinery needed to obtain an estimate for the solution of this equation. This approach suggests a substantial improvement of the current large t-asymptotics estimate for $\zeta\left(\frac{1}{2} + it\right)$.

Keywords: riemann zeta function; asymptotics; exponential sums

---

1. Introduction

It is well known that the leading asymptotics for large t of $\zeta(s)$ can be expressed in terms of a transcendental sum,

$$\zeta(s) \sim \sum_{m=1}^{[t]} \frac{1}{m^s}, \qquad s = \sigma + it, \quad 0 < \sigma < 1, \quad t \to \infty, \tag{1}$$

where throughout this paper [A] denotes the integer part of the positive number A. Lindelöf's hypothesis, one of the most important open problems in the history of mathematics, states that for $\sigma = 1/2$, this sum is of order $O(t^\varepsilon)$ for any $\varepsilon > 0$.

The sum of the rhs of (1) is a particular case of an exponential sum. Pioneering results for the estimation of such sums were obtained in 1916 using methods developed by Weyl [1], and Hardy and Littlewood [2], when it was shown that $\zeta(1/2 + it) = O(t^{1/6+\varepsilon})$. In the last 100 years some slight progress was made using the ingenious techniques of Vinogradov [3]. Currently, the best result is due to Bourgain [4] who has been able to reduce the exponent factor to $13/84 \approx 0.155$.

It is interesting that, in contrast to the usual situation in asymptotics where higher order terms in an asymptotic expansion are more complicated, the higher order terms of the asymptotic expansion of $\zeta(s)$ can be computed *explicitly*. Siegel, in his classical paper [5] presented the asymptotic expansion of $\zeta(s)$ to *all* orders in the important case of $x = y = \sqrt{t/2\pi}$. In [6], analogous results are presented for *any* x and y valid to *all* orders. A similar result for the Hurwitz zeta function is presented in [7]. Some of the results of [6] are used in [8] and the latter results are useful for the estimates presented in this paper.

A major obstacle in trying to prove Lindelöf's hypothesis via the estimation of relevant exponential sums is that in estimates one "loses" something (the more powerful the technique, the less the loss). Here we follow the new formalism for analysing the large t-asymptotics of the Riemann zeta function, introduced in [9]. For the sake of clarity of presentation we restrict our attention to the case $\sigma = 1/2$.

We start with the following integral equation derived in Equation (1.6) of [9]:

$$\frac{t}{\pi} \fint_{-t^{\delta_1-1}}^{1+t^{\delta_4-1}} \Re\left\{\frac{\Gamma(it-i\tau t)}{\Gamma(1/2+it)}\Gamma(1/2+i\tau t)\right\} |\zeta(1/2+i\tau t)|^2 \, d\tau + \mathcal{G}(1/2,t)$$
$$+ O\left(e^{-\pi t^{\delta_{14}}}\right) = 0, \qquad t \to \infty, \qquad (2)$$
$$\delta_1 > 0, \; \delta_4 > 0, \; \delta_{14} = \min(\delta_1, \delta_4),$$

where $\Gamma(z)$ denotes the usual gamma function, the principal value integral is defined with respect to $\tau = 1$ and $\mathcal{G}(1/2,t)$ satisfies the estimate

$$\mathcal{G}(1/2,t) = \ln t + O(1), \qquad t \to \infty. \qquad (3)$$

In [9] the computation of the large t asymptotics of (2) is obtained by first splitting the interval $[-t^{\delta_1-1}, 1+t^{\delta_4-1}]$ into the following four subintervals:

$$L_1 = [-t^{\delta_1-1}, t^{-1}], \; L_2 = [t^{-1}, t^{\delta_2-1}],$$
$$L_3 = [t^{\delta_2-1}, 1-t^{\delta_3-1}], \; L_4 = [1-t^{\delta_3-1}, 1+t^{\delta_4-1}], \qquad (4)$$

with $\delta_j \in (0,1)$, $j=1,2,3,4$. We employ the same splitting in (2) and hence the asymptotic evaluation of (2) reduces to the analysis of the four integrals,

$$I_j(t) = \frac{t}{\pi} \fint_{L_j} \Re\left\{\frac{\Gamma(it-i\tau t)}{\Gamma(1/2+it)}\Gamma(1/2+i\tau t)\right\} |\zeta(1/2+i\tau t)|^2 \, d\tau, \; t > 0, \qquad (5)$$

where I_1, I_2, I_3, I_4 also depend on δ_1, δ_2, (δ_2, δ_3), (δ_3, δ_4), respectively, L_1, L_2, L_3, L_4 are defined in (4), and the principal value integral is needed only for I_4.

Organisation of the Paper

In Section 2 we derive a linear integral equation for $|\zeta(s)|^2$. This equation is given by (23) where S_4^P and S_4^{SD} are defined by (18) and (19), respectively.

In Section 3 we present the methodology for deriving the main result of this paper, namely the linear Volterra-type integral equation for $|\zeta(s)|^2$ given by Equation (8) below. In this connection, we first estimate the double sum S_4^P appearing in the linear integral Equation (23):

$$\Re\left\{S_4^P(t,\delta_3)\right\} = O\left(t^{\frac{\delta_3}{2}} \ln t\right), \qquad t \to \infty. \qquad (6)$$

Then, we present heuristic arguments regarding the estimation of S_4^{SD}, which suggest that

$$S_4^{SD} = O\left(t^{\frac{1}{3}-\frac{\delta_3}{2}}(\ln t)^2\right) + O\left(t^{\frac{\delta_3}{2}} \ln t\right), \qquad t \to \infty. \qquad (7)$$

Replacing in (23) S_4^P and S_4^{SD} by (6) and (7), for $0 < \delta_2 < 1/2$ and $\delta_3 = 1/3$, we find the main result of the paper:

$$\left|\zeta\left(\frac{1}{2}+it\right)\right|^2 = \frac{1}{\pi} \int_{t^{\delta_2}}^{t-t^{1/3}} \mathcal{K}(\rho,t) \left|\zeta\left(\frac{1}{2}+i\rho\right)\right|^2 d\rho + O\left(t^{\frac{1}{6}}(\ln t)^2\right), \qquad t \to \infty, \qquad (8)$$

where the kernel $\mathcal{K}(\rho,t)$ is given by

$$\mathcal{K}(\rho,t) = \Re\left\{\frac{\Gamma(it-i\rho)}{\Gamma(1/2+it)}\Gamma(1/2+i\rho)\right\}. \qquad (9)$$

For the rigorous derivation of (7) we make crucial use of some of the results of [10]. For completeness of presentation, the relevant results of [10] are reviewed in Section 3.

In Section 4 we derive (7). This derivation is based on the following: first, on a lemma for partial summation in two dimensions, which is crucial for the analysis of some parts of the sum S_4^{SD}. Second, on the asymptotic estimates of the function $E_4^{SD}(t, \delta)$ appearing in the definition of S_4^{SD} which are given in [11]. Third, on the splitting of S_4^{SD} into three cases involving certain sums denoted by $S^{(i)}$, $S^{(ii)}$, $S^{(iii)}$. Relatively straightforward estimates yield that both $S^{(ii)}$ and $S^{(iii)}$ are $O\left(t^{\frac{\delta}{2}} \ln t\right)$, $t \to \infty$. The estimation of $S^{(i)}$ is quite complicated; details are given in Section 4.3.

Section 5 summarizes the basic results in this paper and discusses future directions.

2. Derivation of a Linear Integral Equation for $|\zeta(s)|^2$

The main result of this section is the linear integral Equation (23) which is obtained from (2) by computing the contribution of the integrals $\{I_j\}_1^4$. In this direction, we first recall the estimates for I_1 and I_2, and then we introduce a methodology that computes explicitly the leading asymptotic behaviour of I_4. In addition, this methodology avoids the need to compute the asymptotics of I_3.

2.1. The Contribution of I_1 and I_2

Using Lemma 4.1 of [9], it can be shown that for δ_1 sufficiently small, I_1 satisfies the estimate

$$I_1(t, \delta_1) = O\left(t^{-1/2 + \frac{4}{3}\delta_1}\right), \qquad t \to \infty. \tag{10}$$

Furthermore, by employing the classical estimates of Atkinson, and following the steps of Lemma 4.2 of [9], it can be shown that that I_2 satisfies the estimate

$$I_2(t, \delta_2) = O\left(t^{-\frac{1}{2} + \delta_2} \ln t\right), \qquad 0 < \delta_2 < 1, \qquad t \to \infty. \tag{11}$$

Thus, for sufficiently small δ_1 and δ_2 Equations (10) and (11) yield

$$I_1 = o(1) \quad \text{and} \quad I_2 = o(1), \qquad t \to \infty. \tag{12}$$

2.2. The Contribution of the Leading Order Term of I_4

Let $\tilde{I}_4$ denote the contribution of the leading order term of I_4. This term is defined by replacing $\zeta(s)$ with the leading term of its large t-asymptotics in (5) for $j = 4$. Using the change of variables $\tau = 1 - \frac{x}{t}$, $\tilde{I}_4$ becomes

$$\tilde{I}_4(t, \delta_3, \delta_4) = \frac{1}{\pi} \fint_{-t^{\delta_4}}^{t^{\delta_3}} \Re\left\{\Gamma(ix) \frac{\Gamma(1/2 + it - ix)}{\Gamma(1/2 + it)}\right\} |\tilde{\zeta}(1/2 + it - ix)|^2 \, dx, \tag{13}$$

where the principal value integral is with respect to $x = 0$, and

$$|\tilde{\zeta}(1/2 + it - ix)|^2 = \sum_{m_1=1}^{[t]} \sum_{m_2=1}^{[t]} \frac{1}{m_1^s m_2^{\bar{s}}} \left(\frac{m_1}{m_2}\right)^{ix}, \qquad s = \frac{1}{2} + it. \tag{14}$$

Thus, we obtain the following expression for the leading behaviour of I_4:

$$\tilde{I}_4(t, \delta_3, \delta_4) = \Re\left\{\sum_{m_1=1}^{[t]} \sum_{m_2=1}^{[t]} \frac{1}{m_1^s} \frac{1}{m_2^{\bar{s}}} J_4\left(\sigma, t, \delta_2, \delta_3, \frac{m_1}{m_2}\right)\right\}, \qquad s = \frac{1}{2} + it, \tag{15}$$

where J_4 is defined by

$$J_4\left(t,\delta_3,\delta_4,\frac{m_1}{m_2}\right) = \frac{1}{\pi}\fint_{-t^{\delta_4}}^{t^{\delta_3}} \Gamma(ix)\frac{\Gamma(1/2+it-ix)}{\Gamma(1/2+it)}\left(\frac{m_1}{m_2}\right)^{ix} dx,$$
$$t>0,\ 0<\delta_3<1,\ 0<\delta_4<1,\ m_j=1,2,\ldots,[t], \tag{16}$$

with the principal value integral defined with respect to $x = 0$.

Theorem 6.1 of [9] gives the estimate

$$\tilde{I}_4(t,\delta_3,\delta_4) = \left\{ -\sum_{m_1=1}^{[t]}\sum_{m_2=1}^{[t]} \frac{1}{m_1^s m_2^{\bar{s}}} + 2\Re\left\{S_4^P\right\} + \Re\left\{S_4^{SD}\right\} \right\}\left[1+O(t^{2\delta_{34}-1})\right],\quad t\to\infty$$
$$s = \frac{1}{2}+it,\ 0<\delta_3<\frac{1}{2},\ 0<\delta_4<\frac{1}{2},\ \delta_{34}=\max\{\delta_3,\delta_4\}, \tag{17}$$

where S_4^P and S_4^{SD} are defined as follows:

$$S_4^P(t,\delta_3) = \sum_{m_1,m_2\in M_4}\frac{1}{m_1^s m_2^{\bar{s}}}e^{-i\frac{m_2}{m_1}t} \tag{18}$$

with

$$M_4 := M_4(\delta_3,t) = \left\{m_j=1,\ldots,[t],\ j=1,2,\ \frac{m_1}{m_2}\in(t^{1-\delta_3},t)\right\},$$

and

$$S_4^{SD}(t,\delta_3) = \sum_{m_1=1}^{[t]}\sum_{m_2=1}^{[t]}\frac{1}{m_1^s m_2^{\bar{s}}}E_4^{SD}(t,\delta_3); \tag{19}$$

E_4^{SD} satisfies the asymptotic estimate

$$E_4^{SD} \sim -\sqrt{\frac{2}{\pi}}e^{i\frac{\pi}{4}}t^{-\frac{\delta_3}{2}}e^{-it^{\delta_3}}t^{i(\delta_3-1)t^{\delta_3}}\frac{1}{\ln\left(\frac{m_2}{m_1}t^{1-\delta_3}\right)}\left(\frac{m_1}{m_2}\right)^{it^{\delta_3}},\quad t\to\infty, \tag{20}$$

when $\frac{m_2}{m_1}t^{1-\delta_3}\neq 1$.

Remark 1. *According to the analysis of [9], the derivation of (17) involves the computation of the contribution of an integral along the so-called Hankel contour. The function S_4^P is related with the contribution of the pole $w_P = -i\frac{m_2}{m_1}t^{1-\delta_3}$ and S_4^{SD} is related with the contribution of the Hankel contour after deforming it so that it passes through the point of steepest descent $w_{SD} = -i$. Hence, we call S_4^P and S_4^{SD} as the Pole and Steepest Descent contribution, respectively.*

2.3. The Contribution of I_3

Let I_3 be defined by (5), with $j=3$. By making the change of variables $\rho = t\tau$, we obtain

$$I_3(t,\delta_2,\delta_3) = \frac{1}{\pi}\int_{t^{\delta_2}}^{t-t^{\delta_3}}\Re\left\{\frac{\Gamma(it-i\rho)}{\Gamma(1/2+it)}\Gamma(1/2+i\rho)\right\}|\zeta(1/2+i\rho)|^2\, d\rho. \tag{21}$$

2.4. A Volterra-Type Integral Equation

It is shown in Appendix A that the first term of the rhs of (17) is the leading asymptotic term of $|\zeta|^2$. Hence, (17) becomes

$$\tilde{I}_4(t,\delta_3,\delta_4) \sim -|\zeta(s)|^2 + 2\Re\left\{S_4^P\right\} + \Re\left\{S_4^{SD}\right\},\ s=\frac{1}{2}+it,\quad t\to\infty. \tag{22}$$

By replacing in (2), I_1 and I_2 by (12), I_3 by (21) and I_4 by (22) we obtain the following Volterra-type integral equation:

$$|\zeta(1/2+it)|^2 = \frac{1}{\pi}\int_{t^{\delta_2}}^{t-t^{\delta_3}} \Re\left\{\frac{\Gamma(it-i\rho)}{\Gamma(1/2+it)}\Gamma(1/2+i\rho)\right\}|\zeta(1/2+i\rho)|^2 d\rho \qquad (23)$$
$$+ 2\Re\left\{S_4^P\right\} + \Re\left\{S_4^{SD}\right\} + \ln t + O(1), \qquad t \to \infty,$$

where S_4^P and S_4^{SD} are given in (18) and (19), respectively.

3. The Methodology for Deriving the Integral Equation (8)

In this section we derive Equation (6) and we also provide heuristic arguments for supporting the validity of Equation (7). The employment of the estimates (6) and (7), evaluated at $\delta_3 = 1/3$, in Equation (23) yields (8).

3.1. An Estimate for S_4^P

In order to estimate the sum S_4^P, we use (1.30) of [9], namely (See Appendix B)

$$S_3(t,\delta_3) = S_4^P(t,\delta_3)\left[1 + O\left(t^{2\delta_3 - 1}\right)\right], \qquad t \to \infty, \qquad (24)$$

with

$$S_3(t,\delta_3) = \sum\sum_{(m_1,m_2)\in M_3} \frac{1}{m_2^s(m_1+m_2)^s}, \qquad s = \frac{1}{2}+it, \qquad (25)$$

and M_3 is defined by

$$M_3 := M_3(\delta_3, t) = \left\{m_j = 1, \ldots, [t], j = 1,2, \frac{m_2}{m_1} < \frac{1}{t^{1-\delta_3}-1}\right\}. \qquad (26)$$

Using results of [6], it is shown in Theorem 5.1 of [8] that

$$S_3(t,\delta_3) = O\left(t^{\frac{\delta_3}{2}}\ln t\right), \qquad t \to \infty. \qquad (27)$$

Thus, (6) follows.

3.2. An Estimate for S_4^{SD}

The definition of S_4^{SD}, given in (19), implies

$$S_4^{SD} = O\left(\frac{1}{t^{\frac{\delta_3}{2}}}\left|\sum_{m_1=1}^{[t]}\sum_{m_2=1}^{[t]}\frac{1}{m_1^{\frac{1}{2}+i(t-t^{\delta_3})}}\frac{1}{m_2^{\frac{1}{2}-i(t-t^{\delta_3})}}\ln\left(\frac{m_2}{m_1}t^{1-\delta_3}\right)\right|\right). \qquad (28)$$

In order to estimate the sum S_4^{SD}, we employ the classical techniques of [10,12–14]. These techniques can be used for the estimation of the sums of the form

$$\sum\sum_{1<m<n<[t]} \frac{1}{m^{\frac{1}{2}+it}n^{\frac{1}{2}-it}}.$$

In this connection we recall the following well known result (see, for example, Theorem 5.12 of [13]):

$$\sum_{n=1}^{[t]} \frac{1}{n^{\frac{1}{2}+it}} = O\left(t^{\frac{1}{6}}\ln t\right), \qquad t \to \infty. \qquad (29)$$

The above result can be further improved, see, for example, Theorem 5.18 of [13]:

$$\left| \sum_{n=1}^{[t]} \frac{1}{n^{\frac{1}{2}+it}} \right|^2 = O\left(t^{\frac{1}{3}}\right), \qquad t \to \infty. \tag{30}$$

Using similar arguments, it is straightforward to show that

$$\left| \sum_{n=1}^{[t]} \frac{1}{n^{\frac{1}{2}+i(t-t^{\delta_3})}} \right|^2 = O\left(t^{\frac{1}{3}}\right), \qquad t \to \infty. \tag{31}$$

It turns out that the techniques of [10] can be directly applied to estimating sums involving the lhs of (31). In this way it can be shown that

$$\sum_{m_1=1}^{[t]} \sum_{m_2=1}^{[t]} \frac{1}{m_1^{\frac{1}{2}+i(t-t^{\delta_3})} m_2^{\frac{1}{2}-i(t-t^{\delta_3})}} = O\left(t^{\frac{1}{3}}(\ln t)^2\right), \qquad t \to \infty.$$

The sum in the rhs of (28), in comparison to the above sum contains the extra term $\frac{1}{\ln\left(\frac{m_2}{m_1} t^{1-\delta_3}\right)}$. Fortunately, this term satisfies the properties needed for the partial summation procedure. Actually, a slight modification of the partial summation technique used in [10] suggests the estimate (7).

We note that the second term of the rhs of (7) is identical with the estimate of (6). This is due to the fact that the estimation of S_4^{SD} involves the splitting of the relevant set of the summation of in three parts, and in one of these parts the summand has the form of the summand of (18).

3.3. Review of Techniques for Estimating Euler-Zagier Double Sums

We summarise some of the techniques used in [10] that will be needed in the estimation of the sum S_4^{SD}. In what follows we use the terminology of [10].

In [10] estimates of the Euler-Zagier double sums are obtained by employing techniques from [12,14]. Indeed, letting $s_j = \sigma_j + it_j$, with $0 \leq \sigma_j < 1$, $j = 1, 2$ and $|t_j| \sim c_j t$, for some positive constants c_j, estimates as $t \to \infty$ are derived for sums of the form

$$\sum_{1 \leq m < n} \frac{1}{m^{s_1}} \frac{1}{n^{s_2}}.$$

A special case of Theorem 1.1 in [10] yields

$$\sum_{1 \leq m < n} \frac{1}{m^{\frac{1}{2}+it}} \frac{1}{n^{\frac{1}{2}+it}} = O\left(t^{\frac{1}{3}}(\ln t)^2\right).$$

The above result, as well as the estimates of Theorem 1.1 therein, provide a 'sharp' generalisation for double sums of the classical result for the single sum of Theorem 5.12 in [13]. In this sense, the results of [10] improve significantly the analogous results of [15].

Here we are interested only on the part of the analysis of [10] concerning the sums of the form

$$S_1 = \sum_{1 \leq m < n < t} \frac{1}{m^{s_1}} \frac{1}{n^{s_2}},$$

and in particular for the case that $\sigma_1 = \sigma_2 = 1/2$. The above sum is estimated by splitting it into two classes of sums:

$$S_1 = \sum_{j=1}^{\left[\frac{\ln 2t}{\ln 2}\right]} \left[T\left(2^{-j}t\right) + U\left(2^{-j}t\right) \right],$$

where
$$T(M) = \sum_{M<m<n\leq 2M} \frac{1}{m^{s_1}} \frac{1}{n^{s_2}} \quad \text{and} \quad U(M) = \sum_{1\leq m\leq M} \frac{1}{m^{s_1}} \sum_{M<n\leq 2M} \frac{1}{n^{s_2}}.$$

The estimation of the sum $U(M)$ is straightforward since it can be reduced to the estimation of a single sum; this is given by employing the Theorem 5.12 of [13], namely

$$\sum_{1\leq m\leq M} \frac{1}{m^{1/2+it}} = O\left(t^{\frac{1}{6}} \ln t\right) \quad \text{and} \quad \sum_{M<m\leq 2M} \frac{1}{n^{1/2+it}} = O\left(t^{\frac{1}{6}}\right).$$

Thus,
$$\sum_{j=1}^{\left[\frac{\ln 2t}{\ln 2}\right]} U\left(2^{-j}t\right) = O\left(t^{\frac{1}{3}}(\ln t)^2\right), \quad t \to \infty.$$

The estimation of the sum $T(M)$ is more elaborate and is based on Lemmas 3.1–3.5, therein. Lemma 3.1 appears in [14], Lemmas 3.2–3.4 appear in [12], and Lemma 3.5 is a variation of the classical and widely used partial summation technique (see for example [13,14]).

Since the latter Lemma plays an important role in our analysis we find it helpful to restate it:

Lemma 1 (Lemma 3.5 in [10]). *Let M and N be positive integers such that $M < N$, $f(x,y)$ be a C^2-function on $[M,N] \times [M,N]$, $g(m,n)$ be an arithmetical function on the same domain, and*

$$G(x,y) = \sum\sum_{x<m\leq n\leq y} g(m,n).$$

Suppose that
$$|G(x,y)| \leq G, \quad |f_x(x,y)| \leq \kappa_1, \quad |f_y(x,y)| \leq \kappa_2, \quad |f_{xy}(x,y)| \leq \kappa_3,$$

for some positive constants G, κ_1, κ_2, κ_3, and for any $M \leq x, y \leq N$. Then, we have

$$\left|\sum\sum_{M<m\leq n\leq N} f(m,n)g(m,n)\right| \leq G\left[f(M,N) + (\kappa_1 + \kappa_2)(N-M) + \kappa_3(N-M)^2\right]. \quad (32)$$

In order to estimate the sum $T(M)$ the set of summation is divided in three regions corresponding to the following three cases:

(a) $M < t^{\frac{1}{3}}$
(b) $t^{\frac{1}{3}} < M < t^{\frac{2}{3}}$
(c) $t^{\frac{2}{3}} < M < t$.

For case (a) it is sufficient to observe that $T(M) = O(M)$, which yields

$$\sum_{j=\left[\frac{2}{3}\frac{\ln t}{\ln 2}\right]}^{\left[\frac{\ln 2t}{\ln 2}\right]} T\left(2^{-j}t\right) = O\left(t^{\frac{1}{3}} \ln t\right), \quad t \to \infty.$$

For case (c) Lemma 3.4 is used to treat the oscillatory part of the sum, i.e., it is shown that $\sum_{M<m<n\leq 2M} \frac{1}{m^{it_1}} \frac{1}{n^{it_2}} = O(t \ln t)$. Then, applying partial summation using Lemma 3.5 which shows that $T(M) = O\left(\frac{1}{M} t \ln t\right)$, it follows that

$$\sum_{j=1}^{\left[\frac{1}{3}\frac{\ln t}{\ln 2}\right]} T\left(2^{-j}t\right) = O\left(t^{\frac{1}{3}}(\ln t)^2\right), \quad t \to \infty.$$

Case (b) is conceptually the same with case (c) but involves more technicalities: Lemmas 3.1–3.3 are used to treat the oscillatory part of the sum, i.e., to show that $\sum_{M<m<n\leq 2M} \frac{1}{m^{it_1}} \frac{1}{n^{it_2}} = O\left(Mt^{\frac{1}{3}}(\ln t)^{\frac{1}{2}}\right)$. Then, applying partial summation by using Lemma 3.5 in order to obtain that $T(M) = O\left(t^{\frac{1}{3}}(\ln t)^{\frac{1}{2}}\right)$, it follows that

$$\sum_{j=\left[\frac{1}{3}\frac{\ln t}{\ln 2}\right]}^{\left[\frac{2}{3}\frac{\ln t}{\ln 2}\right]} T\left(2^{-j}t\right) = O\left(t^{\frac{1}{3}}(\ln t)^{\frac{3}{2}}\right), \qquad t \to \infty.$$

Summarising the above results it follows that

$$S_1 = \sum_{1\leq m<n<t} \frac{1}{m^{s_1}} \frac{1}{n^{s_2}} = O\left(t^{\frac{1}{3}}(\ln t)^2\right), \qquad t \to \infty. \tag{33}$$

Remark 2. *From the above analysis it follows that it is much more complicated to estimate the sums of the form $T(M)$ in comparison to those of the form $U(M)$; the latter ones correspond to set of summations which can be decoupled, whereas the set of summations corresponding to the former ones cannot be decoupled. The latter observation necessitates the use of the Lemma 3.5 in [10], which is related with the partial summation technique. The sets of summation in our work are more complicated, requiring more general forms of that Lemma. In this connection, in Section 4.1 we state a general form of Abel's summation formula for double sums; its proof is presented in [11].*

4. Derivation of the Estimate (7)

In what follows we let $\delta_3 = \delta$, and throughout the rest of the paper we have $s = 1/2 + it$.

In order to derive (7), we split the sum S_4^{SD} in three different sums $S^{(i)}$, $S^{(ii)}$, $S^{(iii)}$, in accordance with the analysis of Section 4.2, below. We analyse these three sums in Section 4.3:

- The estimation of $S^{(iii)}$ is straightforward.
- The estimation of $S^{(ii)}$ involves the use of partial summation technique described in [14].
- The estimation of $S^{(i)}$ is based on the the analysis of [10], but some of the parts of the sum require the use of a partial summation which is more general than the one derived in [10]. In this direction, we will use a lemma on Abel's summation in two dimensions stated below.

4.1. A Lemma for Partial Summation in Two Dimensions

Lemma 3.5 of [10] is a two-dimensional form of the so-called Abel's summation formula, see Appendix C. The difficulty appearing in the proof of Lemma 3.5 of [10] is due to the fact that the set of summation is given by an expression which does not allow the double sum to be decoupled in two single sums. In the separable case the simple form of the Abel's summation formula for double sums is given in Lemma A1, and is straightforward to derive it by applying twice (A1). However, for our analysis we need to generalise Lemma 3.5 of [10]. This generalisation is given by Lemma 2 below, whose proof is given in [11]. It is this form of Abel's summation formula for double sums that is needed for the analysis of the sums (3b) and (4b) appearing in the sum $S_2^{(i)}$, which is analysed in Section 4.3.

Lemma 2. *Let $\theta(\cdot)$ be a linear function and $\phi(\cdot)$ be its inverse. Particular such functions are $\theta(x) = t^{\delta-1}x$, $\phi(x) = t^{1-\delta}x$. Let $M < N$ be positive integers and $f(x,y)$ be a C^2-function on $[\theta(M), \theta(N)] \times [M, N]$, $g(m,n)$ be an arithmetical function on the same domain, and*

$$G(x,y) = \sum\sum_{x<\phi(m)<n\leq y} g(m,n).$$

Suppose that

$$|G(x,y)| \leq G, \quad |f_x(x,y)| \leq \kappa_1, \quad |f_y(x,y)| \leq \kappa_2, \quad |f_{xy}(x,y)| \leq \kappa_3,$$

for some positive constants G, κ_1, κ_2, κ_3, and for any $(x,y) \in [\theta(M), \theta(N)] \times [M, N]$. Then, we have

$$\left| \sum\sum_{M < \phi(m) < n \leq N} f(m,n) g(m,n) \right|$$
$$\leq G \Big[f(\theta(M), N) + \kappa_1 (\theta(N) - \theta(M)) + \kappa_2(N - M) + \kappa_3 (\theta(N) - \theta(M))(N - M) \Big]. \quad (34)$$

Remark 3. *The above formulation is adapted to the subregion (4b) of the splitting presented in Section 4.3, but the choice of function θ (respectively ϕ) is wider than the particular forms chosen in Lemma 2. The result and the proof is the same if we substitute in the above formulation $\sum\sum_{x < \phi(m) < n \leq y}$ with $\sum\sum_{x < n < \phi(m) \leq y}$, thus this result can be adapted to the subregion (3b).*

4.2. The Different forms of E_4^{SD}

Equation (19) with $\delta_3 = \delta$ becomes

$$S_4^{SD}(t, \delta) = \sum_{m_1=1}^{[t]} \sum_{m_2=1}^{[t]} \frac{1}{m_1^s m_2^s} E_4^{SD}(t, \delta). \quad (35)$$

Let $\alpha = \frac{m_2}{m_1} t^{1-\delta}$. It is shown in [11] that the term E_4^{SD} is given by the expression below.

(i) $|\alpha - 1| > c > 0$, with the constant c independent of t:

$$E_4^{SD} \sim -\sqrt{\frac{2}{\pi}} e^{\frac{i\pi}{4}} e^{-it^\delta} \frac{1}{t^{\frac{\delta}{2}}} \frac{1}{\ln \alpha} \alpha^{it^\delta}, \quad t \to \infty. \quad (36)$$

The condition $|\alpha - 1| > c > 0$ yields that $\frac{1}{\ln \alpha} = O\left(\frac{1}{c}\right)$ is bounded.

(ii) $1 \gg |\alpha - 1| \geq \Gamma t^{-\frac{\delta}{2}}$, for some constant $\Gamma > 0$ independent of t:

$$E_4^{SD} \sim -\sqrt{\frac{2}{\pi}} e^{\frac{i\pi}{4}} e^{-it^\delta} \frac{1}{t^{\frac{\delta}{2}}} \frac{1}{\ln \alpha} \alpha^{it^\delta}, \quad t \to \infty. \quad (37)$$

The condition $1 \gg |\alpha - 1| \geq \Gamma t^{-\frac{\delta}{2}}$, yields that $\frac{1}{\ln \alpha} = O(1) t^{\frac{\delta}{2}}$, thus the term $\frac{1}{t^{\frac{\delta}{2}}} \frac{1}{\ln \alpha}$ is bounded. Furthermore, this condition restricts the set of summation in a sufficiently small set, so that we will use a different technique to estimate the relevant sum compared to the case (i).

(iii) $\alpha = 1 + \gamma t^{-\Delta}$, $\Delta \geq \frac{\delta}{2}$, $\gamma \in \mathbb{R}$, for any constant γ independent of t:

The leading contribution is equal to the pole contribution multiplied by some constant c depending only on γ, with $|c(\gamma)| < 2$ and $c(0^\pm) = \pm 1$.

If $\Delta = \frac{\delta}{2}$, then, using the analysis in [11], we obtain

$$E_4^{SD} \sim c(\gamma) e^{-i \frac{m_2}{m_1} t} e^{-i \frac{\gamma^2}{2}} + \frac{1}{\sqrt{2\pi}} e^{\frac{i\pi}{4}} e^{-it^\delta} \frac{1}{t^{\frac{\delta}{2}}}, \quad t \to \infty. \quad (38)$$

If $\Delta > \frac{\delta}{2}$, then, similarly to the above derivation and using the Plemelj's formulae we obtain

$$E_4^{SD} \sim \text{sign}(\gamma) \, e^{-i \frac{m_2}{m_1} t} e^{-i \frac{\gamma^2}{2} t^{\delta - 2\Delta}} \left(1 + O\left(t^{\frac{\delta}{2} - \Delta}\right)\right), \quad t \to \infty. \quad (39)$$

The sets of summation corresponding to cases (ii) and (iii) are bounded by the two red lines in Figure 1.

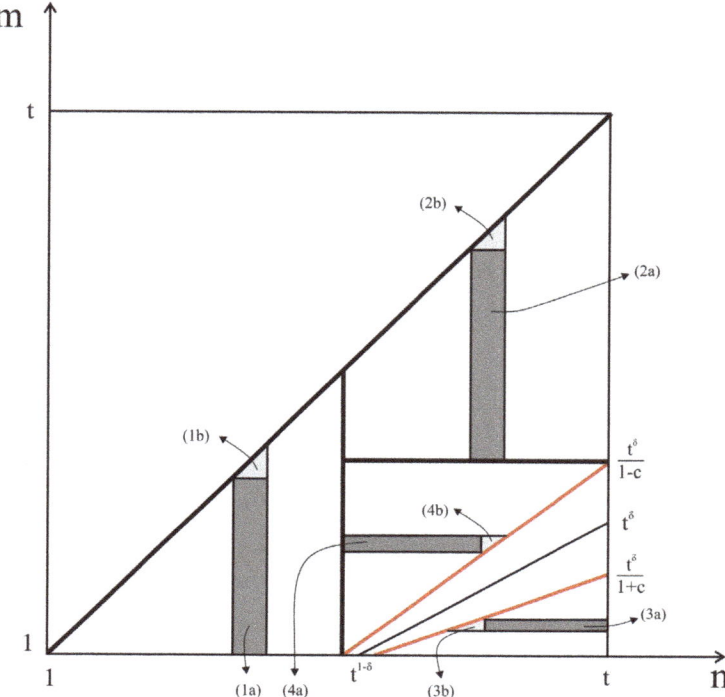

Figure 1. The subregions of the set of summation.

Remark 4. *In Equation* (39) *one observes that for* $\Delta > \frac{\delta}{2}$ *the dominant contribution of* E_4^{SD} *is given by "plus" or "minus" half of the pole contribution (depending on the sign of* γ*), where the pole contribution is given in* (18). *Noting that* $\gamma < 0 \Leftrightarrow m_1, m_2 \in M_4$, *with* M_4 *defined in* (18), *one observes that the dominant contribution of the expression* $2S_4^P + S_4^{SD}$ *appearing in* (17), *is equal to* S_4^P *for all* $\gamma \in \mathbb{R}$ *and* $\Delta > \frac{\delta}{2}$.

The analysis of the case $\Delta = \frac{\delta}{2}$ *is included in (ii). Equation* (38) *elucidates the mechanism responsible for switching the contribution of* E_4^{SD} *from the form* (37) *to the form* (39).

4.3. The Estimation of the Three Parts of S_4^{SD}

In what follows we will estimate the three sums corresponding to the above three forms of the S_4^{SD}.

4.3.1. The Estimate of Case (iii)

Recalling Remark 4, we treat the sum associated with the case (iii) similarly to the derivation of (6), but for a smaller set of summation; hence, it yields the estimate $O\left(t^{\frac{\delta}{2}} \ln t\right)$.

4.3.2. The Estimate of Case (ii)

We treat the sum associated with the case (ii) similarly to Lemma 5.1 in [8], but for a smaller set of summation. Hence, it also yields the estimate $O\left(t^{\frac{\delta}{2}} \ln t\right)$.

It is sufficient to estimate the following sum

$$S^{(ii)} = \frac{1}{t^{\frac{\delta}{2}}} \sum_{m_2=1}^{t^\delta} \sum_{m_1=c_1 m_2 t^{1-\delta}}^{c_2 m_2 t^{1-\delta}} \frac{1}{m_1^{s+it^\delta}} \frac{1}{m_2^{s-it^\delta}} \frac{1}{\ln\left(\frac{m_2}{m_1} t^{1-\delta}\right)}, \qquad (40)$$

where c_1 and c_2 are two positive constants with $c_2 \leq 2c_1$.

Remark 5. *The constraint $c_2 \leq 2c_1$ is satisfied by taking a sufficiently small positive constant c in case (i). Indeed, if $c < \frac{1}{3}$ then the condition $|\alpha - 1| < \frac{1}{3}$, yields*

$$\frac{3}{4} m_2 t^{1-\delta} < m_1 < \frac{3}{2} m_2 t^{1-\delta}.$$

Thus, the condition $1 \gg |\alpha - 1| \geq \Gamma t^{-\frac{\delta}{2}}$ gives rise to two sums of the form (40) with $\frac{3}{4} < c_1 < c_2 < \frac{3}{2}$, thus $c_2 \leq 2c_1$. In particular, we obtain $\frac{3}{4} < c_1 < c_2 < 1$ for the first sum and $1 < c_1 < c_2 < \frac{3}{2}$ for the second sum.

Recalling that $\frac{1}{t^{\frac{\delta}{2}}} \frac{1}{\ln\left(\frac{m_2}{m_1} t^{1-\delta}\right)} = O(1)$, we will first estimate the sum

$$S_A = \sum_{m_2=1}^{t^\delta} \sum_{m_1=c_1 m_2 t^{1-\delta}}^{c_2 m_2 t^{1-\delta}} \frac{1}{m_1^s} \frac{1}{m_2^{\bar{s}}},$$

where c_1 and c_2 are two positive constants with $c_2 \leq 2c_1$. Thus, by using partial summation we will estimate the sum $S^{(ii)}$.

Observing that m_2 takes relatively "small" values in the set of summation of S_A, we use the following inequality without losing crucial information:

$$|S_A| < \sum_{m_2=1}^{[t^\delta]} \frac{1}{m_2^{1/2}} \left| \sum_{m_1=c_1 m_2 t^{1-\delta}}^{c_2 m_2 t^{1-\delta}} \frac{1}{m_1^s} \right|.$$

Then, we estimate the m_1-sum using Theorem 5.9 of [13], namely

$$\sum_{a < n \leq b \leq 2a} n^{it} = O\left(t^{\frac{1}{2}}\right) + O\left(at^{-\frac{1}{2}}\right).$$

Following the partial summation technique appearing in the proof of Theorem 5.12 of [13] and using the fact that $a > c_1 m_2 t^{1-\delta}$, we obtain

$$\sum_{m_1=c_1 m_2 t^{1-\delta}}^{c_2 m_2 t^{1-\delta}} \frac{1}{m_1^s} = O\left(t^{\frac{1}{2}} t^{-\frac{1}{2}(1-\delta)} m_2^{-\frac{1}{2}}\right), \qquad t \to \infty.$$

Thus,

$$S_A = \sum_{m_2=1}^{[t^\delta]} \frac{1}{m_2} O\left(t^{\frac{\delta}{2}}\right) = O\left(t^{\frac{\delta}{2}} \ln t\right), \qquad t \to \infty. \qquad (41)$$

Using the estimate (41), the monotonicity properties of the term $\frac{1}{\ln\left(\frac{m_2}{m_1} t^{1-\delta}\right)}$ appearing in (40), and the fact that $\frac{1}{t^{\frac{\delta}{2}}} \frac{1}{\ln\left(\frac{m_2}{m_1} t^{1-\delta}\right)} = O(1)$, the partial summation technique, as described in [14] and the Appendix B of [8], yields

$$S^{(ii)} = O\left(t^{\frac{\delta}{2}} \ln t\right), \qquad t \to \infty.$$

Remark 6. *The above approaches do not fully exploit the smallness of the set of summation, thus we expect that the above estimates can be sharpened (we recall that the sets of summation corresponding to cases (ii) and (iii) are bounded by the two red lines in Figure 1). In order to exploit the smallness of the set of summation one could follow the techniques presented in [8], which make use of the results of [6]. However, the estimates provided here for $S^{(ii)}$ and $S^{(iii)}$ are sufficient for the purpose of this paper, since they are the same as (and not weaker than) (6).*

4.3.3. The Estimate of Case (i)

In order to estimate this sum we will use techniques similar to the ones used in [10] for the estimation of the double zeta function, but with two main differences: first, we will split the set of summation in more regions, and second, for some of these regions, we will use Lemma 2 needed for the partial summation for double sums, which is a more general form of the Lemma 3.5 in [10].

The term involved in the partial summation is now of the form

$$f(m_1, m_2) = \frac{1}{m_1^{1/2}} \frac{1}{m_2^{1/2}} \frac{1}{\ln\left(\frac{m_2}{m_1} t^{1-\delta}\right)},$$

instead of the term

$$\tilde{f}(m_1, m_2) = \frac{1}{m_1^\sigma} \frac{1}{m_2^\sigma}, \qquad 0 < \sigma < 1,$$

appearing in [10]. However, f shares the same properties with $\tilde{f}$ needed for the application of the partial summation technique, provided that the quantity $\frac{m_2}{m_1} t^{1-\delta}$ is not arbitrarily close to 1; this is ensured by the condition $|\alpha - 1| > c > 0$, with the constant c independent of t. Furthermore, $\{\kappa_j\}_1^3$ remain the same as in the [10], with the exception of the occasional appearance of a logarithmic term, due to $\frac{1}{\ln\left(\frac{m_2}{m_1} t^{1-\delta}\right)}$. However, this term does not affect the relevant estimates; in fact it is slightly helpful since now $\{\kappa_j\}_1^3$ are divided by $\ln t$.

The term involving the exponential sum now has the form

$$g(m_1, m_2) = \frac{1}{m_1^{i(t-t^\delta)}} \frac{1}{m_2^{-i(t-t^\delta)}},$$

instead of the corresponding term of [10]

$$\tilde{g}(m_1, m_2) = \frac{1}{m_1^{it_1}} \frac{1}{m_2^{it_2}}, \qquad t_1 \asymp t_2.$$

Remark 7. *The formalism $t_1 \asymp t_2$ in [10] means that $t_1 = O(t_2)$ and $t_2 = O(t_1)$. This is compatible with the selection of our t_1 and t_2. Furthermore, the fact that $t - t^\delta \sim t$ implies that all relevant estimates are the same. It should be noted that the condition $|t_1 + t_2| \gg 1$ in [10] is imposed because the double Riemann zeta function considered in [10] gives rise to sums which for $t_1 = -t_2$ are not defined. In our work we deal only with sums where the set of summation is $[1, t] \times [1, t]$. In analogy, the single Riemann zeta function $\zeta(s)$ and the relevant single sum are not defined at $s = 1$, however, the sum $\sum_{m=1}^{t} \frac{1}{m}$ can be estimated to be $O(\ln t)$.*

In summary, the analysis in [10] can be applied to the sums appearing to our work.

The Estimate of $S_1^{(i)}$

First, we treat the part of the sum where $m_2 > m_1$: in this case, it is sufficient to estimate the sum

$$S_1^{(i)} = \frac{1}{t^{\frac{\delta}{2}}} \sum_{m_2=1}^{t} \sum_{m_1=1}^{m_2} \frac{1}{m_1^{s+it^\delta}} \frac{1}{m_2^{\bar{s}-it^\delta}} \frac{1}{\ln\left(\frac{m_2}{m_1} t^{1-\delta}\right)}. \qquad (42)$$

First observe that since $t > m_2 > m_1 > 1$, we obtain that $t^{1-\delta} < \frac{m_2}{m_1} t^{1-\delta} < t^{2-\delta}$, thus the quantity $\frac{1}{\ln\left(\frac{m_2}{m_1} t^{1-\delta}\right)}$ is bounded both from above and below by $\frac{1}{\ln t}$ multiplied by some positive constant that depends only on δ. For our purpose it is sufficient to work for $0 < \delta < 1/2$, thus we obtain that $\frac{1}{2}\frac{1}{\ln t} < \frac{1}{\ln\left(\frac{m_2}{m_1} t^{1-\delta}\right)} < \frac{2}{\ln t}$.

The sum $S_1^{(i)}$ is estimated through the analysis provided in [10] with $m_1 = m$, $m_2 = n$. Indeed, we follow the methodology presented in Section 3.3 above by splitting the set of summation in subsets corresponding to the forms $U(M)$ and $T(M)$. For the former case we follow step-by-step the analysis of [10]. Then, we incorporate the contribution of the term $\frac{1}{\ln\left(\frac{n}{m} t^{1-\delta}\right)}$ through the analysis used for the sum $S^{(ii)}$ above. This involves the use of partial summation as described in [14] and the Appendix B of [8]. For the latter case we use $f(m,n) = \frac{1}{m^{1/2}} \frac{1}{n^{1/2}} \frac{1}{\ln\left(\frac{n}{m} t^{1-\delta}\right)}$ and apply the analysis appearing in [10] and described in Section 3.3 above, with the only difference occurring in the application of Lemma 3.5 of [10], where now the bounds will be multiplied by the term $\frac{1}{\ln t}$. Thus, we obtain the estimate

$$\sum_{m_2=1}^{t} \sum_{m_1=1}^{m_2} \frac{1}{m_1^{s+it^\delta}} \frac{1}{m_2^{s-it^\delta}} \frac{1}{\ln\left(\frac{m_2}{m_1} t^{1-\delta}\right)} = O\left(t^{\frac{1}{3}} \ln t\right), \quad t \to \infty, \quad (43)$$

which yields

$$S_1^{(i)} = O\left(t^{\frac{1}{3} - \frac{\delta}{2}} \ln t\right), \quad t \to \infty. \quad (44)$$

Furthermore, the part of the sum where $m_2 = m_1$ becomes the following single sum

$$O\left(\frac{1}{t^{\frac{\delta}{2}}}\right) \sum_{m=1}^{t} \frac{1}{m} \frac{1}{\ln t^{1-\delta}} = O\left(t^{-\frac{\delta}{2}}\right).$$

The Estimate of $S_2^{(i)}$

Next, we will treat the sum in the domain $m_2 < m_1$; this sum presents more difficulties. We first have to split this domain in several subdomains. In each of these subdomains we use the techniques of [10]. Furthermore, in some cases the partial summation requires more general forms of the Lemma involving the partial summation in double sums; for this reason we employ Lemma 2.

Our splitting is motivated by the following observation in the analogue approach of [10]: if the double sum can be decoupled, namely if the domain of summation (in two dimensions) is a rectangle, then estimating this double sum can be reduced to estimating two single sums; this occurs for sums of the form $U(M)$ appearing in [10] (see Section 3.3 above). If the double sum cannot be decoupled, namely if the domain of summation (in two dimensions) is bounded by at least one curve which depends on both the horizontal and the vertical coordinates, then a more sophisticated approach is required, both for the treatment of the double exponential sum and the partial summation technique; this occurs for sums of the form $T(M)$ appearing in [10] (see Section 3.3 above).

Let us use the notation $m_1 = n$, $m_2 = m$. Furthermore, let us denote by D_r the remaining set of summation, i.e., for $(n,m) \in [1,t] \times [1,t]$, let $(n,m) \in D_r$ iff

$$m < n \quad \text{and} \quad n < mt^{1-\delta}(1-c),$$
$$\text{or} \quad (45)$$
$$m < n \quad \text{and} \quad mt^{1-\delta}(1+c) < n,$$

for some sufficiently small constant $c > 0$ (independent of t); these restrictions are induced by the condition $|\alpha - 1| > c > 0$, with $\alpha = \frac{m}{n} t^{1-\delta}$.

In D_r there are two types of regions that correspond to sums of the form $T(M)$ in [10]. The first type is bounded by the line $n = m$ and the second type is bounded by the lines $n = (1 \pm c)mt^{1-\delta}$, for some sufficiently small $c > 0$. For both cases the treatment of the exponential sum follows the arguments presented in [10] (which first appeared in [12,14]). Considering the partial summation, the Lemma 3.5 in [10] is sufficient for the treatment of the first case, however, Lemma 2 is required for the second case.

Thus, in order to estimate the sum

$$S_2^{(i)} = \frac{1}{t^{\frac{\delta}{2}}} \sum\sum_{(n,m)\in D_r} \frac{1}{n^{s+it^\delta}} \frac{1}{m^{\bar{s}-it^\delta}} \frac{1}{\ln\left(\frac{m}{n}t^{1-\delta}\right)}, \qquad (46)$$

we split D_r into four different regions, where, in addition to conditions (45), the following conditions hold:

1. For $1 < M < \frac{1}{2}t^{1-\delta}$, two subregions:

    (1a) $m < M$ and $M < n < 2M$.
    (1b) $M < m < n < 2M$.

2. For $\frac{1}{2}t^{1-\delta} < M < \frac{1}{2}t$, two subregions:

    (2a) $t^\delta < m < M$ and $M < n < 2M$.
    (2b) $M < m < n < 2M$.

3. For $t^{1-\delta} < M < t$, two subregions:

    (3a) $M < mt^{1-\delta} < 2M$ and $t^{1-\delta} < n < M$.
    (3b) $M < n < mt^{1-\delta} < 2M$.

4. For $t^{1-\delta} < M < t$, two subregions:

    (4a) $M < mt^{1-\delta} < 2M$ and $2M < n < t$.
    (4b) $M < mt^{1-\delta} < n < 2M$.

The first subregion of each of the above regions, namely (1a), (2a), (3a) and (4a), are of rectangular shape, see Figure 1. The corresponding sums are treated similarly to the $U(M)$ sums in [10]. It is straightforward to modify the relevant techniques therein according to the discussion of the case $S_1^{(i)}$ and obtain the essential bound of the rhs of (44). In fact, observing that in these regions $\frac{1}{\ln\left(\frac{m}{n}t^{1-\delta}\right)} = O(1)$, one obtains the estimate $O\left(t^{\frac{1}{3}-\frac{\delta}{2}}(\ln t)^2\right)$, $t \to \infty$.

The subregions (1b) and (2b) are of triangular shape, see Figure 1, thus the corresponding sums are treated similarly to the $T(M)$ sums in [10]. The sums in these regions are treated in [10], via Lemma 3.5. It is straightforward to modify accordingly this approach and obtain the same bound as the rhs of (44).

The subregions (3b) and (4b) are also of triangular shape, see Figure 1. In order to analyse these sums we have to modify the approach of estimating the sums $T(M)$ in [10]. It is straightforward to modify the analysis of the oscillatory part of the sum, namely the part which uses the Lemmas 3.1–3.3 therein. For the analogue of the partial summation we need to use Lemma 2 instead of Lemma 3.5 in [10]. Then, we obtain the essential bound of the rhs of (44). In fact, observing that in these regions $\frac{1}{\ln\left(\frac{m}{n}t^{1-\delta}\right)} = O(1)$, one obtains the estimate $O\left(t^{\frac{1}{3}-\frac{\delta}{2}}(\ln t)^2\right)$, $t \to \infty$.

4.4. An Alternative Way to Estimate $S_2^{(i)}$

It is possible to estimate $S_2^{(i)}$ using a different and less technical approach. Let us use the notation $D_2 = \{(n,m) \in [1,t] \times [1,t], \ m < n\}$. Then, we rewrite

$$S_2^{(i)} = \frac{1}{t^{\frac{\delta}{2}}} \sum\sum_{(n,m) \in D_r} \frac{1}{n^{s+it^\delta}} \frac{1}{m^{\bar{s}-it^\delta}} \frac{1}{\ln\left(\frac{m}{n}t^{1-\delta}\right)}, \tag{47}$$

as

$$S_2^{(i)} = \frac{1}{t^{\frac{\delta}{2}}} \sum\sum_{(n,m) \in D_2} \frac{1}{n^{s+it^\delta}} \frac{1}{m^{\bar{s}-it^\delta}} F(n,m) - \frac{1}{t^{\frac{\delta}{2}}} \sum\sum_{(n,m) \in D_2 \setminus D_r} \frac{1}{n^{s+it^\delta}} \frac{1}{m^{\bar{s}-it^\delta}} H(n,m), \tag{48}$$

where the functions F and H are C^2 and are defined as follows

$$F(x,y) = \begin{cases} \frac{1}{\ln\left(\frac{y}{x}t^{1-\delta}\right)}, & (x,y) \in D_r, \\ H(x,y), & (x,y) \in D_2 \setminus D_r, \end{cases} \tag{49}$$

with D_r defined by the conditions (45). Furthermore, the function $P(x,y) : D_2 \to \mathbb{R}$, which is defined by $P(x,y) := \frac{F(x,y)}{x^{1/2}y^{1/2}}$, belongs to C^2 and has the following properties:

$$\begin{aligned} P(x,y) &= O\left(\frac{1}{x^{1/2}y^{1/2}}\right), & P_x(x,y) &= O\left(\frac{1}{x^{3/2}y^{1/2}}\right), \\ P_y(x,y) &= O\left(\frac{1}{x^{1/2}y^{3/2}}\right), & P_{xy}(x,y) &= O\left(\frac{1}{x^{3/2}y^{3/2}}\right). \end{aligned} \tag{50}$$

From (49) the set where we have to assure that $P(x,y) \in C^2(D_2)$ is given by the constraint $\frac{y}{x}t^{1-\delta} = 1 \pm c$, for some sufficiently small positive constant c. Hence, it is sufficient to determine the function $H(x,y) = d\left(\frac{y}{x}t^{1-\delta}\right)$, with the following six properties:

$$\begin{aligned} d(1 \pm c) &= \frac{1}{\ln(1 \pm c)}, & d'(1 \pm c) &= \frac{1}{[\ln(1 \pm c)]^2 (1 \pm c)}, \\ & & d''(1 \pm c) &= \frac{2 + \ln(1 \pm c)}{[\ln(1 \pm c)]^3 (1 \pm c)^2}, \end{aligned} \tag{51}$$

for some fixed and sufficiently small $c > 0$.

Furthermore, the conditions (50) are satisfied if the functions $d(r), d'(r), d''(r)$ are bounded in the interval $r \in (1-c, 1+c)$.

Thus, it is sufficient for $d(r)$ to be a fifth order polynomial which satisfies the conditions (51).

Now, the sum $S_2^{(i)}$ has the appropriate form, so it can be analysed through the following two arguments:

- The first term of the rhs of (48) can be analysed in the same way as the sum $S_1^{(i)}$, with the only difference that in the current analysis we find it more convenient to employ the simpler version of partial summation technique described in Lemma A1.
- The second term of the rhs of (48) can be embedded in the analysis of the sum $S^{(ii)}$. In this case it is more convenient to employ a combination of partial summation techniques, as they are described in Lemmas 2 and A1, respectively.

The resulting estimate remains invariant.

5. Conclusions

The main result of this paper is the derivation of the Volterra-type linear integral Equation (8). In order to derive this equation starting from (1.6) of [9] it is necessary to:

(i)   replace I_4 by $\tilde{I}_4$ defined by (22).
(ii) replace S_4^P by (6).
(iii) replace S_4^{SD} by (7).

The derivation of (i) is based on replacing in the definition of I_4, the term $|\zeta(s)|^2$ by its leading asymptotics. The proof that the error term is indeed small is presented in [11].

The derivation of (ii) is given in Section 3.

The derivation of (iii) is given in Section 4 under the assumption that the function E_4^{SD} appearing in S_4^{SD} is given by Equations (36)–(38); the latter proof is given in [11].

The importance of the derivation of (8) is a consequence of the following considerations: taking into account that the variable ρ appearing in the Γ functions in the integral of (8) satisfies $\rho \geq t^{\delta_2}$ and $t - \rho \geq t^{1/3}$, it follows that these Γ functions can be simplified as $t \to \infty$. Indeed, Equations (4.4), (5.7) and (5.8) on [9] yield

$$\frac{\Gamma(it - i\rho)}{\Gamma(1/2 + it)}\Gamma(1/2 + i\rho) = \sqrt{\frac{2\pi}{t}} e^{-\frac{i\pi}{4}} \frac{1}{\left(1 - \frac{\rho}{t}\right)^{1/2}} e^{it\left[\left(1 - \frac{\rho}{t}\right)\ln\left(1 - \frac{\rho}{t}\right) + \frac{\rho}{t}\ln\left(\frac{\rho}{t}\right)\right]}$$

$$\times \left[1 + O(t^{-\delta_{23}})\right], \quad t \to \infty,$$

with $\delta_{23} = \min\{\delta_2, \delta_3\}$.

Hence, for the specific choice of $\delta_2 = \delta_3 = \frac{1}{3}$, replacing in Equation (8) the combination of the Gamma functions by the rhs of the above equation, we find

$$\left|\zeta\left(\frac{1}{2} + it\right)\right|^2 = \sqrt{\frac{2}{\pi}} \int_{t^{1/3}}^{t - t^{1/3}} \Re\left\{\frac{e^{-\frac{i\pi}{4}}}{(t-\rho)^{1/2}} e^{it\left[\left(1 - \frac{\rho}{t}\right)\ln\left(1 - \frac{\rho}{t}\right) + \frac{\rho}{t}\ln\left(\frac{\rho}{t}\right)\right]}\right\} \left|\zeta\left(\frac{1}{2} + i\rho\right)\right|^2 d\rho \quad (52)$$

$$\times \left[1 + O\left(t^{-1/3}\right)\right] + O\left(t^{\frac{1}{6}}(\ln t)^2\right), \quad t \to \infty,$$

It is straightforward to show that the ansatz $|\zeta(1/2 + it)|^2 = O\left(t^{1/6}(\ln t)^2\right)$ provides a solution of (52). The rigorous proof that the above ansatz provides the unique solution of the linear Volterra integral equation will be presented in [11]. This estimate implies that $\zeta(1/2 + it) = O\left(t^{1/12}\ln t\right)$, which is a dramatic improvement of the current best estimate of the large t behaviour of $\zeta(1/2 + it)$.

Author Contributions: Both authors K.K. and A.S.F. were involved in conceptualization, methodology, formal analysis, investigation, writing and reviewing of the present work.

Funding: This research was funded by EPSRC, grant number 79707.

Acknowledgments: Both authors are supported by the EPSRC, UK. This is part of a large program of study initiated by one of the authors (ASF); in this effort Jonatan Lenells is an indispensable collaborator.

Conflicts of Interest: The authors declare no conflict of interest.

Appendix A. Asymptotics of $|\zeta(s)|^2$

Equation (1.3) of [6] for $s = \frac{1}{2} + it$ and $\eta = 2\pi t$, yields

$$\zeta(s) = \sum_{n=1}^{[t]} \frac{1}{n^s} + O\left(t^{-\frac{1}{2}}\right).$$

Multiplying the above equation with its complex conjugate and using the classical estimate from Theorem 5.12 of [13], which states that

$$\zeta(s) = O\left(t^{\frac{1}{6}} \ln t\right),$$

we obtain

$$|\zeta(s)|^2 = \sum_{m_1=1}^{[t]} \sum_{m_2=1}^{[t]} \frac{1}{m_1^s m_2^s} + O\left(t^{-\frac{1}{3}} \ln t\right).$$

Appendix B. Derivation of (24)

Using the constraint $\frac{m_2}{m_1} < t^{\delta_3 - 1} \ll 1$, we rewrite S_3 as follows:

$$\begin{aligned}
S_3 &= \sum\sum_{(m_1,m_2)\in M_3} \frac{1}{m_2^s(m_1+m_2)^s} = \sum\sum_{(m_1,m_2)\in M_3} \frac{1}{m_2^s m_1^s \left(1+\frac{m_2}{m_1}\right)^s} \\
&= \sum\sum_{(m_1,m_2)\in M_3} \frac{1}{m_2^s m_1^s} \frac{1}{\left(1+\frac{m_2}{m_1}\right)^{1/2}} e^{-it \ln\left(1+\frac{m_2}{m_1}\right)} \\
&= \sum\sum_{(m_1,m_2)\in M_3} \frac{1}{m_2^s m_1^s} \left(1+O\left(t^{\delta_3-1}\right)\right) e^{-it\left[\frac{m_2}{m_1}+O(t^{2\delta_3-2})\right]} \\
&= \sum\sum_{(m_1,m_2)\in M_3} \frac{1}{m_2^s m_1^s} e^{-it\frac{m_2}{m_1}} \left(1+O\left(t^{2\delta_3-1}\right)\right) = S_4^P\left[1+O\left(t^{2\delta_3-1}\right)\right], \quad t\to\infty.
\end{aligned}$$

Appendix C. Abel's Summation

The so-called Abel's summation formula for a single sum is given as follows: let $(a_n)_{n=0}^{\infty}$ be a sequence of real or complex numbers. Define the partial sum function

$$A(y) = \sum_{0\leq n\leq y} a_n, \quad \text{for any real number } y.$$

Fix a real number x, and let ρ be a continuously differentiable function on $[0, x]$. Then,

$$\sum_{0\leq n\leq x} a_n \rho(n) = A(x)\rho(x) - \int_0^x A(u)\rho'(u)\,du. \tag{A1}$$

The simple form of the Abel's summation formula for double sums is given in Lemma A1 below, and is straightforward to derive it by applying twice (A1).

Lemma A1. *Let A, B, C, D be positive integers such that $A < B$, $C < D$ and $f(x,y)$ be a C^2-function on $[A, B] \times [C, D]$, $g(m, n)$ be an arithmetical function on the same domain, and*

$$G(x,y) = \sum_{m=A}^{x} \sum_{n=C}^{y} g(m,n).$$

Suppose that

$$|G(x,y)| \leq G, \quad |f_x(x,y)| \leq \kappa_1, \quad |f_y(x,y)| \leq \kappa_2, \quad |f_{xy}(x,y)| \leq \kappa_3,$$

for some positive constants G, κ_1, κ_2, κ_3, and for any $(x,y) \in [A, B] \times [C, D]$.

Then, we have

$$\left|\sum_{m=A}^{B}\sum_{n=C}^{D} f(m,n)g(m,n)\right|$$
$$\leq G\Big[f(B,D)+\kappa_1(B-A)+\kappa_2(D-C)+\kappa_3(B-A)(D-C)\Big]. \tag{A2}$$

References

1. Weyl, H. Ueber die Gleichverteilung von Zahlen mod. Eins. *Math. Ann.* **1916**, *77*, 313–352. [CrossRef]
2. Hardy, G.H.; Littlewood, J.E. Contributions to the theory of the Riemann zeta-function and the theory of the distribution of primes. *Acta Math.* **1916**, *41*, 119–196. [CrossRef]
3. Vinogradov, I.M. A new method of estimation of trigonometrical sums. *Mat. Sbornik* **1936**, *43*, 175–188.
4. Bourgain, J. Decoupling, exponential sums and the Riemann zeta function. *J. Am. Math. Soc.* **2017**, *30*, 205–224. [CrossRef]
5. Siegel, C.L. Über Riemanns Nachlaßzur analytischen Zahlentheorie. In *Quellen Studien zur Geschichte der Mathematik Astronomie und Physik Abteilung B: Studien 2: 4580, 1932*; Springer: Berlin, Germany, 1966; Volume 1.
6. Fokas, A.S.; Lenells, J. On the asymptotics to all orders of the Riemann Zeta function and of a two-parameter generalization of the Riemann Zeta function. *arXiv* **2012**, to appear, arXiv:1201.2633.
7. Fernandez, A.; Fokas, A.S. Asymptotics to all orders of the Hurwitz zeta function. *J. Math. Anal. Appl.* **2018**, *465*, 423–458. [CrossRef]
8. Kalimeris, K.; Fokas, A.S. Explicit asymptotics for certain single and double exponential sums. *Proc. R. Soc. Edinb. A Math.* **2017**, 1–26. [CrossRef]
9. Fokas, A.S. A novel approach to the Lindelöf hypothesis. *arXiv* **2019**, to appear, arXiv:1708.06607.
10. Kiuci, I.; Tanigawa, Y. Bounds for double zeta functions. *Annali della Scuola Normale Superiore di Pisa-Classe di Scienze* **2006**, *5*, 445–464.
11. Fokas, A.S.; Kalimeris, K.; Lenells, J. A Novel Integral Equation for the Riemann Zeta Function and a Significant Improvement of the Current Large t Estimate. **2019**, in preparation.
12. Krätzel, E. *Lattice Points*; Springer: Berlin, Germany, 1989.
13. Titchmarsh, E.C. *The Theory of the Riemann Zeta-Function*, 2nd ed.; Oxford University Press: Oxford, UK, 1987.
14. Titchmarsh, E.C. On Epstein's Zeta-function. *Proc. Lond. Math. Soc.* **1934**, *2*, 485–500. [CrossRef]
15. Ishikawa, H.; Matsumoto, K. On the estimation of the order of Euler-Zagier multiple zeta-functions. *Ill. J. Math.* **2003**, *47*, 1151–1166. [CrossRef]

© 2019 by the authors. Licensee MDPI, Basel, Switzerland. This article is an open access article distributed under the terms and conditions of the Creative Commons Attribution (CC BY) license (http://creativecommons.org/licenses/by/4.0/).

Article

The Application of Generalized Quasi-Hadamard Products of Certain Subclasses of Analytic Functions with Negative and Missing Coefficients

En Ao * and Shuhai Li

School of Mathematics and Statistics, Chifeng University, Chifeng 02400, Inner Mongolia, China
* Correspondence: cfxyaoen@sina.com

Received: 13 June 2019; Accepted: 9 July 2019; Published: 12 July 2019

Abstract: In this paper, we introduce a new generalized differential operator using a new generalized quasi-Hadamard product, and certain new classes of analytic functions using subordination. We obtain certain results concerning the closure properties of the generalized quasi-Hadamard products and the generalized differential operators for this new subclasses of analytic functions with negative and missing coefficients.

Keywords: analytic functions; quasi-Hadamard; differential operator; closure property

MSC: 30C45

1. Introduction and Motivation

Let $\mathcal{A}(a,k)$ denote the class of functions of the form

$$f(z) = az + \sum_{n=k}^{\infty} a_n z^n \quad (a > 0, k \in N\backslash\{1\} = \{2, 3, \cdots\}), \tag{1}$$

which are analytic in the unit disk $\mathbb{U} = \{z : |z| < 1\}$. Obviously, $\mathcal{A}(1,2) = \mathcal{A}$ denotes the class of functions $f(z)$ normalized by $f(0) = f'(0) - 1 = 0$ which are analytic in $\mathbb{U}$.

Set $\mathcal{T}(a,k)$ be the class of functions of the form

$$f(z) = az - \sum_{n=k}^{\infty} |a_n| z^n \quad (a > 0, k \in N\backslash\{1\} = \{2, 3, \cdots\}).$$

which are analytic in $\mathbb{U}$. It is easy to see that $\mathcal{T}(a,k) \subset \mathcal{A}(a,k)$.

Let $f_i(z) \in \mathcal{T}(a,k) (i = 1,2)$ be given by

$$f_i(z) = az - \sum_{n=k}^{\infty} |a_{n,i}| z^n (i = 1, 2), \tag{2}$$

then the quasi-Hadamard product (or convolution) $f_1 * f_2$ is defined by

$$(f_1 * f_2)(a; z) = a^2 z - \sum_{n=k}^{\infty} |a_{n,1}||a_{n,2}| z^n.$$

For any real numbers p and q, we define the generalized quasi-Hadamard product $f_1 \triangle f_2$ by

$$(f_1 \triangle f_2)(p, q; a, z) = a^2 z - \sum_{n=k}^{\infty} |a_{n,1}|^p |a_{n,2}|^q z^n = (f_2 \triangle f_1)(p, q; a, z). \tag{3}$$

Clearly, for $p = q = 1$, $(f_1 \triangle f_2)(1,1;a,z)$ reduces to the above quasi-Hadamard product $(f_1 * f_2)(a;z)$; for $a = 1$, $(f_1 \triangle f_2)(p,q;1,z)$ reduces to the generalized Hadamard product $(f_1 \triangle f_2)(p,q;z)$ defined by Jae Ho Choi and Yong Chan Kim [1]; and for $p = q = 1, a = 1$, $(f_1 \triangle f_2)(1,1;1,z)$ reduces to the quasi-Hadamard product $(f_1 * f_2)(z)$. For $a = 1, p,q \in N\backslash\{1\}$, $(f_1 \triangle f_2)(p,q;1,z)$ reduces to the quasi-Hadamard product $(\underbrace{f_1 * \cdots * f_1}_{p} * \underbrace{f_2 * \cdots * f_2}_{q})(z)$ (see [2], also see [3,4]).

In 1975, Schild and Silverman [5] studied closure properties of the quasi-Hadamard product $(f_1 * f_2)(z)$ for a starlike function of order α and convex function of order α with negative coefficients in $\mathcal{A}$. In 1983, Owa [2] obtained closure properties of quasi-Hadamard product $(f_1 * f_2 * \cdots * f_m)(z)$ and $(f_1 * f_2 * \cdots * f_m * g_1 * g_2 \cdots * g_l)(z)$ for the same function classes in $\mathcal{A}$. Later Kumar [4] improved some results in 1987. In 1992, Srivastava and Owa [6] studied closure properties of quasi-Hadamard product $(f_1 * f_2 * \cdots * f_m)(z)$ for p-valent starlike function of order α and p-valent convex function of order α class with negative coefficients in $\mathcal{A}$. In 1996, Jae Ho Choi and Yong Chan Kim [1] introduced the generalized Hadamard product $(f_1 \triangle f_2)(p,q;z)$, and obtained the closure properties of $(f_1 \triangle f_2)(p,q;z)$ for a starlike function of order α and convex function of order α with negative coefficients in $\mathcal{A}$. Since then, a lot of authors considered and studied closure properties and characteristics of the quasi-Hadamard product $(f * g)(z)$, $(f_1 * f_2 * \cdots * f_m)(z)$ or $(f_1 * f_2 * \cdots * f_m * g_1 * g_2 * \cdots * g_l)(z)$ for some classes of normalized analytic functions and normalized meromorphic analytic functions, see, for example, [7–15].

Although the closure properties of Hadamard product or quasi-Hadamard product have already been studied in $\mathcal{A}$, our focus is to introduce generalized quasi-Hadamard product, generalized differential operators, and generalized function classes on non-normalized analytic functions, and to discuss the closure properties on generalized analytic function classes.

Now by using the generalized quasi-Hadamard product $(f_1 \triangle f_2)(p,q;a,z)$, we introduce the following differential operator $D^m (m \in N)$ as follows:

$$D^0(f_1 \triangle f_2) = (f_1 \triangle f_2),$$

$$D^1(f_1 \triangle f_2) = D(f_1 \triangle f_2) = z(f_1 \triangle f_2)',$$

$$D^m(f_1 \triangle f_2) = D(D^{m-1}(f_1 \triangle f_2)).$$

We define the generalized differential operator D_μ^m ($\mu \geq 0$) as follows:

$$D_\mu^m(f_1 \triangle f_2) = (1-\mu)D^m(f_1 \triangle f_2) + \mu D^{m+1}(f_1 \triangle f_2).$$

If $f_1 \triangle f_2$ is given by (3), then we can obtain that

$$D^m(f_1 \triangle f_2)(p,q;a,z) = a^2 z - \sum_{n=k}^{\infty} n^m |a_{n,1}|^p |a_{n,2}|^q z^n$$

and

$$D_\mu^m(f_1 \triangle f_2)(p,q;a,z) = a^2 z - \sum_{n=k}^{\infty} [1 + (n-1)\mu] n^m |a_{n,1}|^p |a_{n,2}|^q z^n.$$

Clearly, $D_0^m(f_1 \triangle f_2)(p,q;a,z) = D^m(f_1 \triangle f_2), D_0^0(f_1 \triangle f_2)(p,q;a,z) = (f_1 \triangle f_2)(p,q;a,z)$. For $a = p = q = 1, f_1(z) = z - \sum_{n=k}^{\infty} |a_n| z^n, f_2(z) = \frac{z-z^2-z^k}{1-z}, D^m(f_1 \triangle f_2)(1,1;1,z)$ becomes Sălăgean operator (see [16]). Also, by specializing the parameters μ, p, q, we obtain the following new operators:

$$D_\mu^m(f_1 \triangle f_2)(1,1;a,z) =: D_\mu^m(f_1 * f_2)(a;z) = a^2 z - \sum_{n=k}^{\infty} [1 + (n-1)\mu] n^m |a_{n,1}||a_{n,2}|z^n$$

and
$$D_0^m(f_1 \triangle f_2)(1,1;a,z) =: D^m(f_1 * f_2)(a;z) = a^2 z - \sum_{n=k}^{\infty} n^m |a_{n,1}||a_{n,2}|z^n.$$

For two analytic functions f and g, the function f is subordinate to g in $\mathbb{U}$ (see [17]), written as follows
$$f(z) \prec g(z), z \in \mathbb{U},$$
if there exists an analytic function ω, with $\omega(0) = 0$ and $|\omega(z)| < 1$ such that
$$f(z) = g(\omega(z)).$$

In particular, if the function g is univalent in $\mathbb{U}$, then $f(z) \prec g(z)$ is equivalent to $f(0) = g(0)$ and $f(\mathbb{U}) \subset g(\mathbb{U})$.

We define two generalization classes satisfying the following subordination condition.

Definition 1. $\lambda \geq 0, a > 0, A, B \in R, |A| \leq 1, |B| \leq 1, A \neq B$. A function $f(z) \in \mathcal{A}(a,k)$ is in the class $\mathcal{Q}_\lambda(a,k,A,B)$ if and only if
$$(1-\lambda)\frac{f(z)}{z} + \lambda f'(z) \prec \frac{a(1+Az)}{1+Bz}.$$

For suitable choices λ, a, k, A, B, the class $\mathcal{Q}_\lambda(a,k,A,B)$ reduces the following subclasses.

(1) $\mathcal{Q}_\lambda(a,k,1-2\beta,-1) =: \mathcal{Q}_\lambda(a,k,\beta) = \{f(z) \in \mathcal{A}(a,k) : (1-\lambda)\frac{f(z)}{z} + \lambda f'(z) \prec \frac{a[1+(1-2\beta)z]}{1-z}, \beta < 1\}$. Obviously, $\mathcal{Q}_\lambda(1,2,\beta) =: \mathcal{Q}_\lambda(\beta)$ (see [18]);

(2) $\mathcal{Q}_\lambda(1,2,A,B) =: \mathcal{Q}_\lambda(A,B) = \{f(z) \in \mathcal{A} : (1-\lambda)\frac{f(z)}{z} + \lambda f'(z) \prec \frac{1+Az}{1+Bz}\}$;

(3) $\mathcal{Q}_0(a,k,A,B) =: \mathcal{R}(a,k,A,B) = \{f(z) \in \mathcal{A}(a,k) : \frac{f(z)}{z} \prec \frac{a(1+Az)}{1+Bz}\}$;

(4) $\mathcal{Q}_1(a,k,A,B) =: \mathcal{H}(a,k,A,B) = \{f(z) \in \mathcal{A}(a,k) : f'(z) \prec \frac{a(1+Az)}{1+Bz}\}$. Obviously, $\mathcal{H}(1,k,A,B) =: \mathcal{P}_k(A,B) = \{f(z) \in \mathcal{A}(1,k) : f'(z) \prec \frac{1+Az}{1+Bz}, -1 \leq B < A \leq 1\}$ (see [19]).

Definition 2. Let $\lambda \geq 0, a > 0, A, B \in R, |A| \leq 1, |B| \leq 1, A \neq B$. A function $f(z) \in \mathcal{A}(a,k)$ is in the class $\mathcal{J}_\lambda(a,k,A,B)$ if and only if
$$f'(z) + \lambda z f''(z) \prec \frac{a(1+Az)}{1+Bz}.$$

Clearly, we have the following equivalence:
$$f(z) \in \mathcal{J}_\lambda(a,k,A,B) \iff zf'(z) \in \mathcal{Q}_\lambda(a,k,A,B). \tag{4}$$

Let
$$\mathcal{T}\mathcal{Q}_\lambda(a,k,A,B) = \mathcal{T}(a,k) \bigcap \mathcal{Q}_\lambda(a,k,A,B),$$
$$\mathcal{T}\mathcal{J}_\lambda(a,k,A,B) = \mathcal{T}(a,k) \bigcap \mathcal{J}_\lambda(a,k,A,B).$$

Our object of this paper is to the closure properties of the generalized quasi-Hadamard products, the generalized differential operators for the above generalized classes $\mathcal{T}\mathcal{Q}_\lambda(a,k,A,B)$ and $\mathcal{T}\mathcal{J}_\lambda(a,k,A,B)$. Our results are new in this direction and they give birth to many corollaries.

2. Preliminary Results

Due to derive our main result, we need to talk about the following lemmas.

Lemma 1. $\lambda \geq 0, a > 0, A, B \in R, |A| \leq 1, |B| \leq 1, A \neq B$. If the function $f(z) = az + \sum_{n=k}^{\infty} a_n z^n \in \mathcal{A}(a,k)$ satisfies

$$\sum_{n=k}^{\infty}[1+(n-1)\lambda](1+|B|)|a_n| \leq a|A-B|, \tag{5}$$

then $f(z) \in \mathcal{Q}_\lambda(a,k,A,B)$.

Proof. We assume that the inequality (5) holds true. According to Definition 1, the function $f(z) \in \mathcal{Q}_\lambda(a,k,A,B)$ if and only if there exists an analytic function $\omega(z), \omega(0) = 0, |\omega(z)| < 1 (z \in \mathbb{U})$ such that

$$F(z) = \frac{a(1+A\omega(z))}{1+B\omega(z)} \quad (z \in \mathbb{U}),$$

where

$$F(z) = (1-\lambda)\frac{f(z)}{z} + \lambda f'(z),$$

or equivalently

$$\left|\frac{F(z) - a}{aA - BF(z)}\right| < 1 \ (z \in \mathbb{U}), \tag{6}$$

it suffices to show that

$$|F(z) - a| - |aA - BF(z)| < 0.$$

Therefore, if we let $z \in \partial\mathbb{U} = \{z : z \text{ is complex number and } |z| = 1\}$, we find from (6) that

$$|F(z) - a| - |aA - BF(z)|$$
$$= \left|\sum_{n=k}^{\infty}[1+(n-1)\lambda]a_n z^n\right| - \left|a(A-B) - \sum_{n=k}^{\infty} a(A-B) - \sum_{n=k}^{\infty}[1+(n-1)\lambda]Ba_n z^n\right|$$
$$\leq \sum_{n=k}^{\infty}[1+(n-1)\lambda]|a_n||z|^n - a|A-B| + \sum_{n=k}^{\infty}[1+(n-1)\lambda]|B||a_n||z|^n$$
$$\leq \sum_{n=k}^{\infty}[1+(n-1)\lambda](1+|B|)|a_n| - a|A-B| \leq 0.$$

Hence, by the maximum modulus theorem, we have $f(z) \in \mathcal{Q}_\lambda(a,k,A,B)$. Thus we complete the proof of Lemma 1. □

Lemma 2. Let $\lambda \geq 0, a > 0$, and the function $f(z) = az - \sum_{n=k}^{\infty} |a_n| z^n \in \mathcal{T}(a,k)$.

(1) If $-1 \leq B < A \leq 1, B \leq 0$, then $f(z) \in \mathcal{TQ}_\lambda(a,k,A,B)$ if and only if

$$\sum_{n=k}^{\infty}[1+(n-1)\lambda](1-B)|a_n| \leq a(A-B). \tag{7}$$

(2) If $-1 \leq A < B \leq 1, B \geq 0$, then $f(z) \in \mathcal{TQ}_\lambda(a,k,A,B)$ if and only if

$$\sum_{n=k}^{\infty}[1+(n-1)\lambda](1+B)|a_n| \leq a(B-A).$$

The result is sharp for the function $f(z)$ given by

$$f(z) = az - \frac{a|A-B|}{[1+(k-1)\lambda](1+|B|)} z^k \quad (k \in \mathbb{N}\setminus\{1\} = \{2, 3, \cdots\}).$$

Proof. Since $\mathcal{TQ}_\lambda(a,k,A,B) \subset \mathcal{Q}_\lambda(a,k,A,B)$, according to Lemma 1 we only need to prove the 'only if' part of this Lemma.

Now let us prove the necessity of case (1).

Let $f(z) \in \mathcal{TQ}_\lambda(a,k,A,B), -1 \leq B < A \leq 1, B \leq 0$. Then it satisfies (6) or equivalently

$$\left| \frac{\sum_{n=k}^\infty [1+(n-1)\lambda]|a_n|z^{n-1}}{a(A-B)+\sum_{n=k}^\infty [1+(n-1)\lambda]B|a_n|z^{n-1}} \right| < 1, z \in \mathbb{U}.$$

Since $|\Re(z)| \leq |z|, z \in \mathbb{U}$, we have

$$\Re\left\{ \frac{\sum_{n=k}^\infty [1+(n-1)\lambda]|a_n|z^{n-1}}{a(A-B)+\sum_{n=k}^\infty [1+(n-1)\lambda]B|a_n|z^{n-1}} \right\} < 1, z \in \mathbb{U}. \quad (8)$$

Choose values of z on the real axis so that $(1-\lambda)\frac{f(z)}{z} + \lambda f'(z)$ is real. Upon clearing the denominator in (8) and letting $z \to 1^-$ through real values, we obtain (7).

Similar to the above proof for case (1), we can prove that case (2) is true. Thus we complete the proof of Lemma 2. □

Using arguments similar to those in the proof of Lemmas 1 and 2, we can prove the following Lemmas 3 and 4.

Lemma 3. *Let $\lambda \geq 0, a > 0, A, B \in R, |A| \leq 1, |B| \leq 1, A \neq B$. If the function $f(z) = az + \sum_{n=k}^\infty a_n z^n \in \mathcal{A}(a,k)$ satisfies*

$$\sum_{n=k}^\infty [1+(n-1)\lambda](1+|B|)|n|a_n| \leq a|A-B|,$$

then $f(z) \in \mathcal{J}_\lambda(a,k,A,B)$.

Lemma 4. *Let $\lambda \geq 0, a > 0$, and the function $f(z) = az - \sum_{n=k}^\infty |a_n|z^n \in \mathcal{T}(a,k)$.*

(1) *If $-1 \leq B < A \leq 1, B \leq 0$, then $f(z) \in \mathcal{TJ}_\lambda(a,k,A,B)$ if and only if*

$$\sum_{n=k}^\infty [1+(n-1)\lambda](1-B)n|a_n| \leq a(A-B).$$

(2) *If $-1 \leq A < B \leq 1, B \geq 0$, then $f(z) \in \mathcal{TJ}_\lambda(a,k,A,B)$ if and only if*

$$\sum_{n=k}^\infty [1+(n-1)\lambda](1+B)n|a_n| \leq a(B-A).$$

The result is sharp for the function $f(z)$ given by

$$f(z) = az - \frac{a|A-B|}{k[1+(k-1)\lambda](1+|B|)}z^k \quad (k \in N\backslash\{1\} = \{2,3,\cdots\}).$$

3. Main Results

Theorem 1. *$p > 1$ and the functions $f_i(z)(i=1,2)$ defined by (2) belong to $\mathcal{TQ}_\lambda(a,k,A,B)$.*

(1) *If $-1 \leq B < A \leq 1, B \leq 0, 0 < a < \frac{[1+(k-1)\mu](A-B)k^m}{1-B}$, then $\frac{1}{a}D_\mu^m(f_1 \triangle f_2)(\frac{1}{p}, \frac{p-1}{p}; a, z) \in \mathcal{TQ}_\lambda(a,k,A,\widehat{B})$, where*

$$\frac{a(1-B)A - [1+(k-1)\mu](A-B)k^m}{a(1-B) - [1+(k-1)]\mu(A-B)k^m} \leq \widehat{B} < \min\{A,0\}.$$

(2) If $-1 \leq A < B \leq 1, B \geq 0, 0 < a < \frac{[1+(k-1)\mu](B-A)k^m}{1+B}$, then $\frac{1}{a}D^m_\mu(f_1 \triangle f_2)(\frac{1}{p}, \frac{p-1}{p}; a, z) \in \mathcal{TQ}_\lambda(a, k, A, \widehat{B})$, where

$$\max\{A, 0\} < \widehat{B} \leq \frac{a(1+B)A + [1+(k-1)\mu](B-A)k^m}{a(1+B) - [1+(k-1)\mu](B-A)k^m}.$$

Proof. (1) Suppose that $-1 \leq B < A \leq 1, B \leq 0$. According to Lemma 2, we need to prove

$$\sum_{n=k}^\infty \frac{[1+(n-1)\lambda](1-\widehat{B})[1+(n-1)\mu]}{a(A-\widehat{B})} \frac{n^m}{a} |a_{n,1}|^{\frac{1}{p}} |a_{n,2}|^{\frac{p-1}{p}} \leq 1. \qquad (9)$$

Since $f_i(z) \in \mathcal{TQ}_\lambda(a, k, A, B)$, by Lemma 2 we have

$$\left(\sum_{n=k}^\infty \frac{[1+(n-1)\lambda](1-B)}{a(A-B)} |a_{n,1}|\right)^{\frac{1}{p}} \leq 1$$

and

$$\left(\sum_{n=k}^\infty \frac{[1+(n-1)\lambda](1-B)}{a(A-B)} |a_{n,2}|\right)^{\frac{p-1}{p}} \leq 1.$$

By the Hölder inequality we get

$$\sum_{n=k}^\infty \frac{[1+(n-1)\lambda](1-B)}{a(A-B)} |a_{n,1}|^{\frac{1}{p}} |a_{n,2}|^{\frac{p-1}{p}} \leq 1.$$

Hence the inequality (9) will be satisfied if

$$\frac{[1+(n-1)\mu](1-\widehat{B})n^m}{a(A-\widehat{B})} \leq \frac{1-B}{A-B} \quad (m, n \in N, n \geq k)$$

or if

$$[a(1-B) - [1+(n-1)\mu](A-B)n^m]\widehat{B} \leq a(1-B)A - [1+(n-1)\mu](A-B)n^m \ (m, n \in N, n \geq k). \quad (10)$$

Now define the functions $F_1(n)$ and $G_1(n)$ by

$$F_1(n) = a(1-B) - [1+(n-1)\mu](A-B)n^m$$

and

$$G_1(n) = a(1-B)A - [1+(n-1)\mu](A-B)n^m.$$

When $0 < a < \frac{[1+(n-1)\mu](A-B)k^m}{1-B}$, we obtain that $F_1(n)$ is a decreasing function of $n (n \in N, n \geq k)$ and $F_1(n) < F_1(k) < 0$. Thus the inequality (10) will be satisfied if

$$\widehat{B} \geq \frac{G_1(n)}{F_1(n)} = \frac{a(1-B)A - [1+(n-1)\mu](A-B)n^m}{a(1-B) - [1+(n-1)\mu](A-B)n^m} \quad (m, n \in N, n \geq k). \quad (11)$$

We see that the right hand side of (11) is a decreasing function of $n(n \in N, n \geq k)$. Therefore the inequality (10) is satisfied for all $n(n \in N, n \geq k)$ if

$$\widehat{B} \geq \frac{G_1(k)}{F_1(k)} = \frac{a(1-B)A - [1+(k-1)\mu](A-B)k^m}{a(1-B) - [1+(k-1)\mu](A-B)k^m} \quad (m \in N),$$

which evidently completes the proof of the case (1).

(2) Suppose that $-1 \leq A < B \leq 1, B \geq 0$. According to Lemma 2, we need to prove

$$\sum_{n=k}^{\infty} \frac{[1+(n-1)\lambda](1+\widehat{B})[1+(n-1)\mu]}{a(\widehat{B}-A)} \frac{n^m}{a} |a_{n,1}|^{\frac{1}{p}} |a_{n,2}|^{\frac{p-1}{p}} \leq 1. \tag{12}$$

Similar to case (1), the inequality (12) will be satisfied if

$$\frac{[1+(n-1)\mu](1+\widehat{B})n^m}{a(\widehat{B}-A)} \leq \frac{1+B}{B-A} \quad (m, n \in N, n \geq k)$$

or if

$$[a(1+B) - [1+(n-1)\mu](B-A)n^m]\widehat{B} \geq a(1+B)A + [1+(n-1)\mu](B-A)n^m \quad (m, n \in N, n \geq k). \tag{13}$$

Now define the functions $F_2(n)$ and $G_2(n)$ by

$$F_2(n) = a(1+B) - [1+(n-1)\mu](B-A)n^m$$

and

$$G_2(n) = a(1+B)A + [1+(n-1)\mu](B-A)n^m.$$

When $0 < a < \frac{[1+(n-1)\mu](B-A)k^m}{1+B}$, we obtain that $F_2(n)$ is a decreasing function of $n (n \in N, n \geq k)$ and $F_2(n) < F_2(k) < 0$. Thus the inequality (13) will be satisfied if

$$\widehat{B} \leq \frac{G_2(n)}{F_2(n)} = \frac{a(1+B)A + [1+(n-1)\mu](B-A)n^m}{a(1+B) - [1+(n-1)\mu](B-A)n^m} \quad (m, n \in N, n \geq k). \tag{14}$$

We see that the right hand side of (14) is an increasing function of $n(n \in N, n \geq k)$. Therefore the inequality (13) is satisfied for all $n(n \in N, n \geq k)$ if

$$\widehat{B} \leq \frac{G_2(k)}{F_2(k)} = \frac{a(1+B)A + [1+(n-1)\mu](B-A)k^m}{a(1+B) - [1+(n-1)\mu](B-A)k^m} \quad (m \in N),$$

which evidently completes the proof of the case (2). Thus we complete the proof of Theorem 1. □

Theorem 2. *Let $[1+(n-1)\mu]|A-B|n^m \leq [1+(n-1)\lambda](1+|B|)$. If the functions $f_i(z) (i=1,2)$ defined by (2) belong to $\mathcal{TQ}_\lambda(a,k,A,B)$, then $\frac{1}{a}D_\mu^m(f_1 * f_2)(a;z) \in \mathcal{TQ}_\lambda(a,k,A,B)$.*

Proof. Suppose that $-1 \leq B < A \leq 1, B \leq 0$. According to Lemma 2, we need to prove

$$\sum_{n=k}^{\infty} [1+(n-1)\lambda](1-B)[1+(n-1)\mu] \frac{n^m}{a} |a_{n,1}||a_{n,2}| \leq a(A-B). \tag{15}$$

Since $f_i(z) \in \mathcal{TQ}_\lambda(a,k,A,B)$ $(i=1,2)$, by using Lemma 2 we get

$$\sum_{n=k}^{\infty} [1+(n-1)\lambda](1-B)|a_{n,1}| \leq a(A-B) \tag{16}$$

and

$$\sum_{n=k}^{\infty} [1+(n-1)\lambda](1-B)|a_{n,2}| \leq a(A-B). \tag{17}$$

Therefore, by the Cauchy–Schwarz inequality, we obtain

$$\sum_{n=k}^{\infty}[1+(n-1)\lambda](1-B)\sqrt{|a_{n,1}||a_{n,2}|} \leq a(A-B). \tag{18}$$

This implies that we only need to show that

$$[1+(n-1)\lambda](1-B)[1+(n-1)\mu]\frac{n^m}{a}|a_{n,1}||a_{n,2}| \leq [1+(n-1)\lambda](1-B)\sqrt{|a_{n,1}||a_{n,2}|} \ (n \geq k)$$

or, equivalently, that

$$\sqrt{|a_{n,1}||a_{n,2}|} \leq \frac{a}{[1+(n-1)\mu]n^m} \ (n \geq k). \tag{19}$$

From (18), the inequality (19) is satisfied for all $n(n \in N, n \geq k)$ if

$$[1+(n-1)\mu](A-B)n^m \leq [1+(n-1)\lambda](1-B) \ (n \geq k).$$

Based on the given condition, we get (15).

Also applying Lemma 2 we can prove $\frac{1}{a}D_\mu^m(f_1 * f_2)(a;z) \in \mathcal{T}\mathcal{Q}_\lambda(a,k,A,B)$ for $-1 \leq A < B \leq 1$, $B \geq 0$. Thus we complete the proof of Theorem 2. □

Remark 1. (1) Setting $\mu = 0$ in Theorem 1, we can obtain the closure properties of $\frac{1}{a}\widetilde{D}^m(f_1 * f_2)(\frac{1}{p}, \frac{p-1}{p}; a, z)$ for $\mathcal{T}\mathcal{Q}_\lambda(a,k,A,B)$; (2) Setting $\mu = 0$ in Theorem 2, we can obtain the closure properties of $\frac{1}{a}\widetilde{D}_\mu^m(f_1 * f_2)(a;z)$ for $\mathcal{T}\mathcal{Q}_\lambda(a,k,A,B)$.

Example 1. Let $p > 1, -1 \leq B < A \leq 1, B \leq 0, 0 < a < \frac{A-B}{1-B}$. If the functions $f_i(z)(i = 1,2)$ defined by (2) belong to $\mathcal{T}\mathcal{Q}_\lambda(a,k,A,B)$, then $\frac{1}{a}(f_1 \triangle f_2)(\frac{1}{p}, \frac{p-1}{p}; a, z) \in \mathcal{T}\mathcal{Q}_\lambda(a,k,A,\widehat{B})$, where

$$\frac{a(1-B)A - (A-B)}{a(1-B) - (A-B)} \leq \widehat{B} < \min\{A, 0\}.$$

4. Corollaries and Consequences

On the one hand, by taking special values of parameters A, B, λ, a, k we easily obtain the following closure properties for some important subclasses in $\mathcal{A}(a,k)$.

Putting $A = 1 - 2\beta \ (0 \leq \beta < 1), B = -1$, we obtain the closure properties for the subclass

$$\mathcal{T}\mathcal{Q}_\lambda(a,k,\beta) = \mathcal{T}(a,k) \bigcap \mathcal{Q}_\lambda(a,k,\beta) = \{f(z) \in \mathcal{T}(a,k) : (1-\lambda)\frac{f(z)}{z} + \lambda f'(z) \prec \frac{a[1+(1-2\beta)z]}{1-z}\}.$$

Corollary 1. $p > 1, 0 < \beta < 1$ and the functions $f_i(z)(i = 1,2)$ defined by (2) belong to $\mathcal{T}\mathcal{Q}_\lambda(a,k,\beta)$. If $0 < a < [1+(k-1)\mu](1-\beta)k^m$, then $\frac{1}{a}D_\mu^m(f_1 \triangle f_2)(\frac{1}{p}, \frac{p-1}{p}; a, z) \in \mathcal{T}\mathcal{Q}_\lambda(a,k,1-2\beta,\widehat{B})$, where

$$\frac{a(1-\beta) - [1+(k-1)\mu](1-\beta)k^m}{a - [1+(k-1)\mu](1-\beta)k^m} \leq \widehat{B} < \min\{1-2\beta, 0\}.$$

Corollary 2. Let $0 < \beta < 1, [1+(n-1)\mu](1-\beta)n^m \leq [1+(n-1)\lambda]$. If the functions $f_i(z)(i = 1,2)$ defined by (2) belong to $\mathcal{T}\mathcal{Q}_\lambda(a,k,\beta)$, then $\frac{1}{a}D_\mu^m(f_1 * f_2)(a;z) \in \mathcal{T}\mathcal{Q}_\lambda(a,k,\beta)$. Putting $\lambda = 0$ and $\lambda = 1$, we obtain the closure properties for the subclasses

$$\mathcal{T}\mathcal{R}(a,k,A,B) = \mathcal{T}(a,k) \bigcap \mathcal{H}(a,k,A,B) = \{f(z) \in \mathcal{T}(a,k) : \frac{f(z)}{z} \prec \frac{a(1+Az)}{1+Bz}\}$$

and
$$\mathcal{TH}(a,k,A,B) = \mathcal{T}(a,k) \bigcap \mathcal{H}(a,k,A,B) = \{f(z) \in \mathcal{T}(a,k) : f'(z) \prec \frac{a(1+Az)}{1+Bz}\}.$$

Corollary 3. *Let $[1+(n-1)\mu]|A-B|n^m \leq (1+|B|)$. If the functions $f_i(z)(i=1,2)$ defined by (2) belong to $\mathcal{TR}(a,k,A,B)$, then $\frac{1}{a}D_\mu^m(f_1*f_2)(a;z) \in \mathcal{TR}(a,k,A,B)$.*

Corollary 4. *Let $[1+(n-1)\mu]|A-B|n^m \leq n(1+|B|)$. If the functions $f_i(z)(i=1,2)$ defined by (2) belong to $\mathcal{TH}(a,k,A,B)$, then $\frac{1}{a}D_\mu^m(f_1*f_2)(a;z) \in \mathcal{TH}(a,k,A,B)$. Putting $a=1,k=2$, we obtain the closure properties for the subclass*

$$\mathcal{TQ}_\lambda(A,B) = \mathcal{T}(1,2) \bigcap \mathcal{Q}_\lambda(A,B) = \{f(z) \in \mathcal{T}(1,2) : (1-\lambda)\frac{f(z)}{z} + \lambda f'(z) \prec \frac{a(1+Az)}{1+Bz}\}.$$

Corollary 5. *Let $p > 1$ and the functions $f_i(z)(i=1,2)$ defined by (2) belong to $\mathcal{TQ}_\lambda(A,B)$. If $-1 \leq B < A \leq 1, B \leq 0, (1+\mu)(A-B)2^m - (1-B) > 0$, then $D_\mu^m(f_1 \triangle f_2)(\frac{1}{p}, \frac{p-1}{p}; 1, z) \in \mathcal{TQ}_\lambda(A, \widehat{B})$, where*

$$\frac{a(1-B)A - (1+\mu)(A-B)2^m}{a(1-B) - (1+\mu)(A-B)2^m} \leq \widehat{B} < \min\{A, 0\}.$$

Corollary 6. *Let $[1+(n-1)\mu]|A-B|2^m \leq [1+(n-1)\lambda](1+|B|)$. If the functions $f_i(z)(i=1,2)$ defined by (2) belong to $\mathcal{TQ}_\lambda(A,B)$, then $\frac{1}{a}D_\mu^m(f_1*f_2)(a;z) \in \mathcal{TQ}_\lambda(A,B)$.*

Example 2. *Let $p > 1, 0 < \beta < 1$. If $f_i(z) = z - \sum_{n=2}^\infty |a_{n,i}|z^n \in \mathcal{TQ}_\lambda(1-2\beta, -1), i = 1, 2$, then $(f_1 \triangle f_2)(\frac{1}{p}, \frac{p-1}{p}; 1, z) \in \mathcal{TQ}_\lambda(1-2\beta, \widehat{B})$, where $-1 \leq \widehat{B} < \min\{1-2\beta, 0\}$.*

On the other hand, we can obtain the following closure properties for $\mathcal{TJ}(a,k,A,B)$ according to (4) and Lemma 4.

Corollary 7. *Let $p > 1$ and the functions $f_i(z)(i=1,2)$ defined by (2) belong to $\mathcal{TJ}_\lambda(a,k,A,B)$.*

(1) If $-1 \leq B < A \leq 1, B \leq 0, 0 < a < \frac{[1+(k-1)\mu](A-B)k^m}{1-B}$, then $\frac{1}{a}D_\mu^m(f_1 \triangle f_2)(\frac{1}{p}, \frac{p-1}{p}; a, z) \in \mathcal{TJ}_\lambda(a,k,A,\widehat{B})$, where

$$\frac{a(1-B)A - [1+(k-1)\mu](A-B)k^m}{a(1-B) - [1+(k-1)]\mu(A-B)k^m} \leq \widehat{B} < \min\{A, 0\}.$$

(2) If $-1 \leq A < B \leq 1, B \geq 0, 0 < a < \frac{[1+(k-1)\mu](B-A)k^m}{1+B}$, then $\frac{1}{a}D_\mu^m(f_1 \triangle f_2)(\frac{1}{p}, \frac{p-1}{p}; a, z) \in \mathcal{TJ}_\lambda(a,k,A,\widehat{B})$, where

$$\max\{A,0\} < \widehat{B} \leq \frac{a(1+B)A + [1+(k-1)\mu](B-A)k^m}{a(1+B) - [1+(k-1)\mu](B-A)k^m}.$$

Corollary 8. *Let $[1+(n-1)\mu]|A-B|n^{m-1} \leq [1+(n-1)\lambda](1+|B|)$. If the functions $f_i(z)(i=1,2)$ defined by (2) belong to $\mathcal{TJ}_\lambda(a,k,A,B)$, then $\frac{1}{a}D_\mu^m(f_1*f_2)(a;z) \in \mathcal{TJ}_\lambda(a,k,A,B)$.*

5. Conclusions

In this paper, we mainly study the closure properties of the generalized quasi-Hadamard products, the generalized differential operator and its related special operators for $\mathcal{TQ}_\lambda(a,k,A,B)$ and $\mathcal{TJ}_\lambda(a,k,A,B)$ of analytic functions with negative and missing coefficients. Also, we give two examples and six corollaries to illustrate our results obtained. In the future, we can consider to extend some classical analytic function classes (such as starlike, convex, close-to-convex) in $\mathcal{A}(a,k)$, and discuss the closure properties of the generalized quasi-Hadamard products.

Author Contributions: Conceptualization, E.A. and S.L.; methodology, E.A. and S.L.; software, E.A.; validation, E.A.; formal analysis, S.L.; investigation, E.A.; resources, S.L.; data curation, E.A.; writing—original draft preparation, E.A.; writing—review and editing, E.A.; visualization, S.L.; supervision, E.A.; project administration, E.A.; funding acquisition, E.A.

Funding: This study was supported by the Higher School Foundation of Inner Mongolia of China(Grant No.NJZY19211) and the National Natural Science Foundation of China(Grant No.11561001).

Conflicts of Interest: The authors declare no conflict of interest.

References

1. Choi, J.H.; Kim, Y.C.; Owa, S. Generalizations of Hadamard products of functions with negative coefficients. *J. Math. Anal. Appl.* **1996**, *199*, 495–501. [CrossRef]
2. Owa, S. On the Hadamard products of univalent functions. *Tamkang J. Math.* **1983**, *14*, 15–21.
3. Kumar, V. Hadamard product of certain starlike functions. *J. Math. Anal. Appl.* **1985**, *110*, 425–428. [CrossRef]
4. Kumar, V. Quasi-Hadamard product of certain univalent functions. *J. Math. Anal. Appl.* **1987**, *126*, 70–77. [CrossRef]
5. Schild, A.; Silveman, H. Convluctions of univalent functions with negative coefficients. *Ann. Univ. Mariea Curie-Sklodowsca Sect. A* **1975**, *29*, 99–107.
6. Srivasta, H.M.; Owa, S. *Current Topics in Analytic Function Theory*; World Scientific Publishing Co., Pte. Ltd.: Singapore, 1992; pp. 234–251.
7. Aouf, M.K. The quasi-Hadamard product of certain analytic functions. *Appl. Math. Lett.* **2008**, *21*, 1184–1187. [CrossRef]
8. Frasin, B.A. Quasi-Hadamard product of certain classes of uniformly analytic functions. *Gen. Math.* **2007**, *16*, 29–35.
9. Darwish, H.E. The quasi-Hadamard product of certain starlike and convex functions. *Appl. Math. Lett.* **2007**, *20*, 692–695. [CrossRef]
10. Frasin, B.A.; Aouf, M.K. Quasi-Hadamard product of certain merpmorphic p-valent analytic functions. *Appl. Math. Lett.* **2010**, *23*, 347–350. [CrossRef]
11. Goyal, S.P.; Goswami, P. Quasi-Hadamard product of a generalized class of analytic and univalent functions. *Eur. J. Pure Appl. Math.* **2010**, *6*, 1118–1123.
12. El-Ashwah, R.M.; Aouf, M.K.; Breaz, N. On quasi-Hadamard products of p-valent functions with negative coefficients define by using a differential operator. *Acta Univ. Apulensis* **2011**, *28*, 61–70.
13. Breaz, N.; El-Ashwah, R.M. Quasi-Hadamard product of some uniformly analytic and p-valent functions with negative coefficients. *Carpathian J. Math.* **2014**, *30*, 39–45.
14. EL-Ashwaha, R.M.; Drbuk, M.E. On certain properties for Hadamard product of uniformly univalent meromorphic functions with positive coefficients. *Theory Appl. Math. Comput. Sci.* **2016**, *6*,13–18.
15. Bulut, S. The Quasi-Hadamard Product of Certain Analytic and P-Val. Functions. *J. Math.* **2017**, *15*, 777–781.
16. Sălăgean, G.S. Subclasses of univalent functions. In complex analysis-fifth romanian-finnish seminar, part I (Bucharest, 1981). *Lect. Notes Math* **1983**, *1013*, 362–372.
17. Littlewood, J.E. *Lectures on the Theory of Functions*; Oxford University Press: Oxford, UK, 1944.
18. Ding, S.S.; Ling, Y.; Bao, G.J. Some properties of a class of analytic functions. *J. Math. Anal. Appl.* **1995**, *195*, 71–81. [CrossRef]
19. Kumar, V. On univalent functions with negative and missing coefficients. *J. Math. Res. Expos.* **1984**, *4*, 27–34.

© 2019 by the authors. Licensee MDPI, Basel, Switzerland. This article is an open access article distributed under the terms and conditions of the Creative Commons Attribution (CC BY) license (http://creativecommons.org/licenses/by/4.0/).

Article

Some Properties and Generating Functions of Generalized Harmonic Numbers

Giuseppe Dattoli [1], Silvia Licciardi [1,*], Elio Sabia [1] and Hari M. Srivastava [2,3]

1. ENEA—Frascati Research Center, Via Enrico Fermi 45, 00044 Frascati, Rome, Italy
2. Department of Mathematics and Statistics, University of Victoria, Victoria, BC V8W 3R4, Canada
3. Department of Medical Research, China Medical University Hospital, China Medical University, Taichung 40402, Taiwan
* Correspondence: silviakant@gmail.com or silvia.licciardi@enea.it; Tel.: +39-392-509-6741

Received: 5 May 2019; Accepted: 24 June 2019; Published: 28 Junbe

Abstract: In this paper, we introduce higher-order harmonic numbers and derive their relevant properties and generating functions by using an umbral-type method. We discuss the link with recent works on the subject, and show that the combinations of umbral and other techniques (such as the Laplace and other types of integral transforms) yield a very efficient tool to explore the properties of these numbers.

Keywords: umbral methods; harmonic numbers; special functions; integral representations; laplace and other integral transforms

MSC: 05A40; 44A99; 47B99; 11A99; 33C52; 33C65; 33C99; 33B10; 33B15; 44A05; 44A20; 44A10; 44A15

---

1. Introduction

The properties of many families of special numbers have been profitably studied by the use of methods tracing back to the umbral calculus [1]. Within this context, the definition of the associated polynomials naturally emerges as umbral Newton binomial convolutions (see "The Bernoulli Polynomials §4.2.2" in Reference [1]). The formalism is extremely powerful, and has allowed for the extension of the method to generalized forms of special numbers ([2,3]). The use of umbral techniques has been recently employed in the study of harmonic numbers, whose relationship to Bernoulli numbers has been pointed out in Reference [4]. In this paper, we will extend the use of umbral methods to the case of higher-order harmonic numbers.

In a number of previous papers ([5–7]), different problems concerning harmonic numbers and the relevant generating functions have been touched. The already mentioned use of the umbral-like formalism has allowed for the framing of the theory of harmonic numbers within an algebraic context. Some of the points raised in ([5–7]) have been reconsidered, made rigorous, and generalized by means of different technical frameworks in successive research ([8–15]).

The present investigation concerns the application of the method foreseen in ([5–7]) to generalized forms of harmonic numbers, such as (We use the notation $_m h_n$ instead of $H_n^{(m)}$ recommended in Reference [4] for continuity with previous papers, where it has been adopted to avoid confusion with higher-order Hermite polynomials):

$$_m h_n = \sum_{r=1}^{n} \frac{1}{r^m}, \quad n \geq 1, \qquad (1)$$
$$_m h_0 = 0,$$

namely, "higher-order harmonic numbers" satisfying the property:

$$_m h_{n+1} = {}_m h_n + \frac{1}{(n+1)^m}, \qquad (2)$$

whose associated series is provided by the limit $\lim_{n\to\infty} {}_m h_n$, $m > 1$ is, unlike the ordinary harmonic numbers ($m = 1$), not diverging.

It can be argued that for negative m values, the Harmonic numbers reduce to a finite sum of integers, expressible in terms of Bernoulli numbers, as discussed in the concluding part of the paper (Remark 3). In the following, we will derive a number of apparently new properties and the relevant consequences.

As an introductory example, we provide the following:

Example 1. *We consider the second-order harmonic numbers ($m = 2$) and write*

$$_2 h_n = \int_0^1 \frac{1-x^n}{x-1} \ln(x)\, dx, \quad \forall n \in \mathbb{N}, \qquad (3)$$

which is obtained after setting

$$\frac{1}{r^2} = \int_0^\infty e^{-sr} s\, ds \qquad (4)$$

by noting that

$$_2 h_n = \int_0^\infty \frac{e^{-s(n+1)} - e^{-s}}{e^{-s} - 1}\, s\, ds, \qquad (5)$$

and then by changing the variable of integration.

It is worth stressing that the integral representation allows for the extension of harmonic numbers to non-integer values of the index. The second-order harmonic numbers interpolates between the integer and real values of the index, as shown in the plot given in Figure 1, where it is pointed out that the asymptotic limit of the second-order harmonic numbers is $\frac{\pi^2}{6}$.

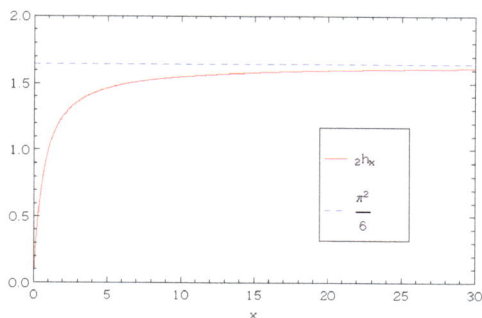

Figure 1. $_2 h_x$ vs x and $\lim_{x \to \infty} {}_2 h_x = \frac{\pi^2}{6}$.

The relevant extension to negative real indices will be considered later in the article.

Let us first consider the generating function associated with the second-order harmonic numbers, which can be cast in the form of an umbral exponential series, as follows.

Definition 1. *We introduce*

$$_{2h}e(t) := 1 + \sum_{n=1}^{\infty} \frac{t^n}{n!} (_2h_n) = e^{_2\hat{h}t} \zeta_0, \qquad (6)$$

where $_2\hat{h}$ is an umbral-like operator acting on the vaccum ζ_0, such that (see [10] for a complete treatment of the umbral method):

$$\begin{aligned} _2\hat{h}^n \zeta_0 &:= \zeta_n = {_2h_n}, \quad n > 0, \\ _2\hat{h}^0 \zeta_0 &= 1 \neq {_2h_0} = 0 \end{aligned} \qquad (7)$$

and

$$_2\hat{h}^\nu {_2\hat{h}^\mu} \zeta_0 = {_2\hat{h}^{\nu+\mu}} \zeta_0, \qquad \forall \nu, \mu \in \mathbb{R}. \qquad (8)$$

(The action of the operator $_2\hat{h}$ should be defined as explained in Reference [10]—namely, as the action of a shift operator on its vacuum, such as here ζ_0, in the following more rigorous way):

$$_2\hat{h} := e^{\partial_z},$$
$$_2\hat{h}^\mu \zeta_0 = e^{\mu \partial_z} \zeta_z \big|_{z=0} = \zeta_{z+\mu}\big|_{z=0} = \int_0^1 \frac{1 - x^{z+\mu}}{x-1} \ln(x)\, dx \bigg|_{z=0} = \int_0^1 \frac{1 - x^\mu}{x-1} \ln(x)\, dx.)$$

Proposition 1. $\forall m \in \mathbb{N}$,

$$_{2h}e(t, m) := \partial_t^m {_{2h}e(t)} = {_2h_m} + \sum_{n=1}^{\infty} \frac{t^n}{n!} (_2h_{n+m}). \qquad (9)$$

Proof. From Equations (6) and (7) it follows that, $\forall m \in \mathbb{N}$,

$$_{2h}e(t, m) = \partial_t^m {_{2h}e(t)} = \partial_t^m e^{_2\hat{h}t} \zeta_0 = {_2\hat{h}^m} e^{_2\hat{h}t} \zeta_0 = {_2h_m} + \sum_{n=1}^{\infty} \frac{t^n}{n!} (_2h_{n+m}).$$

□

Corollary 1. *Limiting ourselves to the first derivative only, it appears evident that the generating function (6) satisfies the identity*

$$\begin{cases} \partial_t {_{2h}e(t)} = {_{2h}e(t)} + f_2(t), \\ _{2h}e(0) = 1 \end{cases} \qquad (10)$$

where

$$f_2(t) = \sum_{n=1}^{\infty} \frac{t^n}{(n+1)(n+1)!} = \frac{1}{t} \int_0^t \frac{e^s - s - 1}{s}\, ds = -\frac{Ein(-t) + t}{t},$$
$$Ein(z) = \int_0^z \frac{1 - e^{-\zeta}}{\zeta}\, d\zeta. \qquad (11)$$

Observation 1. *The problem of specifying the generating function of second-order harmonic numbers is reduced to the solution of the first-order differential Equation (10). The solution writes*

$$_{2h}e(t) = e^t \left(1 + \sum_{n=1}^{\infty} \frac{1}{(n+1)^2} \left(1 - e^{-t} e_n(t)\right) \right), \qquad (12)$$

where

$$e_n(x) = \sum_{r=0}^{n} \frac{x^r}{r!} \qquad (13)$$

are the truncated exponential polynomials [16]. They belong to the family of Appél type polynomials [13] and are defined through the operational identity [17,18]:

$$e_n(x) = \frac{1}{1-\partial_x} \frac{x^n}{n!}. \tag{14}$$

Corollary 2. *We can further elaborate on the previous identities (Equations (13) and (14)), and set*

$$\sum_{n=1}^{\infty} \frac{e_n(t)}{(n+1)^2} = Q_2(t),$$

$$Q_2(t) = \frac{1}{1-\partial_t} f_2(t). \tag{15}$$

Furthermore, since

$$\sum_{n=1}^{\infty} \frac{1}{(n+1)^2} = \frac{\pi^2}{6} - 1, \tag{16}$$

we end up with

$$_2h e(t) = e^t \Sigma_2(t),$$

$$\Sigma_2(t) = \frac{\pi^2}{6} - Q_2(t) e^{-t}. \tag{17}$$

This new result can be viewed as an extension of the generating function for the first-order harmonic numbers derived by Gosper (see below) [19].

It is furthermore evident that the formalism allows the straightforward derivation of other identities, such as

Lemma 1. $\forall t \in \mathbb{R}$

$$\sum_{n=1}^{\infty} \frac{t^n}{n!} (_2h_{n+m}) = e^t \sum_{s=0}^{m} \binom{m}{s} \Sigma_2^{(s)}(t) - {_2h_m}, \tag{18}$$

where the upper index (s) denotes a s-order derivative and is a direct consequence of the identity in Equation (9).

The extension to higher-order harmonic numbers with $m > 2$ follows the same logical steps—namely, the derivation of the associated Cauchy problem.

Corollary 3. *Let $\forall t \in \mathbb{R}, \forall p \in \mathbb{N} : p > 1$,*

$$f_p(t) = \sum_{n=1}^{\infty} \frac{t^n}{(n+1)^{p-1}(n+1)!} \tag{19}$$

and

$$\begin{cases} \partial_t \left({_p}_h e(t) \right) = {_p}_h e(t) + f_p(t) \\ {_p}_h e(0) = 1 \end{cases}, \tag{20}$$

a Cauchy problem. Then, we can write the solution as:

$$_p h e(t) = e^t \left(1 + \sum_{n=1}^{\infty} \frac{1}{(n+1)^p} \left(1 - e^{-t} e_n(t) \right) \right) \tag{21}$$

or

$$_p h e(t) = e^t \Sigma_p(t), \tag{22}$$

with

$$\Sigma_p(t) = \zeta(p) - Q_p(t)e^{-t},$$
$$\zeta(p) = \sum_{n=1}^{\infty} \frac{1}{n^p}, \qquad (23)$$
$$Q_p(t) = \sum_{n=1}^{\infty} \frac{1}{(n+1)^p} e_n(t) = \frac{1}{1-\partial_t} f_p(t).$$

The case $p = 1$ should be treated separately, because the sum on the right-hand side of Equation (21) apparently diverges.

Observation 2. *It is accordingly worth noting that, since*

$$f_1(t) = \sum_{n=1}^{\infty} \frac{t^n}{(n+1)!} = \frac{1}{t}(e^t - t - 1), \qquad (24)$$

we find

$$_1h e(t) = e^t \left(1 + \int_0^t \frac{1-(\tau+1)e^{-\tau}}{\tau} d\tau\right) = e^t \Sigma_1(t),$$
$$\Sigma_1(t) = e^{-t} + Ein(t), \qquad (25)$$

which is a restatement of the Gosper derivation of the generating function of first-order harmonic numbers.

Further comments on the role played by the functions $\Sigma_p(t)$ will be provided in the final section of the paper.

We conclude this introductory section with the inclusion of a further identity.

Definition 2. *Here, we introduce higher-order harmonic number umbral polynomials (for $m = 1$, see Definition 5 in Reference [7]):*

$$_m h_n(x) := (x + {}_m\hat{h})^n \zeta_0 = x^n + \sum_{s=1}^{n} \binom{n}{s} x^{n-s} {}_m h_s,$$
$$_m h_0(x) = 1. \qquad (26)$$

The $_m h_n(x)$ are introduced in umbral form in complete analogy with those associated with the Bernoulli polynomials. They belong to the Sheffer family, and the relevant properties can be studied by means of the techniques discussed in Reference [17].

It has already been noted that (see Equation (31) in Reference [7])

$$_1 h_n(-1) = (-1)^n \left(1 - \frac{1}{n}\right), \qquad (27)$$

which coincides with an analogous identity derived with Mathematica in Reference [20] and, within the present framework from the recurrence

$$_1 h_{n+1}(x) = (x+1) \, _1 h_n(x) + \sum_{s=1}^{n} \binom{n}{s} \frac{1}{(s+1)} x^{n-s} = (x+1) \, _1 h_n(x) + n \int_0^1 dy \int_0^y (x+z)^{n-1} dz. \qquad (28)$$

Analogous results can be derived for the higher-order case, as in

$$_2 h_n(-1) = (-1)^n \left(1 - \frac{{}_1 h_n}{n}\right), \qquad (29)$$

and can be further generalized.

2. Harmonic Numbers and Integral Transforms

The identities we have dealt with in the previous section can be further generalized if the umbral procedure is merged with other techniques, involving things such as methods of an operational nature.

Proposition 2. *We note that an extension of identities of the type reported in Equation (9) is provided by the sum*

$$\sum_{n=0}^{\infty} \frac{t^n}{n!} n^m \,_p h_n = e^t \sum_{k=0}^{m} \left\{ \begin{array}{c} m \\ k \end{array} \right\} t^k \sum_{s=0}^{k} \binom{k}{s} \Sigma_p^{(k)}(t), \qquad (30)$$

where $\left\{ \begin{array}{c} n \\ k \end{array} \right\}$ *are Stirling numbers of the second kind, namely*

$$\left\{ \begin{array}{c} n \\ k \end{array} \right\} = \frac{1}{k!} \sum_{j=0}^{k} (-1)^{k-j} j^n \binom{k}{j}. \qquad (31)$$

Proof. The result in Equation (30) was obtained by merging the umbral formalism with identities of an operational nature. By noting that (see, e.g., Equation (18) in Reference [21] and Equation (1) in [22], and for a more general use of these numbers, see [23–25])

$$(x\partial_x)^n = \sum_{k=0}^{n} \left\{ \begin{array}{c} n \\ k \end{array} \right\} x^k \partial_x^k, \qquad (32)$$

and that if the following sum

$$\sum_{n=0}^{\infty} (t^n a_n) = \sigma(t) \qquad (33)$$

does exist, then the following relation holds true

$$\sum_{n=0}^{\infty} n^m (t^n a_n) = (t\partial_t)^m \sum_{n=0}^{\infty} (t^n a_n) = (t\partial_t)^m \sigma(t) = \sum_{k=0}^{m} \left\{ \begin{array}{c} m \\ k \end{array} \right\} t^k \sigma^{(k)}(t). \qquad (34)$$

The result contained in Equation (30) is, therefore, a consequence of Equations (9) and (34). □

Furthermore, we noticed that by combining the umbral, Laplace transform, and integral representation methods, we could make further progress.

Example 2. *Let us therefore note that (see Corollary 1 in [7])*

$$\frac{1}{1 - {}_1\hat{h} t} \xi_0 = \sum_{r=1}^{\infty} {}_1 h_r \, t^r + 1, \qquad |t| < 1. \qquad (35)$$

The use of the Laplace transform allowed us to write the left-hand side of Equation (35) in the form of

$$\frac{1}{1 - {}_1\hat{h} t} \xi_0 = \int_0^{\infty} e^{-s} e^{s {}_1\hat{h} t} ds \, \xi_0, \qquad (36)$$

which, on account of Equation (25), allows the conclusion controlla se l'eq e' la 25

$$1 + \sum_{n=1}^{\infty} t^n \, (_1h_n) = 1 - \frac{\ln(1-t)}{1-t} = 1 + \frac{Li_1(t)}{1-t}, \quad |t| < 1,$$

$$Li_m(x) = \sum_{r=1}^{\infty} \frac{x^r}{r^m} \equiv \text{Polylogarithm}.$$

(37)

The same procedure applied to higher-order harmonic numbers yields the generating functions:

$$\sum_{r=1}^{\infty} {}_m h_r t^r + 1 = 1 + \frac{Li_m(t)}{1-t}, \quad |t| < 1,$$

(38)

already known for the case $m = 2$.

The following example further underscores the versatility of the procedure we have proposed. We will indeed show that the use of the Gaussian identity

$$e^{b^2} = \frac{1}{\sqrt{\pi}} \int_{-\infty}^{\infty} e^{-\xi^2 + 2b\xi} d\xi$$

(39)

can be exploited to infer further identities on the properties of harmonic numbers.

Example 3. *Now, we consider the following generating function*

$$_{2h}e_2(t) := 1 + \sum_{n=1}^{\infty} \frac{t^n}{n!} \, (_2h_{2n}).$$

(40)

According to our formalism, the corresponding r.h.s. can be written as

$$_{2h}e_2(t) = e^{\left(_2\hat{h}^2\right)t} \tilde{\zeta}_0,$$

(41)

which, on account of the identity (39), can be written as

$$_{2h}e_2(t) = \frac{1}{\sqrt{\pi}} \int_{-\infty}^{+\infty} e^{-s^2 + 2s\sqrt{t}\,_2\hat{h}} ds \; \tilde{\zeta}_0.$$

(42)

The derivation of the sum in Equation (40) is therefore reduced to the evaluation of the following integral

$$_{2h}e_2(t) = \frac{1}{\sqrt{\pi}} \int_{-\infty}^{+\infty} e^{-s^2} \left(\frac{\pi^2}{6} e^{2s\sqrt{t}} - Q_2(2s\sqrt{t}) \right) ds = \frac{\pi^2}{6} e^t - q_2(t),$$

(43)

with

$$q_2(t) = \frac{1}{\sqrt{\pi}} \sum_{n=1}^{\infty} \frac{1}{(n+1)^2} \int_{-\infty}^{+\infty} e^{-s^2} e_n \left(2s\sqrt{t}\right) ds.$$

(44)

3. Final Comments

Before concluding this paper, we show that the umbral formalism we have employed can be pushed even further to infer new properties of the harmonic numbers. To emphasise this point, we start from certain identities established in Reference [20] by the use of the Mathematica code, SIGMA.

Among the examples discussed in Equation (18) in Reference [20], we pick out the following two:

$$
\begin{aligned}
&1)\ \sum_{r=1}^{n}\binom{n}{r}\frac{(-1)^r}{r}=-_1h_n,\\
&2)\ \sum_{r=1}^{n}\binom{n}{r}\frac{(-1)^r}{r}\,_1h_r=-_2h_n.
\end{aligned}
\tag{45}
$$

We can transform the left-hand side of Equations (45) in a Newton binomial by an appropriate definition of umbra.

Case 1

Regarding the first, we define the operator [7]:

$$
\begin{aligned}
\hat{\kappa}^r\psi_0 &= \frac{1}{r},\quad r>0,\\
\hat{\kappa}^0\psi_0 &= 1,
\end{aligned}
\tag{46}
$$

thus casting the first of Equation (45) in the form (see Equation (32) in Reference [7]):

$$
_1h_n = 1 - (1-\hat{\kappa})^n\psi_0,
\tag{47}
$$

which can be exploited to once more derive the Gosper generating function. By multiplying both sides of Equation (47) by $\frac{t^n}{n!}$, and then by summing up on the index n, we obtain:

$$
\sum_{n=0}^{\infty}\frac{t^n}{n!}\,_1h_n = e^t\left(1-e^{-\hat{\kappa}t}\right).
\tag{48}
$$

Keeping the m-th derivative with respect to both sides of Equation (48) yields:

$$
\sum_{n=0}^{\infty}\frac{t^n}{n!}\,_1h_{n+m} = e^t\sum_{r=0}^{m}\binom{m}{r}\partial_t^r\phi(t),
\tag{49}
$$

where

$$
\phi(t) = 1 - e^{-\hat{\kappa}t}\psi_0,
$$

$$
\partial_t^r\phi(t) = -(-\hat{\kappa})^r e^{-\hat{\kappa}t}\psi_0 = -\sum_{s=0}^{\infty}\frac{(-1)^{s+r}\hat{\kappa}^{s+r}}{s!}t^s\psi_0 = \begin{cases} -\sum_{s=0}^{\infty}\dfrac{(-1)^{s+r}}{s!(s+r)}t^s, & r>0,\\ -\sum_{r=1}^{\infty}\dfrac{-t^r}{r!r}, & r=0. \end{cases}
\tag{50}
$$

Case 2

An analogous procedure can be exploited to handle the second:

$$
\hat{\mu}^r\eta_0 = \frac{_1h_r}{r}
\tag{51}
$$

which allows for the derivation of the following identity

$$
\sum_{n=0}^{\infty}\frac{t^n}{n!}\,_2h_{n+m} = e^t\sum_{r=0}^{m}\binom{m}{r}\partial_t^r\,_2\phi(t),
\tag{52}
$$

with

$$_2\phi(t) = 1 - e^{-\hat{\mu} t} \eta_0,$$

$$\partial_t^r {}_2\phi(t) = \begin{cases} -\sum_{s=0}^{\infty} \frac{(-1)^{s+r}}{s!(s+r)} {}_1h_{s+r} \, t^s, & r > 0, \\ -\sum_{r=1}^{\infty} \frac{(-t)^r \, {}_1h_r}{r!r}, & r = 0. \end{cases} \qquad (53)$$

Remark 1. *Let us now use the obvious identity (which holds for all operators):*

$$1^n = [(1 - \hat{\mu}) + \hat{\mu}]^n \, \eta_0. \qquad (54)$$

Expanding the Newton binomial, we find

$$[(1-\hat{\mu}) + \hat{\mu}]^n \, \eta_0 = \sum_{r=0}^{n} \binom{n}{r} \hat{\mu}^{n-r} \sum_{s=0}^{r} \binom{r}{s} (-1)^s \hat{\mu}^s \eta_0 =$$

$$= \left(\sum_{r=0}^{n-1} \binom{n}{r} \sum_{s=1}^{r} \binom{r}{s} (-1)^s \hat{\mu}^{n+s-r} + \sum_{r=0}^{n-1} \binom{n}{r} \hat{\mu}^{n-r} + \sum_{s=1}^{n} \binom{n}{s} (-1)^s \hat{\mu}^s + 1 \right) \eta_0. \qquad (55)$$

Remark 2. *Choosing, for example, the first operator (46) and the property $\hat{\kappa}^r \hat{\kappa}^s \psi_0 = \hat{\kappa}^{s+r} \psi_0 = \frac{1}{s+r}$, we can finally elaborate (54) to get (see [24]):*

$$\sum_{r=1}^{n} \binom{n}{r} \frac{1}{r} = -\sum_{k=1}^{n} \binom{n}{k} \sum_{r=1}^{k} \binom{k}{r} (-1)^r \frac{1}{n+r-k}, \qquad (56)$$

while when repeating the procedure with the realization (51), we end up with

$$\sum_{r=1}^{n} \binom{n}{r} \frac{{}_1h_r}{r} = -\sum_{k=1}^{n} \binom{n}{k} \sum_{r=1}^{k} (-1)^r \binom{k}{r} \frac{{}_1h_{n+r-k}}{n+r-k}. \qquad (57)$$

According to Reference [6] (see Equation (2)), it might also be convenient to use the following umbral definition for the inverse of an integer.

Definition 3. *We introduce the umbral operator*

$$\hat{a}^n \gamma_0 := \frac{1}{n+1}, \qquad \forall n \in \mathbb{N}. \qquad (58)$$

Proposition 3. $\forall n \in \mathbb{N}$ *(see Equation 1 in [6])*

$$\frac{1}{n+1} = \sum_{s=0}^{n} \binom{n}{s} (-1)^s \frac{1}{s+1}. \qquad (59)$$

Proof. It can be proved by induction that from (For $n = 2$ we find $\hat{a}^2 \gamma_0 = (1 - 2\hat{a} + \hat{a}^2) \gamma_0$, according to the prescription in Equation (58), we get

$$\frac{1}{2+1} = 1 - 2\frac{1}{1+1} + \frac{1}{2+1};$$

for $n = 3$, $\hat{a}^3 \gamma_0 = (1 - 3\hat{a} + 3\hat{a}^2 - \hat{a}^3) \gamma_0$, we get

$$\frac{1}{4} = 1 - 3\frac{1}{1+1} + 3\frac{1}{2+1} - \frac{1}{3+1}.)$$

$$\hat{a}^n \gamma_0 = (1-\hat{a})^n \gamma_0. \tag{60}$$

Furthermore, by using the obvious relation

$$\hat{a}^n \gamma_0 = [1-(1-\hat{a})]^n \gamma_0 = \sum_{s=0}^{n} \binom{n}{s} (-1)^s (1-\hat{a})^s, \tag{61}$$

we end up with Equation (59). □

We consider now the square of harmonic numbers, $[_1 h_n]^2$.

Definition 4. *We introduce the umbral operator*

$$_1\hat{h}^n_{(2)}\ _{(2)}\tilde{\xi}_0 := \tilde{h}^n\ \tilde{\xi}_0 = [_1 h_n]^2, \quad \forall n \in \mathbb{N}. \tag{62}$$

Definition 5. *The umbral operator* (62) *can be exploited to define the formal series:*

$$_{1h_{(2)}} e(t) := \tilde{e}(t) = 1 + \sum_{n=1}^{\infty} \frac{t^n}{n!} \tilde{h}^n \tilde{\xi}_0 = e^{\tilde{h}\,t} \tilde{\xi}_0, \tag{63}$$

specifying the associated generating function.

The derivation of Equation (63), according to the previously foreseen method, reduces to the solution of a first-order differential Equation, as it has been shown in the following example.

Example 4. *By noting that*

$$\tilde{h}^{n+1}\tilde{\xi}_0 = [_1 h_n]^2 + 2\,\frac{_1 h_n}{n+1} + \frac{1}{(n+1)^2}, \tag{64}$$

and that

$$\partial_t\, \tilde{e}(t) = \tilde{h}\, e^{\tilde{h}\,t} \tilde{\xi}_0, \tag{65}$$

we easily find

$$\partial_t \tilde{e}(t) = \tilde{e}(t) - 1 + \frac{2}{t} \sum_{n=0}^{\infty} \frac{t^{n+1}}{(n+1)!} \left(_1 h_{n+1} - \frac{1}{n+1}\right) + \sum_{n=0}^{\infty} \frac{t^n}{(n+1)(n+1)!} \tag{66}$$

or

$$\partial_t\, \tilde{e}(t) = \tilde{e}(t) + \frac{2}{t}(_{1h}e(t) - t - 1) - \frac{1}{t}\int_0^t \frac{e^s - s - 1}{s}\, ds \tag{67}$$

which can be solved by the use of the same technique as before, made only slightly more complicated by the non-homogeneous term. The Equation can be straightforwardly solved, and the relevant solution, for $\tilde{e}(0) = 1$, *reads*

$$\tilde{e}(t) = e^t\, (1 + \chi(t)), \tag{68}$$

where

$$\chi(t) = \int_0^t e^{-s} \beta(s)\, ds,$$
$$\beta(t) = \frac{2}{t}(_{1h}e(t) - t - 1) - f_2(t). \tag{69}$$

Remark 3. *In the introductory section we mentioned the link between harmonic and Bernoulli numbers, and they have been paradigmatic examples of applications of umbral methods. Regarding the finite sum of power of integers, we find, for example (according to Chapter 6 in Reference* [4]),

$$\sum_{r=0}^{n} r^m =\ _{(-m)}h_n = \frac{1}{m+1}\left[(\hat{b}+n+1)^{m+1} - \hat{b}^{m+1}\right]\theta_0, \qquad (70)$$

where

$$\hat{b}^r \theta_0 = B_r \qquad (71)$$

is the umbral operator which provides Bernoulli numbers B_r.

Proposition 4. *The use of Equation (70) yields the generating function:*

$$\sum_{m=0}^{\infty} \frac{_{(-m)}h_n\, z^{m+1}}{m!} = \frac{z}{e^z - 1}(e^{(n+1)z} - 1), \qquad \forall z \in \mathbb{R} :|z|< 2\pi. \qquad (72)$$

Proof. $\forall z \in \mathbb{R} :|z|< 2\pi$

$$\sum_{m=0}^{\infty} \frac{_{(-m)}h_n\, z^{m+1}}{m!} = \sum_{m=0}^{\infty} \frac{\left[(\hat{b}+n+1)^{m+1} - \hat{b}^{m+1}\right] z^{m+1}}{(m+1)!} \theta_0 = \left(e^{(\hat{b}+n+1)z} - e^{\hat{b}z}\right)\theta_0 =$$

$$= \left(e^{(n+1)z} - 1\right) e^{\hat{b}z}\theta_0 = \left(e^{(n+1)z} - 1\right) \sum_{r=0}^{\infty} \frac{z^r}{r!} \hat{b}^r \theta_0 =$$

$$= \left(e^{(n+1)z} - 1\right) e^{\hat{b}z}\theta_0 = \left(e^{(n+1)z} - 1\right) \sum_{r=0}^{\infty} \frac{z^r}{r!} B_r$$

which, by taking into account that the generating function of Bernoulli numbers is

$$\sum_{r=0}^{\infty} \frac{z^r}{r!} B_r = \frac{z}{e^z - 1} \qquad (73)$$

yields the statement in Equation (72). □

In this paper, we have provided a hint of the possible interplay between the theory of harmonic numbers and operational methods (be they of umbral ordinary nature). Further progress will be presented in future investigations.

Author Contributions: Conceptualization, G.D.; methodology, G.D., S.L.; data curation: S.L.; validation, G.D., S.L., E.S., H.M.S.; formal analysis, G.D., S.L., E.S., H.M.S.; writing—riginal draft preparation: G.D.; writing—review and editing: S.L.

Funding: The work of S.L. was supported by a *Enea-Research Center Individual Fellowship*

Acknowledgments: The Authors recognizes the kind help of the Referees who pushed for a better organized presentation and for including in the study the Bernoulli numbers.

Conflicts of Interest: The authors declare no conflict of interest.

References

1. Roman, S. *The Umbral Calculus*; Academic Press: New York, NY, USA, 1984; pp. 93–100.
2. Dattoli, G.; Cesarano, C.; Lorenzutta, S. Bernoulli numbers and polynomials from a more general point of view. *Rendiconti Matematica* **2002**, *22*, 193–202.
3. Dattoli, G.; Migliorati, M. *Operational Methods, Harmonic Numbers and Riemann Function*; Internal Report Enea Frascati (Rome) RT/2008/29/FIM; Edizioni Scientifiche—ENEA Centro Ricerche Frascati: Rome, Italy, 2008.
4. Graham, R.L.; Knuth, D.E.; Patashnik, O. *Concrete Mathematics: A Foundation for Computer Science*, 2nd ed.; Addison-Wesley Longman Publishing Co., Inc.: Boston, MA, USA, 1994.
5. Dattoli, G.; Srivastava, H.M. A Note on Harmonic Numbers, Umbral Calculus and Generating Functions. *Appl. Math. Lett.* **2008**, *21*, 686–693. [CrossRef]
6. Dattoli, G.; Zhukovsky, K. Umbral Methods, Combinatorial Identities and Harmonic Numbers. *Appl. Math.* **2011**, *1*, 46–49.

7. Dattoli, G.; Germano, B.; Licciardi, S.; Martinelli, M.R. Umbral Methods and Harmonic Numbers. *Axioms* **2018**, *7*, 62.
8. Coffey, M.W. Expressions for harmonic humber exponential generating functions. In *Contemporary Mathematics, Gems in Experimental Mathematics*; Amdeberhan, T., Simos, E.T., Moll, V.H., Eds.; American Mathematical Society: Washington, DC, USA, 2009; Volume 517, pp. 113–126.
9. Cvijović, D. The Dattoli-Srivastava Conjectures Concerning Generating Functions Involving the Harmonic Numbers. *Appl. Math. Comput.* **2010**, *215*, 4040–4043. [CrossRef]
10. Licciardi, S. Umbral Calculus, a Different Mathematical Language. *arXiv* **2018**, arXiv:1803.03108.
11. Choi, J. Finite summation formulas involving binomial coefficients, harmonic numbers and generalized harmonic numbers. *J. Inequal. Appl.* **2013**, *49*, 11. [CrossRef]
12. Sofo, A.; Srivastava, H.M. A family of shifted harmonic sums. *Ramanujan J.* **2015**, *37*, 89–108. [CrossRef]
13. Sofo, A.; Srivastava, H.M. dentities for the harmonic numbers and binomial coefficients. *Ramanujan J.* **2011**, *25*, 93–113. [CrossRef]
14. Choi, J.; Srivastava, H.M. Some summation formulas involving harmonic numbers and generalized harmonic numbers. *Math. Comput. Model.* **2011**, *54*, 2220–2234. [CrossRef]
15. Mezo, I. Exponential Generating Function of Hyper-Harmonic Numbers Indexed by Arithmetic Progressions. *Cent. Eur. J. Math.* **2013**, *11*, 931–939.
16. Dattoli, G.; Ricci, P.E.; Marinelli, L. Generalized Truncated Exponential Polynomials and Applications, in: Rendiconti dell'Istituto di Matematica dell'Università di Trieste. *Int. J. Math.* **2002**, *34*, 9–18.
17. Dattoli, G.; Migliorati, M.; Srivastava, H.M. Sheffer polynomials, monomiality principle, algebraic methods and the theory of classical polynomials. *Math. Comput. Model.* **2007**, *45*, 1033–1041. [CrossRef]
18. Dattoli, G.; Sabia, E. Generalized Transforms and Special functions. *arXiv* **2010**, arXiv:1010.1679.
19. Gosper R.W. Harmonic summation and exponential gfs. math-fun@cs.arizona.edu posting, 2 August 1996. In *Harmonic Number. MathWorld—A Wolfram Web Resource*; Sondow, J., Weisstein, E.W., Eds.; 1996; Available online: http://mathworld.wolfram.com/HarmonicNumber.html (accessed on 1 May 2019).
20. Schmidt, M.D. Zeta Series Generating Function Transformations Related to Polylogarithm Functions and the k-Order Harmonic Numbers. *Online J. Anal. Comb.* **2017**, *12*, 2.
21. Dattoli, G.; Germano, B.; Martinelli, M.R.; Ricci, P.E. Touchard like polynomials and generalized Stirling numbers. *Appl. Math. Comput.* **2012**, *218*, 6661–6665. [CrossRef]
22. Lang, W. On Generalizations of the Stirling Number Triangles. *J. Integer. Seq.* **2000**, *3*, 3.
23. Srivastava, H.M.; Manocha, H.L. *A Treatise on Generating Functions*; Halsted Press (Ellis Horwood Limited, Chichester): Chichester, UK; John Wiley and Sons: New York, NY, USA, 1984.
24. Comtet, L. *Advanced combinatorics: The Art of Finite and Infinite Expansions*; Springer: Berlin/Heidelberg, Germany, 1974.
25. Johnson, W.P. The Curious History of Faá di Bruno's Formula. *Am. Math. Mon.* **2002**, *109*, 217–234.

 © 2019 by the authors. Licensee MDPI, Basel, Switzerland. This article is an open access article distributed under the terms and conditions of the Creative Commons Attribution (CC BY) license (http://creativecommons.org/licenses/by/4.0/).

Article

Some Classes of Harmonic Mapping with a Symmetric Conjecture Point Defined by Subordination

Lina Ma, Shuhai Li * and Xiaomeng Niu

School of Mathematics and Statistics, Chifeng University, Chifeng 024000, Inner Mongolia, China;
malina00@163.com (L.M.); ndnxm@126.com (X.N.)
* Correspondence: lishms66@163.com

Received: 19 May 2019; Accepted: 12 June 2019; Published: 16 June 2019

Abstract: In the paper, we introduce some subclasses of harmonic mapping, the analytic part of which is related to general starlike (or convex) functions with a symmetric conjecture point defined by subordination. Using the conditions satisfied by the analytic part, we obtain the integral expressions, the coefficient estimates, distortion estimates and the growth estimates of the co-analytic part g, and Jacobian estimates, the growth estimates and covering theorem of the harmonic function f. Through the above research, the geometric properties of the classes are obtained. In particular, we draw figures of extremum functions to better reflect the geometric properties of the classes. For the first time, we introduce and obtain the properties of harmonic univalent functions with respect to symmetric conjugate points. The conclusion of this paper extends the original research.

Keywords: harmonic univalent functions; subordination; with symmetric conjecture point; integral expressions; coefficient estimates; distortion

MSC: 30C45; 30C50; 30C80

---

1. Introduction and Preliminaries

Let $\mathcal{A}$ denote the class of functions in the following form

$$h(z) = z + \sum_{n=2}^{\infty} a_n z^n, \tag{1}$$

where $h(z)$ is analytic in the open unit disk $\mathbb{U} = \{z \in \mathbb{C} : |z| < 1\}$.

$\mathcal{S}, \mathcal{S}^*, \mathcal{K}$ are denoted respectively by the subclasses of $\mathcal{A}$ consisting of univalent, starlike, convex functions (for details, see [1,2]).

Let $\mathcal{P}$ denote the class of functions p satisfying $p(0) = 1$ and $\operatorname{Re} p(z) > 0$, where $z \in \mathbb{U}$.

The function s is subordinate to t in $\mathbb{U}$, written by $s(z) \prec t(z)$, if there exists a Schwarz function σ, analytic in $\mathbb{U}$ with $\sigma(0) = 0$ and $|\sigma(z)| < 1$, satisfying $s(z) = t(\sigma(z))$(see [1]). If the function t is univalent in $\mathbb{U}$ and $s(z) \prec t(z)$, we have the equivalent results as follows,

$$s(0) = t(0) \quad \text{and} \quad s(\mathbb{U}) \subset t(\mathbb{U}).$$

In 1994, Ma and Minda [3] introduce a class $\mathcal{S}^*(\phi)$ of starlike functions defined by subordination, $h(z) \in \mathcal{S}^*(\phi)$ if and only if $\frac{zh'(z)}{h(z)} \prec \phi(z)$, where $h \in \mathcal{A}, \phi \in \mathcal{P}$. The corresponding convex class $\mathcal{K}(\phi)$ was defined in a similar way.

For $\phi(z) = \frac{1+Az}{1+Bz}$ and $-1 \leq B < A \leq 1$, we denote respectively the subclasses of $\mathcal{A}$ by $S^*(A,B)$ and $K(A,B)$ satisfying (see [4]):

$$h \in S^*(A, B) \iff \frac{zh'(z)}{h(z)} \prec \frac{1 + Az}{1 + Bz} \quad (h \in \mathcal{A}, z \in \mathbb{U})$$

and

$$h \in K(A, B) \iff \frac{(zh'(z))'}{h'(z)} \prec \frac{1 + Az}{1 + Bz} \quad (h \in \mathcal{A}, z \in \mathbb{U}).$$

It is easy to see that $h \in K(A, B) \iff zh'(z) \in S^*(A, B)$ and

$$K(A, B) \subset S^*(A, B), \ K(A, B) \subset K \subset \mathcal{S}, \ S^*(A, B) \subset S^* \subset \mathcal{S}.$$

Obviously, $S^*(1 - 2\beta, -1) = S^*(\beta) \ (0 \leq \beta < 1)$ is a starlike function of order β and $K(1 - 2\beta, -1) = K(\beta)$ is a convex function of order β [5]. Especially, $S^*(1, -1) = S^*$ and $K(1, -1) = K$ are well-known starlike functions and convex functions respectively.

In 1959, Sakaguchi [6] introduced the class S_s^* of starlike functions with respect to symmetric points, $f \in S_s^*$ if and only if

$$\operatorname{Re} \frac{zf'(z)}{f(z) - f(-z)} > 0.$$

In 1987, El-Ashwa and Thomas [7] introduced some classes of starlike functions with respect to conjugate points and symmetric conjugate points satisfying the following conditions

$$\operatorname{Re} \frac{zf'(z)}{f(z) + \overline{f(\overline{z})}} > 0 \quad \text{and} \quad \operatorname{Re} \frac{zf'(z)}{f(z) - \overline{f(-\overline{z})}} > 0.$$

In 1933, Fekete and Szegö [8] introduced a classical Fekete-Szegö problem for $f(z) = z + \sum_{n=2}^{\infty} a_n z^n \in \mathcal{S}$ as follows,

$$|a_3 - \mu a_2^2| \leq \begin{cases} 3 - 4\mu, & \mu \leq 0, \\ 1 + 2\exp(\frac{-2\mu}{1-\mu}), & 0 \leq \mu \leq 1, \\ 4\mu - 3, & \mu \geq 1. \end{cases}$$

The result is sharp.

In 1994, Ma and Minda [3] studied the Fekete-Szegö problem of the classes of $S^*(\phi)$ and $K(\phi)$. Many authors studied the problem of Fekete-Szegö and obtained many results (see [9–11]).

A harmonic mapping in $\mathbb{U}$ is a complex valued harmonic function, which maps $\mathbb{U}$ onto the domain $f(\mathbb{U})$. The mapping f has a canonical decomposition $f(z) = h(z) + \overline{g(z)}$ and h and g are analytic in $\mathbb{U}$. h is called the analytic part and g is called the co-analytic part of f. Let S_H denote the class of harmonic mappings with the following form (see [12,13])

$$f = h + \overline{g}, \quad z \in \mathbb{U}, \tag{2}$$

where

$$h(z) = z + \sum_{k=2}^{\infty} a_k z^k \quad \text{and} \quad g(z) = \sum_{k=1}^{\infty} b_k z^k, \ |b_1| = \alpha \in [0, 1). \tag{3}$$

In 1936, Lewy [14] proved that f is univalent and sense-preserving in $\mathbb{U}$ if and only if $J_f(z) > 0$, that is, the second complex dilatation $\omega(z) = g'(z)/h'(z)$ of $f(z)$ satisfying $|\omega(z)| < 1$ in $\mathbb{U}$ (see [12,13]).

Many authors further investigated various subclasses of S_H and obtained some important results. In [15], the authors studied the subclass of S_H with $h \in \mathcal{K}$. Also, Hotta and Michalski [16] studied the properties of a subclass of S_H with h is starlike and obtained the coefficient estimates, distortion estimates and growth estimates of g, and Jacobian estimates of f. Zhu and Huang [17] studied some subclasses of S_H with h is convex, or starlike functions of order β and some sharp estimates of coefficients, distortion, and growth are obtained.

According to the principle of subordination, we introduce the following general subclasses of S_H of harmonic univalent starlike and convex functions with a symmetric conjecture point.

Definition 1. *Let $A, B \in \mathbb{R}, -1 \leq B < A \leq 1$. We denote the function f be in the class $HS_{sc}^{*,\alpha}(A, B)$ of harmonic univalent starlike functions with a symmetric conjecture point if and only if $f \in S_H$ and $h \in S_{cs}^*(A, B)$, that is*

$$\frac{2zh'(z)}{h(z) - \overline{h(-\overline{z})}} \prec \frac{1 + Az}{1 + Bz}. \tag{4}$$

Also, we denote the function f be in the class $HK_{sc}^{\alpha}(A, B)$ of harmonic univalent generalized convex functions with a symmetric conjecture point if and only if $f \in S_H$ and $h \in K_{sc}(A, B)$, that is

$$\frac{2(zh'(z))'}{(h(z) - \overline{h(-\overline{z})})'} \prec \frac{1 + Az}{1 + Bz}, \tag{5}$$

we know that $h \in K_{cs}(A, B) \iff zh' \in S_{cs}^*(A, B)$. Additionally, we define the classes

$$HS_{sc}^*(A, B) = \bigcup_{\alpha \in [0,1)} HS_{sc}^{*,\alpha}(A, B) \quad \text{and} \quad HK_{sc}(A, B) = \bigcup_{\alpha \in [0,1)} HK_{sc}^{\alpha}(A, B). \tag{6}$$

It is clear $HK_{sc}^{\alpha}(A, B) \subset HS_{sc}^{*,\alpha}(A, B)$ and $HK_{sc}(A, B) \subset HS_{sc}^*(A, B)$. Especially, let $S_{cs}^*(1, -1) = S_{cs}^*, K_{cs}(1, -1) = K_{cs}, HS_{sc}^{*,\alpha}(1, -1) = HS_{sc}^{*,\alpha}, HK_{sc}^{\alpha}(1, -1) = HK_{sc}^{\alpha}, S_{cs}^*(1, 1 - 2\beta) = S_{cs}^*(\beta), K_{cs}(1, 1 - 2\beta) = K_{cs}(\beta), HS_{sc}^{*,\alpha}(1, 1 - 2\beta) = HS_{sc}^{*,\alpha}(\beta), HK_{sc}^{\alpha}(1, 1 - 2\beta) = HK_{sc}^{\alpha}(\beta), \beta \in [0, 1)$.

In order to prove our results, we need the following Lemmas.

Lemma 1. *[18] If the function $\omega(z) = c_0 + c_1 z + \ldots + c_n z^n + \ldots$ is analytic with $|\omega(z)| \leq 1$ in $\mathbb{U}$, then*

$$|c_n| \leq 1 - |c_0|^2, n = 1, 2, \ldots, \tag{7}$$

and

$$|c_2 - \gamma c_1^2| \leq \max\{1, |\gamma|\}. \tag{8}$$

Lemma 2. *Let $-1 \leq B < A \leq 1$, $n = 2, 3, \cdots$. (1) If $h(z) = z + \sum\limits_{n=2}^{\infty} a_n z^n \in S_{sc}^*(A, B)$, then*

$$|a_{2n}| \leq F_n(A, B) \quad \text{and} \quad |a_{2n+1}| \leq F_n(A, B), \tag{9}$$

where

$$F_n(A, B) = \frac{\prod\limits_{k=0}^{n-1}(A - B + 2k)}{(2n)!!}. \tag{10}$$

Specially, $F_1(A, B) = \frac{A-B}{2}$, $F_n(1, -1) = 1$. The estimate is sharp if

$$h(z) = \int_0^z \frac{1 + (A - B - 1)t}{(1 - t)(1 - t^2)^{\frac{A-B}{2}}} dt.$$

(2) If $h(z) = z + \sum\limits_{n=2}^{\infty} a_n z^n \in K_{sc}(A, B)$, then

$$|a_{2n}| \leq \frac{F_n(A, B)}{2n} \quad \text{and} \quad |a_{2n+1}| \leq \frac{F_n(A, B)}{2n + 1}, \tag{11}$$

where $F_n(A, B)$ is defined by (10). The estimate is sharp if

$$h(z) = \int_0^z \frac{1}{\eta} \int_0^\eta \frac{1 + (A - B - 1)t}{(1-t)(1-t^2)^{\frac{A-B}{2}}} dt d\eta.$$

Especially, if $A = 1, B = -1$, we have the following results. (i) If $h(z) = z + \sum_{n=2}^\infty a_n z^n \in S_{sc}^*$, then

$$|a_{2n}| \leq 1 \quad \text{and} \quad |a_{2n+1}| \leq 1. \tag{12}$$

The estimate is sharp if $h(z) = \frac{z}{1-z}$. (ii) If $h(z) = z + \sum_{n=2}^\infty a_n z^n \in K_{sc}$, then

$$|a_{2n}| \leq \frac{1}{2n} \quad \text{and} \quad |a_{2n+1}| \leq \frac{1}{2n+1}. \tag{13}$$

The estimate is sharp if $h(z) = -\log(1-z)$.

Proof. Let $h(z) = z + \sum_{n=2}^\infty a_n z^n \in S_{sc}^*(A, B)$, there exists a positive real function $p(z) = 1 + \sum_{k=1}^\infty p_k z^k \in \mathcal{P}$ with $|p_k| \leq A - B$, satisfying

$$\frac{2zh'(z)}{h(z) - \overline{h(-\overline{z})}} = p(z). \tag{14}$$

Comparing the coefficients of the both sides of the equation (14), we have

$$2na_{2n} = p_{2n-1} + a_3 p_{2n-3} + \cdots + a_{2n-1} p_1, \tag{15}$$

and

$$2na_{2n+1} = p_{2n} + a_3 p_{2n-2} + \cdots + a_{2n-1} p_2. \tag{16}$$

It is easy to verify that

$$|a_{2n}| \leq \frac{(A-B)}{2n}(1 + |a_3| + \cdots + |a_{2n-1}|) \tag{17}$$

and

$$|a_{2n+1}| \leq \frac{(A-B)}{2n}(1 + |a_3| + \cdots + |a_{2n-1}|). \tag{18}$$

Let $\phi(n) = 1 + |a_3| + \cdots + |a_{2n-1}|$, from (18), we have

$$\phi(n+1) \leq \frac{\prod_{k=1}^n (A - B + 2k)}{(2n)!!}. \tag{19}$$

According to (17)–(19), we can obtain (9).

If $h(z) = z + \sum_{n=2}^\infty a_n z^n \in K_{sc}(A, B)$, then $zh'(z) \in S_{sc}^*(A, B)$. Using the results in (1), we can obtain (11) easily. □

Lemma 3. Let $A, B \in \mathbb{R}, -1 \leq B < A \leq 1$. (1) If $h(z) = z + \sum_{n=2}^\infty a_n z^n \in S_{sc}^*(A, B), \mu \in \mathbb{C}$, then

$$|a_3 - \mu a_2^2| \leq \frac{A - B}{2} \max\left\{1, \left|B + \frac{\mu(A-B)}{2}\right|\right\}. \tag{20}$$

(2) If $h(z) = z + \sum_{n=2}^{\infty} a_n z^n \in K_{sc}(A, B)$, $\mu \in \mathbb{C}$, then

$$|a_3 - \mu a_2^2| \leq \frac{A-B}{6} \max\left\{1, \left|B + \frac{3\mu(A-B)}{8}\right|\right\}. \tag{21}$$

Proof. Let $h(z) = z + \sum_{n=2}^{\infty} a_n z^n \in S_{sc}^*(A, B)$. By definition 1 and the relationship of subordination, we have

$$\frac{2zh'(z)}{h(z) - \overline{h(-\overline{z})}} = \frac{1 + A\nu(z)}{1 + B\nu(z)}, \tag{22}$$

where $\nu(z) = c_1 z + c_2 z^2 + \cdots$ is analytic in $\mathbb{U}$ satisfying $\nu(0) = 0$ and $|\nu(z)| < 1$.

Comparing the coefficients of the both sides of (22), we obtain

$$a_2 = \frac{A-B}{2} c_1 \quad \text{and} \quad a_3 = \frac{A-B}{2} c_2 - \frac{(A-B)B}{2} c_1^2.$$

Therefore, we have

$$a_3 - \mu a_2^2 = \frac{A-B}{2}\left\{c_2 - \left(B + \frac{\mu(A-B)}{2}\right)c_1^2\right\}.$$

By an application of (8) in Lemma 1, we obtain (20).

The bound is sharp as follows,

$$h(z) = \int_0^z (1 + A\xi)(1 - B\xi)^{\frac{A-B}{2B}} (1 + B\xi)^{\frac{A-3B}{2B}} d\xi$$

or

$$h(z) = \int_0^z (1 + A\xi^2)(1 + B\xi^2)^{\frac{A-3B}{2B}} d\xi.$$

If $h(z) = z + \sum_{n=2}^{\infty} a_n z^n \in K_{sc}(A, B)$, then $zh'(z) \in S_{sc}^*(A, B)$. It is easy to obtain (21) and the bound is sharp as follows,

$$h(z) = \int_0^z \frac{1}{\eta} \int_0^\eta (1 + A\xi)(1 - B\xi)^{\frac{A-B}{2B}} (1 + B\xi)^{\frac{A-3B}{2B}} d\xi d\eta$$

or

$$h(z) = \int_0^z \frac{1}{\eta} \int_0^\eta (1 + A\xi^2)(1 + B\xi^2)^{\frac{A-3B}{2B}} d\xi d\eta.$$

□

Lemma 4. Let $h(z) \in \mathcal{A}$, $0 \leq \beta < 1$, $|z| = r \in [0, 1)$. (1) If $h(z) \in \mathcal{S}^*(\beta)$, then

$$(1 - (1 - 2\beta)r)(1 + r)^{2\beta - 3} \leq |h'(z)| \leq (1 + (1 - 2\beta)r)(1 - r)^{2\beta - 3}, \tag{23}$$

and ([4], Theorem 4 with $A = 1 - 2\beta$, $B = -1$)

$$r(1 + r)^{2\beta - 2} \leq |h(z)| \leq r(1 - r)^{2\beta - 2}, \tag{24}$$

(2) If $h(z) \in \mathcal{K}(\beta)$, then ([19], Theorem 1 with $b = 1$, $A = 1 - 2\beta$, $B = -1$)

$$(1 + r)^{2\beta - 2} \leq |h'(z)| \leq (1 - r)^{2\beta - 2}, \tag{25}$$

and ([19], Theorem 2 with $b = 1$, $A = 1 - 2\beta$, $B = -1$)

$$r(1+r)^{\beta-1} \leq |h(z)| \leq r(1-r)^{\beta-1}. \tag{26}$$

Proof. It suffices to establish the estimate of (23). If $h(z) \in S^*(\beta)$, then

$$\frac{1-(1-2\beta)r}{1+r} \leq \left|\frac{zh'(z)}{h(z)}\right| \leq \frac{1+(1-2\beta)r}{1-r},$$

that is,

$$\frac{1-(1-2\beta)r}{1+r}|h(z)| \leq |zh'(z)| \leq \frac{1+(1-2\beta)r}{1-r}|h(z)|.$$

According to (24), it is not difficult to verify the estimate of (23).

Using the same argument as in the proof of Lemma 2 in [20], we obtain immediately a Lemma as follows. □

Lemma 5. *If $h(z) \in \mathcal{S}^*_{sc}(\beta), 0 \leq \beta < 1$, then $\frac{h(z)-\overline{h}(-\overline{z})}{2} \in S^*(\beta)$. Especially for $\beta = 0$, we get the results of Lemma 2 in [20].*

Lemma 6. *If $h(z) \in \mathcal{K}_{sc}(\beta), 0 \leq \beta < 1$, then $\frac{h(z)-\overline{h}(-\overline{z})}{2} \in \mathcal{K}(\beta)$.*

Lemma 7. *Let $h(z) \in \mathcal{A}, 0 \leq \beta < 1, |z| = r \in [0,1)$. (1) If $h \in \mathcal{S}^*_{sc}(\beta)$, then*

$$\frac{1-(1-2\beta)r}{(1+r)^{3-2\beta}} \leq |h'(z)| \leq \frac{1+(1-2\beta)r}{(1-r)^{3-2\beta}}. \tag{27}$$

(2) If $h \in \mathcal{K}_{sc}(\beta)$, then

$$\frac{1}{(1+r)^{2-2\beta}} \leq |h'(z)| \leq \frac{1}{(1-r)^{2-2\beta}}. \tag{28}$$

Proof. Suppose $h(z) \in \mathcal{S}^*_{sc}(\beta)$, we have

$$\frac{1-(1-2\beta)r}{1+r}\left|\frac{h(z)-\overline{h}(-\overline{z})}{2}\right| \leq |zh'(z)| \leq \frac{1+(1-2\beta)r}{1-r}\left|\frac{h(z)-\overline{h}(-\overline{z})}{2}\right|. \tag{29}$$

According to Lemmas 4 and 5, we have

$$\frac{r}{(1+r)^{2-2\beta}} \leq \left|\frac{h(z)-\overline{h}(-\overline{z})}{2}\right| \leq \frac{r}{(1-r)^{2-2\beta}}. \tag{30}$$

By (29) and (30), we can obtain (27).

If $h(z) \in \mathcal{K}_{sc}(\beta)$, then

$$\frac{1-(1-2\beta)r}{1+r}\left|\frac{(h(z)-\overline{h}(-\overline{z}))'}{2}\right| \leq |(zh'(z))'| \leq \frac{1+(1-2\beta)r}{1-r}\left|\frac{(h(z)-\overline{h}(-\overline{z}))'}{2}\right|. \tag{31}$$

According to Lemmas 4 and 6, we have

$$(1+r)^{2\beta-2} \leq \left|\frac{(h(z)-\overline{h}(-\overline{z}))'}{2}\right| \leq (1-r)^{2\beta-2}. \tag{32}$$

By (31) and (32), we get

$$[1-(1-2\beta)r](1+r)^{2\beta-3} \leq |(zh'(z))'| \leq [1+(1-2\beta)r](1-r)^{2\beta-3}. \tag{33}$$

By (33), integrating along a radial line $\xi = te^{i\theta}$, we obtained immediately,

$$|zh'(z)| \leq \int_0^r [1 + (1-2\beta)t](1-t)^{2\beta-3}dt = \frac{r}{(1-r)^{2-2\beta}}$$

For the left-hand side of (28), we note first that $zh'(z)$ is univalent. Let $H(z) := zh'(z), \Gamma = H(\{z : |z| = r\})$ and let $\xi_1 \in \Gamma$ be the nearest point to the origin. By a rotation we suppose that $\xi_1 > 0$. Let γ be the line segment $0 \leq \xi \leq \xi_1$ and assume that $z_1 = H^{-1}(\xi_1)$ and $L = H^{-1}(\gamma)$. If ς is the variable of integration on L, we have that $d\xi = H'(\varsigma)d\varsigma$ on L. Hence

$$\xi_1 = \int_0^{\xi_1} d\xi = \int_0^{z_1} H'(\varsigma)d\varsigma = \int_0^{z_1} |H'(\varsigma)||d\xi| \geq \int_0^r |H'(te^{i\theta})|dt$$

$$\geq \int_0^r [1 - (1-2\beta)t](1+t)^{2\beta-3}dr = \frac{r}{(1+r)^{2-2\beta}}.$$

So we complete the proof of Lemma 7. □

2. Main Results

Theorem 1. *If $f = h + \overline{g} \in HS_{sc}^{*,\alpha}(A, B)$, then $F = H + \overline{G} \in HK_{sc}^{\alpha}(A, B)$, where $H(z)$ and $G(z)$ satisfy the conditions $zH'(z) = h(z)$ and $zG'(z) = g(z), z \in \mathbb{U}$.*

Proof. Let $f \in HS_{sc}^{*,\alpha}(A, B)$. According to Definition 1 and Alexander's Theorem ([1], p. 43), the function $H(z) \in K_{sc}(A, B)$. Also, $H(0) = 0, H'(0) = \lim_{z \to 0} \frac{h(z)}{z} = h'(0) = 1$, and $|G'(0)| = |\lim_{z \to 0} \frac{g(z)}{z}| = |g'(0)| = \alpha$. Let $\Gamma := [0, h(z)] \subset h(\mathbb{U}), z \in \mathbb{U} - \{0\}$, then

$$|g(z)| = \left|\int_\Gamma d(g \circ h^{-1}(\omega))\right| \leq \int_\Gamma \left|\frac{d(g \circ h^{-1}(\omega))}{d\omega}\right| |d\omega| < \int_\Gamma |d\omega| = |h(z)|.$$

Hence,

$$|G'(z)| = \lim_{t \to z} \left|\frac{g(t)}{t}\right| < \lim_{t \to z} \left|\frac{h(t)}{t}\right| = |H'(z)|.$$

It shows that F is a locally univalent and sense-preserving harmonic function in $\mathbb{U}$. Finally, appealing to ([15], Corollary 2.3), we conclude that $F = H + \overline{G} \in HK_{sc}^{\alpha}(A, B)$. □

Corollary 1. *If $f = h + \overline{g} \in HS_{sc}^{*}(A, B)$, then $F = H + \overline{G} \in HK_{sc}(A, B)$, where $H(z)$ and $G(z)$ satisfy the conditions $zH'(z) = h(z)$ and $zG'(z) = g(z), z \in \mathbb{U}$.*

Next, we give the integral expressions for functions of these classes.

Theorem 2. *If $f = h + \overline{g} \in HS_{sc}^{*,\alpha}(A, B)$, then we have*

$$f(z) = \int_0^z \varphi(\xi)d\xi + \overline{\int_0^z \omega(\xi)\varphi(\xi)d\xi}, \tag{34}$$

where

$$\varphi(\xi) = \frac{1 + Av(\xi)}{1 + Bv(\xi)} \exp \int_0^\xi \frac{(A-B)}{2t}\left\{\frac{v(t)}{1+Bv(t)} + \frac{\overline{v(-\bar{t})}}{1+B\overline{v(-\bar{t})}}\right\}dt, \tag{35}$$

and ω and v are analytic in $\mathbb{U}$ satisfying $|\omega(0)| = \alpha, v(0) = 0, |\omega(z)| < 1, |v(z)| < 1$.

Proof. Let $f = h + \overline{g} \in HS_{sc}^{*,\alpha}(A,B)$. According to Definition 1 and the relationship of subordination, we have

$$g'(z) = \omega(z) h'(z), \tag{36}$$

and

$$\frac{2zh'(z)}{h(z) - \overline{h(-\overline{z})}} = \frac{1 + A\nu(z)}{1 + B\nu(z)}, \tag{37}$$

where ω and ν are analytic in $\mathbb{U}$ satisfying $\omega(0) = b_1, \nu(0) = 0, |\omega(z)| < 1, |\nu(z)| < 1$. Substituting z by $-\overline{z}$ in (37), we get

$$\frac{-2\overline{z}h'(-\overline{z})}{h(-\overline{z}) - \overline{h(z)}} = \frac{1 + A\nu(-\overline{z})}{1 + B\nu(-\overline{z})}. \tag{38}$$

It follows from (37) and (38) that

$$\frac{2z(\overline{h}(z) - h(-\overline{z}))'}{\overline{h}(z) - h(-\overline{z})} = \frac{1 + A\nu(z)}{1 + B\nu(z)} + \frac{1 + A\overline{\nu}(-\overline{z})}{1 + B\overline{\nu}(-\overline{z})}. \tag{39}$$

After integrating the both sides of the equality (39) and calculating it simply, we have

$$\frac{\overline{h}(z) - h(-\overline{z})}{2} = z \exp \int_0^z \frac{(A - B)}{2t} \left\{ \frac{\nu(t)}{1 + B\nu(t)} + \frac{\overline{\nu}(-\overline{t})}{1 + B\overline{\nu}(-\overline{t})} \right\} dt. \tag{40}$$

From (37) and (40), we have

$$h'(z) = \frac{1 + A\nu(z)}{1 + B\nu(z)} \exp \int_0^z \frac{(A - B)}{2t} \left\{ \frac{\nu(t)}{1 + B\nu(t)} + \frac{\overline{\nu}(-\overline{t})}{1 + B\overline{\nu}(-\overline{t})} \right\} dt. \tag{41}$$

Integrating the both sides of the equality (41), we have

$$h(z) = \int_0^z \frac{1 + A\nu(\xi)}{1 + B\nu(\xi)} \exp \int_0^\xi \frac{(A - B)}{2t} \left\{ \frac{\nu(t)}{1 + B\nu(t)} + \frac{\overline{\nu}(-\overline{t})}{1 + B\overline{\nu}(-\overline{t})} \right\} dt d\xi. \tag{42}$$

By (36) and (41), we can obtain

$$g(z) = \int_0^z \omega(\xi) \left(\frac{1 + A\nu(\xi)}{1 + B\nu(\xi)} \right) \exp \int_0^\xi \frac{(A - B)}{2t} \left\{ \frac{\nu(t)}{1 + B\nu(t)} + \frac{\overline{\nu}(-\overline{t})}{1 + B\overline{\nu}(-\overline{t})} \right\} dt d\xi.$$

So, we complete the proof of Theorem 2. □

According to Theorem 2 and $h \in K_{sc}(A, B)$ if and only if $zh'(z) \in S_{sc}^*(A, B)$, we obtain easily the following result.

Theorem 3. *Let $f \in HK_{sc}^\alpha(A, B)$, then we have*

$$f(z) = \int_0^z \frac{1}{\eta} \int_0^\eta \varphi(\xi) d\xi d\eta + \overline{\int_0^z \frac{\omega(\eta)}{\eta} \int_0^\eta \varphi(\xi) d\xi d\eta}. \tag{43}$$

where $\phi(\xi)$ defined by (35), ω and ν are analytic in $\mathbb{U}$ satisfying $|\omega(0)| = \alpha, \nu(0) = 0, |\omega(z)| < 1, |\nu(z)| < 1$.

In the following, we will get the coefficient estimates of the classes.

Theorem 4. *Let $f = h + \overline{g}$, where h and g are given by (3) and $F_k(A, B)$ is defined by (10). If $f \in HS_{sc}^{*,\alpha}(A, B)$, then*

$$|b_{2n}| \leq \begin{cases} \frac{1 - \alpha^2}{2} + \frac{(A - B)\alpha}{2}, & n = 1, \\ \frac{(1 - \alpha^2)}{2n} \left\{ 1 + \sum_{k=1}^{n-1} (4k + 1) F_k(A, B) \right\} + \alpha F_n(A, B), & n \geq 2, \end{cases} \tag{44}$$

and

$$|b_{2n+1}| \leq \begin{cases} \frac{1-\alpha^2}{3}(1+A-B) + \frac{(A-B)\alpha}{2}, & n=1, \\ \frac{(1-\alpha^2)}{2n+1}\left\{1 + \sum_{k=1}^{n-1}(4k+1)F_k(A,B) + 2nF_n(A,B)\right\} + \alpha F_n(A,B), & n \geq 2. \end{cases} \quad (45)$$

The estimate is sharp and the extremal function is

$$f_0^\alpha(z) = \int_0^z \frac{1+(A-B-1)t}{(1-t)(1-t^2)^{\frac{A-B}{2}}}dt + \overline{\int_0^z \frac{(\alpha + (1-\alpha^2-\alpha)t)(1+(A-B-1)t)}{(1-t)^2(1-t^2)^{\frac{A-B}{2}}}dt}. \quad (46)$$

Specially, if $f \in HS_{sc}^{*,\alpha}$, then

$$|b_n| \leq \frac{(n-1)(1-\alpha^2)}{2} + \alpha. \quad (47)$$

The estimate is sharp and the extremal function is

$$f_1^\alpha(z) = \frac{z}{1-z} + \overline{\frac{\alpha z + \frac{1}{2}(1-\alpha^2-2\alpha)z^2}{(1-z)^2}} = z + \sum_{n=2}^\infty z^n + \sum_{n=1}^\infty \overline{\left(\frac{(n-1)(1-\alpha^2)}{2} + \alpha\right)z^n}. \quad (48)$$

Especially, let $\alpha = 0$ and $\alpha = \frac{1}{2}$ in (48) respectively, we have (i) If $f \in HS_{sc}^{*,0}$, then

$$|b_n| \leq \frac{n-1}{2}.$$

The estimate is sharp and the extremal function is

$$f_1^0(z) = \frac{z}{1-z} + \overline{\frac{z^2}{2(1-z)^2}} = z + \sum_{n=2}^\infty z^n + \sum_{n=1}^\infty \overline{\frac{n-1}{2}z^n}.$$

(ii) If $f \in HS_{sc}^{*,\frac{1}{2}}$, then

$$|b_n| \leq \frac{3n+1}{8}.$$

The estimate is sharp and the extremal function is

$$f_1^{\frac{1}{2}}(z) = \frac{z}{1-z} + \overline{\frac{4z-z^2}{8(1-z)^2}} = z + \sum_{n=2}^\infty z^n + \sum_{n=1}^\infty \overline{\frac{3n+1}{8}z^n}.$$

In the following Figure 1, we draw the graph of $f_1^0(z)$ and $f_1^{\frac{1}{2}}(z)$ respectively.

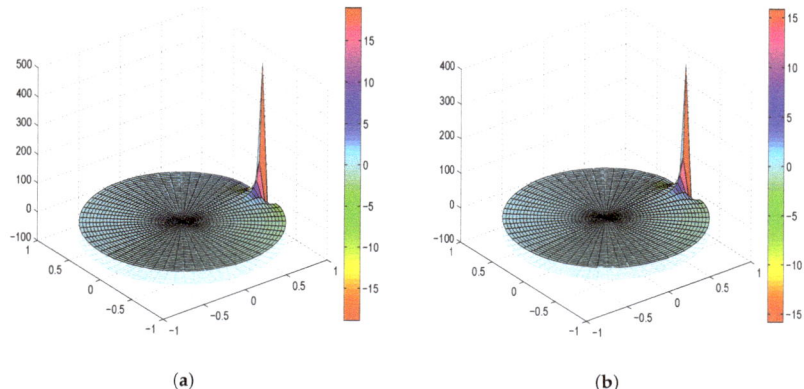

Figure 1. Three dimensional coordinates plus color, the z-axis represents the real part of the function, and the color represents the imaginary part of the function. (**a**) The graph of $f_1^0(z)$; (**b**) The graph of $f_1^{\frac{1}{2}}(z)$.

Proof. By using the relation $g' = \omega h'$, where h and g are given by (3) and $\omega(z) = c_0 + c_1 z + c_2 z^2 + \cdots$ is analytic in $\mathbb{U}$, we obtain

$$2n b_{2n} = \sum_{p=1}^{2n} p a_p c_{2n-p} \quad (a_1 = 1, n \geq 1) \tag{49}$$

and

$$(2n+1) b_{2n+1} = \sum_{p=1}^{2n+1} p a_p c_{2n+1-p} \quad (a_1 = 1, n \geq 1). \tag{50}$$

It is easy to show that

$$2n|b_{2n}| \leq |c_{2n-1}| + 2|a_2||c_{2n-2}| + \ldots + (2n-1)|a_{2n-1}||c_1| + 2n|a_{2n}||c_0| \tag{51}$$

and

$$(2n+1)|b_{2n+1}| \leq |c_{2n}| + 2|a_2||c_{2n-1}| + \ldots + (2n)|a_{2n}||c_1| + (2n+1)|a_{2n+1}||c_0|. \tag{52}$$

Since $g' = \omega h'$, it follows that $c_0 = b_1$. By (7), it can easily be verified that $|c_k| \leq 1 - \alpha^2, k = 1, 2, \cdots, 2n$. Therefore,

$$|b_{2n}| \leq \begin{cases} \frac{1-\alpha^2}{2} + |a_2|\alpha, & n = 1, \\ \frac{(1-\alpha^2)}{2n}(1 + \sum_{k=2}^{2n-1} k|a_k|) + \alpha|a_{2n}|, & n \geq 2, \end{cases} \tag{53}$$

and

$$|b_{2n+1}| \leq \begin{cases} \frac{1-\alpha^2}{3}(1 + 2|a_2|) + |a_3|\alpha, & n = 1, \\ \frac{(1-\alpha^2)}{2n+1}(1 + \sum_{k=2}^{2n} k|a_k|) + \alpha|a_{2n+1}|, & n \geq 2. \end{cases} \tag{54}$$

According to Lemma 2, (53) and (54), by simple calculation, we can obtain (44), (45) and (47). We also obtain the extreme function. Thus, the proof is completed. □

Using the same methods in Theorem 4, we have the following results.

Theorem 5. Let $-1 \leq B < A \leq 1$, $f = h + \overline{g}$, where h and g are given by (3) and $F_k(A, B)$ is defined by (10). If $f \in HK_{sc}^\alpha(A, B)$, then

$$|b_{2n}| \leq \begin{cases} \frac{1-\alpha^2}{2} + \frac{\alpha(A-B)}{4}, & n = 1, \\ \frac{(1-\alpha^2)}{2n}\left(1 + 2\sum_{k=1}^{n-1} F_k(A, B)\right) + \frac{\alpha}{2n}F_n(A, B), & n \geq 2, \end{cases}$$

and

$$|b_{2n+1}| \leq \begin{cases} \frac{(1-\alpha^2)}{3}\left(1 + \frac{A-B}{2}\right) + \frac{\alpha(A-B)}{6}, & n = 1, \\ \frac{(1-\alpha^2)}{2n+1}\left(1 + 2\sum_{k=1}^{n-1} F_k(A, B) + F_n(A, B)\right) + \frac{\alpha}{2n+1}F_n(A, B), & n \geq 2. \end{cases}$$

Specially, if $f \in HK_{sc}^\alpha$, $n = 3, 4, \cdots$, then

$$|b_n| \leq \frac{(n-1)(1-\alpha^2)}{n} + \frac{\alpha}{n}.$$

The estimate is sharp and the extremal function is

$$f_2^\alpha(z) = -\log(1-z) + \overline{(1-\alpha^2)\frac{z}{1-z} - (\alpha^2 + \alpha - 1)\log(1-z)}$$

$$= z + \sum_{n=2}^\infty \frac{1}{n}z^n + \overline{\sum_{n=1}^\infty \left(\frac{(n-1)(1-\alpha^2)}{n} + \frac{\alpha}{n}\right)z^n}.$$

Especially, let $\alpha = 0$ and $\alpha = \frac{1}{2}$ respectively, we have (i) If $f \in HK_{sc}^0$, then

$$|b_n| \leq \frac{n-1}{n}$$

and the estimate is sharp and the extremal function is

$$f_2^0(z) = -\log(1-z) + \overline{\frac{z}{1-z} + \log(1-z)} = z + \sum_{n=2}^\infty \frac{1}{n}z^n + \overline{\sum_{n=1}^\infty \frac{n-1}{n}z^n}.$$

(ii) If $f \in HK_{sc}^{\frac{1}{2}}$, then

$$|b_n| \leq \frac{3n-1}{4n}$$

and the estimate is sharp and the extremal function is

$$f_2^{\frac{1}{2}}(z) = -\log(1-z) + \overline{\frac{3z}{4(1-z)} + \frac{1}{4}\log(1-z)} = z + \sum_{n=2}^\infty \frac{1}{n}z^n + \overline{\sum_{n=1}^\infty \frac{3n-1}{4n}z^n}.$$

In the following Figure 2, we draw the graph of $f_2^0(z)$ and $f_2^{\frac{1}{2}}(z)$ respectively.

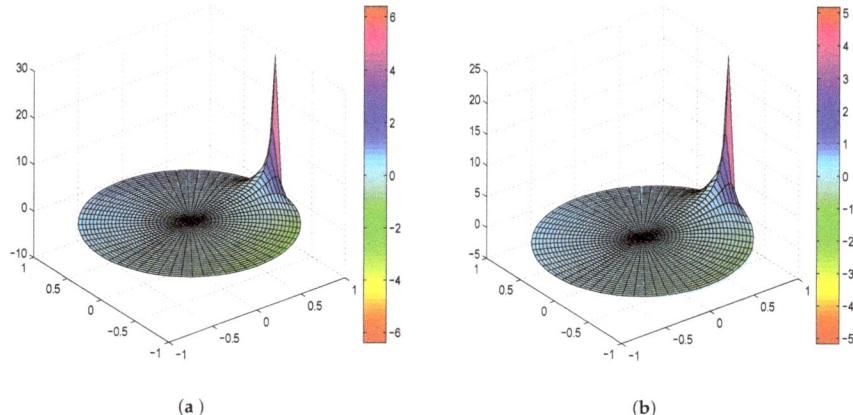

Figure 2. Three dimensional coordinates plus color, the z-axis represents the real part of the function, and the color represents the imaginary part of the function. (**a**) The graph of $f_1^0(z)$; (**b**) The graph of $f_1^{\frac{1}{2}}(z)$.

From Theorems 4 and 5, we have

Corollary 2. *Let* $f = h + \overline{g}$, *where* h *and* g *are given by* (3) *and* $F_k(A, B)$ *is defined by* (10). (1) *If* $f \in HS_{sc}^*(A, B)$, *then*

$$|b_{2n}| \leq \begin{cases} \frac{1}{2} + \frac{(A-B)^2}{8}, & n = 1, \\ \frac{(1+\sum_{k=1}^{n-1}(4k+1)F_k(A,B))^2 + n^2 F_n^2(A,B)}{2n(1+\sum_{k=1}^{n-1}(4k+1)F_k(A,B))}, & n \geq 2, \end{cases}$$

and

$$|b_{2n+1}| \leq \begin{cases} \frac{16+32(A-B)+25(A-B)^2}{48(1+A-B)}, & n = 1, \\ \frac{4(1+\sum_{k=1}^{n-1}(4k+1)F_k(A,B)+2nF_n(A,B))^2 + (2n+1)^2 F_n^2(A,B)}{4(2n+1)(1+\sum_{k=1}^{n-1}(4k+1)F_k(A,B)+2nF_n(A,B))}, & n \geq 2. \end{cases}$$

Especially, if $f \in HS_{sc}^*$, *then* $|b_n| \leq \frac{(n-1)^2+1}{2(n-1)}$. (2) *If* $f \in HK_{sc}(A, B)$, *then*

$$|b_{2n}| \leq \begin{cases} \frac{1}{2} + \frac{(A-B)^2}{32}, & n = 1, \\ \frac{4(1+2\sum_{k=1}^{n-1}F_k(A,B))^2 + F_n^2(A,B)}{8n(1+2\sum_{k=1}^{n-1}F_k(A,B))}, & n \geq 2, \end{cases}$$

and

$$|b_{2n+1}| \leq \begin{cases} \frac{16+16(A-B)+5(A-B)^2}{24(2+A-B)}, & n = 1, \\ \frac{4(1+2\sum_{k=1}^{n-1}F_k(A,B)+F_n(A,B))^2 + F_n^2(A,B)}{4(2n+1)(1+2\sum_{k=1}^{n-1}F_k(A,B)+F_n(A,B))}, & n \geq 2. \end{cases}$$

Especially, if $f \in HK_{sc}$, *then* $|b_n| \leq \frac{4(n-1)^2+1}{4n(n-1)}$.

Also, we give the Fekete-Szegö inequality for functions of these classes.

Theorem 6. Let $f = h + \bar{g}$, where h and g are given by (3), for $\mu \in \mathbb{C}, -1 \leq B < A \leq 1, F_n(A,B)$ is defined by (10). (1) If $f \in HS_{sc}^{*,\alpha}(A,B)$, then

$$|b_3 - \mu b_2^2| \leq \tfrac{(1-\alpha^2)}{3}\left\{1 + \tfrac{3|\mu|(1-\alpha^2)}{4} + \tfrac{(A-B)}{2}|2 - 3\mu b_1|\right\} + \tfrac{(A-B)\alpha}{2}\max\left\{1, |B + \tfrac{\mu b_1}{2}(A-B)|\right\}, \quad (55)$$

$$|b_{2n} - b_{2n-1}| \leq \begin{cases} \tfrac{1}{2}(1-\alpha^2) + (1 + \tfrac{A-B}{2})\alpha, & n = 1, \\ (1-\alpha^2)\left\{(\tfrac{1}{2n} + \tfrac{1}{2n-1})(1 + \sum_{k=1}^{n-1}(4k+1)F_k(A,B)) - F_{n-1}(A,B)\right\} + \\ \alpha(F_n(A,B) + F_{n-1}(A,B)), & n \geq 2, \end{cases} \quad (56)$$

and

$$|b_{2n+1} - b_{2n}| \leq (1-\alpha^2)\left\{(\tfrac{1}{2n+1} + \tfrac{1}{2n})(1 + \sum_{k=1}^{n-1}(4k+1)F_k(A,B)) + \tfrac{2n}{2n+1}F_n(A,B)\right\} \\ + 2\alpha F_n(A,B), \quad n \geq 1. \quad (57)$$

(2) If $f \in HK_{sc}^{\alpha}(A,B)$, then

$$|b_3 - \mu b_2^2| \leq \tfrac{(1-\alpha^2)}{3}\left\{1 + \tfrac{3|\mu|(1-\alpha^2)}{4} + \tfrac{(A-B)}{4}|2 - 3\mu b_1|\right\} + \tfrac{(A-B)\alpha}{6}\max\left\{1, |B + \tfrac{3(A-B)b_1\mu}{8}|\right\}, \quad (58)$$

$$|b_{2n} - b_{2n-1}| \leq \begin{cases} \tfrac{1}{2}(1-\alpha^2) + (1 + \tfrac{A-B}{4})\alpha, & n = 1, \\ (1-\alpha^2)\left\{(\tfrac{1}{2n} + \tfrac{1}{2n-1})(1 + 2\sum_{k=1}^{n-1}F_k(A,B)) - \tfrac{F_{n-1}(A,B)}{2n-1}\right\} + \\ \alpha(\tfrac{F_n(A,B)}{2n} + \tfrac{F_{n-1}(A,B)}{2n-1}), & n \geq 2, \end{cases} \quad (59)$$

and

$$|b_{2n+1} - b_{2n}| \leq (1-\alpha^2)\left\{(\tfrac{1}{2n+1} + \tfrac{1}{2n})(1 + 2\sum_{k=1}^{n-1}F_k(A,B)) + \tfrac{1}{2n+1}F_n(A,B)\right\} + \\ \alpha F_n(A,B)(\tfrac{1}{2n+1} + \tfrac{1}{2n}), \quad n \geq 1. \quad (60)$$

Proof. From the relation (49) and (50), we have

$$2b_2 = c_1 + 2a_2c_0, \quad 3b_3 = c_2 + 2a_2c_1 + 3a_3c_0,$$

and

$$2nb_{2n} = \sum_{p=1}^{2n} pa_p c_{2n-p}, \quad (2n+1)b_{2n+1} = \sum_{p=1}^{2n+1} pa_p c_{2n+1-p} \quad (a_1 = 1, n \geq 1).$$

By (7), we have

$$|b_3 - \mu b_2^2| \leq \tfrac{1-\alpha^2}{3}\left\{1 + \tfrac{3|\mu|(1-\alpha^2)}{4} + |a_2||2 - 3\mu b_1|\right\} + \alpha\left|a_3 - \mu b_1 a_2^2\right|,$$

$$|b_{2n} - b_{2n-1}| \leq \begin{cases} \tfrac{1}{2}(1-\alpha^2) + \alpha(1 + |a_2|), & n = 1, \\ (1-\alpha^2)\left(\tfrac{1}{2n}\sum_{p=1}^{2n-1} p|a_p| + \tfrac{1}{2n-1}\sum_{p=1}^{2n-2} p|a_p|\right) + \alpha(|a_{2n}| + |a_{2n-1}|), & n \geq 2, \end{cases}$$

and

$$|b_{2n+1} - b_{2n}| \leq (1-\alpha^2)\left(\tfrac{1}{2n+1}\sum_{p=1}^{2n} p|a_p| + \tfrac{1}{2n}\sum_{p=1}^{2n-1} p|a_p|\right) + \alpha(|a_{2n+1}| + |a_{2n}|), \quad n \geq 1.$$

According to Lemmas 2 and 3, we can compete the proof of Theorem 6. □

Especially, we let $A = 1, B = -1$, we obtain the following results.

Corollary 3. Let $f = h + \overline{g}$, where h and g are given by (3), for $\mu \in \mathbb{R}$. (1) If $f \in HS_{sc}^{*,\alpha}$, then

$$|b_3 - \mu b_2^2| \leq \frac{(1-\alpha^2)}{3}\left\{1 + \frac{3|\mu|(1-\alpha^2)}{4} + |2 - 3\mu b_1|\right\} + \alpha \max\{1, |b_1\mu - 1|\}, \tag{61}$$

and

$$|b_{n+1} - b_n| \leq \frac{(2n-1)}{2}(1-\alpha^2) + 2\alpha, \quad n \geq 1. \tag{62}$$

Especially, for $f_1^0(z) \in HS_{sc}^{*,0}$ given by Theorem 4, we have $|b_{n+1} - b_n| \leq \frac{1}{2}$. And for $f_1^{\frac{1}{2}}(z) \in HS_{sc}^{*,\frac{1}{2}}$ given by Theorem 4, we have $|b_{n+1} - b_n| \leq \frac{3}{8}$. (2) If $f \in HK_{sc}^{\alpha}$, then

$$|b_3 - \mu b_2^2| \leq \frac{(1-\alpha^2)}{3}\left\{1 + \frac{3|\mu|(1-\alpha^2)}{4} + \frac{1}{2}|2 - 3\mu b_1|\right\} + \frac{\alpha}{3}\max\left\{1, \left|\frac{3b_1\mu}{4} - 1\right|\right\}, \tag{63}$$

$$|b_{n+1} - b_n| \leq (1-\alpha^2)\left(\frac{n}{n+1} + \frac{n-1}{n}\right) + \alpha\left(\frac{1}{n+1} + \frac{1}{n}\right), \quad n \geq 1. \tag{64}$$

Especially, for $f_2^0(z) \in HK_{sc}^0$ given by Theorem 5, we have $|b_{n+1} - b_n| \leq \frac{1}{n(n+1)}$. And for $f_2^{\frac{1}{2}}(z) \in HK_{sc}^{\frac{1}{2}}$ given by Theorem 5, we have $|b_{n+1} - b_n| \leq \frac{1}{4n(n+1)}$.

From Corollary 3, it is easy to obtain the following results.

Corollary 4. Let $f = h + \overline{g}$, where h and g are given by (3). (1) If $f \in HS_{sc}^*$, then

$$|b_{n+1} - b_n| \leq \begin{cases} 2, & n = 1, \\ \frac{4n^2 - 4n + 5}{4n - 2}, & n \geq 2. \end{cases} \tag{65}$$

(2) If $f \in HK_{sc}$, then

$$|b_{n+1} - b_n| \leq \begin{cases} \frac{3}{2}, & n = 1, \\ \frac{16n^4 - 12n^2 + 4n + 5}{4n(n+1)(2n^2-1)}, & n \geq 2. \end{cases} \tag{66}$$

Inspired by Zhu et al. [17], we obtain the distortion estimates and growth estimate of the co-analytic part g, Jacobian estimates, growth estimate and covering theorems of the classes of harmonic mapping with symmetric conjecture point defined by subordination as follows.

Theorem 7. Let $f = h + \overline{g} \in S_H$, $|z| = r \in [0,1)$. (1) If $f \in HS_{sc}^{*,\alpha}(\beta)$, then

$$\frac{\max\{\alpha - r, 0\}[1 - (1-2\beta)r]}{(1-\alpha r)(1+r)^{3-2\beta}} \leq |g'(z)| \leq \frac{(\alpha+r)[1 + (1-2\beta)r]}{(1+\alpha r)(1-r)^{3-2\beta}}. \tag{67}$$

Especially, let $\beta = 0$, for $f_1^0(z) \in HS_{sc}^{*,0}$ given by Theorem 4, we have

$$|g'(z)| \leq \frac{r}{(1-r)^3}.$$

(2) If $f \in HK_{sc}^{\alpha}(\beta)$, then

$$\frac{\max\{\alpha - r, 0\}}{(1-\alpha r)(1+r)^{2-2\beta}} \leq |g'(z)| \leq \frac{(\alpha+r)}{(1+\alpha r)(1-r)^{2-2\beta}}. \tag{68}$$

Especially, let $\beta = 0$, for $f_2^0(z) \in HK_{sc}^0$ given by Theorem 5, we have

$$|g'(z)| \leq \frac{r}{(1-r)^2}.$$

Proof. According to the relation $g' = \omega h'$, $|\omega(0)| = |g'(0)| = |b_1| = \alpha$, it is easy to see $\omega(z)$ such that (see [21]):

$$\left|\frac{\omega(z) - \omega(0)}{1 - \overline{\omega(0)}\omega(z)}\right| \leq |z|, \tag{69}$$

that is,

$$\left|\omega(z) - \frac{\omega(0)(1-r^2)}{1 - |\omega(0)|^2 r^2}\right| \leq \frac{r(1-|\omega(0)|^2)}{1-|\omega(0)|^2 r^2}. \tag{70}$$

From (70), we get

$$\frac{\max\{\alpha - r, 0\}}{1 - \alpha r} \leq |\omega(z)| \leq \frac{\alpha + r}{1 + \alpha r}, \quad z \in \mathbb{U}. \tag{71}$$

Applying (71) and (27), we get (67). Similarly, applying (71) and (28), we get (68). So the proof is completed. □

By using the same method in proof of Lemma 7, it is easy to obtain the following results.

Theorem 8. *Let $f = h + \overline{g} \in S_H$, $|z| = r \in [0, 1)$. (1) If $f \in HS_{sc}^{*,\alpha}(\beta)$, then*

$$\int_0^r \frac{\max\{\alpha - t, 0\}[1 - (1-2\beta)t]}{(1 - \alpha t)(1+t)^{3-2\beta}} dt \leq |g(z)| \leq \int_0^r \frac{(\alpha + t)[1 + (1-2\beta)t]}{(1 + \alpha t)(1-t)^{3-2\beta}} dt. \tag{72}$$

Especially, let $\beta = 0$, for $f_1^0(z) \in HS_{sc}^{*,0}$ given by Theorem 4, we have

$$|g(z)| \leq \frac{r^2}{2(1-r)^2}.$$

(2) If $f \in HK_{sc}^\alpha(\beta)$, then

$$\int_0^r \frac{\max\{\alpha - t, 0\}}{(1 - \alpha t)(1+t)^{2-2\beta}} dt \leq |g(z)| \leq \int_0^r \frac{(\alpha + t)}{(1 + \alpha t)(1-t)^{2-2\beta}} dt. \tag{73}$$

Especially, let $\beta = 0$, for $f_2^0(z) \in HK_{sc}^0$ given by Theorem 5, we have

$$|g(z)| \leq \frac{r}{(1-r)} + \log(1-r).$$

In the following, we can obtain the Jacobian estimates and growth estimates of f.

Theorem 9. *Let $f = h + \overline{g} \in S_H$, $|z| = r \in [0, 1)$. (1) If $f \in HS_{sc}^{*,\alpha}(\beta)$, then*

$$\frac{[1 - (1-2\beta)r]^2(1-\alpha^2)(1-r^2)}{(1+r)^{6-4\beta}(1+\alpha r)^2} \leq J_f(z) \leq \begin{cases} \frac{[1+(1-2\beta)r]^2(1-\alpha^2)(1-r^2)}{(1-r)^{6-4\beta}(1-\alpha r)^2}, & r < \alpha, \\ \frac{[1+(1-2\beta)r]^2}{(1-r)^{6-4\beta}}, & r \geq \alpha. \end{cases}$$

(2) If $f \in HK_{sc}^\alpha(\beta)$, then

$$\frac{(1-\alpha^2)(1-r^2)}{(1+r)^{4-4\beta}(1+\alpha r)^2} \leq J_f(z) \leq \begin{cases} \frac{(1-\alpha^2)(1-r^2)}{(1-r)^{4-4\beta}(1-\alpha r)^2}, & r < \alpha, \\ \frac{1}{(1-r)^{4-4\beta}}, & r \geq \alpha. \end{cases}$$

Proof. We know that the Jacobian of $f = h + \overline{g}$ is in the following form

$$J_f(z) = |h'(z)|^2 - |g'(z)|^2 = |h'(z)|^2(1 - |\omega(z)|^2), \tag{74}$$

where $\omega(z)$ is the dilatation of $f(z)$.

Let $f \in HS_{sc}^{*,\alpha}(\beta)$, applying (27) and (71) to (74), we obtain

$$J_f(z) \geq \frac{[1-(1-2\beta)r]^2}{(1+r)^{6-4\beta}} \cdot \frac{(1-\alpha^2)(1-r^2)}{(1+\alpha r)^2},$$

and

$$J_f(z) \leq \frac{[1+(1-2\beta)r]^2}{(1-r)^{6-4\beta}}\left(1 - \frac{(\max\{(\alpha-r),0\})^2}{(1-\alpha r)^2}\right) = \begin{cases} \frac{[1+(1-2\beta)r]^2}{(1-r)^{6-4\beta}} \cdot \frac{(1-\alpha^2)(1-r^2)}{(1-\alpha r)^2}, & r < \alpha, \\ \frac{[1+(1-2\beta)r]^2}{(1-r)^{6-4\beta}}, & r \geq \alpha. \end{cases}$$

Therefore, we complete the proof of (1). Applying (28) and (71) to (74), (2) of Theorem 9 can be proved by the same method in the same way as shown before. □

Theorem 10. *Let $f = h + \bar{g} \in S_H$, $|z| = r$, $0 \leq r < 1$. (1) If $f \in HS_{sc}^{*,\alpha}(\beta)$, then*

$$\int_0^r \frac{(1-\alpha)(1-\xi)[1-(1-2\beta)\xi]}{(1+\alpha\xi)(1+\xi)^{3-2\beta}} d\xi \leq |f(z)| \leq \int_0^r \frac{(1+\alpha)(1+\xi)[1+(1-2\beta)\xi]}{(1+\alpha\xi)(1-\xi)^{3-2\beta}} d\xi. \quad (75)$$

(2) If $f \in HK_{sc}^{\alpha}(\beta)$, then

$$\int_0^r \frac{(1-\alpha)(1-\xi)}{(1+\alpha\xi)(1+\xi)^{2-2\beta}} d\xi \leq |f(z)| \leq \int_0^r \frac{(1+\alpha)(1+\xi)}{(1+\alpha\xi)(1-\xi)^{2-2\beta}} d\xi. \quad (76)$$

Proof. For any point $z = re^{i\theta} \in \mathbb{U}$, let $\mathbb{U}_r = \mathbb{U}(0,r) = \{z \in \mathbb{U} : |z| < r\}$ and denote

$$d = \min_{z \in \mathbb{U}_r} |f(\mathbb{U}_r)|, \quad (77)$$

and then $\mathbb{U}(0,d) \subseteq f(\mathbb{U}_r) \subseteq f(\mathbb{U})$. Hence, there exists $z_r \in \partial \mathbb{U}_r$ such that $d = |f(z_r)|$. Let $L(t) = tf(z_r)$, $t \in [0,1]$, then $\ell(t) = f^{-1}(L(t))$, $t \in [0,1]$ is a well-defined Jordan arc. For $f = h + \bar{g} \in HS_{sc}^{*,\alpha}(\beta)$, using (27) and (71), we have

$$\begin{aligned} d = |f(z_r)| &= \int_L |d\omega| = \int_\ell |df| = \int_\ell |h'(\eta)d\eta + \overline{g'(\eta)}d\bar{\eta}| \\ &\geq \int_\ell |h'(\eta)|(1 - |\omega(\eta)|)|d\eta| \\ &\geq \int_\ell \frac{(1-\alpha)(1-|\eta|)}{1+\alpha|\eta|} \cdot \frac{[1-(1-2\beta)|\eta|]}{(1+|\eta|)^{3-2\beta}}|d\eta|, \\ &= \int_0^1 \frac{(1-\alpha)(1-|\ell(t)|)}{1+\alpha|\ell(t)|} \cdot \frac{[1-(1-2\beta)|\ell(t)|]}{(1+|\ell(t)|)^{3-2\beta}} dt, \\ &\geq \int_0^r \frac{(1-\alpha)(1-\xi)}{1+\alpha\xi} \cdot \frac{[1-(1-2\beta)\xi]}{(1+\xi)^{3-2\beta}} d\xi \end{aligned}$$

Applying (27) and (71) with a simple calculation, we can obtain the right side of (75). The remainder of the argument is analogous to that in (76) and so is omitted. □

According to (75) and (76), we have the following covering theorems of f.

Corollary 5. *Let $f = h + \bar{g} \in S_H$. (1) If $f \in HS_{sc}^{*,\alpha}(\beta)$, then $\mathbb{U}(0, R_1) \subset f(\mathbb{U})$, where*

$$R_1 = \int_0^1 \frac{(1-\alpha)(1-\xi)[1-(1-2\beta)\xi]}{(1+\alpha\xi)(1+\xi)^{3-2\beta}} d\xi.$$

(2) If $f \in HK_{sc}^{\alpha}(\beta)$, then $\mathbb{U}(0, R_2) \subset f(\mathbb{U})$, where

$$R_2 = \int_0^1 \frac{(1-\alpha)(1-\zeta)}{(1+\alpha\zeta)(1+\zeta)^{2-2\beta}} d\zeta.$$

Note: *In this paper, the geometric properties of the co-analytic part g is obtained by using the analytic part h satisfying certain conditions. Furthermore, the geometric properties of harmonic functions are obtained (see Figures 1 and 2). Using the concepts dealt with in the paper, we can study the geometric properties of the co-analytic part and harmonic function when the analytic part satisfies other conditions. So as to enrich the research field of univalent harmonic mapping.*

Author Contributions: Conceptualization, S.L. and L.M.; methodology, S.L. and L.M.; software, L.M. and X.N.; validation, L.N-M., S.L. and X.N.; formal analysis, S.-H.L; investigation, L.M.; resources, S.L.; data curation, X.N.; writing—original draft preparation, L.M.; writing—review and editing, S.L.; visualization, X.N.; supervision, S.L.; project administration, S.L.; funding acquisition, S.L.

Funding: Supported by the Inner Mongolia autonomous region key institutions of higher learning scientific research projects (No. NJZZ19209), the Natural Science Foundation of the People's Republic of China under Grants (No. 11561001), the Program for Young Talents of Science and Technology in Universities of Inner Mongolia Autonomous Region under Grant (No. NJYT-18-A14), the Natural Science Foundation of Inner Mongolia of the People's Republic of China under Grant (No. 2018MS01026) and the Higher School Research Foundation of Inner Mongolia of China under Grant (No. NJZY18217, No. NJZY17300).

Conflicts of Interest: The authors declare no conflict of interest.

References

1. Duren, P.L. *Univalent Functions: 259 (Grundlehren der MathematischenWissenschaften)*; Springer: New York, NY, USA; Berlin/Heidelberg, Germany; Tokyo, Japan, 1983.
2. Srivastava, H.M.; Owa, S. (Eds.) *Current Topics in Analytic Function Theory*; World Scientific Publishing Company: Singapore; Hackensack, NJ, USA; London, UK; Hong Kong, China, 1992.
3. Ma, W.; Minda, D. A unified treatment of some special classes of univalent functions. In *Proceeding of the Conference on Complex Analysis*; Li, Z., Ren, F., Yang, L., Zhang, S., Eds.; International Press: Boston, MA, USA, 1994; pp. 157–169.
4. Janowski, W. Some extremal problems for certain families of analytic functions. *Ann. Polonici Math.* **1973**, *28*, 297–326. [CrossRef]
5. Robertso, M.I.S. On the theory of univalent functions. *Ann. Math.* **1936**, *37*, 374–408. [CrossRef]
6. Sakaguchi, K. On a certain univalent mapping. *J. Math. Soc. Jpn.* **1959**, *11*, 72–75. [CrossRef]
7. El-Ashwah, R.M.; Thomas, D.K. Some subclasses of closed-to-convex functions. *J. Ramanujan Math. Soc.* **1987**, *2*, 86.
8. Fekete, M.; Szegö, G. Eine bemerkung uber ungerade schlichte funktionen. *J. Lond. Math. Soc.* **1933**, *8*, 85–89. [CrossRef]
9. EL-Ashwah, R.M.; Hassan, A.H. Fekete-Szegö inequalities for certain subclass of analytic functions defined by using Sălăgean operator. *Miskolc Math. Notes* **2017**, *17*, 827–836. [CrossRef]
10. Koepf, W. On the Fekete-Szegö problem for close-to-convex functions. *Proc. Am. Math. Soc.* **1987**, *101*, 89–95.
11. Sakar, F.M.; Aytaş, S.; Güney, H.Ö. On The Fekete-Szegö problem for generalized class $M_{\alpha,\gamma}(\beta)$ defined by differential operator. *Süleyman Demirel Univ. J. Nat. Appl. Sci.* **2016**, *20*, 456–459.
12. Clunie, J.; Sheil-Small, T. Harmonic univalent functions. *Ann. Acad. Sci. Fenn. Ser. AI Math.* **1984**, *39*, 3–25. [CrossRef]
13. Duren, P.L. *Harmonic Mappings in the Plane*; Cambridge University Press: Cambridge, UK, 2004.
14. Lewy, H. On the non-vanishing of the Jacobian in certain one-to-one mappings. *Bull. Am. Math. Soc.* **1936**, *42*, 689–692. [CrossRef]
15. Klimek, D.; Michalski, A. Univalent anti-analytic perturbations of convex analytic mappings in the unit disc. *Ann. Univ. Mariae Curie-Skłodowska Sec. A* **2007**, *61*, 39–49.
16. Hotta, I.; Michalski, A. Locally one-to-one harmonic functions with starlike analytic part. *arXiv* **2014**, arXiv:1404.1826.

17. Zhu, M.; Huang, X. The distortion theorems for harmonic mappings with analytic parts convex or starlike functions of order β. *J. Math.* **2015**, *2015*, 460191. [CrossRef]
18. Kohr, G.; Graham, I. *Geometric Function Theory in one and Higher Dimensions*; Marcel Dekker: New York, NY, USA, 2003.
19. Polatoğlu, Y.; Bolcal, M.; Sen, A. Two-point distortion theorems for certain families of analytic functions in the unit disc. *Int. J. Math. Math. Sci.* **2003**, *66*, 4183–4193. [CrossRef]
20. Chen, M.P.; Wu, Z.R.; Zou, Z.Z. On functions α-starlike with respect to symmetric conjugate points. *J. Math. Anal. Appl.* **1996**, *201*, 25–34. [CrossRef]
21. Goluzin, G.M. *Geometric Theory of Functions of a Complex Variable*; American Mathematical Society: Providence, RI, USA, 1969.

© 2019 by the authors. Licensee MDPI, Basel, Switzerland. This article is an open access article distributed under the terms and conditions of the Creative Commons Attribution (CC BY) license (http://creativecommons.org/licenses/by/4.0/).

Article

A Study of Third Hankel Determinant Problem for Certain Subfamilies of Analytic Functions Involving Cardioid Domain

Lei Shi [1], Izaz Ali [2], Muhammad Arif [2,*], Nak Eun Cho [3,*], Shehzad Hussain [2] and Hassan Khan [2]

[1] School of Mathematics and Statistics, Anyang Normal University, Anyang 455002, China; shimath@163.com
[2] Department of Mathematics, Abdul Wali Khan University Mardan, Mardan 23200, Pakistan; aliizaz719@gmail.com (I.A.); shehzad873822@gmail.com (S.H.); hassanmath@awkum.edu.pk (H.K.)
[3] Department of Applied Mathematics, Pukyong National University, Busan 48513, Korea
* Correspondence: marifmaths@awkum.edu.pk (M.A.); necho@pknu.ac.kr (N.E.C.)

Received: 18 April 2019; Accepted: 6 May 2019; Published: 10 May 2019

Abstract: In the present article, we consider certain subfamilies of analytic functions connected with the cardioid domain in the region of the unit disk. The purpose of this article is to investigate the estimates of the third Hankel determinant for these families. Further, the same bounds have been investigated for two-fold and three-fold symmetric functions.

Keywords: subordinations; starlike functions; convex functions; close-to-convex functions; cardioid domain; Hankel determinant; m-fold symmetric functions

1. Introduction and Definitions

Let $\mathcal{A}$ be the family of all functions that are holomorphic (or analytic) in the open unit disc $\Delta = \{z \in \mathbb{C} : |z| < 1\}$ and having the following Taylor–Maclaurin series form:

$$f(z) = z + \sum_{k=2}^{\infty} a_k z^k \quad (z \in \Delta). \tag{1}$$

Further, let $\mathcal{S}$ represent a subfamily of $\mathcal{A}$, which contains functions that are univalent in Δ. The familiar coefficient conjecture for the function $f \in \mathcal{S}$ of the form (1) was first presented by Bieberbach [1] in 1916 and proven by de-Branges [2] in 1985. In between the years 1916 and 1985, many researchers tried to prove or disprove this conjecture. Consequently, they defined several subfamilies of $\mathcal{S}$ connected with different image domains. Among these, the families $\mathcal{S}^*$, $\mathcal{C}$, and $\mathcal{K}$ of starlike functions, convex functions, and close-to-convex functions, respectively, are the most fundamental subfamilies of $\mathcal{S}$ and have a nice geometric interpretation. These families are defined as:

$$\mathcal{S}^* = \left\{ f \in \mathcal{S} : \frac{zf'(z)}{f(z)} \prec \frac{1+z}{1-z}, \ (z \in \Delta) \right\},$$

$$\mathcal{C} = \left\{ f \in \mathcal{S} : \frac{(zf'(z))'}{f'(z)} \prec \frac{1+z}{1-z}, \ (z \in \Delta) \right\},$$

$$\mathcal{K} = \left\{ f \in \mathcal{S} : \frac{zf'(z)}{g(z)} \prec \frac{1+z}{1-z}, \ \text{for } g(z) \in \mathcal{S}^*, \ (z \in \Delta) \right\},$$

where the symbol "$\prec$" denotes the familiar subordinations between analytic functions and is defined as: the function h_1 is subordinate to a function h_2, symbolically written as $h_1 \prec h_2$ or $h_1(z) \prec h_2(z)$, if we can find a function w, called the Schwarz function, that is holomorphic in Δ with $w(0) = 0$ and

$|w(z)| < 1$ such that $h_1(z) = h_2(w(z))$ $(z \in \Delta)$. In the case of the univalency of h_2 in Δ, then the following relation holds:

$$h_1(z) \prec h_2(z) \quad (z \in \Delta) \quad \Longleftrightarrow \quad h_1(0) = h_2(0) \quad \text{and} \quad h_1(\Delta) \subset h_2(\Delta).$$

In [3], Padmanabhan and Parvatham in 1985 defined a unified family of starlike and convex functions using familiar convolution with the function $z/(1-z)^a$, for $a \in \mathbb{R}$. Later on, Shanmugam [4] generalized this idea by introducing the family:

$$\mathcal{S}_h^*(\phi) = \left\{ f \in \mathcal{A} : \frac{z(f*h)'}{(f*h)} \prec \phi(z), \ (z \in \Delta) \right\},$$

where "$*$" stands for the familiar convolution, ϕ is a convex, and h is a fixed function in $\mathcal{A}$. Furthermore, if we replace h in $\mathcal{S}_h^*(\phi)$ by $z/(1-z)$ and $z/(1-z)^2$, we obtain the families $\mathcal{S}^*(\phi)$ and $\mathcal{C}(\phi)$ respectively. In 1992, Ma and Minda [5] reduced the restriction to a weaker supposition that ϕ is a function, with Re $\phi(z) > 0$ in Δ, whose image domain is symmetric about the real axis and starlike with respect to $\phi(0) = 1$ with $\phi'(0) > 0$ and discussed some properties including distortion, growth, and covering theorems. The family $\mathcal{S}^*(\phi)$ generalizes various subfamilies of the family $\mathcal{A}$, for example;

(i). If $\phi(z) = \frac{1+Az}{1+Bz}$ with $-1 \leq B < A \leq 1$, then $\mathcal{S}^*[A, B] := \mathcal{S}^*\left(\frac{1+Az}{1+Bz}\right)$ is the family of Janowski starlike functions; see [6]. Further, if $A = 1 - 2\alpha$ and $B = -1$ with $0 \leq \alpha < 1$, then we get the family $\mathcal{S}^*(\alpha)$ of starlike functions of order α.

(ii). The family $\mathcal{S}_L^* := \mathcal{S}^*(\sqrt{1+z})$ was introduced by Sokół and Stankiewicz [7], consisting of functions $f \in \mathcal{A}$ such that $zf'(z)/f(z)$ lies in the region bounded by the right-half of the lemniscate of Bernoulli given by $|w^2 - 1| < 1$.

(iii). For $\phi(z) = 1 + \sin z$, the family $\mathcal{S}^*(\phi)$ leads to the family $\mathcal{S}_{\sin}^*$, introduced in [8].

(iv). When we take $\phi(z) = e^z$, then we have $\mathcal{S}_e^* := \mathcal{S}^*(e^z)$ [9].

(v). The family $\mathcal{S}_R^* := \mathcal{S}^*(\phi(z))$ with $\phi(z) = 1 + \frac{z}{k}\frac{k+z}{k-z}$, $k = \sqrt{2} + 1$ was studied in [10].

(vi). By setting $\phi(z) = 1 + \frac{4}{3}z + \frac{2}{3}z^2$, the family $\mathcal{S}^*(\phi)$ reduces to $\mathcal{S}_{car}^*$, introduced by Sharma and his coauthors [11], consisting of functions $f \in \mathcal{A}$ such that $zf'(z)/f(z)$ lies in the region bounded by the cardioid given by:

$$(9x^2 + 9y^2 - 18x + 5)^2 - 16(9x^2 + 9y^2 - 6x + 1) = 0,$$

and also by the Alexandar-type relation, the authors in [11] defined the family $\mathcal{C}_{car}$ by:

$$\mathcal{C}_{car} = \{f \in \mathcal{A} : zf'(z) \in \mathcal{S}_C^* \ (z \in \Delta)\}; \tag{2}$$

see also [12,13]. For more special cases of the family $\mathcal{S}^*(\phi)$, see [14,15]. We now consider the following family connected with the cardioid domain:

$$\mathcal{R}_{car} = \left\{ f \in \mathcal{A} : f'(z) \prec 1 + \frac{4}{3}z + \frac{2}{3}z^2, \ (z \in \Delta) \right\}. \tag{3}$$

For given parameters $q, n \in \mathbb{N} = \{1, 2, \ldots\}$, the Hankel determinant $H_{q,n}(f)$ was defined by Pommerenke [16,17] for a function $f \in \mathcal{S}$ of the form (1) given by:

$$H_{q,n}(f) = \begin{vmatrix} a_n & a_{n+1} & \cdots & a_{n+q-1} \\ a_{n+1} & a_{n+2} & \cdots & a_{n+q} \\ \vdots & \vdots & \cdots & \vdots \\ a_{n+q-1} & a_{n+q} & \cdots & a_{n+2q-2} \end{vmatrix}. \tag{4}$$

The growth of $H_{q,n}(f)$ has been investigated for different subfamilies of univalent functions. Specifically, the absolute sharp bounds of the functional $H_{2,2}(f) = a_2 a_4 - a_3^2$ were found in [18,19] for each of the families $\mathcal{C}$, $\mathcal{S}^*$ and $\mathcal{R}$, where the family $\mathcal{R}$ contains functions of bounded turning. However, the exact estimate of this determinant for the family of close-to-convex functions is still undetermined [20]. Recently, Srivastava and his coauthors [21] found the estimate of the second Hankel determinant for bi-univalent functions involving the symmetric q-derivative operator, while in [22], the authors studied Hankel and Toeplitz determinants for subfamilies of q-starlike functions connected with the conic domain. For more literature, see [23–30].

The Hankel determinant of third order is given as:

$$H_{3,1}(f) = \begin{vmatrix} 1 & a_2 & a_3 \\ a_2 & a_3 & a_4 \\ a_3 & a_4 & a_5 \end{vmatrix} = -a_5 a_2^2 + 2a_2 a_3 a_4 - a_3^3 + a_5 a_3 - a_4^2. \tag{5}$$

The estimation of the determinant $|H_{3,1}(f)|$ is very hard as compared to deriving the bound of $|H_{2,2}(f)|$. The very first paper on $H_{3,1}(f)$ was given in 2010 by Babalola [31], in which he obtained the upper bound of $H_{3,1}(f)$ for the families of $\mathcal{S}^*$, $\mathcal{C}$, and $\mathcal{R}$. Later on, many authors published their work regarding $|H_{3,1}(f)|$ for different subfamilies of univalent functions; see [32–36]. In 2017, Zaprawa [37] improved the results of Babalola as under:

$$|H_{3,1}(f)| \leq \begin{cases} 1, & \text{for } f \in \mathcal{S}^*, \\ \frac{49}{540}, & \text{for } f \in \mathcal{C}, \\ \frac{41}{60}, & \text{for } f \in \mathcal{R}. \end{cases}$$

and claimed that these bounds are still not the best possible. Further, for the sharpness, he examined the subfamilies of $\mathcal{S}^*$, $\mathcal{C}$, and $\mathcal{R}$ consisting of functions with m-fold symmetry and obtained the sharp bounds. Moreover, in 2018, Kwon et al. [38] improved the bound of Zaprawa for $f \in \mathcal{S}^*$ and proved that $|H_{3,1}(f)| \leq 8/9$, but it is not yet the best possible. The authors in [39–41] contributed in a similar direction by generalizing different families of univalent functions with respect to symmetric points. In 2018, Kowalczyk et al. [42] and Lecko et al. [43] obtained the sharp inequalities:

$$|H_{3,1}(f)| \leq 4/135 \quad \text{and} \quad |H_{3,1}(f)| \leq 1/9,$$

for the recognizable families $\mathcal{K}$ and $\mathcal{S}^*(1/2)$, respectively, where the symbol $\mathcal{S}^*(1/2)$ stands for the family of starlike functions of order $1/2$. Furthermore, we would like to cite the work done by Mahmood et al. [44] in which they studied the third Hankel determinant for a subfamily of starlike functions in the q-analogue. Additionally, Zhang et al. [45] studied this determinant for the family $\mathcal{S}_e^*$ and obtained the bound $|H_{3,1}(f)| \leq 0.565$.

In the present article, our aim is to investigate the estimate of $|H_{3,1}(f)|$ for the subfamilies $\mathcal{S}_{car}^*$, $\mathcal{C}_{car}$, and $\mathcal{R}_{car}$ of analytic functions connected with the cardioid domain. Moreover, we also study this problem for families of m-fold symmetric functions connected with the cardioid domain.

2. A Lemma

Let $\mathcal{P}$ denote the family of all functions p that are analytic in Δ with $\Re(p(z)) > 0$ and having the following series representation:

$$p(z) = 1 + \sum_{n=1}^{\infty} c_n z^n \quad (z \in \Delta). \tag{6}$$

Lemma 1. *If $p \in \mathcal{P}$ and it has the form (6), then:*

$$|c_n| \leq 2 \text{ for } n \geq 1, \tag{7}$$

$$|c_m c_n - c_k c_l| \leq 4 \text{ for } m + n = k + l, \tag{8}$$

$$\left| c_{n+2k} - \mu c_n c_k^2 \right| \leq 2(1 + 2\mu); \text{ for } \mu \in \mathbb{R}, \tag{9}$$

$$\left| c_2 - \frac{c_1^2}{2} \right| \leq 2 - \frac{|c_1|^2}{2}, \tag{10}$$

$$|c_{n+k} - \mu c_n c_k| \leq \begin{cases} 2, & 0 \leq \mu \leq 1; \\ 2|2\mu - 1|, & \text{elsewhere.} \end{cases} \tag{11}$$

where the inequalities (7), (10), (11), and (9) are taken from [46].

3. Bound of $|H_{3,1}(f)|$ for the Family $\mathcal{S}^*_{car}$

Theorem 1. *If $f(z)$ of the form (1) belongs to $\mathcal{S}^*_{car}$, then:*

$$|a_2| \leq \tfrac{4}{3}, \quad |a_3| \leq \tfrac{11}{9} \quad \text{and} \quad |a_4| \leq \tfrac{68}{81}.$$

These bounds are the best possible.

Proof. Let $f \in \mathcal{S}^*_{car}$. Then, in the form of the Schwarz function, we have:

$$\frac{zf'(z)}{f(z)} = 1 + \frac{4}{3}w(z) + \frac{2}{3}(w(z))^2 \quad (z \in \Delta).$$

Furthermore, we easily get:

$$\frac{zf'(z)}{f(z)} = 1 + a_2 z + \left(2a_3 - a_2^2\right) z^2 + \left(3a_4 - 3a_2 a_3 + a_2^3\right) z^3$$

$$+ \left(4a_5 - 2a_3^2 - 4a_2 a_4 + 4a_2^2 a_3 - a_2^4\right) z^4 + \cdots. \tag{12}$$

and from series expansion of w with simple calculations, we can write:

$$1 + \frac{4}{3}w(z) + \frac{2}{3}(w(z))^2 = 1 + \frac{2}{3}c_1 z + \left(\frac{2}{3}c_2 - \frac{c_1^2}{6}\right) z^2 + \left(\frac{2}{3}c_3 - \frac{1}{3}c_1 c_2\right) z^3$$

$$+ \left(\frac{2}{3}c_4 + \frac{c_1^4}{24} - \frac{c_2^2}{6} - \frac{c_1 c_3}{3}\right) z^4 + \cdots. \tag{13}$$

By comparing (12) and (13), we get:

$$a_2 = \frac{2}{3}c_1, \tag{14}$$

$$a_3 = \frac{1}{2}\left(\frac{5}{18}c_1^2 + \frac{2}{3}c_2\right), \tag{15}$$

$$a_4 = \frac{1}{3}\left(\frac{c_1 c_2}{3} + \frac{2}{3}c_3 - \frac{c_1^3}{54}\right), \tag{16}$$

$$a_5 = \frac{1}{4}\left(\frac{2}{3}c_4 + \frac{c_2^2}{18} + \frac{7}{27}c_1 c_3 + \frac{7}{486}c_1^4 - \frac{c_1^2 c_2}{9}\right). \tag{17}$$

Applying (7) in (14) and (15), we have:

$$|a_2| \leq \frac{4}{3} \text{ and } |a_3| \leq \frac{11}{9}.$$

Now, reshuffling (16), we get:

$$a_4 = \frac{1}{3}\left\{\frac{2}{3}c_3 + \frac{8}{27}c_1 c_2 + \frac{c_1}{27}\left(c_2 - \frac{c_1^2}{2}\right)\right\}.$$

If we insert $|c_1| = x \in [0,2]$, then we have:

$$|a_4| \leq \frac{1}{3}\left\{\frac{4}{3} + \frac{16}{27}x + \frac{x}{27}\left(2 - \frac{x^2}{2}\right)\right\}.$$

The above function has its maximum value at $x = 2$. Therefore:

$$|a_4| \leq \frac{68}{81}.$$

Equalities are obtained if we take:

$$f(z) = \exp\left(\frac{4}{3}z + \ln z + \frac{1}{3}z^2\right)$$
$$= z + \frac{4}{3}z^2 + \frac{11}{9}z^3 + \frac{68}{81}z^4 + \frac{235}{486}z^5 + \cdots. \tag{18}$$

□

Theorem 2. *If $f \in \mathcal{S}_{car}^*$ and it has the series form (1), then:*

$$|H_{3,1}(f)| \leq \frac{874}{729}.$$

Proof. From (5), the third Hankel determinant can be written as:

$$H_3(1) = -a_2^2 a_5 + 2a_2 a_3 a_4 - a_3^3 + a_3 a_5 - a_4^2.$$

Inserting (14)–(17), we get:

$$H_{3,1}(f) = \frac{7}{729}c_1^4 c_2 + \frac{281}{11664}c_1^3 c_3 + \frac{c_2 c_4}{18} + \frac{23}{324}c_1 c_2 c_3 - \frac{2083}{419904}c_1^6 - \frac{7}{216}c_2^3 - \frac{11}{216}c_1^2 c_4$$
$$- \frac{59}{2592}c_1^2 c_2^2 - \frac{4}{81}c_3^2.$$

Now, rearranging, it yields:

$$H_{3,1}(f) = \frac{2083}{209952}c_1^4\left(c_2 - \frac{c_1^2}{2}\right) + \frac{c_4}{18}\left(c_2 - \frac{c_1^2}{2}\right) + \frac{281}{23328}c_1^3\left(c_3 - \frac{67}{2559}c_1c_2\right) + \frac{5}{216}c_1(c_2c_3 - c_1c_4)$$
$$-\frac{c_1c_3}{648}\left(c_2 - \frac{c_1^2}{2}\right) + \frac{263}{23328}c_1^2\left(c_1c_3 - c_2^2\right) - \frac{4}{81}c_3(c_3 - c_1c_2) - \frac{67}{5832}c_1^2c_2^2 - \frac{7}{216}c_2^3.$$

Applying the triangle inequality:

$$|H_{3,1}(f)| \leq \frac{2083}{209952}|c_1|^4\left|c_2 - \frac{c_1^2}{2}\right| + \frac{|c_4|}{18}\left|c_2 - \frac{c_1^2}{2}\right| + \frac{281}{23328}|c_1|^3\left|c_3 - \frac{67}{2559}c_1c_2\right| + \frac{5}{216}|c_1||c_2c_3 - c_1c_4|$$
$$+\frac{|c_1||c_3|}{648}\left|c_2 - \frac{c_1^2}{2}\right| + \frac{263}{23328}|c_1|^2\left|c_1c_3 - c_2^2\right| + \frac{4}{81}|c_3||c_3 - c_1c_2| + \frac{67}{5832}|c_1|^2|c_2|^2 + \frac{7}{216}|c_2|^3;$$

besides, (7), (10), (11) and (8) lead us to:

$$|H_{3,1}(f)| \leq \frac{2083}{209952}|c_1|^4\left(2 - \frac{|c_1|^2}{2}\right) + \frac{1}{9}\left(2 - \frac{|c_1|^2}{2}\right) + \frac{281}{11664}|c_1|^3 + \frac{5}{54}|c_1| + \frac{|c_1|}{324}\left(2 - \frac{|c_1|^2}{2}\right)$$
$$+\frac{263}{5832}|c_1|^2 + \frac{16}{81} + \frac{67}{1458}|c_1|^2 + \frac{7}{27}.$$

If we insert $|c_1| = x \in [0,2]$, then we have:

$$|H_3(f)| \leq \frac{2083}{209952}x^4\left(2 - \frac{x^2}{2}\right) + \frac{1}{9}\left(2 - \frac{x^2}{2}\right) + \frac{281}{11664}x^3 + \frac{5}{54}x + \frac{x}{324}\left(2 - \frac{x^2}{2}\right)$$
$$+\frac{263}{5832}x^2 + \frac{16}{81} + \frac{67}{1458}x^2 + \frac{7}{27} = \Phi(x), \text{ say.}$$

Then, the function $\Phi(x)$ is increasing. Therefore, we get its maximum value by putting $x = 2$,

$$|H_{3,1}(f)| \leq \frac{874}{729}.$$

Thus, the proof follows. □

From the function given by (18), we conclude the following conjecture.

Conjecture 3.1. Let $f \in \mathcal{S}_{car}^*$ and in the form (1). Then, the sharp bound is:

$$|H_{3,1}(f)| \leq \frac{827}{13122}.$$

4. Bound of $|H_{3,1}(f)|$ for the Family $\mathcal{C}_{car}$

Theorem 3. If $f \in \mathcal{C}_{car}$ and has the series form (1), then:

$$|a_2| \leq \tfrac{2}{3}, \quad |a_3| \leq \tfrac{11}{27} \quad \text{and} \quad |a_4| \leq \tfrac{17}{81}.$$

These bounds are the best possible.

Proof. Let the function $f \in \mathcal{C}_{car}$. Then, by the Alexandar-type relation, we say that $zf' \in \mathcal{S}_{car}^*$, and hence, using the coefficient bounds of the family $\mathcal{S}_{car}^*$, which was proven in the last Theorem, we get the needed bounds. □

Theorem 4. *Let f have the form (1) and belong to $\mathcal{C}_{car}$. Then:*

$$|H_{3,1}(f)| \leq \frac{319}{4374}.$$

Proof. From (5), the third Hankel determinant can be obtained as:

$$H_{3,1}(f) = -a_2^2 a_5 + 2a_2 a_3 a_4 - a_3^3 + a_3 a_5 - a_4^2.$$

Utilizing the definition of the family $\mathcal{C}_{car}$, we easily have:

$$\begin{aligned} H_{3,1}(f) &= \frac{97}{174960} c_1^4 c_2 + \frac{61}{58320} c_1^3 c_3 + \frac{1}{270} c_2 c_4 + \frac{1}{405} c_1 c_2 c_3 - \frac{617}{3149280} c_1^6 - \frac{31}{29160} c_2^3 \\ &\quad - \frac{7}{3240} c_1^2 c_4 - \frac{143}{116640} c_1^2 c_2^2 - \frac{1}{324} c_3^2. \end{aligned}$$

After reordering, it yields:

$$\begin{aligned} H_{3,1}(f) &= \frac{97}{349920} c_1^4 \left(c_2 - \frac{617}{873} c_1^2\right) - \frac{143}{116640} c_1^2 c_2 \left(c_2 - \frac{97}{429} c_1^2\right) - \frac{7}{3240} c_1^2 \left(c_4 - \frac{61}{126} c_1 c_3\right) \\ &\quad + \frac{c_2}{270} \left(c_4 - \frac{31}{108} c_2^2\right) - \frac{c_3}{324} \left(c_3 - \frac{324}{405} c_1 c_2\right). \end{aligned}$$

Using the triangle inequality, we get:

$$\begin{aligned} |H_{3,1}(f)| &\leq \frac{97}{349920} |c_1|^4 \left|c_2 - \frac{617}{873} c_1^2\right| + \frac{143}{116640} |c_1|^2 |c_2| \left|c_2 - \frac{97}{429} c_1^2\right| + \frac{7}{3240} |c_1|^2 \left|c_4 - \frac{61}{126} c_1 c_3\right| \\ &\quad + \frac{|c_2|}{270} \left|c_4 - \frac{31}{108} c_2^2\right| + \frac{|c_3|}{324} \left|c_3 - \frac{324}{405} c_1 c_2\right|. \end{aligned}$$

The application of (7) and (11) leads us to:

$$\begin{aligned} |H_{3,1}(f)| &\leq \frac{97}{10935} + \frac{143}{7290} + \frac{7}{405} + \frac{4}{270} + \frac{4}{324} \\ &= \frac{319}{4374}. \end{aligned}$$

Thus, the proof is completed. □

5. Bound of $|H_{3,1}(f)|$ for the Family $\mathcal{R}_{car}$

Theorem 5. *Let $f \in \mathcal{R}_{car}$ and be given in the form (1). Then:*

$$|a_2| \leq \tfrac{2}{3}, \quad |a_3| \leq \tfrac{4}{9}, \quad |a_4| \leq \tfrac{1}{3}.$$

These results are the best possible.

Proof. Let $f \in \mathcal{R}_{car}$. Then, we can write (3), in the form of the Schwarz function, as:

$$f'(z) = 1 + \frac{4}{3} w(z) + \frac{2}{3} (w(z))^2, \quad (z \in \Delta).$$

Since:

$$f'(z) = 1 + 2a_2 z + 3a_3 z^2 + 4a_4 z^3 + 5a_5 z^4 + \cdots, \tag{19}$$

by comparing (19) and (13), we may get:

$$a_2 = \frac{c_1}{3}, \tag{20}$$

$$a_3 = \frac{2}{9}\left(c_2 - \frac{c_1^2}{4}\right), \tag{21}$$

$$a_4 = \frac{1}{6}\left(c_3 - \frac{c_1 c_2}{2}\right), \tag{22}$$

$$a_5 = \frac{1}{15}\left(2c_4 + \frac{c_1^4}{8} - \frac{c_2^2}{2} - c_1 c_3\right). \tag{23}$$

Using (7) in (20), we get:

$$|a_2| \leq \frac{2}{3}.$$

Applying (11) in (21) and (22), we obtain:

$$|a_3| \leq \frac{4}{9} \text{ and } |a_4| \leq \frac{1}{3}.$$

Thus, the proof is completed.

Equalities in each coefficient $|a_2|$, $|a_3|$, and $|a_4|$ are obtained respectively by taking:

$$f_1(z) = z + \frac{2}{3}z^2 + \frac{2}{9}z^3,$$
$$f_2(z) = z + \frac{4}{9}z^3 + \frac{2}{15}z^5,$$
$$f_3(z) = z + \frac{1}{3}z^4 + \frac{2}{21}z^7.$$

□

Theorem 6. *Let $f \in \mathcal{R}_{car}$ and be given in the form (1). Then:*

$$|H_{3,1}(f)| \leq \frac{754}{1215}.$$

Proof. From (5), the third Hankel determinant can be written as:

$$H_3(1) = -a_2^2 a_5 + 2 a_2 a_3 a_4 - a_3^3 + a_3 a_5 - a_4^2.$$

Utilizing (20)–(23), we have:

$$H_{3,1}(f) = \frac{7}{2430}c_1^4 c_2 + \frac{2}{405}c_1^3 c_3 + \frac{4}{135}c_2 c_4 + \frac{61}{1620}c_1 c_2 c_3 - \frac{71}{58320}c_1^6 - \frac{67}{1620}c_2^3$$
$$- \frac{c_1^2 c_4}{45} - \frac{107}{19440}c_1^2 c_2^2 - \frac{c_3^2}{36}.$$

By rearranging, it yields:

$$H_{3,1}(f) = \frac{7}{4860}c_1^4\left(c_2 - \frac{71}{84}c_1^2\right) - \frac{107}{19440}c_1^2 c_2\left(c_2 - \frac{28}{107}c_1^2\right) - \frac{c_1^2}{45}\left(c_4 - \frac{2}{9}c_1 c_3\right)$$
$$- \frac{c_3}{36}\left(c_3 - \frac{61}{45}c_1 c_2\right) + \frac{4}{135}c_2\left(c_4 - \frac{67}{108}c_2^2\right).$$

Implementing the triangle inequality, we have:

$$|H_{3,1}(f)| \leq \frac{7}{4860}|c_1|^4 \left|c_2 - \frac{71}{84}c_1^2\right| + \frac{107}{19440}|c_1|^2|c_2|\left|c_2 - \frac{28}{107}c_1^2\right| + \frac{|c_1|^2}{45}\left|c_4 - \frac{2}{9}c_1 c_3\right|$$
$$+ \frac{|c_3|}{36}\left|c_3 - \frac{61}{45}c_1 c_2\right| + \frac{4}{135}|c_2|\left|c_4 - \frac{67}{108}c_2^2\right|.$$

(7) and (11) lead us to:

$$|H_{3,1}(f)| \leq \frac{224}{4860} + \frac{1712}{19440} + \frac{8}{45} + \frac{77}{405} + \frac{16}{135}.$$
$$= \frac{754}{1215}.$$

Thus, the proof of this result is completed. □

6. Bounds of $|H_{3,1}(f)|$ for Two-Fold and Three-Fold Functions

Let $m \in \mathbb{N} = \{1, 2, \ldots\}$. If a rotation Ω about the origin through an angle $2\pi/m$ carries Ω on itself, then such a domain Ω is called m-fold symmetric. An analytic function f is m-fold symmetric in Δ, if:

$$f\left(e^{2\pi i/m}z\right) = e^{2\pi i/m}f(z) \ (z \in \Delta).$$

By $\mathcal{S}^{(m)}$, we define the family of m-fold univalent functions having the following Taylor series form:

$$f(z) = z + \sum_{k=1}^{\infty} a_{mk+1} z^{mk+1} \ (z \in \Delta). \tag{24}$$

The subfamilies $\mathcal{S}_{car}^{*(m)}$, $\mathcal{C}_{car}^{(m)}$, and $\mathcal{R}_{car}^{(m)}$ of $\mathcal{S}^{(m)}$ are the families of the m-fold symmetric starlike, convex, and bounded turning functions, respectively, associated with the cardioid functions. More intuitively, an analytic function f of the form (24) belongs to the families $\mathcal{S}_{car}^{*(m)}$, $\mathcal{C}_{car}^{(m)}$, and $\mathcal{R}_{car}^{(m)}$ if and only if:

$$\frac{zf'(z)}{f(z)} = 1 + \frac{4}{3}\left(\frac{p(z)-1}{p(z)+1}\right) + \frac{2}{3}\left(\frac{p(z)-1}{p(z)+1}\right)^2, \ p \in \mathcal{P}^{(m)}, \tag{25}$$

$$1 + \frac{zf''(z)}{f'(z)} = 1 + \frac{4}{3}\left(\frac{p(z)-1}{p(z)+1}\right) + \frac{2}{3}\left(\frac{p(z)-1}{p(z)+1}\right)^2, \ p \in \mathcal{P}^{(m)}, \tag{26}$$

$$f'(z) = 1 + \frac{4}{3}\left(\frac{p(z)-1}{p(z)+1}\right) + \frac{2}{3}\left(\frac{p(z)-1}{p(z)+1}\right)^2, \ p \in \mathcal{P}^{(m)}, \tag{27}$$

where the family $\mathcal{P}^{(m)}$ is defined by:

$$\mathcal{P}^{(m)} = \left\{ p \in \mathcal{P} : p(z) = 1 + \sum_{k=1}^{\infty} c_{mk} z^{mk}, \ (z \in \mathbf{D}) \right\}. \tag{28}$$

Now, we prove some theorems concerned with two-fold and three-fold symmetric functions.

Theorem 7. *If $f \in \mathcal{S}_{car}^{*(2)}$ and it has the form given in (24), then:*

$$|H_{3,1}(f)| \leq \frac{2}{9}.$$

Proof. Let $f \in \mathcal{S}_{car}^{*(2)}$. Then, there exists a function $p \in \mathcal{P}^{(2)}$ such that:

$$\frac{zf'(z)}{f(z)} = 1 + \frac{4}{3}\left(\frac{p(z)-1}{p(z)+1}\right) + \frac{2}{3}\left(\frac{p(z)-1}{p(z)+1}\right)^2.$$

Using the series form (24) and (28), when $m = 2$ in the above relation, we can get:

$$a_3 = \frac{c_2}{3}, \tag{29}$$

$$a_5 = \frac{1}{4}\left(\frac{c_2^2}{18} + \frac{2}{3}c_4\right). \tag{30}$$

Now:

$$H_3(f) = a_3 a_5 - a_3^3.$$

Utilizing (29) and (30), we get:

$$H_{3,1}(f) = -\frac{7}{216}c_2^3 + \frac{c_2 c_4}{18}.$$

By reordering, it yields:

$$H_{3,1}(f) = \frac{c_2}{18}\left(c_4 - \frac{7}{12}c_2^2\right).$$

Using the triangle inequality long with (11) and (7), we have:

$$|H_{3,1}(f)| \leq \frac{2}{9}.$$

Hence, the proof is done. □

Theorem 8. *If $f \in \mathcal{S}_{car}^{*(3)}$ and it has the form (24), then:*

$$|H_{3,1}(f)| \leq \frac{16}{81}.$$

The result is sharp for the function:

$$f(z) = \exp\left(\ln z + \frac{4}{9}z^3 + \frac{1}{9}z^6\right) = z + \frac{4}{9}z^4 + \frac{17}{81}z^7 + \cdots. \tag{31}$$

Proof. Let $f \in \mathcal{S}_{car}^{*(3)}$. Then, there exists a function $p \in \mathcal{P}^{(3)}$ such that:

$$\frac{zf'(z)}{f(z)} = 1 + \frac{4}{3}\left(\frac{p(z)-1}{p(z)+1}\right) + \frac{2}{3}\left(\frac{p(z)-1}{p(z)+1}\right)^2.$$

Utilizing the series form (24) and (28), when $m = 3$ in the above relation, we can obtain:

$$a_4 = \frac{2}{9}c_3.$$

Then,

$$H_{3,1}(f) = -a_4^2 = -\frac{4}{81}c_3^2.$$

Utilizing (7) along with triangle inequality, we have:

$$|H_{3,1}(f)| \leq \frac{16}{81}.$$

Thus, the proof is completed. □

Theorem 9. *Let $f \in \mathcal{C}_{car}^{(2)}$, and it has the form (24), then:*

$$|H_{3,1}(f)| \leq \frac{2}{135}.$$

Proof. Let $f \in \mathcal{C}_{car}^{(2)}$. Then, there exists a function $p \in \mathcal{P}^{(2)}$ such that:

$$1 + \frac{zf''(z)}{f'(z)} = 1 + \frac{4}{3}\left(\frac{p(z)-1}{p(z)+1}\right) + \frac{2}{3}\left(\frac{p(z)-1}{p(z)+1}\right)^2.$$

Utilizing the series form (24) and (28), when $m = 2$ in the above relation, we can obtain:

$$a_3 = \frac{c_2}{9}, \tag{32}$$

$$a_5 = \frac{1}{20}\left(\frac{c_2^2}{18} + \frac{2}{3}c_4\right). \tag{33}$$

$$H_{3,1}(f) = a_3 a_5 - a_3^3.$$

Using (32) and (33), we have:

$$H_{3,1}(f) = -\frac{31}{29160}c_2^3 + \frac{c_2 c_4}{270}.$$

Now, reordering the above equation, we obtain:

$$H_3(f) = \frac{c_2}{270}\left(c_4 - \frac{31}{108}c_2^2\right).$$

Application of (7), (11), and the triangle inequality leads us to:

$$|H_{3,1}(f)| \leq \frac{2}{135}.$$

Thus, the required result is completed. □

Theorem 10. *If $f \in \mathcal{C}_{car}^{(3)}$ and it has the form given in (24), then:*

$$|H_{3,1}(f)| \leq \frac{1}{81}.$$

The result is sharp for the function:

$$f(z) = \int_0^z \frac{\exp\left(\ln x + \frac{4}{9}x^3 + \frac{1}{9}x^6\right)}{x}dx = z + \frac{1}{9}z^4 + \frac{17}{657}z^7 + \cdots.$$

Proof. Let $f \in \mathcal{C}_{car}^{(3)}$. Then, there exists a function $p \in \mathcal{P}^{(3)}$ such that:

$$1 + \frac{zf''(z)}{f'(z)} = 1 + \frac{4}{3}\left(\frac{p(z)-1}{p(z)+1}\right) + \frac{2}{3}\left(\frac{p(z)-1}{p(z)+1}\right)^2.$$

Utilizing the series form (24) and (28), when $m = 3$ in the above relation, we obtain:

$$a_4 = \frac{c_3}{18}.$$

Then:
$$H_{3,1}(f) = -a_4^2 = -\frac{c_3^2}{324}.$$

Implementing (7) and the triangle inequality, we have:
$$|H_{3,1}(f)| \leq \frac{1}{81}.$$

Hence, the proof is done. □

Theorem 11. *Let* $f \in \mathcal{R}_{car}^{(2)}$ *be of the form* (24). *Then:*
$$|H_{3,1}(f)| \leq \frac{16}{135}.$$

Proof. Since $f \in \mathcal{R}_{car}^{(2)}$, therefore there exists a function $p \in \mathcal{P}^{(2)}$ such that:
$$f'(z) = 1 + \frac{4}{3}\left(\frac{p(z)-1}{p(z)+1}\right) + \frac{2}{3}\left(\frac{p(z)-1}{p(z)+1}\right)^2.$$

For $f \in \mathcal{R}_{car}^{(2)}$, using the series form (24) and (28), when $m = 2$ in the above relation, we can write:
$$a_3 = \frac{2}{6}c_2, \tag{34}$$
$$a_5 = \frac{1}{5}\left(\frac{2}{3}c_4 - \frac{c_2^2}{6}\right). \tag{35}$$

It is clear that for $f \in \mathcal{R}_{car}^{(2)}$,
$$H_{3,1}(f) := a_3 a_5 - a_3^3.$$

Applying (34) and (35), we have:
$$H_{3,1}(f) = \frac{4}{135}c_2 c_4 - \frac{67}{3645}c_2^3.$$

By rearrangement, we have:
$$H_{3,1}(f) = \frac{4}{135}c_2(c_4 - \frac{67}{108}c_2^2).$$

Using Lemma (7), (10), and triangle inequality, we get:
$$|H_{3,1}(f)| \leq \frac{16}{135}.$$

Hence, the proof is completed. □

Theorem 12. *If* $f \in \mathcal{R}_{car}^{(3)}$ *and it is of the form* (24), *then:*
$$|H_{3,1}(f)| \leq \frac{1}{9}.$$

This result is sharp for the function:
$$f(z) = \int_0^z \left(1 + \frac{4}{3}x^3 + \frac{2}{3}x^6\right) dx = z + \frac{1}{3}z^4 + \frac{2}{21}z^7.$$

Proof. Since $f \in \mathcal{R}_{car}^{(3)}$, there exists a function $p \in \mathcal{P}^{(3)}$ such that:

$$f'(z) = 1 + \frac{4}{3}\left(\frac{p(z)-1}{p(z)+1}\right) + \frac{2}{3}\left(\frac{p(z)-1}{p(z)+1}\right)^2.$$

For $f \in \mathcal{R}_{car}^{(3)}$, using the series form (24) and (28), when $m = 2$ in the above relation, we can write:

$$a_4 = \frac{c_3}{6}.$$

Then:

$$H_{3,1}(f) := -a_4^2 = -\frac{c_3^2}{36}.$$

Implementing (7), we have:

$$|H_{3,1}(f)| \leq \frac{1}{9}.$$

Hence, the proof is completed. □

7. Conclusions

In this article, we studied the Hankel determinant $H_{3,1}(f)$ for the subfamilies $\mathcal{S}_{car}^*$, $\mathcal{C}_{car}$, and $\mathcal{R}_{car}$ of the analytic function using a very simple technique. Further, these bounds were also discussed for two-fold symmetric and three-fold symmetric functions.

Author Contributions: The authors have equally contributed to accomplish this research work.

Funding: This article is supported financially by the Anyang Normal University, Anyang 455002, Henan, China. The fourth author was supported by the Basic Science Research Program through the National Research Foundation of Korea (NRF) funded by the Ministry of Education, Science and Technology (No. 2016R1D1A1A09916450).

Conflicts of Interest: The authors have no conflict of interest.

References

1. Bieberbach, L. Über dié koeffizienten derjenigen Potenzreihen, welche eine schlichte Abbildung des Einheitskreises vermitteln. *Sitzungsberichte Preussische Akademie der Wissenschaften* **1916**, *138*, 940–955.
2. De-Branges, L. A proof of the Bieberbach conjecture. *Acta Math.* **1985**, *154*, 137–152. [CrossRef]
3. Padmanabhan, K.S.; Parvatham, R. Some applications of differential subordination. *Bull. Aust. Math. Soc.* **1985**, *32*, 321–330. [CrossRef]
4. Shanmugam, T.N. Convolution and differential subordination. *Int. J. Math. Math. Sci.* **1989**, *12*, 333–340. [CrossRef]
5. Ma, W.; Minda, D. A unified treatment of some special classes of univalent functions. In *Proceeding of the Conference on Complex Analysis*; Li, Z., Ren, F., Yang, L., Zhang, S., Eds.; International Press: Cambridge, MA, USA, 1992; pp. 157–169.
6. Janowski, W. Extremal problems for a family of functions with positive real part and for some related families. *Ann. Pol. Math.* **1971**, *23*, 159–177. [CrossRef]
7. Sokół, J.; Stankiewicz, J. Radius of convexity of some subclasses of strongly starlike functions. *Zeszyty Nauk. Politech. Rzeszowskiej Mat.* **1996**, *19*, 101–105.
8. Cho, N.E.; Kowalczyk, B.; Kwon, O.S.; Lecko, A.; Sim, Y.J. Some coefficient inequalities related to the Hankel determinant for strongly starlike functions of order alpha. *J. Math. Inequal.* **2017**, *11*, 429–439. [CrossRef]
9. Mendiratta, R.; Nagpal, S.; Ravichandran, V. On a subclass of strongly starlike functions associated with exponential function. *Bull. Malays. Math. Sci. Soc.* **2015**, *38*, 365–386. [CrossRef]
10. Kumar, S.; Ravichandran, V. A subclass of starlike functions associated with a rational function. *Southeast Asian Bull. Math.* **2016**, *40*, 199–212.
11. Sharma, K.; Jain, N.K.; Ravichandran, V. Starlike functions associated with a cardioid. *Afrika Matematika* **2016**, *27*, 923–939. [CrossRef]

12. Ravichandran, V.; Sharma, K. Sufficient conditions for starlikeness. *J. Korean Math. Soc.* **2015**, *52*, 727–749. [CrossRef]
13. Sharma, K.; Ravichandran, V. Application of subordination theory to starlike functions. *Bull. Iran. Math. Soc.* **2016**, *42*, 761–777.
14. Kargar, R.; Ebadian, A.; Sokół, J. On Booth lemniscate of starlike functions. *Anal. Math. Phys.* **2019**, *9*, 143–154. [CrossRef]
15. Raina, R.K.; Sokol, J. On coefficient estimates for a certain class of starlike functions. *Hacet. J. Math. Stat.* **2015**, *44*, 1427–1433. [CrossRef]
16. Pommerenke, C. On the coefficients and Hankel determinants of univalent functions. *J. Lond. Math. Soc.* **1966**, *1*, 111–122. [CrossRef]
17. Pommerenke, C. On the Hankel determinants of univalent functions. *Mathematika* **1967**, *14*, 108–112. [CrossRef]
18. Janteng, A.; Halim, S.A.; Darus, M. Coefficient inequality for a function whose derivative has a positive real part. *J. Inequal. Pure Appl. Math.* **2006**, *7*, 1–5.
19. Janteng, A.; Halim, S.A.; Darus, M. Hankel determinant for starlike and convex functions. *Int. J. Math. Anal.* **2007**, *1*, 619–625.
20. Răducanu, D.; Zaprawa, P. Second Hankel determinant for close-to-convex functions. *Compt. Rendus Math.* **2017**, *355*, 1063–1071. [CrossRef]
21. Srivastava, H.M.; Altınkaya, S.; Yalçın, S. Hankel determinant for a subclass of bi-univalent functions defined by using a symmetric q-derivative operator. *Filomat* **2018**, *32*, 503–516. [CrossRef]
22. Srivastava, H.M.; Ahmad, Q.Z.; Khan, N.; Khan, B. Hankel and Toeplitz determinants for a subclass of q-starlike functions associated with a general conic domain. *Mathematics* **2019**, *7*, 181. [CrossRef]
23. Çaglar, M.; Deniz, E.; Srivastava, H.M. Second Hankel determinant for certain subclasses of bi-univalent functions. *Turk. J. Math.* **2017**, *41*, 694–706. [CrossRef]
24. Bansal, D. Upper bound of second Hankel determinant for a new class of analytic functions. *Appl. Math. Lett.* **2013**, *26*, 103–107. [CrossRef]
25. Hayman, W.K. On second Hankel determinant of mean univalent functions. *Proc. Lond. Math. Soc.* **1968**, *3*, 77–94. [CrossRef]
26. Lee, S.K.; Ravichandran, V.; Supramaniam, S. Bounds for the second Hankel determinant of certain univalent functions. *J. Inequal. Appl.* **2013**, *2013*, 281. [CrossRef]
27. Altınkaya, Ş.; Yalçın, S. Upper bound of second Hankel determinant for bi-Bazilevic functions. *Mediterr. J. Math.* **2016**, *13*, 4081–4090. [CrossRef]
28. Liu, M.S.; Xu, J.F.; Yang, M. Upper bound of second Hankel determinant for certain subclasses of analytic functions. *Abstr. Appl. Anal.* **2014**, *2014*, 603180. [CrossRef]
29. Noonan, J.W.; Thomas, D.K. On the second Hankel determinant of areally mean p-valent functions. *Trans. Am. Math. Soc.* **1976**, *223*, 337–346.
30. Orhan, H.; Magesh, N.; Yamini, J. Bounds for the second Hankel determinant of certain bi-univalent functions. *Turk. J. Math.* **2016**, *40*, 679–687. [CrossRef]
31. Babalola, K.O. On $H_3(1)$ Hankel determinant for some classes of univalent functions. *Inequal. Theory Appl.* **2010**, *6*, 1–7.
32. Altınkaya, Ş.; Yalçın, S. Third Hankel determinant for Bazilevič functions. *Adv. Math.* **2016**, *5*, 91–96.
33. Bansal, D.; Maharana, S.; Prajapat, J.K. Third order Hankel Determinant for certain univalent functions. *J. Korean Math. Soc.* **2015**, *52*, 1139–1148. [CrossRef]
34. Krishna, D.V.; Venkateswarlu, B.; RamReddy, T. Third Hankel determinant for bounded turning functions of order alpha. *J. Niger. Math. Soc.* **2015**, *34*, 121–127. [CrossRef]
35. Raza, M.; Malik, S.N. Upper bound of third Hankel determinant for a class of analytic functions related with lemniscate of Bernoulli. *J. Inequal. Appl.* **2013**, *2013*, 412. [CrossRef]
36. Shanmugam, G.; Stephen, B.A.; Babalola, K.O. Third Hankel determinant for α-starlike functions. *Gulf J. Math.* **2014**, *2*, 107–113.
37. Zaprawa, P. Third Hankel determinants for subclasses of univalent functions. *Mediterr. J. Math.* **2017**, *14*, 19. [CrossRef]
38. Kwon, O.S.; Lecko, A.; Sim, Y.J. The bound of the Hankel determinant of the third kind for starlike functions. *Bull. Malays. Math. Sci. Soc.* **2019**, *42*, 767–780. [CrossRef]

39. Mahmood, S.; Khan, I.; Srivastava, H.M.; Malik, S.N. Inclusion relations for certain families of integral operators associated with conic regions. *J. Inequal. Appl.* **2019**, *2019*, 59. [CrossRef]
40. Mahmood, S.; Srivastava, H.M.; Malik, S.N. Some subclasses of uniformly univalent functions with respect to symmetric points. *Symmetry* **2019**, *11*, 287. [CrossRef]
41. Mahmood, S.; Jabeen, M.; Malik, S.N.; Srivastava, H.M.; Manzoor, R.; Riaz, S.M. Some coefficient inequalities of q-starlike functions associated with conic domain defined by q-derivative. *J. Funct. Spaces* **2018**, *2018*, 8492072. [CrossRef]
42. Kowalczyk, B.; Lecko, A.; Sim, Y.J. The sharp bound of the Hankel determinant of the third kind for convex functions. *Bull. Aust. Math. Soc.* **2018**, *97*, 435–445. [CrossRef]
43. Lecko, A.; Sim, Y.J.; Śmiarowska, B. The sharp bound of the Hankel determinant of the third kind for starlike functions of order 1/2. *Complex Anal. Oper. Theory* **2018**, 1–8. [CrossRef]
44. Mahmood, S.; Srivastava, H.M.; Khan, N.; Ahmad, Q.Z.; Khan, B.; Ali, I. Upper bound of the third Hankel determinant for a subclass of q-starlike functions. *Symmetry* **2019**, *11*, 347. [CrossRef]
45. Zhang, H.-Y.; Tang, H.; Niu, X.-M. Third-order Hankel determinant for certain class of analytic functions related with exponential function. *Symmetry* **2018**, *10*, 501. [CrossRef]
46. Pommerenke, C. *Univalent Functions*; Vandenhoeck and Ruprecht: Gottingen, Germany, 1975.

© 2019 by the authors. Licensee MDPI, Basel, Switzerland. This article is an open access article distributed under the terms and conditions of the Creative Commons Attribution (CC BY) license (http://creativecommons.org/licenses/by/4.0/).

Article

Remarks on the Generalized Fractional Laplacian Operator

Chenkuan Li [1,*], Changpin Li [2], Thomas Humphries [1] and Hunter Plowman [1]

1. Department of Mathematics and Computer Science, Brandon University, Brandon, MB R7A 6A9, Canada; humphrte65@brandonu.ca (T.H.); plowmahh10@brandonu.ca (H.P.)
2. Department of Mathematics, Shanghai University, Shanghai 200444, China; lcp@shu.edu.cn
* Correspondence: lic@brandonu.ca

Received: 25 February 2019; Accepted: 25 March 2019; Published: 29 March 2019

Abstract: The fractional Laplacian, also known as the Riesz fractional derivative operator, describes an unusual diffusion process due to random displacements executed by jumpers that are able to walk to neighbouring or nearby sites, as well as perform excursions to remote sites by way of Lévy flights. The fractional Laplacian has many applications in the boundary behaviours of solutions to differential equations. The goal of this paper is to investigate the half-order Laplacian operator $(-\Delta)^{\frac{1}{2}}$ in the distributional sense, based on the generalized convolution and Temple's delta sequence. Several interesting examples related to the fractional Laplacian operator of order $1/2$ are presented with applications to differential equations, some of which cannot be obtained in the classical sense by the standard definition of the fractional Laplacian via Fourier transform.

Keywords: distribution; fractional Laplacian; Riesz fractional derivative; delta sequence; convolution

MSC: 46F10; 26A33

In recent years, the fractional Laplacian operator has gained considerable attention due to its applications in many disciplines, such as partial differential equations, long-range interactions, anomalous diffusions and non-local quantum theories. There is also the physical meaning of the fractional Laplacian operator in bounded domains through its associated stochastic processes. However, the half-order Laplacian operator $(-\Delta)^{\frac{1}{2}}$, often appearing in various literature works and applications, needs to be studied carefully as the first-order Riesz derivative is undefined in the classical sense. The goal of this work is to use a new distributional approach to defining operator $(-\Delta)^{\frac{1}{2}}$ in the generalized sense by Temple's delta sequence, as well as present fresh techniques in computing examples of the fractional Laplacian operator of order $1/2$ and applications to solving partial differential equations related to this operator.

1. Introduction

Let $s \in (0,1)$ and $\Delta = \partial^2/\partial x_1^2 + \cdots + \partial^2/\partial x_n^2$. The fractional Laplacian of a function $u: R^n \to R$ is defined as:

$$(-\Delta)^s u(x) = C_{n,s} \text{P.V.} \int_{R^n} \frac{u(x) - u(\zeta)}{|x - \zeta|^{n+2s}} d\zeta, \tag{1}$$

where P.V. stands for the Cauchy principal value, and the constant $C_{n,s}$ is given by:

$$C_{n,s} = \left(\int_{R^n} \frac{1 - \cos y_1}{|y|^{n+2s}} dy \right)^{-1} = \pi^{-n/2} 2^{2s} \frac{\Gamma(\frac{n+2s}{2})}{\Gamma(1-s)} s,$$

and $y = (y_1, y_2, \cdots, y_n) \in R^n$.

On the other hand, the fractional Laplacian in R^n can be written by the Fourier transform:

$$(-\Delta)^s u(x) = \frac{1}{(2\pi)^n} \int_{R^n} |\zeta|^{2s} (u, e^{-i\zeta x})_{L^2} e^{i\zeta x} d\zeta = \mathcal{F}^{-1}\{|\zeta|^{2s} \mathcal{F}(u)(\zeta)\}(x)$$

where:

$$(u, e^{-i\zeta x})_{L^2} = \int_{R^n} u(x) e^{-i\zeta x} dx, \quad \hat{u}(\zeta) = \mathcal{F}\{u\}(\zeta) = \frac{1}{(2\pi)^{n/2}} \int_{R^n} u(x) e^{-i\zeta x} dx,$$

$$\mathcal{F}^{-1}\{\hat{u}\}(x) = \frac{1}{(2\pi)^{n/2}} \int_{R^n} \hat{u}(\zeta) e^{i\zeta x} d\zeta.$$

Hence, the fractional Laplacian is really a pseudo-differential operator with symbol $|\zeta|^{2s}$.

Let L^p be the Lebesgue space with $p \in [1, \infty)$, B_0 be the space of continuous functions vanishing at infinity, and B_{bu} be the space of bounded uniformly-continuous functions. M.Kwaśnicki recently presented ten equivalent definitions in [1] for defining $(-\Delta)^s$ over these three spaces, including the above Fourier definition.

There have been many studies, including numerical analysis approaches, on the fractional Laplacian with applications in solving certain differential equations on bounded domains and in the theory of stochastic processes and anomalous diffusion [2–10]. For example, the work in [11] used the fractional Laplacian for linear and nonlinear lossy media, as well as studying a linear integro-differential equation wave model. The work in [12] studied a finite difference method of solving parabolic partial integro-differential equations, with possibly singular kernels. These arise in option pricing theory when the random evolution of the underlying asset is driven by a Lévy process related to the fractional Laplacian or, more generally, a time-inhomogeneous jump-diffusion process. Using the fractional Laplacian operator, Araci et al. [13] investigated the following q-difference boundary value problem:

$$D_q^\gamma(\phi_p(D_q^\delta y(t))) + f(t, y(t)) = 0, \quad (0 < t < 1; \ 3 < \delta < 4)$$

with the boundary conditions:

$$y(0) = (D_q y)(0) = (D_q^2 y)(0),$$
$$a_1(D_q y)(1) + a_2(D_q^2 y)(1) = 0, \quad \text{and} \quad D_{0+}^\gamma y(t)|_{t=0} = 0.$$

They proved the existence and uniqueness of a positive and nondecreasing solution for this problem by means of a fixed point theorem involving partially-ordered sets.

Let $\Omega \subset R^n$ denote a bounded, open domain. For $u(x): \Omega \to R$, D'Elia and Gunzburger [14] investigated the action of the nonlocal diffusion operator $\mathcal{L}$ on the function $u(x)$ as:

$$\mathcal{L} u(x) = 2 \int_{R^n} (u(y) - u(x)) \gamma(x, y) dy \quad \forall x \in \Omega \subseteq R^n,$$

where the volume of Ω is non-zero and the kernel $\gamma(x, y): \Omega \times \Omega \to R$ is a non-negative symmetric mapping, as well as the nonlocal, steady-state diffusion equation:

$$-\mathcal{L} u = f \quad \text{on } \Omega,$$
$$u = 0 \quad \text{on } R^n \setminus \Omega.$$

An example of $\gamma(x, y)$ is given by:

$$\gamma(x, y) = \frac{\sigma(x, y)}{|y - x|^{n+2s}}$$

with $\sigma(x, y)$ bounded from above and below by positive constants. This nonlocal diffusion operator has, as special cases, the fractional Laplacian and fractional differential operators that arise in several applications. The corresponding evolution model was further studied in [15]. Recently, Hu et al. [16] studied the following high-dimensional Caputo-type parabolic equation with the fractional Laplacian by using the finite difference method:

$$_C D_{0,t}^\alpha u = -(-\Delta)^s u + f, \quad x \in \Omega, \ t > 0,$$
$$u(x, 0) = u_0(x),$$
$$u(x, t) = 0 \quad \text{on} \quad x \in R^n \setminus \Omega,$$

where $\alpha \in (0,1), s \in (0,1)$ and $\Omega \subset R^n$. In particular, this involves the half-order Laplacian operator $(-\Delta)^{\frac{1}{2}}$ when $s = 1/2$. The convergence and error estimate of the established finite difference scheme are shown with several examples.

On the other hand, the Riesz fractional derivative is generally given as:

$$_{RZ} D_x^\alpha u(x) = -\frac{(_{RL} D_{-\infty, x}^\alpha + _{RL} D_{\infty, x}^\alpha) u(x)}{2 \cos(\alpha \pi/2)} \quad (2)$$

where $0 < \alpha < 2$ and $\alpha \neq 1$, and:

$$_{RL} D_{-\infty, x}^\alpha u(x) = \frac{1}{\Gamma(n-\alpha)} \frac{d^n}{dx^n} \int_{-\infty}^x (x-\zeta)^{n-\alpha-1} u(\zeta) d\zeta,$$

$$_{RL} D_{\infty, x}^\alpha u(x) = \frac{1}{\Gamma(n-\alpha)} \frac{d^n}{dx^n} \int_\infty^x (x-\zeta)^{n-\alpha-1} u(\zeta) d\zeta$$

for $n - 1 < \alpha < n \in Z^+$.

Note that in space fractional quantum mechanics, the $\alpha = 2$ case corresponds to the Schrödinger equation for a massive non-relativistic particle, while the $\alpha = 1$ case needs to be examined carefully, both on physical and mathematical grounds, since Equation (2) is undefined for $\alpha = 1$.

It follows from [10,17–22] that:

$$_{RL} D_{-\infty, x}^\alpha u(x) = \frac{d^2}{dx^2} [I_{-\infty, x}^{2-\alpha} u(x)], \quad 1 < \alpha < 2$$

and:

$$_{RL} D_{-\infty, x}^\alpha u(x) = I_{-\infty, x}^{-\alpha} u(x) = \frac{1}{\Gamma(2-\alpha)} \frac{d^2}{dx^2} \int_{-\infty}^x \frac{u(t)}{(x-t)^{\alpha-1}} dt$$
$$= \frac{1}{\Gamma(2-\alpha)} \frac{d^2}{dx^2} \int_0^\infty \zeta^{-\alpha+1} u(x-\zeta) d\zeta$$

by making the variable change $\zeta = x - t$.

Applying two identities:

$$\zeta^{-\alpha+1} = (\alpha - 1) \int_\zeta^\infty \frac{d\eta}{\eta^\alpha},$$

$$\frac{\partial^2 u(x-\zeta)}{\partial x^2} = \frac{\partial^2 u(x-\zeta)}{\partial \zeta^2}$$

we get:

$$\begin{aligned}
{}_{RL}D^\alpha_{-\infty,x}u(x) &= \frac{\alpha-1}{\Gamma(2-\alpha)}\int_0^\infty \frac{\partial^2 u(x-\zeta)}{\partial \zeta^2}\left[\int_\zeta^\infty \frac{d\eta}{\eta^\alpha}\right]d\zeta \\
&= \frac{1-\alpha}{\Gamma(2-\alpha)}\int_0^\infty \left[\int_\zeta^\infty \frac{d\eta}{\eta^\alpha}\right]d\frac{\partial u(x-\zeta)}{\partial \zeta} \\
&= \frac{1}{\Gamma(1-\alpha)}\frac{\partial u(x-\zeta)}{\partial \zeta}\int_\zeta^\infty \frac{d\eta}{\eta^\alpha}\bigg|_{\zeta=0}^\infty + \frac{1}{\Gamma(1-\alpha)}\int_0^\infty \frac{\partial u(x-\zeta)}{\partial \zeta}\frac{1}{\zeta^\alpha}d\zeta \\
&= -\frac{1}{\Gamma(1-\alpha)}\int_0^\infty \frac{1}{\zeta^\alpha}du(x-\zeta) \\
&= -\frac{1}{\Gamma(1-\alpha)}\frac{u(x-\zeta)}{\zeta^\alpha}\bigg|_{\zeta=0}^\infty - \frac{\alpha}{\Gamma(1-\alpha)}\int_0^\infty \frac{u(x-\zeta)}{\zeta^{\alpha+1}}d\zeta \\
&= \frac{\alpha}{\Gamma(1-\alpha)}\int_0^\infty \frac{u(x)}{\zeta^{\alpha+1}}d\zeta - \frac{\alpha}{\Gamma(1-\alpha)}\int_0^\infty \frac{u(x-\zeta)}{\zeta^{\alpha+1}}d\zeta \\
&= -\frac{\alpha}{\Gamma(1-\alpha)}\int_0^\infty \frac{u(x-\zeta)-u(x)}{\zeta^{\alpha+1}}d\zeta, \quad 1<\alpha<2
\end{aligned}$$

by removing the term:

$$\int_0^\infty \frac{d\eta}{\eta^\alpha},$$

due to the meaning of the finite part integral as the integral is divergent and the finite part:

$$\frac{\eta^{-\alpha+1}}{-\alpha+1}\bigg|_{\eta=\infty} = 0.$$

With the same argument, we come to:

$${}_{RL}D^\alpha_{\infty,x}u(x) = -\frac{\alpha}{\Gamma(1-\alpha)}\int_0^\infty \frac{u(x+\zeta)-u(x)}{\zeta^{\alpha+1}}d\zeta, \quad 1<\alpha<2.$$

Hence, we have another representation of the Riesz fractional derivative from Equation (2):

$${}_{RZ}D^\alpha_x u(x) = \frac{\Gamma(1+\alpha)\sin\alpha\pi/2}{\pi}\int_0^\infty \frac{u(x+\zeta)-2u(x)+u(x-\zeta)}{\zeta^{1+\alpha}}d\zeta, \quad 1<\alpha<2.$$

Similarly, we can claim that this representation still holds for the entire range $0<\alpha\leq 2$ [23]. In particular,

$${}_{RZ}D^1_x u(x) = \frac{1}{\pi}\int_0^\infty \frac{u(x+\zeta)-2u(x)+u(x-\zeta)}{\zeta^2}d\zeta = \frac{1}{\pi}(\text{P.V.}\frac{1}{\zeta^2}, u(x+\zeta))$$

based on the formula:

$$(\text{P.V.}\frac{1}{x^2}, \phi(x)) = \int_0^\infty \frac{\phi(x)-2\phi(0)+\phi(-x)}{x^2}dx.$$

Clearly, Equation (1) becomes:

$$
\begin{aligned}
(-\Delta)^{1/2} u(x) &= C_{1,1/2} \, \text{P.V.} \int_R \frac{u(x) - u(\zeta)}{|x - \zeta|^2} d\zeta \\
&= \frac{1}{\pi} \text{P.V.} \int_R \frac{u(x) - u(\zeta)}{(x - \zeta)^2} d\zeta \\
&= -\frac{1}{\pi} \int_0^\infty \frac{u(x + \zeta) - 2u(x) + u(x - \zeta)}{\zeta^2} d\zeta \quad (3)\\
&= -{}_{RZ} D_x^1 u(x) \quad (4)
\end{aligned}
$$

for $s = 1/2$ and $n = 1$.

Therefore, investigations of the half-order Laplacian operator $(-\Delta)^{\frac{1}{2}}$ on R are equivalent to studies of the first-order Riesz derivative. We can define $-(-\Delta)^{\frac{1}{2}} u$ as the Riesz derivative ${}_{RZ} D_x^\alpha u(x)$ in the case of $\alpha = 1$, which is undefined in Equation (2) in the classical sense. The aim of this work is to study the operator $(-\Delta)^{\frac{1}{2}}$ on R in distribution explicitly and implicitly, using a particular delta sequence and the generalized convolution. We also present several interesting examples, such as $(-\Delta)^{\frac{1}{2}} \delta(x)$ and $(-\Delta)^{\frac{1}{2}} \theta(x)$, which are undefined in the classical sense. At the end of this work, we describe applications of such studies to solving the differential equations involving the half-order Laplacian operator.

2. The Explicit Approach to $(-\Delta)^{1/2} u$

In order to extend the fractional Laplacian $(-\Delta)^{1/2}$ distributionally, we briefly introduce the following basic concepts of distributions. Let $\mathcal{D}(R)$ be the Schwartz space (testing function space) [24,25] of infinitely-differentiable functions with compact support in R and $\mathcal{D}'(R)$ the (dual) space of distributions defined on $\mathcal{D}(R)$. A sequence $\phi_1, \phi_2, \cdots, \phi_n, \cdots$ goes to zero in $\mathcal{D}(R)$ if and only if these functions vanish outside a certain fixed bounded set and converge to zero uniformly together with their derivatives of any order.

The functional $\delta^{(n)}(x - x_0)$ is defined as:

$$(\delta^{(n)}(x - x_0), \phi(x)) = (-1)^n \phi^{(n)}(x_0)$$

where $\phi \in \mathcal{D}(R)$. Clearly, $\delta^{(n)}(x - x_0)$ is a linear and continuous functional on $\mathcal{D}(R)$, and hence, $\delta^{(n)}(x - x_0) \in \mathcal{D}'(R)$.

Define the unit step function $\theta(x)$ as:

$$\theta(x) = \begin{cases} 1 & \text{if } x > 0, \\ 0 & \text{if } x < 0. \end{cases}$$

Then,

$$(\theta(x), \phi(x)) = \int_0^\infty \phi(x) dx \quad \text{for} \quad \phi \in \mathcal{D}(R),$$

which implies $\theta(x) \in \mathcal{D}'(R)$.

Let $f \in \mathcal{D}'(R)$. The distributional derivative of f, denoted by f' or df/dx, is defined as:

$$(f', \phi) = -(f, \phi')$$

for $\phi \in \mathcal{D}(R)$.

Clearly, $f' \in \mathcal{D}'(R)$ and every distribution has a derivative. As an example, we are going to show that $\theta'(x) = \delta(x)$ distributionally, although $\theta(x)$ is not defined at $x = 0$ in the classical sense. Indeed,

$$(\theta'(x), \phi(x)) = -(\theta(x), \phi'(x)) = -\int_0^\infty \phi'(x)dx = \phi(0) = (\delta(x), \phi(x)),$$

which claims:
$$\theta'(x) = \delta(x).$$

It can be shown that the ordinary rules of differentiation also apply to distributions. For instance, the derivative of a sum is the sum of the derivatives, and a constant can be commuted with the derivative operator.

Definition 1. *Let f and g be distributions in $\mathcal{D}'(R)$ satisfying either of the following conditions:*

(a) *either f or g has bounded support (set of all essential points).*
(b) *the supports of f and g are bounded on the same side (either on the left or right).*

Then, the convolution $f * g$ is defined by the equation:

$$((f * g)(x), \phi(x)) = (g(x), (f(y), \phi(x+y)))$$

for $\phi \in \mathcal{D}$. Clearly, we have:
$$f * g = g * f$$

from Definition 1.

It follows from the definition above that:

$$\delta^{(n)}(x - x_0) * f(x) = f^{(n)}(x - x_0)$$

for any distribution $f \in \mathcal{D}'(R)$.

Let $\delta_n(x) = n\rho(nx)$ be Temple's δ-sequence for $n = 1, 2, \cdots$, where $\rho(x)$ is a fixed, infinitely-differentiable function on R, having the following properties [26,27]:

(i) $\rho(x) \geq 0$,
(ii) $\rho(x) = 0$ for $|x| \geq 1$,
(iii) $\rho(x) = \rho(-x)$,
(iv) $\int_{-\infty}^\infty \rho(x)dx = 1$.

An example of such a $\rho(x)$ function is given as:

$$\rho(x) = \begin{cases} ce^{-\frac{1}{1-x^2}} & \text{if } |x| < 1, \\ 0 & \text{otherwise} \end{cases}$$

where:
$$c^{-1} = \int_{-1}^1 e^{-\frac{1}{1-x^2}} dx.$$

This delta-sequence plays an important role in defining products of distributions [28,29]. Let f be a continuous function on R. Then,

$$(f * \delta_n)(x) = \int_{-\infty}^\infty f(x-t)\delta_n(t)dt = \int_{-\infty}^\infty f(t)\delta_n(x-t)dt$$

uniformly converges to f on any compact subset of R. Indeed, we assume that f is continuous on R, and L is any compact subset of R. Then, f is uniformly continuous on L, and for all $\epsilon > 0$, there exists $\delta_1 > 0$ such that:
$$|f(x-t) - f(x)| < \epsilon$$
for all $x \in L$ and $|t| < \delta_1$. This implies that:
$$|(f * \delta_n)(x) - f(x)| \leq \int_{-\infty}^{\infty} |f(x-t) - f(x)| \delta_n(t) dt < \epsilon$$
for all $x \in L$ and $1/n < \delta_1$ by noting that:
$$\int_{-\infty}^{\infty} \delta_n(t) dt = 1.$$

Furthermore, if $f \in \mathcal{D}'(R)$, then $(f * \delta_n)(x)$ converges to f in $\mathcal{D}'(R)$. Indeed,
$$\lim_{n \to \infty} ((f * \delta_n)(x), \phi(x)) = \lim_{n \to \infty} (f(x), (\delta_n(y), \phi(x+y))) = (f(x), \phi(x))$$
by noting that:
$$\lim_{n \to \infty} (\delta_n(y), \phi(x+y)) = \phi(x)$$
in $\mathcal{D}(R)$.

In order to study the half-order Laplacian operator in the distribution, we introduce an infinitely-differentiable function $\tau(x)$ satisfying the following conditions:

(i) $\tau(x) = \tau(-x)$,
(ii) $0 \leq \tau(x) \leq 1$,
(iii) $\tau(x) = 1$ if $|x| \leq 1/2$,
(iv) $\tau(x) = 0$ if $|x| \geq 1$.

Define:
$$\phi_m(x) = \begin{cases} 1 & \text{if } |x| \leq m, \\ \tau(m^m x - m^{m+1}) & \text{if } x > m, \\ \tau(m^m x + m^{m+1}) & \text{if } x < -m, \end{cases}$$
for $m = 1, 2, \cdots$. Then, $\phi_m(x) \in \mathcal{D}(R)$ with the support $[-m - m^{-m}, m + m^{-m}]$.

From Equation (3), we have:
$$(-\Delta)^{1/2} u(x) = -\frac{1}{\pi}(\text{P.V.}\frac{1}{t^2}, u(x+t)) = -\frac{1}{\pi}(\text{P.V.}\frac{1}{t}, u'(x+t))$$
if $u \in \mathcal{D}(R)$ as:
$$(\text{P.V.}\frac{1}{t^2}, \phi(t)) = \int_0^{\infty} \frac{\phi(t) - 2\phi(0) + \phi(-t)}{t^2} dt,$$
$$\frac{d}{dt}\text{P.V.}\frac{1}{t} = -\text{P.V.}\frac{1}{t^2}.$$

This suggests the following explicit definition for defining $(-\Delta)^{\frac{1}{2}} u(x)$. This explicit definition directly evaluates the half-order fractional Laplacian of $u(x)$ as a function of x, without relating to any testing function in the Schwartz space.

Definition 2. Let $u \in \mathcal{D}'(R)$ and $u'_n = u' * \delta_n = (u'(t), \delta_n(x-t))$ for $n = 1, 2, \cdots$. We define the half-order Laplacian operator $(-\Delta)^{\frac{1}{2}}$ on $\mathcal{D}'(R)$ as:

$$\begin{aligned}(-\Delta)^{\frac{1}{2}} u(x) &= -\frac{1}{\pi} \lim_{m \to \infty} \lim_{n \to \infty} \left(P.V. \frac{1}{t}, \phi_m(t) u'_n(x+t) \right) \\ &= -\frac{1}{\pi} \lim_{m \to \infty} \lim_{n \to \infty} \int_0^\infty \phi_m(t) \frac{u'_n(x+t) - u'_n(x-t)}{t} dt \end{aligned} \quad (5)$$

if it exists.

Clearly, the integral:

$$\int_0^\infty \phi_m(t) \frac{u'_n(x+t) - u'_n(x-t)}{t} dt$$

is well defined as $\phi_m(t)$ has a bounded support and:

$$\lim_{t \to 0^+} \frac{u'_n(x+t) - u'_n(x-t)}{t} = u''_n(x).$$

Theorem 1.
$$(-\Delta)^{\frac{1}{2}} \delta(x) = -\frac{1}{\pi} P.V. \frac{1}{x^2}.$$

Proof. Assuming $x > 0$, we choose a large n such that $1/n < x$. This infers that $\delta'_n(x+t) = 0$, and:

$$\begin{aligned}(-\Delta)^{\frac{1}{2}} \delta(x) &= \lim_{m \to \infty} \lim_{n \to \infty} -\frac{1}{\pi} \int_0^\infty \phi_m(t) \frac{\delta'_n(x+t) - \delta'_n(x-t)}{t} dt \\ &= \lim_{m \to \infty} \lim_{n \to \infty} \frac{1}{\pi} \int_0^\infty \phi_m(t) \frac{\delta'_n(x-t)}{t} dt \end{aligned}$$

from Definition 2. Making the substitution $w = x - t$, we get:

$$\begin{aligned}(-\Delta)^{\frac{1}{2}} \delta(x) &= \lim_{m \to \infty} \lim_{n \to \infty} \frac{1}{\pi} \int_{-\infty}^x \delta'_n(w) \frac{\phi_m(x-w)}{x-w} dw \\ &= \lim_{m \to \infty} \lim_{n \to \infty} \frac{1}{\pi} \int_{-\infty}^\infty \delta'_n(w) \frac{\phi_m(x-w)}{x-w} dw \\ &= \lim_{m \to \infty} -\frac{1}{\pi} \frac{\partial}{\partial w} \left[\frac{\phi_m(x-w)}{x-w} \right] \bigg|_{w=0} \\ &= \lim_{m \to \infty} \frac{\phi'_m(x) x - \phi_m(x)}{\pi x^2} \end{aligned}$$

by noting that $\delta_n(w)$ is a delta sequence, and:

$$\frac{\phi_m(x-w)}{x-w}$$

is a testing function of w if $w < x$, and:

$$\delta'_n(w) \frac{\phi_m(x-w)}{x-w}$$

is identical to zero if $w \geq x$ as $\delta'_n(w) = 0$. Choosing m such that $x < m$, we derive that:

$$\phi'_m(x) = 0 \quad \text{and} \quad \phi_m(x) = 1$$

as $\phi_m(x) = 1$ if $|x| < m$. Hence,

$$(-\Delta)^{\frac{1}{2}}\delta(x) = -\frac{1}{\pi}\text{P.V.}\frac{1}{x^2}.$$

If $x < 0$, we set $y = -x$ and:

$$-\frac{1}{\pi}\int_0^\infty \phi_m(t)\frac{\delta_n'(x+t) - \delta_n'(x-t)}{t}dt = -\frac{1}{\pi}\int_0^\infty \phi_m(t)\frac{\delta_n'(-y+t) - \delta_n'(-y-t)}{t}dt$$

$$= -\frac{1}{\pi}\int_0^\infty \phi_m(t)\frac{\delta_n'(y+t) - \delta_n'(y-t)}{t}dt,$$

which implies that:

$$(-\Delta)^{\frac{1}{2}}\delta(x) = \lim_{m\to\infty}\frac{\phi_m'(y)y - \phi_m(y)}{\pi y^2} = -\frac{1}{\pi}\text{P.V.}\frac{1}{x^2}.$$

Finally, we have for $x = 0$ by making the variable change $u = nt$:

$$\int_0^\infty \phi_m(t)\frac{\delta_n'(t) - \delta_n'(-t)}{t}dt = 2\int_0^\infty \phi_m(t)\frac{\delta_n'(t)}{t}dt$$

$$= 2n^2\int_0^\infty \phi_m(t)\frac{\rho'(nt)}{t}dt$$

$$= 2n^2\int_0^1 \phi_m(u/n)\frac{\rho'(u)}{u}du = 2n^2\int_0^1 \frac{\rho'(u)}{u}du$$

$$= O(n^2)$$

by noting that the term $\phi_m(u/n) = 1$, and:

$$\int_0^1 \frac{\rho'(u)}{u}du = \int_0^1 \frac{\rho'(u) - \rho'(0)}{u}du$$

is well defined as $\rho(u)$ is an even function. This completes the proof of Theorem 1. □

From [24], we have the distributions P.V.x^{-2m} for $m = 1, 2, \cdots$ and P.V.x^{-2m-1} for $m = 0, 1, \cdots$ given as:

$$(\text{P.V.}x^{-2m}, \phi) = \int_0^\infty x^{-2m}\{\phi(x) + \phi(-x)$$

$$-2\left[\phi(0) + \frac{x^2}{2!} + \cdots + \frac{x^{2m-2}}{(2m-2)!}\phi^{(2m-2)}(0)\right]\}dx,$$

$$(\text{P.V.}x^{-2m-1}, \phi) = \int_0^\infty x^{-2m-1}\{\phi(x) - \phi(-x)$$

$$-2\left[x\phi'(0) + \frac{x^3}{3!} + \cdots + \frac{x^{2m-1}}{(2m-1)!}\phi^{(2m-1)}(0)\right]\}dx.$$

Theorem 2.

$$(-\Delta)^{\frac{1}{2}}\delta^{(m)}(x) = \frac{(-1)^{m+1}(m+1)!}{\pi}\text{P.V.}\frac{1}{x^{m+2}}$$

for $m = -1, 0, 1, \ldots$. In particular, we have for $m = -1$ that:

$$(-\Delta)^{\frac{1}{2}}\theta(x) = \frac{1}{\pi}\text{P.V.}\frac{1}{x}.$$

Proof. We start with the case $m = -1$. Then, by Definition 2,

$$\begin{aligned}
(-\Delta)^{\frac{1}{2}}\theta(x) &= \lim_{m \to \infty} \lim_{n \to \infty} -\frac{1}{\pi} \int_0^\infty \phi_m(t) \frac{\delta_n(x+t) - \delta_n(x-t)}{t} dt \\
&= \lim_{m \to \infty} \lim_{n \to \infty} \frac{1}{\pi} \int_0^\infty \phi_m(t) \frac{\delta_n(x-t)}{t} dt
\end{aligned}$$

for $x > 0$ and a large n such that $1/n < x$. Following the proof of Theorem 1, we come to:

$$\begin{aligned}
(-\Delta)^{\frac{1}{2}}\theta(x) &= \lim_{m \to \infty} \lim_{n \to \infty} \frac{1}{\pi} \int_{-\infty}^x \delta_n(w) \frac{\phi_m(x-w)}{x-w} dw \\
&= \lim_{m \to \infty} \lim_{n \to \infty} \frac{1}{\pi} \int_{-\infty}^\infty \delta_n(w) \frac{\phi_m(x-w)}{x-w} dw \\
&= \lim_{m \to \infty} \frac{\phi_m(x)}{\pi x} \\
&= \frac{1}{\pi} \text{P.V.} \frac{1}{x}.
\end{aligned}$$

The case $x < 0$ follows similarly. To compute $(-\Delta)^{\frac{1}{2}} \delta^{(m)}(x)$, we note that for a large n:

$$(-\Delta)^{\frac{1}{2}} \delta^{(m)}(x) = \lim_{m \to \infty} \lim_{n \to \infty} \frac{1}{\pi} \int_{-\infty}^\infty \delta_n^{(m+1)}(w) \frac{\phi_m(x-w)}{x-w} dw$$

from the proof of Theorem 1 and:

$$\begin{aligned}
&\lim_{m \to \infty} \frac{(-1)^{m+1}}{\pi} \frac{\partial^{m+1}}{\partial w^{m+1}} \left[\frac{\phi_m(x-w)}{x-w}\right]\bigg|_{w=0} \\
&= \lim_{m \to \infty} \frac{(-1)^{m+1}}{\pi} \sum_{k=0}^{m+1} \binom{m+1}{k} (-1)^k \phi_m^{(k)}(x-w) \left(\frac{1}{x-w}\right)^{(m+1-k)}\bigg|_{w=0} \\
&= \lim_{m \to \infty} \frac{(-1)^{m+1}}{\pi} \binom{m+1}{0} \phi_m(x-w) \left(\frac{1}{x-w}\right)^{(m+1)}\bigg|_{w=0} \\
&= \frac{(-1)^{m+1}(m+1)!}{\pi} \text{P.V.} \frac{1}{x^{m+2}}
\end{aligned}$$

since $\lim_{m \to \infty} \phi_m(x) = 1$ and:

$$\lim_{m \to \infty} \frac{(-1)^{m+1}}{\pi} \sum_{k=1}^{m+1} \binom{m+1}{k} (-1)^k \phi_m^{(k)}(x-w) \left(\frac{1}{x-w}\right)^{(m+1-k)}\bigg|_{w=0} = 0$$

from the definition of $\phi_m(x)$. This completes the proof of Theorem 2. □

We should note that Theorem 2 cannot be derived by the standard definition of the fractional Laplacian via Fourier transform, and:

$$(-\Delta)^{\frac{1}{2}}\theta(x) = \frac{1}{\pi} \text{P.V.} \frac{1}{x}$$

holds in the distributional sense and:

$$\frac{d}{dx}\theta(x) = \delta(x),$$

although $\theta(x)$ is discontinuous at $x = 0$ in the classical sense.

Theorem 3.

$$(-\Delta)^{\frac{1}{2}} \sin x = \sin x, \tag{6}$$
$$(-\Delta)^{\frac{1}{2}} \cos x = \cos x, \tag{7}$$
$$(-\Delta)^{\frac{1}{2}} (ax + b) = 0. \tag{8}$$

where a and b are arbitrary constants.

Proof. Clearly, $\sin x$ is an infinitely-differentiable function on R. This claims that:

$$(\sin x)' * \delta_n = \cos x * \delta_n$$

uniformly converges to $\cos x$ on any compact subset of R. Therefore,

$$\begin{aligned}
(-\Delta)^{\frac{1}{2}} \sin x &= \lim_{m \to \infty} -\frac{1}{\pi} \int_0^\infty \phi_m(t) \frac{\cos(x+t) - \cos(x-t)}{t} dt \\
&= \lim_{m \to \infty} -\frac{1}{\pi} \int_0^\infty \phi_m(t) \frac{-2 \sin x \sin t}{t} dt \\
&= \frac{2 \sin x}{\pi} \int_0^\infty \frac{\sin t}{t} dt = \sin x
\end{aligned}$$

by using:

$$\int_0^\infty \frac{\sin t}{t} dt = \pi/2.$$

Similarly,

$$\begin{aligned}
(-\Delta)^{\frac{1}{2}} \cos x &= \lim_{m \to \infty} -\frac{1}{\pi} \int_0^\infty \phi_m(t) \frac{-\sin(x+t) + \sin(x-t)}{t} dt \\
&= \lim_{m \to \infty} -\frac{1}{\pi} \int_0^\infty \phi_m(t) \frac{-2 \cos x \sin t}{t} dt \\
&= \frac{2 \cos x}{\pi} \int_0^\infty \frac{\sin t}{t} dt = \cos x.
\end{aligned}$$

Finally,

$$\begin{aligned}
(-\Delta)^{\frac{1}{2}} (ax + b) &= \lim_{m \to \infty} \lim_{n \to \infty} -\frac{1}{\pi} \int_0^\infty \phi_m(t) \frac{(b*\delta_n)(x+t) - (b*\delta_n)(x-t)}{t} dt \\
&= \lim_{m \to \infty} -\frac{1}{\pi} \int_0^\infty \phi_m(t) \frac{b - b}{t} dt = 0.
\end{aligned}$$

This completes the proof of Theorem 3. □

We would like to mention that Theorem 3 extends several classical results obtained in [30] to distributions by Definition 2 with a new approach.

Theorem 4.

$$(-\Delta)^{\frac{1}{2}} \arctan x = \frac{x}{1 + x^2}.$$

Proof. Clearly, $(\arctan x)' = \frac{1}{1 + x^2}$ is continuous on R, and

$$(\arctan x)' * \delta_n = \left(\frac{1}{1 + t^2}, \delta_n(x - t) \right)$$

uniformly converges to $1/(1+x^2)$ on any compact subset of R. Therefore,

$$
\begin{aligned}
(-\Delta)^{\frac{1}{2}} \arctan x &= \lim_{m \to \infty} -\frac{1}{\pi} \int_0^\infty \phi_m(t) \frac{\frac{1}{1+(x+t)^2} - \frac{1}{1+(x-t)^2}}{t} dt \\
&= \frac{4x}{\pi} \lim_{m \to \infty} \int_0^\infty \phi_m(t) \frac{dt}{(1+(x-t)^2)(1+(x+t)^2)} \\
&= \frac{4x}{\pi} \int_0^\infty \frac{dt}{(1+(x-t)^2)(1+(x+t)^2)}
\end{aligned}
$$

by noting that the integral:

$$
\int_0^\infty \frac{dt}{(1+(x-t)^2)(1+(x+t)^2)}
$$

is well defined for every point $x \in R$. It remains to show that:

$$
\int_0^\infty \frac{dt}{(1+(x-t)^2)(1+(x+t)^2)} = \frac{\pi}{4+4x^2}
$$

for all $x \in R$. First, we note that:

$$
R(z) = \frac{1}{(1+(x-z)^2)(1+(x+z)^2)}
$$

is even with respect to z and has two singular points $z_1 = x + i$ and $z_2 = i - x$ in the upper half-plane. Clearly, we have for $x \neq 0$ that:

$$
\begin{aligned}
Res\{R(z), x+i\} &= \lim_{z \to x+i} \frac{z-x-i}{(1+(x-z)^2)(1+(x+z)^2)} = \frac{1}{2i(1+(2x+i)^2)}, \\
Res\{R(z), i-x\} &= \lim_{z \to i-x} \frac{z+x-i}{(1+(x-z)^2)(1+(x+z)^2)} = \frac{1}{2i(1+(2x-i)^2)}.
\end{aligned}
$$

By Cauchy's residue theorem, we get:

$$
\begin{aligned}
\int_{-\infty}^\infty \frac{dt}{(1+(x-t)^2)(1+(x+t)^2)} &= 2\pi i [Res\{R(z), x+i\} + Res\{R(z), i-x\}] \\
&= \frac{2\pi i}{2i(1+(2x+i)^2)} + \frac{2\pi i}{2i(1+(2x-i)^2)} \\
&= \pi \frac{8x^2}{(1+(2x+i)^2)(1+(2x-i)^2)} \\
&= \frac{\pi}{2+2x^2}
\end{aligned}
$$

using:

$$
\left(1+(2x+i)^2\right)\left(1+(2x-i)^2\right) = 16x^2 + 16x^4.
$$

This implies that:

$$
\int_0^\infty \frac{dt}{(1+(x-t)^2)(1+(x+t)^2)} = \frac{\pi}{4+4x^2}
$$

for all nonzero x. Furthermore, we derive for $x = 0$ that:

$$
\int_0^\infty \frac{dt}{(1+t^2)^2} = \frac{\pi}{4}
$$

by using the identity [31]:

$$\int_{-\infty}^{\infty} \frac{dt}{(1+t^2)^n} = \frac{(2n-3)(2n-5)\cdots 1}{(2n-2)(2n-4)\cdots 2}\pi.$$

This completes the proof of Theorem 4. □

3. The Implicit Approach to $(-\Delta)^{\frac{1}{2}}u$

It seems infeasible to calculate directly the fractional Laplacian operator of some functions or distributions by Definition 2. For example,

$$\begin{aligned}
(-\Delta)^{\frac{1}{2}}e^x &= \lim_{m\to\infty}\lim_{n\to\infty} -\frac{1}{\pi}\int_0^\infty \phi_m(t)\frac{u_n'(x+t)-u_n'(x-t)}{t}dt \\
&= \lim_{m\to\infty} -\frac{1}{\pi}\int_0^\infty \phi_m(t)\frac{e^{x+t}-e^{x-t}}{t}dt \\
&= -\frac{e^x}{\pi}\lim_{m\to\infty}\int_0^\infty \phi_m(t)\frac{e^t-e^{-t}}{t}dt
\end{aligned}$$

where:

$$u_n(x) = (e^t, \delta_n(x-t))$$

uniformly converges to e^x as it is continuous on R. Clearly, the right-hand side of the above integral is divergent as:

$$\lim_{t\to\infty}\frac{e^t-e^{-t}}{t} = \infty.$$

In this section, we are going to provide another definition for dealing with $(-\Delta)^{\frac{1}{2}}u(x)$ efficiently, based on a testing function with compact support. This definition is implicit and only used to define the meaning of:

$$((-\Delta)^{\frac{1}{2}}u(x), \phi(x)),$$

rather than finding an explicit function of x. It clearly makes sense in the distribution as we regard $(-\Delta)^{\frac{1}{2}}u(x)$ as a functional (not a function) on the Schwartz testing space $\mathcal{D}(R)$. We must point out that this implicit-definition, using a different generalization, is independent of the explicit one provided in Section 2.

Definition 3. *Let $u \in \mathcal{D}'(R)$ and $u_n' = u' * \delta_n = (u'(t), \delta_n(x-t))$ for $n = 1, 2, \cdots$. We define the half-order Laplacian operator $(-\Delta)_i^{\frac{1}{2}}$ on $\mathcal{D}'(R)$ (adding the index i to distinguish from $(-\Delta)^{\frac{1}{2}}$) for $\phi \in \mathcal{D}(R)$ as:*

$$\begin{aligned}
((-\Delta)_i^{\frac{1}{2}}u(x), \phi(x)) &= -\frac{1}{\pi}\lim_{n\to\infty}\left(P.V.\frac{1}{t}, \phi(t)u_n'(x+t)\right) \\
&= -\frac{1}{\pi}\lim_{n\to\infty}\int_0^\infty \frac{u_n'(x+t)\phi(t)-u_n'(x-t)\phi(-t)}{t}dt \quad (9)
\end{aligned}$$

if it exists.

Clearly, the integral:

$$\int_0^\infty \frac{u_n'(x+t)\phi(t)-u_n'(x-t)\phi(-t)}{t}dt$$

is well defined, as $\phi(t)$ has a bounded support and:

$$\lim_{t\to 0^+}\frac{u_n'(x+t)\phi(t)-u_n'(x-t)\phi(-t)}{t} = 2u_n''(x)\phi(0)+2u_n'(x)\phi'(0).$$

It follows from Definition 3 that:

$$((-\Delta)_i^{\frac{1}{2}} e^x, \phi(x)) = -\frac{e^x}{\pi}(\text{P.V.}\frac{1}{x}, e^x\phi(x))$$

by noting that:

$$(\text{P.V.}\frac{1}{x}, e^x\phi(x)) = \int_0^\infty \frac{e^x\phi(x) - e^{-x}\phi(-x)}{x}dx$$

is well defined and is a number for each ϕ, since $e^x\phi(x)$ is also a testing function in the Schwartz space.

Furthermore,

$$((-\Delta)_i^{\frac{1}{2}} \cosh x, \phi(x)) = -\frac{e^x}{2\pi}(\text{P.V.}\frac{1}{x}, e^x\phi(x)) + \frac{e^{-x}}{2\pi}(\text{P.V.}\frac{1}{x}, e^{-x}\phi(x)).$$

Theorem 5. *Let $u(x)$ be an infinitely-differentiable function satisfying:*

$$u(x) = \sum_{k=0}^\infty \frac{u^{(k)}(0)}{k!}x^k.$$

Then,

$$((-\Delta)_i^{\frac{1}{2}} u(x), \phi(x)) = -\frac{u'(x)}{\pi}(\text{P.V.}\frac{1}{x}, \phi(x)) - \frac{1}{\pi}\left(\frac{u'(x+t) - u'(x)}{t}, \phi(t)\right). \tag{10}$$

Proof. Clearly,

$$u(x+t) = \sum_{k=0}^\infty \frac{u^{(k)}(x)}{k!}t^k,$$

$$u'(x+t) = \sum_{k=0}^\infty \frac{u^{(k+1)}(x)}{k!}t^k$$

and:

$$u_n'(x) = (u' * \delta_n)(x) = (u'(t), \delta_n(x-t))$$

uniformly converges to $u(x)$ on any compact subset of R. By Definition 3,

$$\begin{aligned}((-\Delta)_i^{\frac{1}{2}} u(x), \phi(x)) &= \lim_{n\to\infty} -\frac{1}{\pi}\int_0^\infty \frac{u_n'(x+t)\phi(t) - u_n'(x-t)\phi(-t)}{t}dt \\ &= -\frac{1}{\pi}\int_0^\infty \frac{u'(x+t)\phi(t) - u'(x-t)\phi(-t)}{t}dt \\ &= -\frac{1}{\pi}\int_0^\infty \frac{\sum_{k=0}^\infty \frac{u^{(k+1)}(x)}{k!}t^k\phi(t) - \sum_{k=0}^\infty \frac{u^{(k+1)}(x)}{k!}(-t)^k\phi(-t)}{t}dt \\ &= -\frac{1}{\pi}\int_0^\infty \frac{\phi(t)u'(x) - \phi(-t)u'(x)}{t}dt \\ &\quad -\frac{1}{\pi}\int_0^\infty \frac{\sum_{k=1}^\infty \frac{u^{(k+1)}(x)}{k!}t^k\phi(t) - \sum_{k=1}^\infty \frac{u^{(k+1)}(x)}{k!}(-t)^k\phi(-t)}{t}dt \\ &= -\frac{u'(x)}{\pi}(\text{P.V.}\frac{1}{x}, \phi(x)) - \frac{1}{\pi}\left(\frac{u'(x+t) - u'(x)}{t}, \phi(t)\right)\end{aligned}$$

since

$$\sum_{k=1}^\infty \frac{u^{(k+1)}(x)}{k!}t^k = u'(x+t) - u'(x).$$

We should note that the term:
$$\left(\frac{u'(x+t) - u'(x)}{t}, \phi(t)\right)$$
is well defined for every point $x \in R$, and it is indeed not Cauchy's principal value as:
$$\lim_{t \to 0} \frac{u'(x+t) - u'(x)}{t} = u''(x).$$

This completes the proof of Theorem 5. □

It follows from Theorem 5 that:
$$((-\Delta)_i^{\frac{1}{2}}(ax+b), \phi(x)) = -\frac{a}{\pi}(\text{P.V.}\frac{1}{x}, \phi(x))$$

which implies that:
$$(-\Delta)_i^{\frac{1}{2}}(ax+b) = -\frac{a}{\pi}\text{P.V.}\frac{1}{x} \neq (-\Delta)^{\frac{1}{2}}(ax+b) = 0.$$

Furthermore,
$$((-\Delta)_i^{\frac{1}{2}} \sin x, \phi(x)) = -\frac{\cos x}{\pi}\left(\text{P.V.}\frac{\cos t}{t}, \phi(t)\right) + \frac{\sin x}{\pi}\left(\frac{\sin t}{t}, \phi(t)\right).$$

Remark 1. *To end this section, we must point out that if we replace $\phi(t)$ by $\phi_m(t)$ used in Section 2 and add the limit, then the two operators $(-\Delta)_i^{\frac{1}{2}}$ and $(-\Delta)^{\frac{1}{2}}$ are identical for some functions. For instance,*
$$\lim_{m \to \infty}((-\Delta)_i^{\frac{1}{2}} \sin x, \phi_m(x)) = \frac{\sin x}{\pi}\int_{-\infty}^{\infty}\frac{\sin t}{t}dt = \sin x$$

as:
$$\left(\text{P.V.}\frac{\cos t}{t}, \phi_m(t)\right) = 0.$$

4. Conclusions

This paper introduced two independent definitions for defining the fractional Laplacian of the half-order $(-\Delta)^{\frac{1}{2}}$ in the distribution, both explicitly and implicitly. We demonstrate several examples, such as $(-\Delta)^{\frac{1}{2}}\delta^{(m)}(x)$ and $(-\Delta)^{\frac{1}{2}}\arctan x$, some of which are undefined in the classical sense. The results obtained have potential applications in solving the differential equations involving the half-order Laplacian operator. For example, the differential equation:
$$(-\Delta)^{\frac{1}{2}}u(x) = \frac{x}{1+x^2}$$
has a solution:
$$u(x) = \arctan x + ax + b$$
on any non-empty subset of R, and the differential equation:
$$(-\Delta)^{\frac{1}{2}}u(x) = \text{P.V.}\frac{1}{x}$$
has a solution:
$$u(x) = \pi\theta(x) + ax + b$$
where a and b are arbitrary constants.

Author Contributions: The order of the author list reflects contributions to the paper.

Funding: This work is partially supported by NSERC (Canada 2017-00001) and NSFC (China 11671251).

Acknowledgments: The authors are grateful to the reviewers and editor for the careful reading of the paper with several productive suggestions and corrections, which certainly improved its quality.

Conflicts of Interest: The authors declare no conflict of interest.

References

1. Kwaśnicki, M. Ten equivalent definitions of the fractional Laplace operator. *Fract. Calc. Appl. Anal.* **2017**, *20*, 7–51. [CrossRef]
2. Pozrikidis, C. *The Fractional Laplacian*; CRC Press: Boca Raton, FL, USA, 2016.
3. Yang, Q.; Liu, F.; Turner, I. Numerical methods for fractional partial differential equations with Riesz space fractional derivatives. *Appl. Math. Model.* **2010**, *34*, 200–218. [CrossRef]
4. Huang, Y.; Oberman, A. Numerical methods for the fractional Laplacian: A finite difference-quadrature approach. *SIAM J. Numer. Anal.* **2014**, *52*, 3056–3084. [CrossRef]
5. Zoia, A.; Rosso, A.; Kardar, M. Fractional Laplacian in bounded domains. *Phys. Rev. E* **2007**, *76*, 1116–1126. [CrossRef] [PubMed]
6. Laskin, N. Fractional Schrödinger equation. *Phys. Rev. E* **2002**, *66*, 056108. [CrossRef] [PubMed]
7. Caffarelli, L.; Silvestre, L. An extension problem related to the fractional Laplacian. *Commun. Part. Differ. Equ.* **2007**, *32*, 1245–1260. [CrossRef]
8. Hu, Y.; Li, C.P.; Li, H.F. The finite difference method for Caputo-type parabolic equation with fractional Laplacian: One-dimension case. *Chaos Solitons Fract.* **2017**, *102*, 319–326. [CrossRef]
9. Barrios, B.; Colorado, E.; Servadei, R.; Soria, F. A critical fractional equation with concave–convex power nonlinearities. *Annales de l'I.H.P. Analyse non Linéaire* **2015**, *32*, 875–900. [CrossRef]
10. Bayin, S.Ş. Definition of the Riesz derivative and its application to space fractional quantum mechanics. *J. Math. Phys.* **2016**, *57*, 123501. [CrossRef]
11. Chen, W.; Holm, S. Fractional Laplacian time-space models for linear and nonlinear lossy media exhibiting arbitrary frequency power-law dependency. *J. Acoust. Soc. Am.* **2004**, *115*, 1424–1430. [CrossRef]
12. Cont, R.; Voltchkova, E. A finite difference scheme for option pricing in jump diffusion and exponential Lévy models. *SIAM J. Numer. Anal.* **2005**, *43*, 1596–1626. [CrossRef]
13. Araci, S.; Şen, E.; Açikgöz, M.; Srivastava, H.M. Existence and uniqueness of positive and nondecreasing solutions for a class of fractional boundary value problems involving the *p*-Laplacian operator. *Adv. Differ. Equ.* **2015**, *2015*, 40. [CrossRef]
14. D'Elia, M.; Gunzburger, M. The fractional Laplacian operator on bounded domains as a special case of the nonlocal diffusion operator. *Comput. Math. Appl.* **2013**, *66*, 1245–1260. [CrossRef]
15. Chen, A.; Du, Q.; Li, C.P.; Zhou, Z. Asymptotically compatible schemes for space-time nonlocal diffusion equations. *Chaos Solitons Fract.* **2017**, *102*, 361–371. [CrossRef]
16. Hu, Y.; Li, C.P.; Li, H.F. The finite difference method for Caputo-type parabolic equation with fractional Laplacian: More than one space dimension. *Int. J. Comput. Math.* **2018**, *95*, 1114–1130. [CrossRef]
17. Kilbas, A.A.; Srivastava, H.M.; Trujillo, J.J. *Theory and Applications of Fractional Differential Equations*; Elsevier/North-Holland: Amsterdam, The Netherlands, 2006.
18. Podlubny, I. *Fractional Differential Equations*; Academic Press: New York, NY, USA, 1999.
19. Herrmann, R. *Fractional Calculus: An Introduction for Physicists*; World Scientific: Singapore, 2011.
20. Li, C.P.; Zeng, F.H. *Numerical Methods for Fractional Calculus*; CRC Press: Boca Raton, FL, USA, 2015.
21. Bayin, S.S. *Mathematical Methods in Science and Engineering*; John Wiley & Sons: New York, NY, USA, 2006.
22. Li, C.P.; Chen, A. Numerical methods for fractional partial differential equations. *Int. J. Comput. Math.* **2018**, *95*, 1048–1099. [CrossRef]
23. Gorenflo, R.; Mainardi, F. Essentials of Fractional Calculus. Available online: http://citeseerx.ist.psu.edu/viewdoc/download?doi=10.1.1.28.1961&rep=rep1&type=pdf (accessed on 10 July 2018).
24. Gel'fand, I.M.; Shilov, G.E. *Generalized Functions*; Academic Press: New York, NY, USA, 1964; Volume I.
25. Stein, E. *Functional Analysis: Introduction to Further Topics in Analysis*; Princeton Lectures in Analysis; Book 4; Princeton University Press: Princeton, NJ, USA, 2011.

26. Temple, G. The theory of generalized functions. *Proc. R. Soc. Ser. A* **1955**, *28*, 175–190.
27. Li, C. *A Review on the Products of Distributions*; Taş, K., Tenreiro Machado, J.A., Baleanu, D., Eds.; Math. Methods Eng.; Springer: Dordrecht, The Netherlands, 2007; pp. 71–96.
28. Li, C. The products on the unit sphere and even-dimension spaces. *J. Math. Anal. Appl.* **2005**, *305*, 97–106. [CrossRef]
29. Cheng, L.; Li, C. A commutative neutrix product of distributions on R^m. *Math. Nachr.* **1991**, *151*, 345–355.
30. Barrios, B.; García-Melián, J.; Quaas, A. Periodic solutions for the one-dimensional fractional Laplacian. *arXiv* **2018**, arXiv:1803.08739v2.
31. Gradshteyn, I.S.; Ryzhik, I.M. *Tables of Integrals, Series, and Products*; Academic Press: New York, NY, USA, 1980.

© 2019 by the authors. Licensee MDPI, Basel, Switzerland. This article is an open access article distributed under the terms and conditions of the Creative Commons Attribution (CC BY) license (http://creativecommons.org/licenses/by/4.0/).

Article

On Geometric Properties of Normalized Hyper-Bessel Functions

Khurshid Ahmad [1], Saima Mustafa [2], Muhey U. Din [3], Shafiq ur Rehman [4], Mohsan Raza [5,*] and Muhammad Arif [1]

1. Department of Mathematics, Abdul Wali Khan University, Mardan 23200, Pakistan; khurshidahmad410@gmail.com (K.A.); marifmaths@awkum.edu.pk (M.A.)
2. Department of Mathematics, PMAS Arid Agriculture University, Rawalpindi 46000, Pakistan; saimamustafa28@gmail.com
3. Department of Mathematics, Islamia College Faisalabad, Faisalabad 38000, Pakistan; muheyudin@yahoo.com
4. Department of Mathematics, COMSATS University Islamabad, Attock 43600, Pakistan; shafiq_ur_rahman2@yahoo.com
5. Department of Mathematics, Government College University Faisalabad, Faisalabad 38000, Pakistan
* Correspondence: mohsan976@yahoo.com

Received: 18 February 2019; Accepted: 21 March 2019; Published: 28 March 2019

Abstract: In this paper, the normalized hyper-Bessel functions are studied. Certain sufficient conditions are determined such that the hyper-Bessel functions are close-to-convex, starlike and convex in the open unit disc. We also study the Hardy spaces of hyper-Bessel functions.

Keywords: univalent functions; starlikeness; convexity; close-to-convexity; hyper-Bessel functions; Hardy space

MSC: Primary 30C45, 33C10; Secondary 30C20, 30C75

---

1. Introduction

Let $\mathcal{H}$ denote the class of functions that are analytic in $\mathcal{U} = \{z : |z| < 1\}$, and $\mathcal{A}$ denote the class of functions f that are analytic in $\mathcal{U}$ having the Taylor series form

$$f(z) = z + \sum_{n=2}^{\infty} a_n z^n, \ z \in \mathcal{U}. \tag{1}$$

The class $\mathcal{S}$ of univalent functions f is the class of those functions in $\mathcal{A}$ that are one-to-one in $\mathcal{U}$. Let $\mathcal{S}^*$ denote the class of all functions f such that $f(\mathcal{U})$ is star-shaped domain with respect to origin while $\mathcal{C}$ denotes the class of functions f such that f maps $\mathcal{U}$ onto a domain which is convex. A function f in $\mathcal{A}$ belongs to the class $\mathcal{S}^*(\alpha)$ of starlike functions of order α if and only if $Re\left(zf'(z)/f(z)\right) > \alpha$, $\alpha \in [0,1)$. For $\alpha \in [0,1)$, a function $f \in \mathcal{A}$ is convex of order α if and only if $Re\left(1 + zf''(z)/f'(z)\right) > \alpha$ in $\mathcal{U}$. This class of functions is dented by $\mathcal{C}(\alpha)$. It is clear that $\mathcal{S}^*(0) = \mathcal{S}^*$ and $\mathcal{C}(0) = \mathcal{C}$ are the usual classes of starlike and convex functions respectively. A function f in $\mathcal{A}$ is said to be close-to-convex function in $\mathcal{U}$, if $f(\mathcal{U})$ is close-to-convex. That is, the complement of $f(\mathcal{U})$ can be expressed as the union of non-intersecting half-lines. In other words a function f in $\mathcal{A}$ is said to be close-to-convex if and only if $Re\left(zf'(z)/g(z)\right) > 0$ for some starlike function g. In particular if $g(z) = z$, then $Re\left(f'(z)\right) > 0$. The class of close-to-convex functions is denoted by $\mathcal{K}$. The functions in class $\mathcal{K}$ are univalent in $\mathcal{U}$. For some details about these classes of functions one can refer to [1]. Consider the class $\mathcal{P}_\delta(\alpha)$ of functions p such that $p(0) = 1$ and

$$\operatorname{Re}\left\{e^{i\delta} p(z)\right\} > \alpha, \quad z \in \mathcal{U}, \alpha \in [0,1), \delta \in \mathbb{R}.$$

Also consider the class $\mathcal{R}_\delta(\alpha)$ of functions $f \in \mathcal{A}$ such that

$$\operatorname{Re}\left\{e^{i\delta} f'(z)\right\} > \alpha, \quad z \in \mathcal{U}, \alpha \in [0,1), \delta \in \mathbb{R}.$$

These classes were introduced and investigated by Baricz [2]. For $\delta = 0$, we have the classes $\mathcal{P}_0(\alpha)$ and $\mathcal{R}_0(\alpha)$. Also for $\delta = 0$ and $\alpha = 0$, we have the classes $\mathcal{P}$ and $\mathcal{R}$.

Special functions have great importance in pure and applied mathematics. The wide use of these functions have attracted many researchers to work on the different directions. Geometric properties of special functions such as Hypergeometric functions, Bessel functions, Struve functions, Mittag-Leffler functions, Wright functions and some other related functions is an ongoing part of research in geometric function theory. We refer for some geometric properties of these functions [2–6] and references therein.

We consider the hyper-Bessel function in the form of the hypergeometric functions defined as

$$J_{\gamma_c}(z) = \frac{\left(\frac{z}{c+1}\right)^{\gamma_1 + \gamma_2 + \ldots + \gamma_c}}{\prod_{i=1}^{c} \Gamma(\gamma_i + 1)} \, {}_0F_c\left((\gamma_c + 1); -\left(\frac{z}{c+1}\right)^{c+1}\right), \tag{2}$$

where the notation

$$_mF_n\left(\binom{(\eta)_m}{(\delta)_n}; x\right) = \sum_{k=0}^{\infty} \frac{(\eta_1)_k (\eta_2)_k \ldots (\eta_m)_k}{(\delta_1)_k (\delta_2)_k \ldots (\delta_n)_k} \frac{x^k}{k!}, \tag{3}$$

represents the generalized Hypergeometric functions and γ_c represents the array of c parameters $\gamma_1, \gamma_2, \ldots, \gamma_c$. By combining Equations (2) and (3), we get the following infinite representation of the hyper-Bessel functions

$$J_{\gamma_c}(z) = \sum_{n=0}^{\infty} \frac{(-1)^n}{n! \prod_{i=1}^{c} \Gamma(\gamma_i + n + 1)} \left(\frac{z}{c+1}\right)^{n(c+1) + \gamma_1 + \gamma_2 + \ldots + \gamma_c}, \tag{4}$$

since J_{γ_c} is not in class $\mathcal{A}$. Therefore, consider the hyper-Bessel function $\mathcal{J}_{\gamma_c}$ which is defined by

$$\mathcal{J}_{\gamma_c}(z) = 1 + \sum_{n=2}^{\infty} \frac{(-1)^{n-1}}{(n-1)!(c+1)^{(n-1)(c+1)} \prod_{i=1}^{c} (\alpha_i + 1)_{n-1}} z^{(n-1)(c+1)}. \tag{5}$$

It is observed that the function $\mathcal{J}_{\gamma_c}$ defined in (5) is not in the class $\mathcal{A}$. Here, we consider the following normalized form of the hyper-Bessel function for our own convenience.

$$\mathcal{H}_{\gamma_c}(z) = z\mathcal{J}_{\gamma_c}(z) = z + \sum_{n=2}^{\infty} \frac{(-1)^{n-1}}{(n-1)!(c+1)^{(n-1)(c+1)} \prod_{i=1}^{c} (\gamma_i + 1)_{n-1}} z^{(n-1)(c+1)+1}. \tag{6}$$

For some details about the hyper-Bessel functions one can refer to [7–9]. Recently Aktas et al. [8] studied some geometric properties of hyper-Bessel function. In particular, they studied radii of starlikeness, convexity, and uniform convexity of hyper-Bessel functions. Motivated by the above works, we study the geometric properties of hyper-Bessel function $\mathcal{H}_{\gamma_c}$ given by the power series (6). We determine the conditions on parameters that ensure the hyper-Bessel function to be starlike of order α, convex of order α, close-to-convex of order $(\frac{1+\alpha}{2})$. We also study the convexity and starlikeness in the domain $\mathcal{U}_{1/2} = \{z : |z| < \frac{1}{2}\}$. Sufficient conditions on univalency of an integral operator defined by hyper-Bessel function is also studied. We find the conditions on normalized hyper-Bessel function to belong to the Hardy space $\mathcal{H}^p$.

To prove our results, we require the following.

Lemma 1. *If $f \in \mathcal{A}$ satisfy $|f'(z) - 1| < 1$ for each $z \in \mathcal{U}$, then f is convex in $\mathcal{U}_{1/2} = \{z : |z| < \frac{1}{2}\}$ [10].*

Lemma 2. *If $f \in \mathcal{A}$ satisfy $\left|\frac{f(z)}{z} - 1\right| < 1$ for each $z \in \mathcal{U}$, then f is starlike in $\mathcal{U}_{1/2} = \{z : |z| < \frac{1}{2}\}$ [11].*

Lemma 3. *Let $\beta \in \mathbb{C}$ with $Re(\beta) > 0$, $c \in \mathbb{C}$ with $|c| \leq 1$, $c \neq -1$ [12]. If $h \in \mathcal{A}$ satisfies*

$$\left| c|z|^{2\beta} + \left(1 - |z|^{2\beta} \frac{zh''(z)}{\beta h'(z)}\right) \right| \leq 1, \quad z \in \mathcal{U},$$

then the integral operator

$$C_\beta(z) = \left\{ \beta \int_0^z t^{\beta-1} h'(t) dt \right\}^{1/\beta}, \quad z \in \mathcal{U},$$

is analytic and univalent in $\mathcal{U}$.

Lemma 4. *If $f \in \mathcal{A}$ [13] satisfies the inequality*

$$|zf''(z)| < \frac{1-\alpha}{4}, \quad (z \in \mathcal{U}, 0 \leq \alpha < 1),$$

then

$$Re f'(z) > \frac{1+\alpha}{4}, \quad (z \in \mathcal{U}, 0 \leq \alpha < 1).$$

Lemma 5. *If $f \in \mathcal{A}$ satisfies $\left|\frac{zf''(z)}{f'(z)}\right| < \frac{1}{2}$ [14], then $f \in \mathcal{UCV}$.*

2. Geometric Properties of Normalized Hyper-Bessel Function

Theorem 1. *Let $i \in \{1, 2, 3, ..., c\}$, $\gamma_i > -1$ with $\alpha \in [0, 1)$ and $z \in \mathcal{U}$. Then the following results are true:*

(i) *If $\prod_{i=1}^{c} (\gamma_i + 1) > \frac{(2c+7-5\alpha) + \sqrt{(2c+7-5\alpha)^2 - 8(1-\alpha)(c+4-3\alpha)}}{4\varsigma(1-\alpha)}$, then $\mathcal{H}_{\gamma_c} \in \mathcal{S}^*(\alpha)$.*

(ii) *If $\prod_{i=1}^{c} (\gamma_i + 1) > \frac{\{(2c+7)(1-\alpha) - (2c^2+5c+3)\} + \Psi}{2\varsigma\{2(1-\alpha) - (4c^2+10c+6)\}}$, where*

$$\Psi = \sqrt{\{(2c+7)(1-\alpha) - (2c^2+5c+3)\}^2 - 4\{2(1-\alpha) - (4c^2+10c+6)\}(c+4)(1-\alpha)},$$

then $\mathcal{H}_{\gamma_c} \in \mathcal{C}(\alpha)$.

(iii) *If $\prod_{i=1}^{c} (\gamma_i + 1) > \frac{1-\alpha}{\varsigma(1-\alpha) - 4(c+1)(2c+3)}$, then $\mathcal{H}_{\gamma_c} \in \mathcal{K}(\frac{1+\alpha}{2})$.*

(iv) *If $\prod_{i=1}^{c} (\gamma_i + 1) > \frac{3-\alpha}{2\varsigma(1-\alpha)}$, then $\frac{\mathcal{H}_{\gamma_c}}{z} \in \mathcal{P}(\alpha)$.*

Proof. (i) By using the inequalities

$$n! \geq n, \ (\gamma_i + 1)_n \geq (\gamma_i + 1)^n, \ \forall\, n \in \mathbb{N},$$

we obtain

$$\left| \mathcal{H}'_{\gamma_c}(z) - \frac{\mathcal{H}_{\gamma_c}(z)}{z} \right| \leq \frac{(c+1)}{\zeta\eta} \sum_{n\geq 1} \left(\frac{1}{\zeta\eta} \right)^{n-1},$$

where

$$\zeta = (c+1)^{c+1} \text{ and } \eta = \prod_{i=1}^{c} (\gamma_i + 1).$$

This implies that

$$\left| \mathcal{H}'_{\gamma_c}(z) - \frac{\mathcal{H}_{\gamma_c}(z)}{z} \right| \leq \frac{c+1}{\zeta\eta - 1}. \tag{7}$$

Furthermore, if we use the inequality

$$n! \geq 2^{n-1}, \ (\gamma_i + 1)_n \geq (\gamma_i + 1)^n, \ \forall\, n \in \mathbb{N},$$

then

$$\left| \frac{\mathcal{H}_{\gamma_c}(z)}{z} \right| = \left| 1 + \sum_{n\geq 1} \frac{(-1)^{n-1}}{(n-1)!(c+1)^{(n-1)(c+1)} \prod_{i=1}^{c} (\gamma_i + 1)_{n-1}} z^{(n-1)(c+1)} \right|$$

$$\geq 1 - \frac{1}{\zeta\eta} \sum_{n\geq 1} \left(\frac{1}{2\zeta\eta} \right)^{n-1}$$

$$= \frac{2\zeta\eta - 3}{2\zeta\eta - 1}. \tag{8}$$

By combining Equations (7) and (8), we obtain

$$\left| \frac{z\mathcal{H}'_{\gamma_c}(z)}{\mathcal{H}_{\gamma_c}(z)} - 1 \right| \leq \frac{(c+1)(2\zeta\eta - 1)}{(\zeta\eta - 1)(2\zeta\eta - 3)}. \tag{9}$$

For $\mathcal{H}_{\gamma_c} \in \mathcal{S}^*(\alpha)$, we must have

$$\frac{(c+1)(2\zeta\eta - 1)}{(\zeta\eta - 1)(2\zeta\eta - 3)} < 1 - \alpha.$$

So, $\mathcal{H}_{\gamma_c} \in \mathcal{S}^*(\alpha)$, where $0 \leq \alpha < 1 - \frac{(c+1)(2\zeta\eta-1)}{(\zeta\eta-1)(2\zeta\eta-3)}$.

(ii) To prove that the function $\mathcal{H}_{\gamma_c} \in \mathcal{C}(\alpha)$, we have to show that $\left|\frac{z\mathcal{H}''_{\gamma_c}(z)}{\mathcal{H}'_{\gamma_c}(z)}\right| < 1 - \alpha$. Consider

$$\mathcal{H}_{\gamma_c}(z) = \sum_{n \geq 0} \frac{(-1)^n}{n!(c+1)^{n(c+1)} \prod_{i=1}^{c}(\gamma_i+1)_n} z^{n(c+1)+1},$$

$$z\mathcal{H}''_{\gamma_c}(z) = \sum_{n \geq 0} \frac{\{n^2(c+1)^2 + n(c+1)\}(-1)^n}{n!(c+1)^{n(c+1)} \prod_{i=1}^{c}(\gamma_i+1)_n} z^{n(c+1)},$$

$$= \sum_{n \geq 1} \frac{(n-1)^2(c+1)^2}{(n-1)!(c+1)^{(n-1)(c+1)} \prod_{i=1}^{c}(\gamma_i+1)_{n-1}} z^{n(c+1)}$$

$$+ \sum_{n \geq 1} \frac{(n-1)(c+1)}{(n-1)!(c+1)^{(n-1)(c+1)} \prod_{i=1}^{c}(\gamma_i+1)_{n-1}} z^{n(c+1)}.$$

By using the inequalities

$$(n-1)! \geq \frac{(n-1)^2}{2}, \; (n-1)! \geq n-1, \; (\alpha_i+1)_n \geq (\alpha_i+1)^n, \; \forall\, n \in \mathbb{N},$$

we have

$$\left|z\mathcal{H}''_{\alpha_c}(z)\right| = \left| \begin{array}{c} \sum_{n \geq 1} \frac{(n-1)^2(c+1)^2}{(n-1)!(c+1)^{(n-1)(c+1)} \prod_{i=1}^{c}(\alpha_i+1)_{n-1}} z^{n(c+1)} \\ + \sum_{n \geq 1} \frac{(n-1)(c+1)}{(n-1)!(c+1)^{(n-1)(c+1)} \prod_{i=1}^{c}(\alpha_i+1)_{n-1}} z^{n(c+1)} \end{array} \right|$$

$$\leq 2(c+1)^2 \sum_{n \geq 1} \left(\frac{1}{\zeta \eta}\right)^{n-1} + (c+1) \sum_{n \geq 1} \left(\frac{1}{\zeta \eta}\right)^{n-1},$$

where

$$\zeta = (c+1)^{c+1} \text{ and } \eta = \prod_{i=1}^{c}(\gamma_i+1).$$

This implies that

$$\left|z\mathcal{H}''_{\gamma_c}(z)\right| \leq \frac{\zeta \eta (c+1)(2c+3)}{(\zeta \eta - 1)} \tag{10}$$

Furthermore, if we use the inequalities

$$n! \geq n, \; n! \geq 2^{n-1}, \; (\gamma_i+1)_n \geq (\gamma_i+1)^n, \; \forall\, n \in \mathbb{N},$$

then we get

$$\left|\mathcal{H}'_{\gamma_c}(z)\right| \geq 1 - \sum_{n\geq 1} \frac{n(c+1)+1}{n!\zeta^n \eta^n}$$

$$\geq 1 - \frac{c+1}{\zeta\eta}\sum_{n\geq 1}\left(\frac{1}{\zeta\eta}\right)^{n-1} + \frac{1}{\zeta\eta}\sum_{n\geq 1}\left(\frac{1}{2\zeta\eta}\right)^{n-1}$$

$$= \frac{(\zeta\eta-1)(2\zeta\eta-3)-(2\zeta\eta-1)(c+1)}{(\zeta\eta-1)(2\zeta\eta-1)}. \tag{11}$$

By combining Equations (10) and (11), we get

$$\left|\frac{z\mathcal{H}''_{\gamma_c}(z)}{\mathcal{H}'_{\gamma_c}(z)}\right| \leq \frac{\zeta\eta(2\zeta\eta-1)(c+1)(2c+3)}{(\zeta\eta-1)(2\zeta\eta-3)-(2\zeta\eta-1)(c+1)} < 1-\alpha.$$

This implies that $\mathcal{H}_{\gamma_c} \in \mathcal{C}(\alpha)$, where $0 \leq \alpha < 1 - \frac{\zeta\eta(2\zeta\eta-1)(c+1)(2c+3)}{(\zeta\eta-1)(2\zeta\eta-3)-(2\zeta\eta-1)(c+1)}$.

(iii) Using the inequality (10) and Lemma 4, we have

$$\left|z\mathcal{H}''_{\gamma_c}(z)\right| \leq \frac{\zeta\eta(c+1)(2c+3)}{(\zeta\eta-1)} < \frac{1-\alpha}{4},$$

where $0 \leq \alpha < 1 - 4\frac{\zeta\eta(c+1)(2c+3)}{(\zeta\eta-1)}$. This shows that $\mathcal{H}_{\gamma_c} \in \mathcal{K}(\frac{1+\alpha}{2})$. Therefore $Re\left(\mathcal{H}'_{\gamma_c}(z)\right) > \frac{1+\alpha}{2}$.

(iv) To prove that $\frac{\mathcal{H}_{\gamma_c}}{z} \in \mathcal{P}(\alpha)$, we have to show that $|h(z)-1| < 1$, where $h(z) = \frac{\mathcal{H}_{\gamma_c}(z)/z - \alpha}{1-\alpha}$. By using the inequality

$$2^{n-1} \leq n!, n \in \mathbb{N},$$

we have

$$|h(z)-1| = \left|\frac{1}{1-\alpha}\sum_{n=1}^{\infty}\frac{(-1)^n}{n!\,(c+1)^{n(c+1)}\prod_{i=1}^{c}(\gamma_i+1)_n}z^{n(c+1)+1}\right|$$

$$\leq \frac{1}{(1-\alpha)}\frac{1}{\zeta\eta}\sum_{n=1}^{\infty}\left(\frac{1}{2\zeta\eta}\right)^{n-1}$$

$$= \frac{1}{(1-\alpha)}\frac{2}{2\zeta\eta-1}.$$

Therefore, $\frac{\mathcal{H}_{\gamma_c}}{z} \in \mathcal{P}(\alpha)$ for $0 < \alpha < 1 - \frac{2}{2(c+1)^{n(c+1)}\prod_{i=1}^{c}(\gamma_i+1)_n - 1}$. □

Putting $\alpha = 0$ in Theorem 1, we have the following results.

Corollary 1. *Let $i \in \{1,2,3,...,c\}, \gamma_i > -1$ and $z \in \mathcal{U}$. Then the followings are true:*

(i) *If $\prod_{i=1}^{c}(\gamma_i+1) > \frac{(2c+7)+\sqrt{(2c+7)^2-8(c+4)}}{4\varsigma}$, then $\mathcal{H}_{\gamma_c} \in \mathcal{S}^*$.*

(ii) *If $\prod_{i=1}^{c}(\gamma_i+1) > \frac{\{(2c+7)-(2c^2+5c+3)\}+\sqrt{\{(2c+7)-(2c^2+5c+3)\}^2-4(c+4)\{2-(4c^2+10c+6)\}}}{2\varsigma\{2-(4c^2+10c+6)\}}$,*

then $\mathcal{H}_{\gamma_c} \in \mathcal{C}$.

(iii) If $\prod_{i=1}^{c}(\gamma_i+1) > \frac{1}{\varsigma\{1-4(c+1)(2c+3)\}}$, then $\mathcal{H}_{\gamma_c} \in \mathcal{K}(\frac{1}{2})$.

(iv) If $\prod_{i=1}^{c}(\gamma_i+1) > \frac{3}{2\varsigma}$, then $\frac{\mathcal{H}_{\gamma_c}}{z} \in \mathcal{P}$.

3. Starlikeness and Convexity in $\mathcal{U}_{1/2}$

Theorem 2. *Let $i \in \{1,2,3,...,c\}, \gamma_i > -1$ and $z \in \mathcal{U}$. Then the following assertions are true:*

(i) If $\prod_{i=1}^{c}(\gamma_i+1) > \frac{3}{2\varsigma}$, then $\mathcal{H}_{\gamma_c}$ is starlike in $\mathcal{U}_{1/2}$.

(ii) If $\prod_{i=1}^{c}(\gamma_i+1) > \frac{(c+3)+\sqrt{c^2+4c+2}}{2\varsigma}$, then $\mathcal{H}_{\gamma_c}$ is convex in $\mathcal{U}_{1/2}$.

Proof. (i) By using the inequality $2^{n-1} \leq n!$, $n \in \mathbb{N}$, we obtain

$$\left|\frac{\mathcal{H}_{\gamma_c}(z)}{z} - 1\right| \leq \left|\sum_{n=1}^{\infty} \frac{(-1)^n}{n!\,(c+1)^{n(c+1)} \prod_{i=1}^{c}(\gamma_i+1)_n} z^{n(c+1)}\right|$$

$$\leq \sum_{n=1}^{\infty}\left(\frac{1}{n!\left\{(c+1)^{(c+1)}\right\}^n \left\{\prod_{i=1}^{c}(\gamma_i+1)\right\}^n}\right)$$

$$\leq \frac{1}{\varsigma\eta}\sum_{n=1}^{\infty}\left(\frac{1}{2\varsigma\eta}\right)^{n-1} = \frac{2}{2\varsigma\eta - 1}.$$

In view of Lemma 2, $\mathcal{H}_{\gamma_c}$ is starlike in $\mathcal{U}_{1/2}$, if $\frac{2}{2\varsigma\eta-1} < 1$, which is true under the given hypothesis.

(ii) Consider,

$$\left|\mathcal{H}'_{\gamma_c}(z) - 1\right| \leq \sum_{n=1}^{\infty} \frac{n(c+1)+1}{n!\,(c+1)^{n(c+1)} \prod_{i=1}^{c}(\gamma_i+1)_n}$$

$$\leq \sum_{n=1}^{\infty} \frac{n(c+1)}{n!\varsigma^n\eta^n} + \sum_{n=1}^{\infty} \frac{1}{n!\varsigma^n\eta^n}.$$

Since, $n! \geq n$, for all $n \in \mathbb{N}$ and $n! \geq 2^{n-1}$, for all $n \in \mathbb{N}$. Therefore

$$\left|\mathcal{H}'_{\gamma_c}(z) - 1\right| \leq \frac{c+1}{\varsigma\eta}\sum_{n=1}^{\infty}\left(\frac{1}{\varsigma\eta}\right)^{n-1} + \frac{1}{\varsigma\eta}\sum_{n=1}^{\infty}\left(\frac{1}{2\varsigma\eta}\right)^{n-1}$$

$$= \frac{(c+1)(2\varsigma\eta-1)+2(\varsigma\eta-1)}{(\varsigma\eta-1)(2\varsigma\eta-1)}.$$

In view of Lemma 1, $\mathcal{H}_{\gamma_c}$ is convex in $\mathcal{U}_{1/2}$, if $\frac{(c+1)(2\varsigma\eta-1)+2(\varsigma\eta-1)}{(\varsigma\eta-1)(2\varsigma\eta-1)} < 1$, but this is true under the hypothesis. □

Consider the integral operator $\mathcal{F}_\beta : \mathcal{U} \to \mathbb{C}$, where $\beta \in \mathbb{C}, \beta \neq 0$,

$$\mathcal{F}_\beta(z) = \left\{ \beta \int_0^z t^{\beta-2} \mathcal{H}_{\gamma_c}(t) \, d(t) \right\}^{\frac{1}{\beta}}, \quad z \in \mathcal{U}.$$

Here $\mathcal{F}_\beta \in \mathcal{A}$. In the next theorem, we obtain the conditions so that $\mathcal{F}_\beta$ is univalent in $\mathcal{U}$.

Theorem 3. *Let* $i \in \{1,2,3,...,c\}, \gamma_i > -1$ *and* $z \in \mathcal{U}$. *Let* $\prod_{i=1}^{c}(\gamma_i + 1) > \frac{(2c+7)+\sqrt{(7+2c)^2-8(2c+3)}}{4\varsigma}$ *and suppose that* $M \in \mathbb{R}^+$ *such that* $|\mathcal{H}_{\gamma_c}(z)| \leq M$ *in the open unit disc. If*

$$|\beta - 1| + \frac{(c+1)(2\zeta\eta - 1)}{(\zeta\eta - 1)(2\zeta\eta - 3)} + \frac{M}{|\beta|} \leq 1,$$

then $\mathcal{F}_\beta$ *is univalent in* $\mathcal{U}$.

Proof. A calculations gives us

$$\frac{z\mathcal{F}_\beta''(z)}{\mathcal{F}_\beta'(z)} = \frac{z\mathcal{H}_{\gamma_c}'(z)}{\mathcal{H}_{\gamma_c}(z)} + \frac{z^{\beta-1}}{\beta}\mathcal{H}_{\gamma_c}(z) + \beta - 2, \, z \in \mathcal{U}.$$

Since $\mathcal{H}_{\gamma_c} \in \mathcal{A}$, then by the Schwarz Lemma, triangle inequality and Equation (9), we obtain

$$\left(1-|z|^2\right)\left|\frac{z\mathcal{F}_\beta''(z)}{\mathcal{F}_\beta'(z)}\right| \leq \left(1-|z|^2\right)\left\{|\beta-1| + \left|\frac{z\mathcal{H}_{\gamma_c}'(z)}{\mathcal{H}_{\gamma_c}(z)} - 1\right| + \frac{|z|^{\mathcal{R}(\beta)}}{|\beta|}\left|\frac{\mathcal{H}_{\gamma_c}(z)}{z}\right|\right\}$$

$$\leq \left(1-|z|^2\right)\left\{|\beta-1| + \frac{(c+1)(2\zeta\eta-1)}{(\zeta\eta-1)(2\zeta\eta-3)} + \frac{M}{|\beta|}\right\}$$

$$\leq 1.$$

This shows that the given integral operator satisfying the Becker's criterion for univalence [12], hence $\mathcal{F}_\beta$ is univalent in $\mathcal{U}$. □

4. Uniformly Convexity of Hyper-Bessel Functions

Theorem 4. *If* $i \in \{1,2,3,...,c\}, \gamma_i > -1$ *and* $z \in \mathcal{U}$. *If* $\prod_{i=1}^{c}(\gamma_i+1) > \frac{(4c^2+8c-1)+\sqrt{(4c^2+8c-1)^2-4(c+4)(8c^2+20c+10)}}{2(8c^2+20c+10)\varsigma}$, *then* $\mathcal{H}_{\gamma_c} \in \mathcal{UCV}$.

Proof. Since

$$\left|\frac{z\mathcal{H}_{\gamma_c}''(z)}{\mathcal{H}_{\gamma_c}'(z)}\right| \leq \frac{\zeta\eta(2\zeta\eta-1)(c+1)(2c+3)}{(\zeta\eta-1)(2\zeta\eta-3)-(2\zeta\eta-1)(c+1)}.$$

By using Lemma 5, we have

$$\left|\frac{z\mathcal{H}_{\gamma_c}''(z)}{\mathcal{H}_{\gamma_c}'(z)}\right| < \frac{1}{2},$$

if

$$\frac{\zeta\eta(2\zeta\eta-1)(c+1)(2c+3)}{(\zeta\eta-1)(2\zeta\eta-3)-(2\zeta\eta-1)(c+1)} < \frac{1}{2}.$$

This implies that

$$\prod_{i=1}^{c}(\gamma_i+1) > \frac{(4c^2+8c-1)+\sqrt{(4c^2+8c-1)^2-4(c+4)(8c^2+20c+10)}}{2(8c^2+20c+10)\varsigma}.$$

Hence, we obtain the required result. □

5. Hardy Spaces Of Hyper-Bessel Functions

Let $\mathcal{H}^\infty$ denote the space of all bounded functions on $\mathcal{H}$. Let $f \in \mathcal{H}$, set

$$M_p(r,f) = \begin{cases} \left(\frac{1}{2\pi}\int_0^{2\pi}|f(re^{i\theta})|^p \, d\theta\right)^{1/p}, & 0 < p < \infty, \\ \sup\{|f(z)| : |z| \le r\}, & p = \infty. \end{cases}$$

Then the function $f \in \mathcal{H}^p$ if $M_p(r,f)$ is bounded for all $r \in [0,1)$. It is clear that

$$\mathcal{H}^\infty \subset \mathcal{H}^q \subset \mathcal{H}^p, \, 0 < q < p < \infty.$$

For some details, see [15] (page 2). It is also known [16] (page 64, Section 4.5) (see also [15]) that for $Re(f'(z)) > 0$ in $\mathcal{U}$, then

$$\begin{cases} f' \in \mathcal{H}^q, & q < 1, \\ f \in \mathcal{H}^{q/(1-q)}, & 0 < q < 1. \end{cases}$$

We require the following results to prove our results.

Lemma 6. $\mathcal{P}_0(\alpha) * \mathcal{P}_0(\beta) \subset \mathcal{P}_0(\gamma)$, where $\gamma = 1 - 2(1-\alpha)(1-\beta)$ with $\alpha, \beta < 1$ and the value of γ is best possible [17].

Lemma 7. For $\alpha, \beta < 1$ and $\gamma = 1 - 2(1-\alpha)(1-\beta)$, we have $\mathcal{R}_0(\alpha) * \mathcal{R}_0(\beta) \subset \mathcal{R}_0(\gamma)$ or equivalently $\mathcal{P}_0(\alpha) * \mathcal{P}_0(\beta) \subset \mathcal{P}_0(\gamma)$ [18].

Lemma 8. If the function $f \in \mathcal{C}(\alpha)$ [19], where $\alpha \in [0,1)$ is not of the form

$$f(z) = \begin{cases} \theta + \eta(1-ze^{i\gamma})^{2\alpha-1}, & \alpha \ne \frac{1}{2}, \\ \theta + \eta\log(1-ze^{i\gamma}), & \alpha \ne \frac{1}{2}, \end{cases}$$

for $\zeta, \eta \in \mathbb{C}$ and $\gamma \in \mathbb{R}$, then the following statements hold:

(i) There exists $\delta = \delta(f) > 0$, such that $f' \in \mathcal{H}^{\delta + \frac{1}{2(1-\alpha)}}$.
(ii) If $\alpha \in [1, 1/2)$, then there exists $\tau = \tau(f) > 0$, such that $f \in \mathcal{H}^{\tau + 1/(1-2\alpha)}$.
(iii) If $\alpha \ge 1/2$, then $f \in \mathcal{H}^\infty$.

Theorem 5. Let $i \in \{1, 2, 3, ..., c\}, \gamma_i > -1$ with $\alpha \in [0, 1)$ and $z \in \mathcal{U}$. Let

$$\prod_{i=1}^{c}(\gamma_i+1) > \frac{\{(2c+7)(1-\alpha)-(2c^2+5c+3)\}+\Phi}{2\varsigma\{2(1-\alpha)-(4c^2+10c+6)\}},$$

where

$$\Phi = \sqrt{\{(2c+7)(1-\alpha) - (2c^2+5c+3)\}^2 - 4\{2(1-\alpha) - (4c^2+10c+6)\}(c+4)(1-\alpha)}.$$

Then

(i) $\mathcal{H}_{\gamma_c} \in \mathcal{H}^{1/1-2\alpha}$ for $\alpha \in [0, 1/2)$.
(ii) $\mathcal{H}_{\gamma_c} \in \mathcal{H}^{\infty}$ for $\alpha \geq 1/2$.

Proof. By using the definition of Hypergeometric function

$$_2F_1(a,b,c;z) = \sum_{n=0}^{\infty} \frac{(a)_n (b)_n}{(c)_n} \frac{z^n}{n!},$$

we have

$$\theta + \frac{\vartheta z}{(1-ze^{i\psi})^{1-\alpha}} = \theta + \vartheta z \,_2F_1(1, 1-2\alpha, 1; ze^{i\psi})$$

$$= \theta + \vartheta \sum_{n=0}^{\infty} \frac{(1-2\alpha)_n}{n!} e^{i\psi n} z^{n+1},$$

for $\theta, \vartheta \in \mathbb{C}$, $\alpha \neq 1/2$ and for $\psi \in \mathbb{R}$. On the other hand

$$\theta + \vartheta \log(1 - ze^{i\psi}) = \theta - \vartheta z \,_2F_1(1, 1, 2; ze^{i\psi})$$

$$= \theta - \vartheta \sum_{n=0}^{\infty} \frac{1}{n+1} e^{i\psi n} z^{n+1}.$$

This implies that $\mathcal{H}_{\gamma_c}$ is not of the form $\theta + \vartheta z(1-ze^{i\psi})^{2\alpha-1}$ for $\alpha \neq 1/2$ and $\theta + \vartheta \log(1-ze^{i\psi})$ for $\alpha = 1/2$ respectively. Also from part (ii) of Theorem 1, $\mathcal{H}_{\gamma_c}$ is convex of order α. Hence by using Lemma 8, we have required result. □

Theorem 6. Let $i \in \{1, 2, 3, ..., c\}$, $\gamma_i > -1$ with $\alpha \in [0, 1)$ and $z \in \mathcal{U}$. If $\prod_{i=1}^{c}(\gamma_i + 1) > \frac{3-\alpha}{2\zeta(1-\alpha)}$, then $\frac{\mathcal{H}_{\gamma_c}}{z} \in \mathcal{P}(\alpha)$. If $f \in \mathcal{R}(\varrho)$, with $\varrho < 1$, then $\mathcal{H}_{\gamma_c} * f \in \mathcal{R}(\tau)$, where $\tau = 1 - 2(1-\alpha)(1-\varrho)$.

Proof. Let $h(z) = \mathcal{H}_{\gamma_c}(z) * f(z)$, then $h'(z) = \frac{\mathcal{H}_{\gamma_c}(z)}{z} * f'(z)$. Now from Theorem 1 of part (iv), we have $\frac{\mathcal{H}_{\gamma_c}}{z} \in \mathcal{P}(\alpha)$. By using Lemma 6 and the fact that $f' \in \mathcal{P}(\varrho)$, we have $h'(z) \in \mathcal{P}(\tau)$, where $\tau = 1 - 2(1-\alpha)(1-\varrho)$. Consequently, we have $h \in \mathcal{R}(\tau)$. □

Corollary 2. Let $i \in \{1, 2, 3, ..., c\}$, $\gamma_i > -1$ with $\alpha \in [0, 1)$ and $z \in \mathcal{U}$. If $\prod_{i=1}^{c}(\gamma_i + 1) > \frac{3-\alpha}{2\zeta(1-\alpha)}$, then $\frac{\mathcal{H}_{\gamma_c}}{z} \in \mathcal{P}(\alpha)$. If $f \in \mathcal{R}(\varrho)$, $\varrho = (1-2\alpha)(2-2\alpha)$, then $\mathcal{H}_{\gamma_c} * f \in \mathcal{R}(0)$.

Author Contributions: Conceptualization, K.A.; Formal analysis, S.M. and M.R.; Funding acquisition, M.U.D. and S.u.R.; Investigation, K.A.; Methodology, M.A. and M.R.; Supervision, M.A.; Validation, S.M. and M.R.; Visualization, K.A.;Writing—original draft, K.A.;Writing—review and editing, M.R.

Funding: The work here is partially supported by HEC grant: 5689/Punjab/NRPU/R&D/HEC/2016.

Acknowledgments: The authors thank the referees for their valuable suggestions to improve the paper.

Conflicts of Interest: The authors declare no conflict of interest.

References

1. Goodman, A.W. *Univalent Functions*; Mariner: Tampa, FL, USA, 1983; Volumes 1–2.
2. Baricz, Á. Bessel transforms and Hardy space of generalized Bessel functions. *Mathematica* **2006**, *48*, 127–136.
3. Baricz, Á.; Ponnusamy, S.; Singh, S. Modified Dini functions: Monotonicity patterns and functional inequalities. *Acta Math. Hungrica* **2016**, *149*, 120–142. [CrossRef]
4. Baricz, Á.; Szász, R. The radius of convexity of normalized Bessel functions. *Anal. Math.* **2015**, *41*, 141–151. [CrossRef]
5. Raza, M.; Din, M.U.; Malik, S.N. Certain geometric properties of normalized Wright Functions. *J. Funct. Spaces* **2016**, *2016*, 1896154. [CrossRef]
6. Yagmur, N.; Orhan, H. Hardy space of generalized Struve functions. *Complex Var. Elliptic Equ.* **2014**, *59*, 929–936. [CrossRef]
7. Aktaş, I. Partial sums of hyper-Bessel function with applications. *arXiv* **2018**, arXiv:1806.09813.
8. Aktaş, I.; Baricz, Á.; Singh, S. Geometric and monotonic properties of hyper-Bessel functions. *arXiv* **2018**, arXiv:1802.05226.
9. Chaggara, H.; Romdhane, N.B. On the zeros of the hyper-Bessel function. *Integr. Trans. Spec. Funct.* **2015**, *26*, 96–101. [CrossRef]
10. Mocanu, P.T. Some starlike conditions for analytic functions. *Rev. Roum. Math. Pures Appl.* **1988**, *33*, 117–124.
11. MacGregor, T.H. The radius of univalence of certain analytic functions II. *Proc. Am. Math. Soc.* **1963**, *14*, 521–524. [CrossRef]
12. Pescar, V. A new generalization of Ahlfors's and Becker's criterion of univalence. *Bull. Malaysian Math. Soc. (Second Ser.)* **1996**, *19*, 53–54.
13. Owa, S.; Nunokawa, M.; Saitoh, H.; Srivastava, H.M. Close-to-convexity, starlikeness, and convexity of certain analytic functions. *Appl. Math. Lett.* **2002**, *15*, 63–69. [CrossRef]
14. Ravichandran, V. On uniformly convex functions. *Ganita* **2002**, *53*, 117–124.
15. Duren, P.L. Theory of H^p Spaces; A series of Monographs and Textbooks in Pure and Applied Mathematics Volume 38; Academic Press: New York, NY, USA; London, UK, 1970.
16. Priwalow, I.I. *Randeigenschaften Analytischer Functionen*; Deutscher Verlag Der Wissenschaften: Berlin, Germany, 1956.
17. Stankiewicz, J.; Stankiewicz, Z. Some applications of Hadamard convolutions in the theory of functions. *Ann. Univ. Mariae Curie-Sklodowska* **1986**, *40*, 251–265.
18. Ponnusamy, S. The Hardy space of hypergeometric functions. *Complex Var. Theor. Appl.* **1996**, *29*, 83–96. [CrossRef]
19. Eenigenburg, P.J.; Keogh, F.R. The Hardy class of some univalent functions and their derivatives. *Mich. Math J.* **1970**, *17*, 335–346. [CrossRef]

© 2019 by the authors. Licensee MDPI, Basel, Switzerland. This article is an open access article distributed under the terms and conditions of the Creative Commons Attribution (CC BY) license (http://creativecommons.org/licenses/by/4.0/).

Article

Some Reciprocal Classes of Close-to-Convex and Quasi-Convex Analytic Functions

Shahid Mahmood [1], Janusz Sokół [2], Hari Mohan Srivastava [3,4] and Sarfraz Nawaz Malik [5,*]

[1] Department of Mechanical Engineering, Sarhad University of Science & I.T, Peshawar 25000, Pakistan; shahidmahmood757@gmail.com
[2] Department of Mathematics, Rzeszów University of Technology, Al. Powstańców, Warszawy 12, 35-959 Rzeszów, Poland; jsokol@prz.edu.pl
[3] Department of Mathematics and Statistics, University of Victoria, Victoria, BC V8W 3R4, Canada; harimsri@math.uvic.ca
[4] Department of Medical Research, China Medical University Hospital, China Medical University, Taichung 40402, Taiwan
[5] Department of Mathematics, COMSATS University Islamabad, Wah Campus 47040, Pakistan
* Correspondence: snmalik110@yahoo.com

Received: 18 December 2018; Accepted: 22 March 2019; Published: 27 March 2019

Abstract: The present paper comprises the study of certain functions which are analytic and defined in terms of reciprocal function. The reciprocal classes of close-to-convex functions and quasi-convex functions are defined and studied. Various interesting properties, such as sufficiency criteria, coefficient estimates, distortion results, and a few others, are investigated for these newly defined sub-classes.

Keywords: subordination; functions with positive real part; reciprocals

MSC: 30C45; 30C50

1. Introduction

We denote by $\mathcal{A}$ the class of analytic functions on the unit disc $\mathbb{U} = \{z \in \mathbb{C} : |z| < 1\}$ having the following taylor series representation:

$$f(z) = z + \sum_{n=2}^{\infty} a_n z^n. \tag{1}$$

The analytic function f will be subordinate to an analytic function g, if there exists an analytic function w, known as a Schwarz function, with $w(0) = 0$ and $|w(z)| < |z|$, such that $f(z) = g(w(z))$. Moreover, if the function g is univalent in $\mathbb{U}$, then we have the following (see [1,2]):

$$f(z) \prec g(z), \quad z \in \mathbb{U} \iff f(0) = g(0) \quad \text{and} \quad f(\mathbb{U}) \subset g(\mathbb{U}).$$

Uralegaddi et al. [3] introduced the reciprocal classes $\mathcal{M}(\gamma)$ of starlike and $\mathcal{N}(\gamma)$ of convex functions for $1 \leq \gamma \leq \dfrac{4}{3}$, which were further studied by Owa et al. [4–6] for the values $\gamma \geq 1$.

The classes $\mathcal{M}(\gamma)$ of starlike functions and $\mathcal{N}(\gamma)$ of reciprocal order convex functions γ, $(\gamma > 1)$ are defined as follows:

$$\mathcal{M}(\gamma) = \left\{ f \in \mathcal{A} : \mathfrak{Re}\frac{zf'(z)}{f(z)} < \gamma,\, z \in \mathbb{U} \right\},$$

$$\mathcal{N}(\gamma) = \left\{ f \in \mathcal{A} : \mathfrak{Re}\left\{1 + \frac{zf''(z)}{f'(z)}\right\} < \gamma,\, z \in \mathbb{U} \right\}.$$

Using the same concept, together with the idea of k-uniformly starlike and γ ordered convex functions, Nishiwaki and Owa [7] defined the reciprocal classes of uniformly starlike $\mathcal{MD}(k, \gamma)$ and convex functions $\mathcal{ND}(k, \gamma)$. The class $\mathcal{MD}(k, \gamma)$ denotes the subclass of $\mathcal{A}$ consisting of functions f satisfying the inequality

$$\mathfrak{Re}\frac{zf'(z)}{f(z)} < k\left|\frac{zf'(z)}{f(z)} - 1\right| + \gamma,\quad (z \in \mathbb{U}),$$

for some γ $(\gamma > 1)$ and k $(k \leq 0)$ and the class $\mathcal{ND}(k, \gamma)$ denotes the subclass of $\mathcal{A}$ consisting of functions $f(z)$ satisfying the inequality

$$\mathfrak{Re}\frac{(zf'(z))'}{f'(z)} < \gamma + k\left|\frac{(zf'(z))'}{f'(z)} - 1\right|,\quad (z \in \mathbb{U}),$$

for some γ $(\gamma > 1)$ and k $(k \leq 0)$. They also proved that the well-known Alexander relation holds between $\mathcal{MD}(k, \gamma)$ and $\mathcal{ND}(k, \gamma)$. This means that

$$f \in \mathcal{ND}(k, \gamma) \iff zf' \in \mathcal{MD}(k, \gamma).$$

For a more detailed and recent study on uniformly convex and starlike functions, we refer the reader to [8–12].

Considering the above defined classes, we introduce the following classes.

Definition 1. *Let f belong to $\mathcal{A}$. Then, it will belong to the class $\mathcal{KD}(\beta, \gamma)$ if there exists $g \in \mathcal{MD}(\gamma)$ such that*

$$\mathfrak{Re}\left\{\frac{zf'(z)}{g(z)}\right\} < \beta,\quad (z \in \mathbb{U}),\tag{2}$$

for some $\beta, \gamma > 1$.

Definition 2. *Let f belong to $\mathcal{A}$. Then, it will belong to the class $\mathcal{QD}(\beta, \gamma)$ if there exists $g \in \mathcal{ND}(\gamma)$ such that*

$$\mathfrak{Re}\left\{\frac{(zf'(z))'}{g'(z)}\right\} < \beta,\quad (z \in \mathbb{U}),\tag{3}$$

for some $\beta, \gamma > 1$.

It is clear, from (2) and (3), that

$$f(z) \in \mathcal{QD}(\beta, \gamma) \iff zf'(z) \in \mathcal{KD}(\beta, \gamma).$$

Definition 3. *Let f belong to $\mathcal{A}$. Then, it will belong to the class $\mathcal{KD}(k, \beta, \gamma)$ if there exists $g \in \mathcal{MD}(k, \gamma)$ such that*

$$\mathfrak{Re}\left\{\frac{zf'(z)}{g(z)}\right\} < k\left|\frac{zf'(z)}{g(z)} - 1\right| + \beta,\quad (z \in \mathbb{U}),\tag{4}$$

for some $k \leq 0$ and $\beta, \gamma > 1$.

Definition 4. Let f belong to $\mathcal{A}$. Then, it is said to be in the class $\mathcal{QD}(k, \beta, \gamma)$ if there exists $g \in \mathcal{ND}(k, \gamma)$ such that

$$\mathfrak{Re}\left\{\frac{(zf'(z))'}{g'(z)}\right\} < k\left|\frac{(zf'(z))'}{g'(z)} - 1\right| + \beta, \quad (z \in \mathbb{U}), \tag{5}$$

for some $k \leq 0$ and $\beta, \gamma > 1$.

We can see, from (4) and (5), that the well-known relation of Alexander type holds between the classes $\mathcal{KD}(k, \beta, \gamma)$ and $\mathcal{QD}(k, \beta, \gamma)$, which means that

$$f(z) \in \mathcal{QD}(k, \beta, \gamma) \quad \Leftrightarrow \quad zf'(z) \in \mathcal{KD}(k, \beta, \gamma).$$

2. Preliminary Lemmas

Lemma 1. *For positive integers t and σ, we have*

$$\sigma \sum_{j=1}^{t} \frac{(\sigma)_{j-1}}{(j-1)!} = \frac{(\sigma)_t}{(t-1)!}, \tag{6}$$

where $(\sigma)_t$ is the Pochhammer symbol, defined by

$$(\sigma)_t = \frac{\Gamma(\sigma+t)}{\Gamma(\sigma)} = \sigma(\sigma+1)(\sigma+2)(\sigma+3)\cdots(\sigma+t-1).$$

Proof. Consider

$$\begin{aligned}
&\sigma \sum_{j=1}^{t} \frac{(\sigma)_{j-1}}{(j-1)!} \\
&= \sigma\left(1 + \frac{\sigma}{1} + \frac{(\sigma)_2}{2!} + \frac{(\sigma)_3}{3!} + \frac{(\sigma)_4}{4!} + \cdots + \frac{(\sigma)_{t-1}}{(t-1)!}\right) \\
&= \sigma(1+\sigma)\left(1 + \frac{\sigma}{2} + \frac{\sigma(\sigma+2)}{2 \times 3} + \cdots + \frac{\sigma(\sigma+2)\cdots(\sigma+t-2)}{2 \times \cdots \times (t-1)}\right) \\
&= \sigma(1+\sigma)\frac{(\sigma+2)}{2}\left(1 + \frac{\sigma}{3} + \cdots + \frac{\sigma(\sigma+3)\cdots(\sigma+t-2)}{3 \times 4 \times \cdots \times (t-1)}\right) \\
&= \sigma(1+\sigma)\frac{(\sigma+2)}{2}\frac{(\sigma+3)}{3}\left(1 + \frac{\sigma}{4} + \cdots + \frac{\sigma(\sigma+4)\cdots(\sigma+t-2)}{4 \times \cdots \times (t-1)}\right) \\
&= \sigma(1+\sigma)\frac{(\sigma+2)}{2}\frac{(\sigma+3)}{3}\frac{(\sigma+4)}{4}\left(1 + \frac{\sigma}{5} + \cdots + \frac{\sigma\cdots(\sigma+t-2)}{5 \times 6 \times \cdots \times (t-1)}\right) \\
&= \sigma(1+\sigma)\frac{(\sigma+2)}{2}\frac{(\sigma+3)}{3}\frac{(\sigma+4)}{4}\cdots\left(1 + \frac{\sigma}{t-1}\right) \\
&= \sigma(1+\sigma)\frac{(\sigma+2)}{2}\frac{(\sigma+3)}{3}\frac{(\sigma+4)}{4}\cdots\left(\frac{\sigma+(t-1)}{t-1}\right) \\
&= \frac{(\sigma)_t}{(t-1)!}.
\end{aligned}$$

□

Lemma 2. *If $f(z) \in \mathcal{MD}(k, \gamma)$, then*

$$f(z) \in \mathcal{MD}\left(\frac{\gamma-k}{1-k}\right).$$

Proof. Using the definition, we write

$$\Re\frac{zf'(z)}{f(z)} < k\left|\frac{zf'(z)}{f(z)} - 1\right| + \gamma$$

$$\leq k\Re\frac{zf'(z)}{f(z)} + \gamma - k,$$

which implies that

$$(1-k)\Re\frac{zf'(z)}{f(z)} < \gamma - k.$$

After simplification, we obtain

$$\Re\frac{zf'(z)}{f(z)} < \frac{\gamma - k}{1 - k}, \quad (k \leq 0, \ \gamma > 1).$$

As $\frac{\gamma - k}{1 - k} > 1$, we have $f(z) \in \mathcal{MD}\left(\frac{\gamma - k}{1 - k}\right)$. With this, we obtain the required result. □

Lemma 3. *If f belongs to the class $\mathcal{MD}(k, \gamma)$, then*

$$|a_n| \leq \frac{(\delta_{k,\gamma})_{n-1}}{(n-1)!}, \tag{7}$$

where

$$\delta_{k,\gamma} = \frac{2(\gamma - 1)}{1 - k}. \tag{8}$$

Proof. Let us define a function

$$p(z) = \frac{(\gamma - k) - (1-k)\left(\frac{zf'(z)}{f(z)}\right)}{\gamma - 1}, \tag{9}$$

where $p \in \mathcal{P}$, the class of Caratheodory functions (see [1]). One may write

$$\frac{zf'(z)}{f(z)} = \frac{(\gamma - k) - (\gamma - 1)p(z)}{1 - k}, \tag{10}$$

or

$$zf'(z) = \left(\frac{\gamma - k}{1 - k} - \frac{\gamma - 1}{1 - k}p(z)\right)f(z). \tag{11}$$

Let us write $p(z)$ as $p(z) = 1 + \sum_{n=1}^{\infty} p_n z^n$ and let f have the series form, as in (1). Then, (11) can be written as

$$\sum_{n=1}^{\infty} n a_n z^n = \left(\sum_{n=1}^{\infty} a_n z^n\right)\left(\frac{\gamma - k}{1 - k} - \frac{\gamma - 1}{1 - k}\left(1 + \sum_{n=1}^{\infty} p_n z^n\right)\right), \quad a_1 = 1$$

which reduces to

$$\sum_{n=1}^{\infty} n a_n z^n = \left(\sum_{n=1}^{\infty} a_n z^n\right)\left(1 - \frac{\gamma - 1}{1 - k}\sum_{n=1}^{\infty} p_n z^n\right)$$

$$= \sum_{n=1}^{\infty} a_n z^n - \frac{\gamma - 1}{1 - k}\left(\sum_{n=1}^{\infty} a_n z^n\right)\left(\sum_{n=1}^{\infty} p_n z^n\right).$$

This implies that

$$\sum_{n=1}^{\infty}(n-1)a_n z^n = -\frac{\gamma-1}{1-k}\sum_{n=1}^{\infty}\left(\sum_{j=0}^{n-1}a_j p_{n-j}\right)z^n.$$

After comparing the nth term's coefficients, appearing on both sides, combined with the fact that $a_0 = 0$, we obtain

$$a_n = \frac{-(\gamma-1)}{(n-1)(1-k)}\sum_{j=1}^{n-1}a_j p_{n-j}.$$

Now, we take the absolute value and then apply the triangle inequality to get

$$|a_n| \leq \frac{\gamma-1}{(n-1)(1-k)}\sum_{j=1}^{n-1}|a_j||p_{n-j}|.$$

Applying the coefficient estimates, such that $|p_n| \leq 2$ $(n \geq 1)$ for Caratheodory functions [1], we obtain

$$|a_n| \leq \frac{2(\gamma-1)}{(n-1)(1-k)}\sum_{j=1}^{n-1}|a_j|.$$

$$|a_n| \leq \frac{\delta_{k,\gamma}}{n-1}\sum_{j=1}^{n-1}|a_j|, \tag{12}$$

where $\delta_{k,\gamma} = \dfrac{2(\gamma-1)}{1-k}$. We prove (7) by induction on n. Thus, first for $n=2$, we obtain the following from (12):

$$|a_2| \leq \frac{\delta_{k,\gamma}}{1} = \frac{(\delta_{k,\gamma})_{2-1}}{(2-1)!}. \tag{13}$$

This proves that, for $n=2$, (7) is true. For $n=3$, we obtain

$$|a_3| \leq \frac{\delta_{k,\gamma}}{2}(1+|a_2|) = \frac{\delta_{k,\gamma}(1+\delta_{k,\gamma})}{2} = \frac{(\delta_{k,\gamma})_{3-1}}{(3-1)!}.$$

This proves that when $n=3$, (7) holds true. Now, we assume that for $t \leq n$, (7) is true, that means

$$|a_t| \leq \frac{(\delta_{k,\gamma})_{t-1}}{(t-1)!} \quad t=1,2,\ldots,n. \tag{14}$$

Using (12) and (14), we have

$$|a_{t+1}| \leq \frac{\delta_{k,\gamma}}{t}\sum_{j=1}^{t}|a_j| \leq \frac{\delta_{k,\gamma}}{t}\sum_{j=1}^{t}\frac{(\delta_{k,\gamma})_{j-1}}{(j-1)!}.$$

After applying (6), we obtain

$$|a_{t+1}| \leq \frac{1}{t}\frac{(\delta_{k,\gamma})_t}{(t-1)!} = \frac{(\delta_{k,\gamma})_t}{t!}.$$

As a result of mathematical induction, it is shown that (7) is true for all $n \geq 2$. Hence, the required bound is obtained. □

Lemma 4 ([13]). *Let w be analytic in $\mathbb{U}$ with $w(0) = 0$. If there exists $z_0 \in \mathbb{U}$ such that*

$$\max_{|z| \leq |z_0|}|w(z)| = |w(z_0)|,$$

then
$$z_0 w'(z_0) = cw(z_0),$$
where c is real and $c \geq 1$.

3. Main Results

Theorem 1. *If $f(z) \in \mathcal{KD}(k, \beta, \gamma)$, then*
$$f(z) \in \mathcal{KD}\left(\frac{\beta - k}{1 - k}, \gamma\right).$$

Proof. If $f(z) \in \mathcal{KD}(k, \beta, \gamma)$, then $k \leq 0$, $\beta > 1$, and so we obtain
$$\mathfrak{Re}\left\{\frac{zf'(z)}{g(z)}\right\} < k\left|\frac{zf'(z)}{g(z)} - 1\right| + \beta$$
$$\leq \beta + k\mathfrak{Re}\left\{\frac{zf'(z)}{g(z)} - 1\right\},$$
which leads to
$$\mathfrak{Re}\left\{\frac{zf'(z)}{g(z)}\right\} - k\mathfrak{Re}\left\{\frac{zf'(z)}{g(z)}\right\} < -k + \beta.$$

After simplification, we obtain
$$\mathfrak{Re}\left\{\frac{zf'(z)}{g(z)}\right\} < \frac{\beta - k}{1 - k}, \quad (k \leq 0, \beta > 1). \tag{15}$$

This completes the proof. □

In a similar way, one can easily prove the following important result.

Theorem 2. *If $f \in \mathcal{QD}(k, \beta, \gamma)$, then*
$$f \in \mathcal{QD}\left(\frac{\beta - k}{1 - k}, \gamma\right).$$

Theorem 3. *If $f(z) \in \mathcal{KD}(k, \beta, \gamma)$, then*
$$|a_n| \leq \frac{(\delta_{k,\gamma})_{n-1}}{n!} + \frac{|\delta_{k,\beta}|}{n}\sum_{j=1}^{n-1}\frac{(\delta_{k,\gamma})_{j-1}}{(j-1)!},$$
where $\delta_{k,\gamma}$ is given by (8) and
$$\delta_{k,\beta} = \frac{2(\beta - 1)}{1 - k}. \tag{16}$$

Proof. If f is in the class $\mathcal{KD}(k, \beta, \gamma)$, then there exists $g(z) \in \mathcal{MD}(k, \gamma)$ such that the function
$$p(z) = \frac{(\beta - k) - (1 - k)\left(\frac{zf'(z)}{g(z)}\right)}{\beta - 1} \tag{17}$$
belongs to $\mathcal{P}$. Therefore, we write
$$zf'(z) = \frac{\beta - k}{1 - k}g(z) - \frac{\beta - 1}{1 - k}g(z)p(z). \tag{18}$$

Let us write $p(z)$ as $p(z) = 1 + \sum_{n=1}^{\infty} p_n z^n$, $g(z)$ as $g(z) = z + \sum_{n=2}^{\infty} b_n z^n$, and let $f(z)$ have the series form as in (1). Then, (18) can be written as

$$z + \sum_{n=2}^{\infty} n a_n z^n = \frac{\beta - k}{1 - k}\left(z + \sum_{n=2}^{\infty} b_n z^n\right) - \frac{\beta - 1}{1 - k}\left(1 + \sum_{n=1}^{\infty} p_n z^n\right)\left(z + \sum_{n=2}^{\infty} b_n z^n\right).$$

Comparing the nth term's coefficients on both sides, we obtain

$$n a_n = b_n - \frac{\beta - 1}{1 - k}\left[p_{n-1} + p_{n-2} b_2 + p_{n-3} b_3 + \ldots + p_1 b_{n-1}\right].$$

By taking the absolute value, we get

$$n|a_n| = \left|b_n - \frac{\beta - 1}{1 - k}[p_{n-1} + p_{n-2} b_2 + p_{n-3} b_3 + \ldots + p_1 b_{n-1}]\right|$$

$$\leq |b_n| + \frac{\beta - 1}{1 - k}|p_{n-1} + p_{n-2} b_2 + p_{n-3} b_3 + \ldots + p_1 b_{n-1}|.$$

Applying the triangle inequality, we obtain

$$n|a_n| \leq |b_n| + \frac{\beta - 1}{1 - k}\{|p_{n-1}| + |p_{n-2} b_2| + |p_{n-3} b_3| + \ldots + |p_1 b_{n-1}|\}. \tag{19}$$

As $\mathfrak{Re}\{p(z)\} > 0$ in $\mathbb{U}$, we have $|p_n| \leq 2$ $(n \geq 1)$ (see [1]). Then, from (19), we have

$$n|a_n| \leq |b_n| + \frac{2(\beta - 1)}{1 - k} \sum_{j=1}^{n-1} |b_j|,$$

where $b_1 = 1$. Using Lemma (3), we obtain

$$n|a_n| \leq \frac{(\delta_{k,\gamma})_{n-1}}{(n-1)!} + \delta_{k,\beta} \sum_{j=1}^{n-1} \frac{(\delta_{k,\gamma})_{j-1}}{(j-1)!},$$

where $\delta_{k,\beta} = \dfrac{2(\beta - 1)}{1 - k}$ and $\delta_{k,\gamma}$ is defined by (8). This can be written as

$$|a_n| \leq \frac{(\delta_{k,\gamma})_{n-1}}{n!} + \frac{\delta_{k,\beta}}{n} \sum_{j=1}^{n-1} \frac{(\delta_{k,\gamma})_{j-1}}{(j-1)!}.$$

This completes the proof. □

From Definition 4 and Theorem 2, we immediately get the following corollary.

Corollary 1. *If $f(z) \in \mathcal{QD}(k, \beta, \gamma)$, then*

$$|a_n| \leq \frac{1}{n}\left[\frac{(\delta_{k,\gamma})_{n-1}}{n!} + \frac{\delta_{k,\beta}}{n} \sum_{j=1}^{n-1} \frac{(\delta_{k,\gamma})_{j-1}}{(j-1)!}\right],$$

where $\delta_{k,\beta}$ and $\delta_{k,\gamma}$ are given by (16) and (8), respectively.

By taking $k = 0$ in the above results, we obtain the coefficient inequality for the classes $\mathcal{KD}(\beta, \gamma)$ and $\mathcal{QD}(\beta, \gamma)$.

Theorem 4. *If a function $f \in \mathcal{KD}(k, \beta, \gamma)$, then there exists $g \in \mathcal{MD}(k, \gamma)$ such that*

$$\frac{zf'(z)}{g(z)} \prec 1 + 2(\beta_1 - 1) - \frac{2(\beta_1 - 1)}{1 - z}, \quad (z \in \mathbb{U}), \tag{20}$$

where

$$\beta_1 = \frac{\beta - k}{1 - k}. \tag{21}$$

Proof. Let $f(z) \in \mathcal{KD}(k, \beta, \gamma)$. Then, there exists $g(z)$ in $\mathcal{MD}(k, \gamma)$ and a Schwarz function $w(z)$ such that

$$\frac{\beta_1 - \left(\frac{zf'(z)}{g(z)}\right)}{\beta_1 - 1} = \frac{1 + w(z)}{1 - w(z)}, \tag{22}$$

as $w(z)$ is analytic $\mathbb{U}$ with $w(0) = 0$ and

$$\Re\left(\frac{1 + w(z)}{1 - w(z)}\right) > 0, \quad (z \in \mathbb{U}).$$

So, from (22), we obtain

$$\begin{aligned}
\frac{zf'(z)}{g(z)} &= \beta_1 - (\beta_1 - 1)\left(\frac{1 + w(z)}{1 - w(z)}\right) \\
&= \frac{\beta_1(1 - w(z)) - (\beta_1 - 1)(1 + w(z))}{1 - w(z)} \\
&= \frac{1 + w(z) - 2\beta_1 w(z)}{1 - w(z)} \\
&= \frac{1 - w(z) - 2(\beta_1 - 1)w(z)}{1 - w(z)} \\
&= \frac{1 - w(z) + 2(\beta_1 - 1) - 2(\beta_1 - 1)w(z) - 2(\beta_1 - 1)}{1 - w(z)} \\
&= \frac{1 - w(z) + 2(\beta_1 - 1)(1 - w(z)) - 2(\beta_1 - 1)}{1 - w(z)}.
\end{aligned}$$

This implies that

$$\frac{zf'(z)}{g(z)} = 1 + 2(\beta_1 - 1) - \frac{2(\beta_1 - 1)}{1 - w(z)},$$

and hence

$$\frac{zf'(z)}{g(z)} \prec 1 + 2(\beta_1 - 1) - \frac{2(\beta_1 - 1)}{1 - z}, \quad (z \in \mathbb{U}),$$

which is as required in (20). □

Corollary 2. *If $f \in \mathcal{QD}(k, \beta, \gamma)$, then there exists $g \in \mathcal{ND}(k, \gamma)$ such that*

$$\frac{(zf'(z))'}{g'(z)} \prec 1 + 2(\beta_1 - 1) - \frac{2(\beta_1 - 1)}{(1 - z)}, \quad (z \in \mathbb{U}), \tag{23}$$

where β_1 is given by (21).

Theorem 5. *If $f \in \mathcal{KD}(k, \beta, \gamma)$, then there exists a function $g \in \mathcal{MD}(k, \gamma)$ such that*

$$\frac{1 - (2\beta_1 - 1)r}{1 - r} \leq \Re \frac{zf'(z)}{g(z)} \leq \frac{1 + (2\beta_1 - 1)r}{1 + r}, \tag{24}$$

where $|z| = r < 1$ and β_1 is given by (21).

Proof. Using Theorem 4, we define the function ϕ as follows

$$\phi(z) = 1 + 2(\beta_1 - 1) + \frac{2(1 - \beta_1)}{1 - z}, (z \in \mathbb{U}).$$

Letting $z = re^{i\theta} (0 \leq r < 1)$, we observe that

$$\Re\phi(z) = 1 + 2(\beta_1 - 1) + \frac{2(1 - \beta_1)(1 - r\cos\theta)}{1 + r^2 - 2r\cos\theta}.$$

Let us define

$$\psi(t) = \frac{1 - rt}{1 + r^2 - 2rt}, (t = \cos\theta).$$

As $\psi'(t) = \dfrac{r(1 - r^2)}{(1 + r^2 - 2rt)^2} \geq 0$ (since $r < 1$), we get

$$1 + 2(\beta_1 - 1) + \frac{2(1 - \beta_1)}{1 - r} \leq \Re\phi(z) \leq 1 + 2(\beta_1 - 1) + \frac{2(1 - \beta_1)}{1 + r}.$$

After simplification, we have

$$\frac{1 - (2\beta_1 - 1)r}{1 - r} \leq \Re\phi(z) \leq \frac{1 + (2\beta_1 - 1)r}{1 + r}.$$

With the fact that $\dfrac{zf'(z)}{g(z)} \prec \phi(z), (z \in \mathbb{U})$ and as ϕ is univalent in $\mathbb{U}$, by using (22), we get the required result. □

Corollary 3. *If $f \in \mathcal{QD}(k, \beta, \gamma)$, then there exists $g \in \mathcal{ND}(k, \gamma)$ such that*

$$\frac{1 - (2\beta_1 - 1)r}{1 - r} \leq \Re\frac{(zf'(z))'}{g'(z)} \leq \frac{1 + (2\beta_1 - 1)r}{1 + r}, \quad (25)$$

where $|z| = r < 1$ and β_1 is given by (21).

Theorem 6. *Assume that a function $f \in \mathcal{A}$ satisfies*

$$\Re\left(\frac{zg'(z)}{g(z)} - \frac{zf''(z)}{f'(z)}\right) > \frac{\beta_1 + 1}{2\beta_1}, \quad (z \in \mathbb{U}), \quad (26)$$

for some $g(z) \in \mathcal{MD}(k, \gamma)$ and for real β_1 given by (21). If

$$\phi(z) = \frac{zf'(z)}{g(z)}$$

is analytic in $\mathbb{U}$ and $\phi(z) \neq 0$ and $\phi(z) \neq 2\beta_1 - 1$ in $\mathbb{U}$, then $f \in \mathcal{KD}(k, \beta_1)$.

Proof. Let us define a function $w(z)$ by

$$w(z) = \frac{\phi(z) - 1}{\phi(z) + (1 - 2\beta_1)}, \quad z \in \mathbb{U}.$$

Then, $w(z)$ is analytic in $\mathbb{U}$ as $\phi(z) \neq 2\beta_1 - 1$ and

$$\phi(z) = \frac{zf'(z)}{g(z)} = \frac{1 + (1 - 2\beta_1) w(z)}{1 - w(z)}. \tag{27}$$

Because $\phi(z) \neq 0$, we use logarithmic differentiation to get

$$\frac{1}{z} + \frac{f''(z)}{f'(z)} - \frac{g'(z)}{g(z)} = \frac{(1 - 2\beta_1) w'(z)}{1 + (1 - 2\beta_1) w(z)} + \frac{w'(z)}{1 - w(z)},$$

which further yields

$$\frac{zg'(z)}{g(z)} - \frac{zf''(z)}{f'(z)} = 1 - \frac{(1 - 2\beta_1) zw'(z)}{1 + (1 - 2\beta_1) w(z)} - \frac{zw'(z)}{1 - w(z)}. \tag{28}$$

Then, we note that w is analytic in open unit disk and $w(0) = 0$. Therefore, from (28), we obtain

$$\Re \left(\frac{zg'(z)}{g(z)} - \frac{zf''(z)}{f'(z)} \right) = \Re \left(1 - \frac{(1 - 2\beta_1) zw'(z)}{1 + (1 - 2\beta_1) w(z)} - \frac{zw'(z)}{1 - w(z)} \right)$$
$$> \frac{\beta_1 + 1}{2\beta_1}.$$

Suppose there exists a point $z_0 \in \mathbb{U}$ such that

$$\max_{|z| \leq |z_0|} |w(z)| = |w(z_0)| = 1,$$

then, by Lemma 4, we can write $w(z_0) = e^{i\theta}$ and $z_0 w'(z_0) = c e^{i\theta}$ for a point z_0, and we have

$$\Re \left(\frac{z_0 g'(z_0)}{g(z_0)} - \frac{z_0 f''(z_0)}{f'(z_0)} \right)$$
$$= \Re \left(1 - \frac{(1 - 2\beta_1) c e^{i\theta}}{1 + (1 - 2\beta_1) e^{i\theta}} - \frac{c e^{i\theta}}{1 - e^{i\theta}} \right)$$
$$= \Re \left(1 - \frac{c (1 - 2\beta_1) \left(e^{i\theta} + (1 - 2\beta_1) \right)}{1 + (1 - 2\beta_1)^2 + 2 (1 - 2\beta_1) \cos \theta} + \frac{c (1 - e^{i\theta})}{2 (1 - \cos \theta)} \right)$$
$$= 1 + \frac{c (2\beta_1 - 1) [\cos \theta + (1 - 2\beta_1)]}{1 + (1 - 2\beta_1)^2 + 2 (1 - 2\beta_1) \cos \theta} + \frac{c}{2}$$
$$\leq 1 - \frac{c (2\beta_1 - 1)}{2\beta_1} + \frac{c}{2}$$
$$= 1 - \frac{c (\beta_1 - 1)}{2\beta_1}$$
$$\leq 1 - \frac{\beta_1 - 1}{2\beta_1}, \text{ as } c < 1$$
$$= \frac{\beta_1 + 1}{2\beta_1},$$

which gives that

$$\Re \left\{ \frac{z_0 g'(z_0)}{g(z_0)} - \frac{z_0 f''(z_0)}{f'(z_0)} \right\} \leq \frac{\beta_1 + 1}{2\beta_1},$$

which is the contradiction to the supposed condition (26). Hence, there is no $z_0 \in \mathbb{U}$ such that $|w(z_0)| = 1$. This implies that $|w(z)| < 1, (z \in \mathbb{U})$ and, therefore, by (27), we have

$$\frac{zf'(z)}{g(z)} \prec \frac{1-(2\beta_1-1)z}{1-z}$$

or

$$\mathfrak{Re}\left\{\frac{zf'(z)}{g(z)}\right\} < \beta_1, \ z \in \mathbb{U}.$$

Hence, we conclude that $f(z) \in \mathcal{KD}(k, \beta_1)$. □

Theorem 7. Assume that $k \leq 0$ and $\beta > 1$. If $f \in \mathcal{A}$ and if there exists $g \in \mathcal{MD}(k, \gamma)$ such that

$$\left|\frac{zf'(z)}{g(z)} - 1\right| < \frac{\beta-1}{1-k} \quad z \in \mathbb{U}, \tag{29}$$

then $f \in \mathcal{KD}(k, \beta, \gamma)$.

Proof. We have

$$\left|\frac{zf'(z)}{g(z)} - 1\right| < \frac{\beta-1}{1-k}$$

$$\Rightarrow (1-k)\left|\frac{zf'(z)}{g(z)} - 1\right| + 1 < \beta$$

$$\Rightarrow \left|\frac{zf'(z)}{g(z)} - 1\right| + 1 < k\left|\frac{zf'(z)}{g(z)} - 1\right| + \beta$$

$$\Rightarrow \mathfrak{Re}\frac{zf'(z)}{g(z)} < k\left|\frac{zf'(z)}{g(z)} - 1\right| + \beta$$

$$\Rightarrow f \in \mathcal{KD}(k, \beta, \gamma).$$

□

Corollary 4. Let $f \in \mathcal{A}$ have the form (1). Assume that $g = z + b_2 z^2 + \cdots$ belongs to the class $\mathcal{MD}(k, \gamma)$ and satisfies

$$\left|\frac{\sum_{n=2}^{\infty}(na_n - b_n)z^{n-1}}{1 + \sum_{n=2}^{\infty} b_n z^{n-1}}\right| < \frac{\beta-1}{1-k} \quad z \in \mathbb{U}, \tag{30}$$

for some k $(k \leq 0)$, β $(\beta > 1)$.
Then, $f(z) \in \mathcal{KD}(k, \beta, \gamma)$.

Proof. We have

$$\left|\frac{zf'(z)}{g(z)} - 1\right|$$

$$= \left|\frac{z + \sum_{n=2}^{\infty} na_n z^n}{z + \sum_{n=2}^{\infty} b_n z^n} - 1\right|$$

$$= \left|\frac{\sum_{n=2}^{\infty}(na_n - b_n)z^{n-1}}{1 + \sum_{n=2}^{\infty} b_n z^{n-1}}\right|$$

$$< \frac{\beta-1}{1-k},$$

and hence (29) follows immediately from (30). □

Theorem 8. Let $f \in \mathcal{A}$ have the form (1) and let $g = z + \sum_{n=2}^{\infty} b_n z^n$, belonging to the class $\mathcal{MD}(k,\gamma)$, satisfy

$$1 + \sum_{n=2}^{\infty} (n|a_n| + y|b_n|) < y \quad z \in \mathbb{U}, \tag{31}$$

for some k ($k \leq 0$), β ($\beta > 1$) and where

$$y = \frac{(\beta - 1)}{(1 - k)} > 0.$$

Then, $f(z) \in \mathcal{KD}(k, \beta, \gamma)$.

Proof. Consider

$$1 + \sum_{n=2}^{\infty} (n|a_n| + y|b_n|) < y \tag{32}$$

$$\Rightarrow 1 + \sum_{n=2}^{\infty} n|a_n| < y - y\sum_{n=2}^{\infty} |b_n|$$

$$\Rightarrow 0 < y - y\sum_{n=2}^{\infty} |b_n|$$

$$\Rightarrow 0 < y - y\sum_{n=2}^{\infty} |b_n||z^{n-1}|$$

$$\Rightarrow 0 < y \left| 1 + \sum_{n=2}^{\infty} b_n z^{n-1} \right|. \tag{33}$$

We have

$$1 + \sum_{n=2}^{\infty} (n|a_n| + y|b_n|) < y$$

$$\Rightarrow 1 + \sum_{n=2}^{\infty} n|a_n| < y - y\sum_{n=2}^{\infty} |b_n|$$

$$\Rightarrow 1 + \sum_{n=2}^{\infty} n|a_n||z^{n-1}| < y - y\sum_{n=2}^{\infty} |b_n||z^{n-1}|$$

$$\Rightarrow \left| 1 + \sum_{n=2}^{\infty} na_n z^{n-1} \right| < y \left| 1 + \sum_{n=2}^{\infty} b_n z^{n-1} \right|$$

$$\Rightarrow \left| \frac{1 + \sum_{n=2}^{\infty} na_n z^{n-1}}{1 + \sum_{n=2}^{\infty} b_n z^{n-1}} \right| < y,$$

from (33). By (30), it follows that $f \in \mathcal{KD}(k, \beta, \gamma)$. $\square$

Author Contributions: Conceptualization, S.M.; Formal analysis, S.N.M. and J.S.; Funding acquisition, S.M.; Investigation, S.M.; Methodology, S.M. and S.N.M.; Supervision, H.M.S. and J.S.; Validation, H.M.S.; Visualization, S.M.;Writing—original draft, S.M.;Writing—review and editing, S.M. and S.N.M.

Funding: This research is supported by Sarhad University of Science & I.T, Peshawar 25000, Pakistan.

Acknowledgments: The authors are grateful to referees for their valuable comments which improved the quality of work and presentation of paper.

Conflicts of Interest: The authors declare no conflict of interest.

References

1. Goodman, A.W. *Univalent Functions, Vol. I, II*; Polygnal Publishing House: Washington, NJ, USA, 1983.

2. Miller, S.S.; Mocanu, P.T. *Differential Subordinations Theory and Applications*; Marcel Decker Inc.: New York, NY, USA, 2000.
3. Uralegaddi, B.A.; Ganigi, M.D.; Sarangi, S.M. Univalent functions with positive coefficients. *Tamkang J. Math.* **1994**, *25*, 225–230.
4. Nishiwaki, J.; Owa, S. Coefficient estimates for certain classes of analytic functions. *J. Ineq. Pure Appl. Math.* **2002**, *3*, 1–5.
5. Nishiwaki, J.; Owa, S. Coefficient inequalities for analytic functions. *Int. J. Math. Math. Sci.* **2002**, *29*, 285–290. [CrossRef]
6. Owa, S.; Srivastava, H.M. Some generalized convolution properties associated with cerain subclasses of analytic functions. *J. Ineq. Pure Appl. Math.* **2002**, *3*, 1–13.
7. Nishiwaki, J.; Owa, S. Certain classes of analytic functions concerned with uniformly starlike and convex functions. *Appl. Math. Comput.* **2007**, *187*, 350–355. [CrossRef]
8. Arif, M.; Ayaz, M.; Sokół, J. On a subclass of multivalent close to convex functions. *Publ. Inst. Math.* **2017**, *101*, 161–168. [CrossRef]
9. Arif, M.; Umar, S.; Mahmood, S.; Sokół, J. New reciprocal class of analytic functions associated with linear operator. *Iran. J. Sci. Technol. Trans. Sci.* **2016**, *42*, 881–886. [CrossRef]
10. Mahmood, S.; Malik, S.N.; Farman, S.; Riaz, S.M.J.; Farwa, S. Uniformly alpha-quasi convex functions defined by Janowski functions. *J. Funct. Spaces* **2018** , *2018*, 6049512. [CrossRef]
11. Malik, S.N.; Mahmood, S.; Raza, M.; Farman, S.; Zainab, S.; Muhammad, N. Coefficient Inequalities of Functions Associated with Hyperbolic Domain. *Mathematics* **2019**, *7*, 88. [CrossRef]
12. Malik, S.N.; Mahmood, S.; Raza, M.; Farman, S.; Zainab, S. Coefficient Inequalities of Functions Associated with Petal Type Domain. *Mathematics* **2018**, *6*, 298. [CrossRef]
13. Jack, I.S. Functions starlike and convex of order α. *J. Lond. Math. Soc.* **1971**, *2*, 469–474. [CrossRef]

© 2019 by the authors. Licensee MDPI, Basel, Switzerland. This article is an open access article distributed under the terms and conditions of the Creative Commons Attribution (CC BY) license (http://creativecommons.org/licenses/by/4.0/).

MDPI
St. Alban-Anlage 66
4052 Basel
Switzerland
Tel. +41 61 683 77 34
Fax +41 61 302 89 18
www.mdpi.com

Mathematics Editorial Office
E-mail: mathematics@mdpi.com
www.mdpi.com/journal/mathematics

www.ingramcontent.com/pod-product-compliance
Lightning Source LLC
LaVergne TN
LVHW070205100526
838202LV00015B/1999